The Institute of Mathematics
and its Applications
Conference Series

The Institute of Mathematics
and its Applications
Conference Series

Previous volumes in this series were published by Academic Press to whom all enquiries should be addressed. Forthcoming volumes will be published by Oxford University Press throughout the world.

NEW SERIES

1. *Supercomputers and parallel computation* Edited by D. J. Paddon
2. *The mathematical basis of finite element methods* Edited by David F. Griffiths
3. *Multigrid methods for integral and differential equations* Edited by D. J. Paddon and H. Holstein
4. *Turbulence and diffusion in stable environments* Edited by J. C. R. Hunt
5. *Wave propagation and scattering* Edited by B. J. Uscinski
6. *The mathematics of surfaces* Edited by J. A. Gregory
7. *Numerical methods for fluid dynamics II* Edited by K. W. Morton and M. J. Baines
8. *Analysing conflict and its resolution* Edited by P. G. Bennett
9. *The state of the art in numerical analysis* Edited by A. Iserles and M. J. D. Powell
10. *Algorithms for approximation* Edited by J. C. Mason and M. G. Cox
11. *The mathematics of surfaces II* Edited by R. R. Martin
12. *Mathematics in signal processing* Edited by T. S. Durrani, J. B. Abbiss, J. E. Hudson, R. N. Madan, J. G. McWhirter, and T. A. Moore

Mathematics in signal processing

Based on the proceedings of a conference organized by the Institute of Mathematics and its Applications on Mathematics in Signal Processing, held at University of Bath in September 1985

Edited by

T. S. DURRANI
University of Strathclyde, Scotland

J. B. ABBISS
Spectron Development Labs, Inc., USA

J. E. HUDSON
University of Technology, Loughborough

R. N. MADAN
Office of Naval Research, USA

J. G. McWHIRTER
Royal Signals and Radar Establishment

T. A. MOORE
British Broadcasting Corporation

CLARENDON PRESS · OXFORD · 1987

Oxford University Press, Walton Street, Oxford OX2 6DP

Oxford New York Toronto
Delhi Bombay Calcutta Madras Karachi
Petaling Jaya Singapore Hong Kong Tokyo
Nairobi Dar es Salaam Cape Town
Melbourne Auckland

and associated companies in
Beirut Berlin Ibadan Nicosia

Oxford is a trade mark of Oxford University Press

Published in the United States
by Oxford University Press, New York

British Library Cataloguing in Publication Data

Mathematics in signal processing: based on the proceedings of a conference organized by the Institute of Mathematics and its Applications on Mathematics in signal processing, held at University of Bath in September, 1985.—(The Institute of mathematics and its Applications conference series. New series; no. 12).
1. Signal processing—Digital techniques —Mathematics
I. Durrani, Tariq S. II. Institute of Mathematics and its Applications III. Series
621.38'043'0151 TK5102.5
ISBN 0–19–853613–5

Library of Congress Cataloging-in-Publication Data
Data available

Printed in Great Britain by
The Alden Press, Oxford

PREFACE

Signal processing represents one of the major growth areas in the application of advanced analytical techniques which come from a variety of mathematical sources such as approximation theory, number theory, linear algebra, optimisation theory, time series analysis, and others.

With the rapid development of new tools and concepts in the subject area it was felt that a Conference which brings together leading experts in the field and the practitioners of signal processing, should lead to a stimulating exchange of experience and ideas, and provide a platform for the identification of current trends and future directions. A Conference on 'Mathematics in Signal Processing' was organised by the Institute of Mathematics and its Applications at Bath on 17-19 September 1985, in association with the Institute of Physics, Institution of Electrical Engineers, and the Institute of Electrical and Electronic Engineers (UK Section).

The response to the Call for Papers was overwhelming and after long deliberations, the Conference Organising Committee selected 81 papers for presentation. The submissions fell naturally into the mainstream categories of the subject, and the Conference was organised to have 6 paper sessions and 6 poster sessions. The Conference was held on the pleasant and leafy campus of the University of Bath, and was attended by some two hundred delegates from over eighteen countries.

The proceedings represent papers selected after a comprehensive refereeing process following their presentation at the Conference. The Proceedings provides an interesting insight into Signal Processing as it includes papers from leading mathematicians, scientists and practising engineers. It is a state of the art record of the subject, and an impressive collection of new results which strongly endorse the infusion of Mathematics in Signal Processing.

The undersigned are grateful to Miss C. Richards and her staff at the Institute of Mathematics and its Applications for their assistance with the organisation of the Conference, and these Proceedings.

T.S. Durrani
J.B. Abbiss
J.E. Hudson
R.N. Madan
J.G. McWhirter
T.A. Moore

ACKNOWLEDGEMENTS

The Institute thanks the authors of the papers, the editors, Dr. J.B. Abbiss (Spectron Development Labs. Inc., USA), Professor T.S. Durrani, (University of Strathclyde), Dr. J.E. Hudson (University of Technology, Loughborough), Dr. R.N. Madan (Office of Naval Research, USA), Dr. J.G. McWhirter (Royal Signals and Radar Establishment) and Mr. T.A. Moore (British Broadcasting Corporation) and also Miss P. Irving and Miss Deborah Brown for typing the papers.

CONTENTS

CONTRIBUTORS

J.B. ABBISS; *Spectron Development Laboratories Inc., 3303 Harbor Boulevard, Costa Mesa, California 92626, U.S.A.*

M. AHMADI; *Department of Electrical Engineering, University of Windsor, Windsor, Ontario, Canada.*

J. ALCAZAR-FERNANDEZ; *ETSI Telecommunicacion-UPM, Cdad. Universitaria, 28040 Madrid, Spain.*

B.C. BARBER; *Space Department, Royal Aircraft Establishment, Farnborough, Hants, GU14 6TD.*

G.R. BARNES; *Department of Mathematics, University of Louisville, Louisville, Kentucky, 40292, U.S.A.*

S. BARNETT; *School of Mathematical Sciences, University of Bradford, West Yorkshire, BD7 1DP.*

M. BERTERO; *Dipartimento di Fisica dell' Universita & Instituto Nazionale di Fisica Nucleare, Genoa, Italy.*

J.M. BLACKLEDGE; *Physics Department, King's College, London University, Strand, London, WC2R 2LS.*

J.F. BOHME; *Lehrstuhl fur Signaltheorie, Ruhr-Universitat, 4630 Bochum, Federal Republic of Germany.*

J.R. CALDAS PINTO; *Departamento de Engenharia Mecanica, Instituto Superior Techico, Lisbon, Portugal.*

A. CANTONI; *Department of Electrical & Computer Engineering, University of Newcastle, Shortland 2305, New South Wales, Australia.*

A.S. CARASSO; *Mathematical Anaylsis Division, National Bureau of Standards, Gaithersburg, Maryland 20899, U.S.A.*

S. CASTAN; *Universite Paul Sabatier 118, route de Narbonne, 31062 Toulouse, France.*

W.K. CHAM; *Science Centre, Chinese University of Hong Kong, Shatin, N.T., Hong Kong.*

R.J. CLARKE; *Department of Electronic & Electrical Engineering, Loughborough University of Technology, Loughborough, Leicestershire, LE11 3TU.*

P.M. CLARKSON; *Institute of Sound & Vibration Research, University of Southampton, Southampton, SO9 5NJ.*

M.A. CLEMENTS; *School of Electrical Engineering, Georgia Institute of Technology, Atlanta, Georgia, U.S.A.*

G. CORSINI; *Istituto di Elettronica e Telecomunicazioni, Facolta di Ingegneria, Universita di Pisa, Via Diotisalvi 2, Pisa 56100, Italy.*

C.F.N. COWAN; *Department of Electrical Engineering, University of Edinburgh, Edinburgh, Scotland.*

E. DALLE MESE; *Istituto di Elettronica e Telecomunicazioni, Facolta di Ingegneria, Universita di Pisa, Via Diotisalvi 2, Pisa 56100, Italy.*

M. DEFRISE; *Theoretische Natuurkunde, Vrije Universiteit Brussel, Belgium.*

C. DE MOL; *Department of Mathematics, Universite Libre de Bruxelles, Belgium.*

G. DEMOMENT; *Laboratoire des Signaux & Systemes, Ecole Superieure d'Electricite, Plateau du Moulon, 91190 Gif-sur-Yvette, France.*

B. DE MOOR; *ESAT, Electrical Engineering Department, Katholieke Universiteit Leuven, Belgium.*

J. DE SOUSA PINTO; *Department of Mathematics, University of Aveiro, Portugal.*

P. DEWILDE; *Department of Electrical Engineering, Delft University of Technology, The Netherlands.*

X.C. DU; *Laboratoire des Signaux et Systemes, Plateau du Moulon, 91190 Gif-sur-Yvette, France.*

A.M. DUFFIELD; *Department of Physics, University of Surrey, Guildford, GU2 5XH.*

T.S. DURRANI; *Department of Electronic & Electrical Engineering, University of Strathclyde, 204 George Street, Glasgow, G1 1XW, Scotland.*

P.G. EARWICKER; *Space Department, Royal Aircraft Establishment, Farnborough, Hants, GU14 6TD.*

D.R. FARRIER; *Department of Electrical Engineering, The University, Southampton, SO9 5NH.*

M.A. FIDDY; *Physics Department, King's College, London University, Strand, London WC2R 2LS.*

A.R. FIGUEIRAS-VIDAL; *Department of Signal Theory & Processing, Polytechnic University of Madrid, Ciudad Universitaria, 28040 Madrid, Spain.*

P. FLANDRIN; *Laboratoire de Traitement du Signal, ICPI, Lyon, 25 Rue de Plat, 69288 Lyon Cedex 02, France.*

R.N. GORGUI-NAGUIB; *Imperial College of Science & Technology, Department of Electrical Engineering, Exhibition Road, London, SW7 2BT.*

C.E. GOUTIS; *Department of Electrical Engineering, Patras University, Patras, Greece.*

M.J.C. GOVER; *Department of Mathematical Sciences, University of Bradford, West Yorkshire, BD7 1DP.*

S. HAMMARLING; *Numerical Algorithms Group Ltd., NAG Central Office, 256 Banbury Road, Oxford, OX2 7DE.*

A. HOUACINE; *Laboratoire des Signaux et Systemes, Plateau du Moulon, 91190 Gif-sur-Yvette, France.*

M.H. HAYES; *School of Electrical Engineering, Georgia Institute of Technology, Atlanta, Georgia 30332, U.S.A.*

J.R. HIGGINS; *Cambridgeshire College of Arts & Technology, East Road, Cambridge, CB1 1PT.*

F. HLAWATSCH; *Institut fur Nachrichtentechnik und Hochfrequenztechnik, Technische Universitat Wien, A-1040 Wien, Austria.*

R.F. HOSKINS; *Department of Applied Computing & Mathematics, Cranfield Institute of Technology, Cranfield, Bedford, MK43 0AL.*

M. IBRAHIM SEZAN; *Department of Electrical & Electronic Engineering, University of Nottingham, Nottingham, NG7 2RD.*

M.K. IBRAHIM; *Department of Electrical & Electronic Engineering, University of Nottingham, Nottingham, NG7 2RD.*

D.J. JEFFRIES; *Department of Electrical Engineering, The University, Southampton, SO9 5NH.*

M.C. JONES; *Department of Physics, University of Surrey, Guildford, GU2 5XH.*

N.B. JONES; *Department of Engineering, University of Leicester, Leicester, LE1 7RH.*

M. KAVEH; *University of Minnesota, 123 Church Street South East, Minneapolis, MN 55455, U.S.A.*

J.T. KENT; *Department of Statistics, University of Leeds, Leeds, LS2 9JT.*

M. KIMBERLEY; *Department of Mathematical Sciences, Goldsmiths' College, University of London, New Cross, London, SE14 6NW.*

R.A. KING; *Department of Electrical Engineering, Imperial College of Science and Technology, Exhibition Road, London, SW7 2BT.*

R. KNOTSCH; *Abteilung fur Elektrotechnik, Ruhr-Universitat, 4630 Bochum, West Germany.*

P.J. LAGO; *R.D. Manuel II, Oporto, Portugal.*

R.M. LEAHY; *Signal & Image Processing Institute, University of Southern California, Los Angeles, California, CA 90089, U.S.A.*

K.V. MARDIA; *Department of Statistics, University of Leeds, Leeds, LS2 9JT.*

J.G. McWHIRTER; *Royal Signals & Radar Establishment, St. Andrews Road, Great Malvern, Worcs, WR14 3PS.*

D.H. MUGLER; *Department of Mathematics, University of Santa Clara, California, U.S.A.*

J.M. PAEZ-BORRALLO; *ETSI Telecommunicacion-UPM, Ciudad, Universitaria, 28040 Madrid, Spain.*

A. PAREKH; *Graduate Division of Biomedical Engineering, EAPS III, University of Sussex, Falmer, Brighton, Sussex, BN1 9QT.*

E.R. PIKE; *Royal Signals & Radar Establishment, St. Andrews Road, Great Malvern, Worcs, WR14 3PS.*

F. PIPER; *Department of Mathematics, Royal Holloway & Bedford New College, University of London, Egham, Surrey.*

C.V.K. PRABHAKARA RAO; *Department of ECE, Drexel University, Philadelphia, PA 19104, U.S.A.*

R. REGGIANNINI; *Istituto di Elettronica e Telecomunicazioni, Universita di Pisa, Via Diotisalvi 2, I-56100, Pisa, Italy.*

B. RUIZ-MEZCUA; *ETSI Telecommunicacion UPM, Cdad., Universitaria, 28040 Madrid, Spain.*

D. SAINT-FELIX; *Laboratoire des Signaux et Systemes, Plateau du Moulon, 91190 Gif-sur- Yvette, France.*

A.S. SETHI; *Department of Electrical Engineering, University of Windsor, Windsor, Ontario, Canada.*

D.B. SHARMAN; *Department of Electronic & Electrical Engineering, University of Strathclyde, 204 George Street, Glasgow, Scotland, G1 1XW.*

J. SHEN; *Universite Paul Sabatier 118, route de Narbonne, 31062 Toulouse, France.*

T.J. SHEPHERD; *Royal Signals & Radar Establishment, St. Andrews Road, Great Malvern, Worcs, WR14 3PS.*

V. SHVARTSMAN; *Electronic Design & Research, Inc., Louisville, KY, USA.*

J. SKILLING; *Department of Applied Mathematics & Theoretical Physics, University of Cambridge, Silver Street, Cambridge, CB3 9EW.*

M.J. SMITH; *Department of Electrical Engineering, University of Edinburgh, The King's Buildings, Edinburgh, EH9 3JL.*

M. SOUMEKH; *Department of Electrical Engineering, Worcester Polytechnic Institute, Worcester, MA, U.S.A.*

W. SPLETTSOSSER; *Lehrstuhl A fur Mathematik, RWTH Aachen, Aachen, West Germany.*

H. STARK; *School of Engineering, Rensselaer Polytechnic Institute, Troy, New York 12180-3590, U.S.A.*

K. STEWART; *Signal Processing Group, University of Strathclyde, 204 George Street, Glasgow, G1 1XW, Scotland.*

D.M. TITTERINGTON; *Department of Statistics, University of Glasgow, Scotland, G12 8QQ.*

E.A. TRACHTENBERG; *Department of ECE, Drexel University, Philadelphia, Pennsylvania 19104, U.S.A.*

J. VANDEWALLE; *Electrical Engineering Department, ESAT Laboratory, Kardinnaal Mecierlaan 94, B-3030 Heverlee, Belgium.*

E.I. VERRIEST; *Georgia Institute of Technology, School of Electrical Engineering, Atlanta, Georgia 30332, U.S.A.*

W.A. WARD; *Earth Sciences Department, University of Leeds, Leeds, LS2 9JT.*

P.E. WELLSTEAD; *Department of Mathematics, UMIST, PO Box 88, Manchester, M60 1QD.*

D.M. WILKES; *School of Electrical Engineering, Georgia Institute of Technology, Atlanta 30332, U.S.A.*

I SIGNAL ANALYSIS AND MODELLING

SIGNAL REPRESENTATIONS GEOMETRY AND CATASTROPHES IN THE TIME-FREQUENCY PLANE

P. Flandrin
(Laboratoire de Traitement du Signal, ICPI, Lyon, France)
and
F. Hlawatsch
(Institut für Nachrichtentechnik und Hochfrequenztechnik, Technische Universität Wien, Wien, Austria)

ABSTRACT

Generalized Wigner Distribution (GWD) is a class of time-frequency signal representations with particularly interesting properties. These properties being well known, we here take a more descriptive standpoint and discuss the "shape" of GWD. It is shown that GWD is in principle governed by the characteristic laws of interference geometry. For the important class of amplitude and frequency modulated signals, this result is confirmed and rendered more concrete by approximations using the stationary phase method and catastrophe theory. Optimality of Wigner Distribution, among the whole GWD class, is clearly revealed by its ideally concentrated and symmetric interference structure.

1. INTRODUCTION

Joint time-frequency representations of signals, like Wigner Distribution, Spectrogram, and Ambiguity Function, are valuable tools for signal analysis. This paper is concerned with Generalized Wigner Distribution (GWD), which, among the time-frequency representations of the energy density type, satisfies a great number of desirable properties and is thus most attractive from a theoretical standpoint (see Claasen and Mecklenbräuker (1980)).

Let $x(t) \leftrightarrow X(f)$, $y(t) \leftrightarrow Y(f)$ denote two signals with their spectra, where t is time and f is frequency. GWD $W_{x,y}^{(\alpha)}(t,f)$ is a family of time-frequency representations of x,y depending on a parameter $\alpha \in \mathbb{R}$,

$$W_{x,y}^{(\alpha)}(t,f) = \int_\tau x(t+(\tfrac{1}{2}+\alpha)\tau)\ y^*(t-(\tfrac{1}{2}-\alpha)\tau) e^{-j2\pi f\tau} d\tau$$
$$= \int_\nu X(f+(\tfrac{1}{2}-\alpha)\nu)\ Y^*(f-(\tfrac{1}{2}+\alpha)\nu)\ e^{j2\pi t\nu} d\nu. \qquad (1.1)$$

In most applications, a single signal is analyzed,

$$W_x^{(\alpha)}(t,f) \stackrel{\Delta}{=} W_{x,x}^{(\alpha)}(t,f). \qquad (1.2)$$

Two special cases of GWD are noteworthy: $\alpha=0$ and $\alpha=\tfrac{1}{2}$ yield Wigner Distribution (WD) and Rihaczek Distribution (RD), respectively. GWD will be complex-valued unless $\alpha=0$.

In this paper, rather than discussing formal properties of GWD, we shall take the user's standpoint and ask what GWD will look like in typical cases. It will be demonstrated that the shape of GWD is determined by a characteristic property which we call interference geometry. The general formulation of interference geometry will then clearly show that the choice $\alpha=0$, i.e. WD, is optimum for practical applications. The interference phenomenon has been previously studied in Janssen (1982), Flandrin (1984), Hlawatsch (1984).

2. OUTER INTERFERENCE

GWD is a bilinear signal transformation and is thus governed by a bilinear superposition law: if

$$z(t) = a.x(t) + b.y(t) \qquad (2.1)$$

then

$$W_z^{(\alpha)}(t,f) = |a|^2\ W_x^{(\alpha)}(t,f) + |b|^2\ W_y^{(\alpha)}(t,f) + I_{x,y}^{(\alpha)}(t,f), \qquad (2.2)$$

with the cross term or outer interference term

$$I_{x,y}^{(\alpha)}(t,f) = ab^*\ W_{x,y}^{(\alpha)}(t,f) + ba^*\ W_{y,x}^{(\alpha)}(t,f). \qquad (2.3)$$

Note that only for $\alpha=0$ (WD) this reduces to

$$I_{x,y}^{(0)}(t,f) = 2\ \mathrm{Re}\left\{ab^*\ W_{x,y}^{(0)}(t,f)\right\}. \qquad (2.4)$$

It is obvious that the occurrence of outer interference terms will strongly influence the shape of GWD in the case of multicomponent signals. To consider a simple test case, suppose that in (2.1) x and y are two signals x_+ and x_- which are derived from a given signal x_o through time-frequency shifts by $(\frac{\tau}{2},\frac{\nu}{2})$ and $(-\frac{\tau}{2},-\frac{\nu}{2})$ respectively. With the operator $s_\zeta^{(\eta)}$ denoting a shift by ζ with respect to the variable η, the components of (2.2) are then (assuming a = b = 1)

$$W_{x_\pm}^{(\alpha)}(t,f) = s_{\pm\tau/2}^{(t)}\, s_{\pm\nu/2}^{(f)} \left\{W_{x_o}^{(\alpha)}(t,f)\right\} \qquad (2.5)$$

$$W_{x_\pm,x_\mp}^{(\alpha)}(t,f) = s_{\mp\alpha\tau}^{(t)}\, s_{\pm\alpha\nu}^{(f)} \left\{W_{x_o}^{(\alpha)}(t,f)\; e^{-j2\pi(\nu t-\tau f \mp \alpha.\tau\nu)}\right\} \qquad (2.6)$$

While the overall position of the two "signal terms" (2.5) is thus independent from α, this is not at all true for the interference term components (2.6). In fact, (2.6) already expresses the interference geometry of GWD: the interference term consists in general of two distinct subterms which are derived from $W_{x_o}^{(\alpha)}$ by a plane-wave modulation and subsequent shifts by $(\mp\alpha\tau, \pm\alpha\nu)$. The modulation causes an oscillation in a direction orthogonal to the line connecting the two "signal terms" $W_{x_\pm}^{(\alpha)}$. Oscillation "frequency" is given by the distance between the two signal terms, it is independent of α.

A schematic illustration of interference geometry is given by Fig. 1.a. For $\alpha = \frac{1}{2}$ (RD) and $\alpha = -\frac{1}{2}$, signal and interference terms form the corners of a rectangle with lateral lengths τ,ν (Fig. 1.b). For α=0 (WD), the two interference subterms merge into a single real-valued interference term

$$I_{x_+,x_-}^{(0)}(t,f) = 2.W_{x_o}^{(0)}(t,f) \;.\; \cos 2\pi(\nu t - \tau f) \qquad (2.7)$$

located at the centre between the two signal terms (Fig. 1.c). The uniqueness of WD is hence stressed by interference geometry through an ideal concetration and symmetry of interference terms.

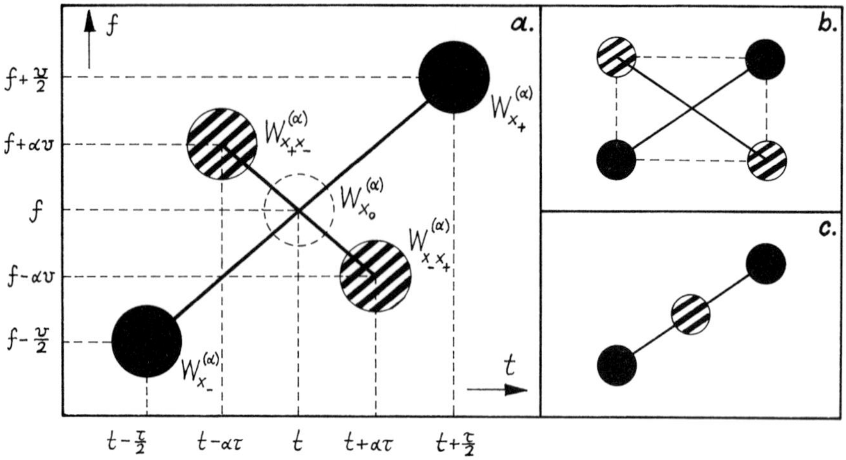

Fig. 1

The laws of interference geometry, although derived above by a simple example, are in fact a general property of GWD. A general indication of the position of outer interference terms (not, however, of their oscillations), is given by GWD's "outer interference formula" (comp. Janssen (1982))

$$|W^{(\alpha)}_{x,y}(t,f)|^2 =$$

$$\iint_{\tau\nu} W^{(\alpha)}_{x}(t+(\tfrac{1}{2}+\alpha)\tau,\ f+(\tfrac{1}{2}-\alpha)\nu)\ .\ W^{(\alpha)*}_{y}(t-(\tfrac{1}{2}-\alpha)\tau, f-(\tfrac{1}{2}+\alpha)\nu)\ d\tau d\nu : \tag{2.8}$$

if the signal terms $W^{(\alpha)}_{x}$, $W^{(\alpha)}_{y}$ are located around points $(t\pm{}^{\tau}/_{2},\ f\pm{}^{\nu}/_{2})$ in the (t,f) -plane, then the interference term $I^{(\alpha)}_{x,y}$ is located around points $(t\mp\alpha\tau,\ f\pm\alpha\nu)$. This agrees with interference geometry.

3. INNER INTERFERENCE

So far, the interference character of GWD has been considered as a direct consequence of GWD's bilinear superposition law. Interference terms, however, are not always explicit bilinear cross terms of distinct signal components (outer interference). Rather, GWD also shows an inherent interference structure which is equally present in the GWD of monocomponent signals (inner interference).

Consider the GWD of a signal x which may be monocomponent. With y=x, (2.8) yields the "inner interference formula"

$$|W_x^{(\alpha)}(t,f)|^2 =$$

$$\iint_{\tau\nu} W_x^{(\alpha)}(t+(\tfrac{1}{2}+\alpha)\tau, f+(\tfrac{1}{2}-\alpha)\nu) \,.\, W_x^{(\alpha)*}(t-(\tfrac{1}{2}-\alpha)\tau, f-(\tfrac{1}{2}+\alpha)\nu)\, d\nu. \qquad (3.1)$$

If $W_x^{(\alpha)}$ is large and non-oscillatory around points $(t\pm{}^{\tau}/_{2}, f\pm{}^{\nu}/_{2})$ (we could then say that some part of signal energy is located there), then $W_x^{(\alpha)}$ is also large around the points $(t\mp\alpha\tau, f\pm\alpha\nu)$. These, then, are points of inner interference, in perfect agreement with interference geometry.

Inner interference terms also show the characteristic oscillations described by interference geometry. This will be demonstrated for an important class of signals in the next section.

4. GWD OF AM-FM SIGNALS

We consider a complex signal x(t) with amplitude and frequency modulation (AM-FM)

$$x(t) = a(t)\, e^{j\phi(t)}, \quad a \geqslant 0, \ \phi \in \mathbb{R}, \qquad (4.1)$$

where the AM term a(t) is slowly varying and the instantaneous phase $\phi(t)$ is determined by instantaneous frequency $f_i(t)$,

$$f_i(t) = \frac{1}{2\pi}\, \phi'(t). \qquad (4.2)$$

For such a signal, GWD can be written as

$$W_x^{(\alpha)}(t,f) = \int_{\tau} L(t,\tau)\, e^{j\Phi(t,f;\tau)} d\tau \qquad (4.3)$$

with

$$L(t,\tau) = a(t+(\tfrac{1}{2}+\alpha)\tau).a(t-(\tfrac{1}{2}-\alpha)\tau) \qquad (4.4)$$

$$\Phi(t,f;\tau) = \phi(t+(\tfrac{1}{2}+\alpha)\tau)-\phi(t-(\tfrac{1}{2}-\alpha)\tau) - 2\pi f.\tau. \qquad (4.5)$$

GWD (4.3) is an oscillatory integral for which an approximate solution is obtained by the method of stationary phase (see e.g. Papoulis (1977)): for a specific point (t_o, f_o), let $\{\tau_i\}$,

$i=1,\ldots,n$ be the values of τ for which

$$\Phi_o'(\tau_i) = 0 \qquad \text{and} \qquad \Phi_o''(\tau_i) \neq 0, \tag{4.6}$$

where $\Phi_o(\tau)$ is short for $\Phi(t_o,f_o;\tau)$. Then each of the τ_i will yield a distinct contribution to $W_x^{(\alpha)}(t_o,f_o)$:

$$W_x^{(\alpha)}(t_o,f_o) \simeq \sum_{i=1}^{n} \frac{L(t_o;\tau_i)}{\sqrt{|\Phi_o''(\tau_i)|}} e^{j[\Phi_o(\tau_i) + \frac{\pi}{4}\,\text{sign}\,\Phi_o''(\tau_i)]}. \tag{4.7}$$

Eqs. (4.6) amd (4.5) yield a relation between t_o,f_o and τ,

$$f_o = \frac{1}{2}[f_i(t_1) + f_i(t_2)] + \alpha\,[f_i(t_1) - f_i(t_2)] \tag{4.8}$$

and the condition

$$(\tfrac{1}{2}+\alpha)^2 . f_i'(t_1) \neq (\tfrac{1}{2}-\alpha)^2 . f_i'(t_2) \tag{4.9}$$

where $t_1 = t_o + (\frac{1}{2}+\alpha)\tau$, $t_2 = t_o - (\frac{1}{2}-\alpha)\tau$. Defining t,f and ν by

$$t = \tfrac{1}{2}[t_1+t_2],\ \tau = t_1-t_2\ ;\ f = \tfrac{1}{2}[f_i(t_1)+f_i(t_2)],\ \nu = f_i(t_1)-f_i(t_2), \tag{4.10}$$

this can be formulated as follows: two points $f_i(t_1)$, $f_i(t_2)$ of the instantaneous frequency curve $f_i(t)$ interfere and create two subterms of inner interference at points $(t_o,f_o) = (t\mp\alpha\tau, t\pm\alpha\nu)$, according to whether $+\tau$ or $-\tau$ is inserted in (4.7). This is shown by Fig. 2. The connection with interference geometry is easily found by regarding the instantaneous frequency curve $f_i(t)$ as signal term; different points of this one signal term then interfere according to the laws of interference geometry (comp. Fig. 1).

The interference terms' oscillation, too, can be derived from the stationary phase approximation (4.7). Consider, for a specific point (t_o,f_o), a single contribution to (4.7) given by some τ. In local surroundings of (t_o, f_o), i.e. for small $|\Delta t|$, $|\Delta f|$, (4.7) then yields

$$W_x^{(\alpha)}(t_o+\Delta t, f_o+\Delta f) \simeq C(t_o,f_o).a(t_1)a(t_2).e^{j2\pi[\nu.(t_o+\Delta t)-\tau(f_o+\Delta f)]}, \tag{4.11}$$

where $f_i(t_1)$ and $f_i(t_2)$ are the interfering points of the instantaneous frequency curve (see above) and $\tau=t_1-t_2$, $\nu=f_i(t_1)-f_i(t_2)$ are the time and frequency differences between these points. This is exactly the plane-wave oscillation described by interference geometry.

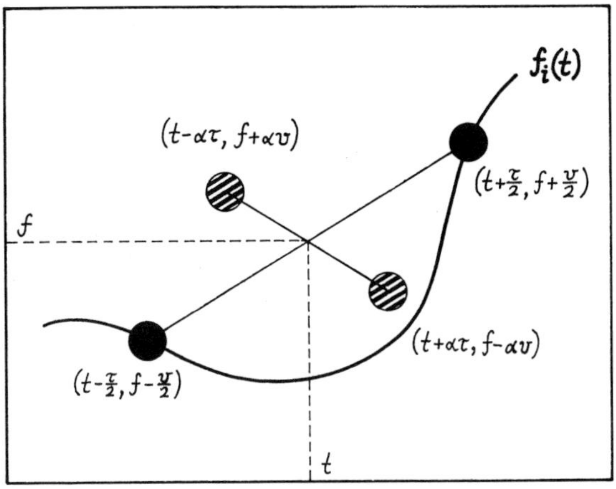

Fig. 2

The strong dependence of the interference terms' position on α will sometimes lead to dramatic differences between the shapes of different GWDs. Fig. 3 compares WD ($\alpha=0$) and real part of RD ($\alpha=\frac{1}{2}$) of a chirp signal (FM linear, AM according to raised-cosine window). Note that these results can be predicted from interference geometry (comp. Fig. 1.c, Fig. 1.b).

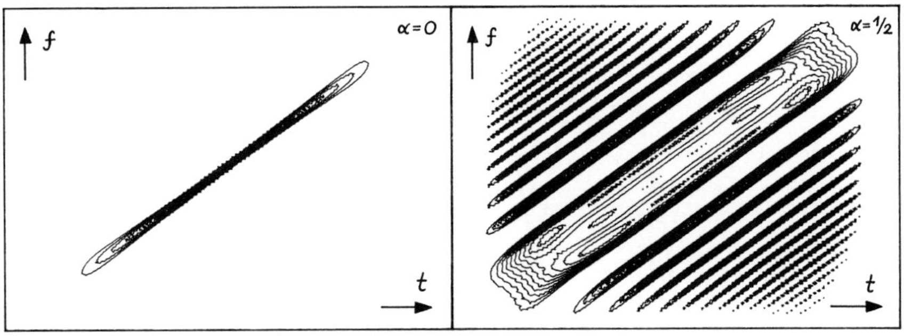

Fig. 3

5. CATASTROPHES OF WD

We consider again the AM-FM signal (4.1) but restrict to the WD case ($\alpha=0$) for simplicity. For a specific (t_o,f_o), the stationary points are given by (4.8)

$$f_o = \frac{1}{2}[f_i(t_o + \tau/2) + f_i(t_o - \tau/2)], \tag{5.1}$$

provided that (4.9)

$$f_i'(t_o + \tau/2) \neq f_i'(t_o - \tau/2). \tag{5.2}$$

For a geometrical interpretation, consider the simple example of a sinusoidal FM shown in Fig. 4. Evidently, there are three cases:

i) For (t_o, f_o) inside the shaded area (point P_1), there are two stationary points $\pm\tau$ whose combination (see (4.7)) yields an oscillatory structure (the local cosine terms of inner WD interference). According to (5.1), the two stationary points $\pm\tau$ are defined by the chord whose endpoints lie on $f_i(t)$ and whose midpoint is P_1 (interference geometry, comp. Fig. 2 with $\alpha=0$).

ii) On $f_i(t)$ (point P_2), the two stationary points coalesce ($\tau \to 0$, zero length chord). Since (5.2) is no longer valid, the stationary phase approximation (4.7) diverges.

iii) Outside the shaded area (point P_3), there is no τ satisfying (5.1), and the stationary phase approximation vanishes.

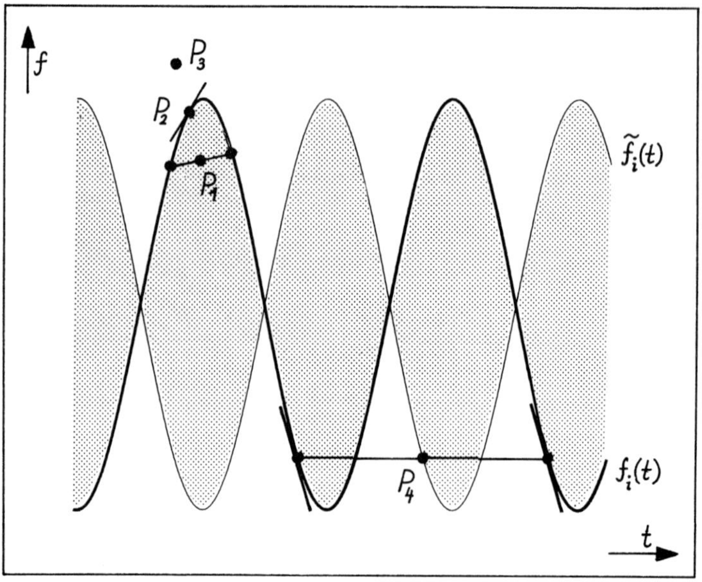

Fig. 4

The sudden change in behaviour which occurs when passing through the boundaries of the shaded area in Fig. 4 can be viewed as a "catastrophe" of WD. In fact catastrophe theory (CT) can be applied to yield a qualitative description of these changes, see Poston and Stewart (1978), Berry (1977), Flandrin and Escudié with contributions of Bouachache (1981). According to CT, the occurrence of "catastrophic" changes of oscillatory integrals like (4.3) depends only on the singularities (vanishing of derivatives) of the phase function $\Phi_o(\tau)$. This, in a sense, generalizes the stationary phase method. The specific interest of CT lies in the fact that it yields a classification of the basic geometrical structures (morphologies) which typically occur. In our case, only two such typical structures ("catastrophes") exist, namely, the "fold" and the "cusp" catastrophe (see Poston and Stewart (1978)).

It follows from this that, in the qualitative behaviour of WD, four cases can be distinguished, according to the derivatives of the phase function $\Phi_o(\tau)$ (see (4.5) with α=0):

i) *Stationary point areas* (e.g. the shaded area in Fig. 4) consist of those points (t_o, f_o) for which there exist two τ's such that $\Phi_o'(\tau) = 0$, $\Phi_o''(\tau) \neq 0$. An approximation of WD is here given by the stationary phase method (cosine function).

ii) *Fold lines* consist of those points (t_o, f_o) for which $\Phi_o'(\tau) = \Phi_o''(\tau) = 0$, $\Phi_o'''(\tau) \neq 0$. This, in particular, implies

$$f_i'(t_o + \tau/2) = f_i'(t_o - \tau/2) , \qquad (5.3)$$

so that here the slopes of $f_i(t)$ at the chord endpoints (interfering points) are equal. As shown by Fig. 4, there are two types of fold lines: a) the instantaneous frequency curve $f_i(t)$ itself (e.g. point P_2) for which τ=0; an approximation of WD is here given by the Airy function; b) a "ghost" curve $\tilde{f}_i(t)$ (e.g. point P_4) for which $\tau \neq 0$.

iii) *Cusp points* (t_o, f_o) are characterized by $\Phi_o'(\tau)=\Phi_o''(\tau)= \Phi_o'''(\tau) = 0$, $\Phi_o''''(\tau) \neq 0$. They are points of fold lines (ghost type) for which, in addition to equal slopes (see (5.3)), the curvatures of $f_i(t)$ at the chord endpoints are equal in magnitude but opposite in sign (see Fig. 5, point P),

$$f_i''(t_o + {}^{\tau}/_2) = - f_i''(t_o - {}^{\tau}/_2). \qquad (5.4)$$

At cusp points, the WD is peaked and can be approximated by a Pearcey function.

iv) *Points of higher-order singularities* are midpoints of an infinity of chords. They occur only in the case of a perfectly (anti-) symmetrical $f_i(t)$. At these points, the WD is highly peaked. Fig. 6 gives three examples (point P).

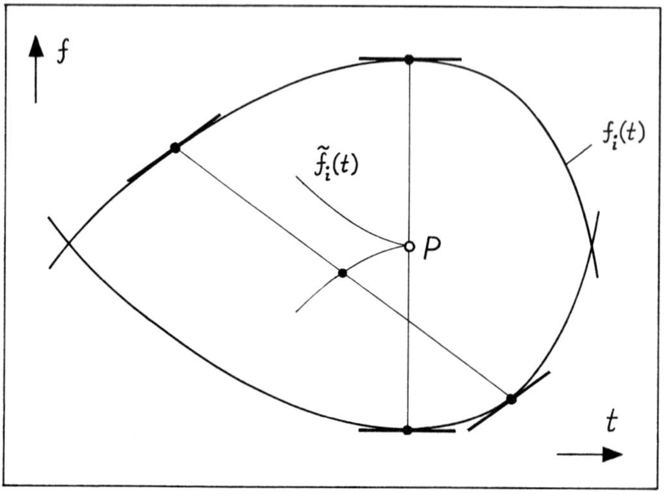

Fig. 5

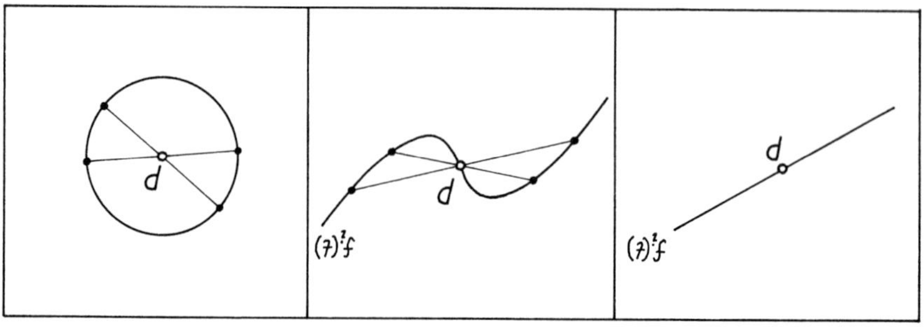

Fig. 6

6. CONCLUSION

GWD is in principle governed by the simple laws of interference geometry. Variation of GWD's parameter α will essentially lead to a variation of the interference terms' position which may have a dramatic influence on GWD's shape. Thus different GWDs of the same signal may look totally different even though most of the mathematical properties are the same. It is thus from the standpoint of interference geometry that "optimality" of WD (α=0) is most evident.

The description of the GWD and WD of AM-FM signals which was based on the stationary phase method and catastrophe theory is consistent with interference geometry but at the same time yields more concrete results. We stress that the approach taken here could be extended to bilinear signal representations other than GWD. In particular, the class of Generalized Ambiguity Functions (which is dual to GWD by Fourier transform, see Hlawatsch (1985)) shows an interference geometry which is analogous to that of GWD (see Flandrin and Escudié (1981), Flandrin (1984)).

REFERENCES

Berry, M.V., (1977) "Semi-classical mechanics in phase-space: a study of Wigner's function", *Phil. Trans. Roy. Soc.* London A 287, pp. 237-271.

Claasen, T.A.C.M. and Mecklenbräuker, W.F.G., (1980) The Wigner Distribution - a Tool for Time-Frequency Signal Analysis", Part III. *Philips J. Res.* **35**, pp. 372-389.

Flandrin, P. and Escudié, B., (1981) "Géométrie des fonctions d'ambiguité et des représentations conjointes de Ville: l'approche de la théorie des catastrophes". 8è Coll. Trait. Signal GRETSI, Nice, pp. 69-74.

Flandrin, P., (1984) "Some Features of Time-Frequency Representations of Multicomponent Signals", Proc. ICASSP 1984, San Diego, Calif., pp. 41.B.4.1-41.B.4.4.

Hlawatsch, F., (1984) "Interference Terms in the Wigner Distribution". Int. Conf. on Digital Signal Processing, Florence, pp. 363-367.

Hlawatsch, F., (1985) "Duality of Time-Frequency Signal Representations: Energy Density Domain and Correlation Domain". IASTED Int. Sympos., Paris.

Janssen, A.J.E.M., (1982) "On the Locus and Spread of Pseudo-Density Functions in the Time-Frequency Plane". *Philips J. Res.* **37**, pp. 79-110.

Papoulis, A., (1977) "Signal Analysis". Mc Graw-Hill.

Poston, T. and Stewart, I., (1978) "Catastrophe Theory and its Applications". Pitman, London.

A UNIFIED THEORY OF MODEL REDUCTION VIA GLEASON MEASURES

E.I. Verriest
*(School of Electrical Engineering,
Georgia Institute of Technology, Atlanta, Georgia)*

ABSTRACT

Earlier work has cast the stochastic realization and approximation problem in the framework of the RV-coefficient. This allowed the introduction of a common measure for the "goodness of fit" for the different realization algorithms. This paper explores the deeper geometrical basis for this common measure in a unified theory for the data driven and exact covariance approaches.

1. INTRODUCTION

1.1 Scope of the Paper

In the theory of identification, signal processing, and digital filtering, a problem of fundamental importance is that of finding a finite dimensional Markovian representation of a stochastic process from the covariance information. This problem is known as the Stochastic Realization Problem, and has received a great deal of attention. Whenever a finite set of real data is gathered, all processing is done over finite sets, and an underlying probabilistic description is absent in most cases. As a result, covariances must be estimated by sample covariances, and a "degradation" of the theoretical realization solutions results. A more direct, data driven approach is needed. Moreover, for many applications, the Markovian representation or state space model may be too complex, due to its high dimensionality, thus barring efficient computational management. This motivates the quest for approximate lower order models, and the need for common measures to evaluate and compare different approaches.

1.2 *Historical Background*

Akaike (1975), Faurre (1976), and Baram (1981) developed a stochastic realization theory based on the information interface between the past and the future of a time series and the concepts of canonical correlation analysis (CCA). Desai and Pal (1982) extended the results. They obtain forward-backward dual models with state covariances which are equal and diagonal. They are the stochastic counterpart of the deterministic balanced realizations. See Moore (1981) and Verriest and Kailath (1983).

Arun and Kung (1983) proposed the Karhunen-Loeve method (KLM) as a basis for the stochastic realization.

Ramos and Verriest (1984) unified the theory by showing that both CRA and KLM, given the exact covariances, are special cases of a more general optimization problem, using the RV-coefficient introduced by Escoufier (1973). Verriest (1985) explored the connection of the exact covariance and real data case further by relating the RV-coefficient to certain operators in a tensor product space.

2. STOCHASTIC REALIZATION PROBLEM

2.1 *Problem Formulation*

Given the covariance sequence $\Lambda(k)$ of a rational stationary, zero mean, discrete time vector sequence $\{y_k\}$, the stochastic realization problem consists in finding a Markovian representation of the form

$$x_{k+1} = Fx_k + w_k \tag{2.1.1}$$

$$\bar{y}_k = Hx_k + v_k \tag{2.1.2}$$

where $\{w_k\}$ and $\{v_k\}$ are White Gaussian noises with

$$E\begin{bmatrix} w_k \\ v_k \end{bmatrix} [w'_e \quad v'_e] = \begin{bmatrix} Q & S \\ S' & R \end{bmatrix} \delta_{k,\ell} \;;\; \forall k,\ell \tag{2.1.3}$$

such that $E(\bar{y}_{k+n}\bar{y}'_k) = \Lambda(k)$. $\delta(k,\ell)$ is the Kronecker delta.

The solution to this problem is described by Faurre (1976).

2.2 *Information Interface Between Past and Future*

Assume that the stochastic time-series $\{y_k\}$ is Gaussian (with zero mean). The relevant random variables are then in the Hilbert space $L_2(\Omega,B,P)$ and conditional expectations can be interpreted as orthogonal projections onto subspaces $L_2(\Omega, F_k^{\{y_k\}}, P)$.

For the time-series $\{y_k\}$ define the infinite vectors

$$y_k^+ = \begin{bmatrix} y_k \\ y_{k+1} \\ \vdots \end{bmatrix}, \text{ the future} \qquad y_k^- = \begin{bmatrix} y_{k-1} \\ y_{k-2} \\ \vdots \end{bmatrix}, \text{ the past} \tag{2.2.1}$$

and define the semi-infinite covariance matrices

$$\hat{H} = E\{Y_k^+(Y_k^-)'\},\ R^+ = E\{Y_k^+(Y_k^+)'\},\ R^- = E\{Y_k^-(Y_k^-)'\}\ . \tag{2.2.2}$$

Within this representation, the forward and backward predictor subspaces are

$$X_k = \mathrm{Span}(Y_k^+ \mid Y_k^-),\ Z_{k-1} = \mathrm{Span}(Y_k^- \mid Y_k^+) \tag{2.2.3}$$

$(A|B)$ denotes the projection of span (A) onto the Hilbert space spanned by the components of B. These two spaces form the information interface between R^+ and R^-. Either one can be used to define a minimal Markovian representation (forwards or backwards). The canonical correlations lead to a natural distance measure between the past and the future, which for the Gaussian case is exactly the Kullback-Leibler information.

Alternatively, the past can be treated as the instrumental variables for predicting the future. See Arun and Kung (1983).

Ramos and Verriest (1984) and Ramos (1985) resolved the two methods by putting them in a common framework, optimizing Escoufier's RV-coefficient (1973) under different constraints. If for random vectors X and Y

$$\operatorname{cov}(X,Y)=\begin{bmatrix}\Sigma_{11} & \Sigma_{12}\\ \Sigma_{21} & \Sigma_{22}\end{bmatrix},\ \text{then}\quad RV(X,Y)=\frac{\operatorname{Tr}(\Sigma_{12}\Sigma_{21})}{\sqrt{(\Sigma_{11}^2)\ \operatorname{Tr}(\Sigma_{22}^2)}}\geqslant 0 \tag{2.2.4}$$

This measure also allows the computation of a "figure of merit" for each algorithm in a consistent way. Our new Gleason approach gives a natural interpretation for (2.2.4) (and other) measures.

3. THE LATTICE OF SUBSPACES AND GLEASON MEASURES

Let H be a Hilbert space. The set of all closed subspaces of H has the structure of an orthocomplemented complete lattice, also called a logic. A one-to-one correspondence exists between the lattice of all closed subspaces of H and the lattice Proj H of all orthoprojectors on H.

In his study of the mathematical foundations of quantum mechanics, Mackey posed the problem of finding all positive measures on the closed subspaces of a Hilbert space. Such a measure must have the property that for any countable collection $\{S_i\}$ of mutually orthogonal closed subspaces the mapping is σ-additive, i.e.,

$$\sum_i \mu(S_i) = \mu(\sum_i S_i) \tag{3.1}$$

A measure satisfying the above property is for instance obtained by selecting a vector v in the Hilbert space H, and for each subspace A of H defining

$$\mu_v(A) = \| P^A(v) \|^2 \tag{3.2}$$

where P^A is the projection operation on A. Clearly, finite convex conbinations of such measures also satisfy the conditions for such measures, and passing to the limit, any positive semidefinite trace class operator T also defines such a measure via

$$\mu(A) = \operatorname{Tr}(TP^A) \tag{3.3}$$

Gleason (1957) has shown that in a separable Hilbert space of dimension at least three, every measure on the closed subspaces can be represented as above, with T a positive definite operator of trace class. Further extensions of Gleason measures which are vector and operator valued, have been surveyed by Jajte (1979).

Let H be a separable Hilbert space, and Proj H the lattice of all orthogonal projectors in H. Let also E be some Banach space, then:

Definition: A mapping ξ : Proj H $\to$ E is said to be an E-valued Gleason measure if

(1) For any sequence of pairwise orthogonal projectors $P_1, P_2, \ldots$ from Proj H

$$\sum_i \xi P_i = \xi(\sum_i P_i) \qquad (3.4)$$

the series on the left hand side being weakly convergent,

(2) $\sup\{\| \xi P_i \| : P_i \varepsilon \text{ Proj H}\} < \infty$ (3.5)

An important class of Gleason measures taking values in a Hilbert space are the Orthogonally Scattered Measures (OSG). Let H and K be two Hilbert spaces, with dimension at least three.

Definition: A Gleason measure ξ : Proj H $\to$ K is said to be an orthogonally scattered measure (OSG-measure) if for any orthogonal projectors P,Q in Proj H the following implication holds.

$$P \perp Q \dashrightarrow \xi p \perp \xi Q \qquad (3.6)$$

Note that automatically for all P $\| \xi P \| \leqslant \| \xi I \|$ is implied, where I is the identity operator, corresponding with the (sub) space H.

Any OSG-measure defines a positive Gleason measure by

$$\mu P = \| \xi P \|^2, \; P \varepsilon \text{ Proj H} \qquad (3.7)$$

By Gleason's theorem, there exists then a nonnegative self-adjoint trace-class operator T such that

$$\mu P = \text{Tr TP} \;, \; P \;\varepsilon \text{ Proj H} \qquad (3.8)$$

The above can be interpreted as a "variance"; if we similarly define a "covariance" by $\text{COV}\{P,Q\} = (\xi P, \xi Q)_K$, then for any commuting projectors P, Q ε Proj H one has

$$(\xi P,\xi Q)_K = \text{Tr } TPQ \tag{3.9}$$

This formula is not true in general for arbitrary projectors in a complex Hilbert space. However, if H and K are real Hilbert spaces, then it can be shown that (for T given by Gleason's theorem (3.9))

$$(\xi P,\xi Q)_K = \text{Tr } TPQ = \text{Tr } TQP, \text{ all } P, Q \,\varepsilon\, \text{Proj } H \tag{3.10}$$

4. APPLICATIONS TO REALIZATION THEORY

4.1 A Correlation Measure for Subspaces

As shown by Verriest (1985), the (exact) stochastic realization and the (real) signal modelling benefit from the use of the RV-coefficient. In the first case, the formalism is used in the comparison of random variables, while in the second it compares data matrices. We present here a geometric point of view, motivated by the observation that for the stochastic realization problem, the underlying space $L_2^p(\Omega,B,m)$ and in the real data case, the space R^{pxN} are isomorphic with the tensor product spaces, respectively

$$L_2^p(\Omega,B,m) \sim R^p \otimes L_2(\Omega,B,m) \tag{4.1.1}$$

$$R^{pxN} \sim R^p \otimes R^N \tag{4.1.2}$$

In general, let G and H be separable Hilbert spaces. Let $\{\phi_i\}$ be a Complete Orthonormal Set (CONS) in G, and $\{\psi_i\}$ a CONS in H. Any vector x in the tensor product space $G \otimes H$ has a decomposition

$$x = \sum_i |\, x_i > < \psi_i \,| \tag{4.1.3}$$

where $x_i \,\varepsilon\, G$. The vector x in the tensor product space will be referred to as a "prior." Introduce now the superposition of measures on Proj G induced by the prior.

$$\mu_x = \sum_i \mu_i \tag{4.1.4}$$

The μ_i are the Gleason measures corresponding to x_i. For all subspaces of G, it follows that

$$\mu_x(A) = \text{Tr } T_x P^A \tag{4.1.5}$$

where

$$T_x = \sum_i | x_i > < x_i | = xx' \tag{4.1.6}$$

is interpreted as a gramian or covariance operator.

The measure $\mu_x(A)$ gives a numeric value to the closeness of A to G, given the prior x.

The problem of finding the subspace of fixed dimension which "looks most like H from the point of view of x" is then solved by letting P^A be the projector on the eigenspace of T_x with the largest principal components. See Aragon and Couot (1976), who also stated several equivalent problems relating to the principal component analysis. Note that $\mu_x(H) = \text{Tr } T_x$.

However, this measure does not lead to a useful definition of the correlation between subspaces. Indeed, consistent with the above "variance" μ_x we have the covariance (using 3.10)

$$(\xi_x(A), \xi_x(B)) = \text{Tr } T_x P^A P^B \tag{4.1.7}$$

But for $A \perp B$, we get $(\xi_x(A), \xi_x(B)) = 0$. There is no interface between A and B. This situation is displeasing, but can be resolved. The operator $T_x : G \to G$ is a characteristic for the given x in $G \otimes H$ (in fact, a "sufficient statistic"), and one can think of T (or μ) as conditioned by the vector $x \in G \otimes H$. In this sense, the extended projectors $\tilde{P}^B \in \text{Proj } G \otimes H$ defined by

$$\tilde{P}^B x = \sum_i P^B | x_i > < \phi_i | = \sum_i \xi_{x_i}(B) < \phi_i | \tag{4.1.8}$$

yield a "coherent" addition of OSG measures, conditioned on x (i.e., a posterior measure). The posterior variance of $A \in \text{Proj } G$, given x is then the operator from $G \to G$

$$(\tilde{P}^A x)(\tilde{P}^A x)' = \sum_i P^A |x_i><x_i| P^A = P^A T_x P^A \tag{4.1.9}$$

and the covariance

$$(\tilde{P}^B x)(\tilde{P}^A x)' = \sum_i P^B |x_i\rangle\langle x_i| P^A = P^B T_x P^A \qquad (4.1.10)$$

This displays the coupling or interface between A and B given x. In order to attach a numerical value to this interface, any norm on the various restrictions $P^B T_x P^A$ can be chosen. The following natural definition follows.

Definition: The "Correlation" between subspaces A and B in Proj G is

$$\rho(A,B|x) = \frac{\| P^A T_x P^B \|}{\sqrt{\| P^A T_x P^A \| \; \| P^A T_x P^A \|}} \qquad (4.1.11)$$

Let K be the fixed subspace span $(\phi_1, \ldots, \phi_k)$ of G, then the principal component analysis and canonical correlation analysis are respectively (in Frobenius norm) (O(K) is the orthogonal group on K)

$$\text{PCA} : \max_{M \in O(K)} \frac{\mathrm{Tr}\,(MT_{12}T_{21}M')}{\sqrt{\mathrm{Tr}\,(MT_{11}M')^2 \; \mathrm{Tr}\,(T_{22})^2}} \qquad (4.1.12)$$

$$\text{CCA} : \max_{\substack{M \in O(K) \\ N \in O(K^{\perp})}} \frac{\mathrm{Tr}\,(MT_{12}NN'T_{21}M')}{\sqrt{\mathrm{Tr}\,(MT_{11}M')^2 \; \mathrm{Tr}\,(NT_{22}N')^2}} \qquad (4.1.13)$$

which are the formulas obtained by Robert and Escoufier (1976). If $G \otimes H = L^2_p(\Omega, B, m)$ or R^{pN} then T is respectively the covariance Σ or sample covariance S.

4.2 Data Driven Stochastic Realization Solution

Assuming that a data stream $\{y_k, |k| \leq N\}$ is observed, a data matrix Y can be formed by considering $(Y_{-N}, 0, \ldots, 0)'$, $(Y_{-N+1}, Y_{-N}, 0, \ldots, 0)'$ ... $((Y_N, Y_{N-1}, \ldots, Y_{-N})'$, ..., $(0, \ldots, 0, Y_N)'$

as consecutive samples of the vector in $G = R^{2N+1}$ (the "pure states"). In order to avoid the nasty end effects due to the substitution of zeros where data is missing, a linear superposition of these states, weighted by the sequence $\{q_j \geq 0;\ j=1,\ldots,4N+1\}$ may be used. Let $Q = \text{diag}\{q_i\}$. The Gleason operator is $T_{(Y,q)} = YQY'$. The "past" is span $\{|\phi_{-N}\rangle \ldots |\phi_{-1}\rangle\}$, and the future span $\{|\phi_0\rangle \ldots |\phi_N\rangle\}$. A recursive realization algorithm, which optimally uses all the data (in real time) would necessarily involve the update of $T_{(Y,q)_N}$.

5. CONCLUSIONS

By determining well motivated measures for the correlation between subspaces of a Hilbert space, based on the available prior information, it was possible to unify several tools from multivariate analysis, and introduce common measures for their evaluation. We have only discussed the principal component and the canonical correlation analyses. Discriminant analysis and a rational way for discarding variables can also be treated. Our inspiration for this work came from the desire to better motivate and explain the use of the RV-statistic problems. In particular, exact realization theory and its signal processing counterpart (i.e., the real data case) are unified. The fact that the data should come first looks natural from this viewpoint. Deterministic modelling, cluster analysis (in pattern recognition), and quantization of random fields are other applications of our abstract framework. The variation lies in the choice of the spaces G and H, and the constraints that are natural for the problem.

6. REFERENCES

Akaike, A., (1975) Markovian Representation of Stochastic Processes by Canonical Variables, *SIAM J. Control* **13**, No 4, 162-173.

Aragon, Y. and Couot, J., (1976) Une Definition de l'Operateur d'Escoufier, *Comptes rendus,* 283, series A, 867-869.

Arun, K.S., Bhaskar Rao, D.V. and Kung, S.Y., (1983) A New Prediction Efficiency Criterion for Approximate Stochastic Realization, Proc. 22nd Conf. on Decision and Control, 1353-1365.

Baram, Y., (1981) Realization and Reduction of Markovian Models from Nonstationary Data, *IEEE Trans. Automatic Control AC-26,* no. 6, 1225-1231.

Desai, U.B. and Pal, D., (1982) A Realization Approach to Stochastic Model Reduction and Balanced Stochastic Realizations, Proc. 21st Conf. on Decision and Control, 1105-1111.

Escoufier, Y., (1973) Le Traitement des Variables Vectorielles, *Biometrics* 29, 751-760.

Faurre, P.L., (1976) Stochastic Realization Algorithms, in: Mehra, R.K. and Lainiotis, D.G., (eds.), System Identification: Advances and Case Studies, (Academic Press).

Gleason, A.M., (1957) Measures on the Closed Subspaces of a Hilbert Space, *J. Math. Mech.* **6**, 885-893.

Jajte, R., (1979) Gleason Measure, in: Bharucha-Reid (ed.), Probabilistic Analysis and Related Topics, Vol. 2 (Academic Press).

Moore, B.C., (1981) Principal Component Analysis in Linear Systems: Controllability, Observability, and Model Reduction, *IEEE Trans. Automatic Control,* AC-26, No. 5 (1) 17-32.

Ramos, J.A., (1985) A Stochastic Approach to Streamflow Modelling, Ph.D. Dissertation, School of Civil Engineering, Georgia Institute of Technology.

Ramos, J.A. and Verriest, E.I., (1984) A Unifying Tool for Comparing Stochastic Realization Algorithms and Model Reduction Techniques, Proc. 1984 ACC, San Diego.

Robert, P. and Escoufier, Y., (1976) A Unifying Tool for Linear Multivariate Statistical Methods: The RV- Coefficient, *Appl. Statist.* **25**, no. 3, 257-265.

Verriest, E.I., (1985) Projection Techniques for Model Reduction, in Lindquist, A., and Byrnes, C., (eds.) Modelling, Identification and Robust Control, (North Holland).

Verriest, E.I. and Kailath, T., (1983) On Generalized Balanced Realizations, *IEEE Trans. Automatic Control, AC-28,* no. 8, 833-844.

A FRESH APPROACH TO THE DERIVATIVE SAMPLING THEOREM

J.R. Higgins
*(Department of Science,
Cambridgeshire College of Arts and Technology, Cambridge)*

ABSTRACT

Convergence properties of the derivative sampling series are addressed. Two forms of the series are discussed, one which allows for over-sampling, the other in which sampling is at Nyquist rate. Some attention is given to the way in which the two parts of the series partition the signal to be reconstructed, and to which subsets of the Hilbert space of finite energy band-limited signals these parts belong.

1. INTRODUCTION

The reconstruction of a band-limited signal from knowledge of its samples together with samples of its derivatives was suggested by C.E. Shannon (see, e.g., Higgins (1985) p. 60 and the references cited there). The formula

$$f(t) = \sum_{n=-\infty}^{\infty} \left\{ \frac{f'(n)}{\pi} \frac{\sin^2 \pi(t-n)/2}{\pi(t-n)/2} + \frac{f(n)}{2} \left[\frac{\sin\pi(t-n)/2}{\pi(t-n)/2} \right]^2 \right\} \tag{1.1}$$

is known to hold for every f belonging to the Hilbert space PW of Paley-Wiener functions, that is, the class of finite energy signals band-limited to $[-\pi,\pi]$. The convergence is in the norm of PW and also pointwise, uniformly over all of $\mathbb{R}$.

For the sake of brevity let

$$s_p^q(t) \triangleq \frac{\sin^q(\pi t/q)}{(\pi t/q)^p}, \quad 0 < p \leq q.$$

Then the series in (1.1) is

$$\Sigma \left\{ \frac{f'(n)}{\pi} S_1^2(t - n) + \frac{f(n)}{2} S_2^2(t - n) \right\} .$$

Running indices on summation signs will continue to be omitted without confusion.

Sampling in the series above is at twice Nyquist rate; so this scheme allows for oversampling and the series can be called the "oversampled form". The same scheme allows sampling at Nyquist rate; indeed in §4 it will be shown that the "Nyquist rate form"

$$\Sigma \left\{ \frac{2f'(2n)}{\pi} S_1^2(t - 2n) + f(2n) S_2^2(t - 2n) \right\} \tag{1.2}$$

represents every $f \in PW$ (this was asserted in Higgins (1985) p. 66 but the discussion there is obscure).

There is nothing special about the even integers, and one can sample the function and its derivative at the odd integers using the same scheme. This scheme does not, however, allow the function and its derivative to be sampled at interlaced sets of points. There are other schemes which do allow this (Kwan Fai Cheung et al. (1985)), but they are more complicated and there is also the danger of encountering ill posed problems.

In §2 a new proof will be given of (1.1) which starts from the fact that $\{S_1^2(t - n)\}$ is a complete orthonormal (CON) set in PW. While it is well known that the ordinary Whittaker-Shannon, or cardinal, series is an orthogonal expansion in PW, it is not clear what the analogous situation for derivative sampling should be. One can hardly expect things to be so clear cut; indeed it is a routine matter to show (by taking Fourier transforms, for example) that $\{S_2^2(t - n)\}$ is not orthogonal over $\mathbb{R}$. Nevertheless there is a little to be said about the Hilbert space structure associated with (1.1), and in §3 it will be shown that the two parts of the series effect a kind of partition of PW.

2. THE OVERSAMPLED FORM IN PW

Let us denote the Fourier transform of f by $\mathcal{F}f$ or $f^{\wedge}$, the Hilbert transform by $\mathcal{H} f$ or $f^{\sim}$ and the derivative by $\mathcal{D} f$ or f'.

Let

$$T_1 : f \mapsto \Sigma \frac{f'(n)}{\pi} S_1^2(t - n) \tag{2.1}$$

$$T_2 : f \mapsto \Sigma \frac{f(n)}{2} S_2^2(t - n). \tag{2.2}$$

Lemma 1. If $f \in PW$ the series in (2.1) converges to $f'^{\sim}(t)/\pi$ in norm, and uniformly over all of $\mathbb{R}$.

Proof. The set $\{S_1^2(t - n) : n \in \mathbb{Z}\}$ is CON in PW with coefficient functionals $(-f^{\sim}(n))$ (Higgins (1985)). Also, since $D: PW \mapsto PW$ we have $(f'(n)) \in \ell^2$. Hence (2.1) converges in norm to an element, g say, of PW such that $-g^{\sim}(n) = f'(n)/\pi$. Then by the ordinary sampling theorem in PW we must have $-g^{\sim}(t) \equiv f'^{\sim}(t)/\pi$. Taking Hilbert transforms, $-g^{\sim\sim}(t) = g(t) = f'^{\sim}(t)/\pi$.

The pointwise convergence and uniformity follow from the convergence in norm by a standard principle in Hilbert spaces with reproducing kernel.

Lemma 2. If $f \in PW$ the series in (2.2) converges to $f(t) - f'^{\sim}(t)/\pi$ in norm and uniformly over all of $\mathbb{R}$.

Proof. Let $F \in PW$ and consider an expansion

$$F(t) = \Sigma\, c_n S_2^2(t - n)$$

for some coefficients (c_n). Suppose also that $F^{\wedge}(x)/\tau(x) \in L^2(-\pi,\pi)$, where τ is the "tent" or "triangle" function

$$\tau(x) = \begin{cases} 1 - |x|/\pi \;; & |x| \leqslant \pi \\ 0 \;; & |x| > \pi. \end{cases}$$

Then

$$\begin{aligned} &\| F(t) - \sum^{N} c_n S_2^2(t - n)\|_{PW} \\ &= \| F^{\wedge}(x) - \sqrt{\tfrac{2}{\pi}} \sum^{N} c_n \tau(x) e^{-inx}\|_{L^2(-\pi,\pi)} \\ &\leqslant \| \frac{F^{\wedge}(x)}{\tau(x)} - \sqrt{\tfrac{2}{\pi}} \sum^{N} c_n e^{-inx}\|_{L^2(-\pi,\pi)} \end{aligned}$$

since $|\tau(x)| \leqslant 1$. This last expression is vanishingly small for large N if $(\sqrt{\tfrac{2}{\pi}}\, c_n)$ is chosen to be the sequence of Fourier coefficients of $F^{\wedge}(x)/\tau(x)$. Now choose $F(t) \triangleq f(t) - f'^{\sim}(t)/\pi$, $f \in PW$. Then

$$F^{\wedge}(x) = f^{\wedge}(x) - f'^{\sim\wedge}(x)/\pi$$

$$= f^{\wedge}(x) - (\text{sgn } x)\ xf^{\wedge}(x)/\pi$$

$$= \tau(x)\ f^{\wedge}(x).$$

Hence this F is a candidate for expansion. Finally,

$$c_n = \sqrt{\frac{\pi}{2}}\frac{1}{2\pi}\int_{-\pi}^{\pi} \tau(x)\ [f^{\wedge}(x)/\tau(x)]\ e^{inx}\ dx$$

$$= \frac{1}{2\sqrt{(2\pi)}}\int_{-\pi}^{\pi} f^{\wedge}(x)\ e^{inx}\ dx$$

$$= f(n)/2.$$

Since $T_1 + T_2$ is the identity operator on PW, we have the

THEOREM The representation (1.1) holds for every $f \in PW$ in norm and pointwise, uniformly over all of $\mathbb{R}$.

3. THE RANGES OF T_1 AND T_2

It is easy to see that both T_1 and T_2 are bounded linear operators on PW. The first maps PW to $A = \{f'^{\sim}(t)/\pi : f \in PW\}$. The second maps PW to $\{f(t) - f'^{\sim}(t)/\pi : f \in PW\}$.

Since H is unitary on PW the invertibility of $T_1 = HD$ is equivalent to that of D, and that of $T_2 = I - (1/\pi)HD = I - (1/\pi)DH$ (I is the identity operator) is equivalent to that of $\pi I - D$. But neither D nor $\pi I - D$ has a bounded inverse in PW (Weston (1949)). Hence neither A nor B is all of PW; otherwise the "bounded inverse" theorem would be contradicted.

In fact both A and B must be open subsets of PW. At the same time both A and B are dense in PW, since $F \in A$ if $F^{\wedge}$ vanishes in a neighbourhood of the origin, and belongs to B if $F^{\wedge}$ vanishes in neighbourhoods of $\pm\ \pi$. Since these two criteria are not mutually exclusive we find that not only is $A \cap B \neq \emptyset$, but in fact $A \cap B$ is dense in PW.

Although A and B make a strange kind of partition for PW, we do have from §2 that

$$f \in PW \Rightarrow f = \alpha + \beta,\ \alpha \in A,\ \beta \in B.$$

4. THE NYQUIST RATE FORM

Our object in this section is to prove the following

THEOREM <u>The series (1.2) converges to f in norm and pointwise, uniformly over all of $\mathbb{R}$, for every $f \in PW$.</u>

Some tedious but routine calculational detail has been omitted for the sake of clarity.

At this point it will be natural to approach our proof via projection operators. Let

$$S_1 : f \mapsto \Sigma \frac{2f'(2n)}{\pi} s_1^2(t - 2n)$$

$$S_2 : f \mapsto \Sigma f(2n) s_2^2(t - 2n).$$

$$P_e : PW \to PW_e = \text{closed linear span of } \{s_1^2(t - 2n) : n \in \mathbb{Z}\}.$$

Then evidently $S_1 = P_e T_1$; because $\{s_1^2(t - n)\}$ is CON in PW, P_e is a projection. Hence S_1 is bounded.

Two further steps will reveal the method of proof. They are

<u>Step 1</u>. It can be shown by direct calculation that each member of the "cardinal" basis $\{s_1^1(t - k) : k \in \mathbb{Z}\}$ for PW is faithfully expanded by the operator $S = S_1 + S_2$.

<u>Step 2</u>. If an operator on a separable Hilbert space H is bounded and linear and acts as the identity operator on a basis for H, it is the identity on all of H.

Thus, if we knew that S_2 were bounded the required result would follow immediately. Unfortunately S_2 is harder to deal with than S_1 because $\{s_2^2(t - n)\}$ is not CON in PW.

We require five further steps to complete the proof.

<u>Step 3</u>. Decompose PW into a direct sum $\Sigma \oplus S_n$ as follows. Let $S_n \subset PW$ be defined so that $f_n \in S_n$ if $\hat{f}$ has support on

$$\left[-\frac{n\pi}{n+1}, -\frac{(n-1)\pi}{n}\right) \cup \left[\frac{(n-1)\pi}{n}, \frac{n\pi}{n+1}\right); \quad n = 1,2, \ldots .$$

Note that if $n \neq m$, $\text{supp}\, f_n^{\wedge} \cap \text{supp}\, f_m^{\wedge} = \emptyset$ so that (with the appropriate inner products) $\langle f_n, f_m \rangle = \langle f_n^{\wedge}, f_m^{\wedge} \rangle = 0$. Also, (null function) $\in S_n$ for every n, so that we can write, for $f \in PW$,

$$f = \Sigma f_n \; ; \; f_n \in S_n.$$

DEFINITION

$$B_n \triangleq \{f_n - f_n'^{\sim}/\pi : f_n \in S_n\} \subset B.$$

Step 4. $T_2|S_n$ (T_2 with domain restricted to S_n) has a bounded inverse on B_n.

This implies that B_n is a subspace of PW.

DEFINITIONS

$$Q_e : \Sigma \frac{f(n)}{2} s_2^2(t - n) \mapsto \Sigma f(2n)\; s_2^2(t - 2n)$$

$$Q_o : \Sigma \frac{f(n)}{2} s_2^2(t - n) \mapsto \Sigma f(2n + 1)\; s_2^2(t - (2n + 1)).$$

Step 5. It can be shown that

$$B_n = (\text{range } Q_e |B_n) \oplus (\text{range } Q_o|B_n)$$

in the Banach space sense, so that $Q_e|B_n$ and $Q_o|B_n$ are Banach space projections (they are not projections in the Hilbert space sense). As such they are bounded.

Step 6. $Q_e|B_n$ is bounded on B_n, hence $S|S_n = (S_1 + Q_e T_2)|S_n$ is the identity operator restricted to S_n (by Steps 1. and 2., since for every n, some subset of $\{s_1^1(t - n)\}$ spans S_n).

Step 7. Finally we can define $S = S_1 + S_2$ on all of PW by

$$S f = \lim_{N\to\infty} S \sum^{N} f_n$$

and require to show that $Sf = f$.

First, the limit exists since

$$S \sum^{N} f_n = \sum^{N} f_n \to f \text{ in norm,}$$

hence $(S \sum^{N} f_n)$ is a Cauchy sequence.

Second,

$$\| \lim S \sum^{N} f_n - f \|$$

$$= \lim \| S \sum^{N} f_n - f \|$$

$$= \lim \| \sum^{N} f_n - f \|$$

$$= 0.$$

Clearly a parallel proof treats the case of sampling at the odd integers.

5. REFERENCES

Higgins, J.R., (1985) "Five short stories about the cardinal series". *Bull. Amer. Math. Soc.* **12**, 45-89.

Kwan Fai Cheung and Marks, R.J., (1985) "Ill posed sampling theorems". *IEEE Trans. Circuits and Systems CAS* - **32**, 481-484.

Weston, J.D., (1949) "The Cardinal Series in Hilbert space". *Proc. Cambridge Philos. Soc.* **45**, 335-341.

SOME NEW DIFFERENCE SCHEMES FOR THE PREDICTION OF BANDLIMITED SIGNALS FROM PAST SAMPLES

D.H. Mugler
(Department of Mathematics, University of Santa Clara, California, USA)

and

W. Splettstößer
(Lehrstuhl A für Mathematik, RWTH Aachen, West Germany)

ABSTRACT

Difference formulas, with modified differences which extend the usual Newton formulas for interpolation, have been shown to provide a means to predict the value of a bandlimited, finite-energy signal from its samples at a discrete set of past times. Difference formulas have an advantage over other formulas of adjusting easily to the number of signal samples available.

In this paper, a correction term for the previous formulas is presented. When this factor is combined with the former theory, a predictor - corrector formula results which allows for another level of accuracy to be gained. A different means of extending the previous methods involves a formula that is shown to fit into the context of difference methods and which provides an optimal prediction formula for a wider class of signals.

1. INTRODUCTION

The fundamental problem considered in this paper is to determine, i.e. predict, the value f(t) of a bandlimited signal based on its past samples at equally-spaced points, $t_k = t - kT$ for $k \geq 1$, using a linear prediction formula

$$f(t) = \sum_{k=1}^{n} a_{kn} f(t-kT) = \sum_{k=1}^{n} a_{kn} f(t_k). \qquad (1.1)$$

Unless otherwise stated, we will assume that f(t) has the form

$$f(t) = \frac{1}{\sqrt{2\pi}} \int_{-\pi W}^{\pi W} F(\omega) e^{i\omega t} \, d\omega ,$$

with $F \in L^2(-\pi W, \pi W)$. Existence of prediction formulas of form (1.1) is known under the assumption that $TW < 1$, where $W = 1/T$ is the Nyquist rate of sampling, see e.g. Splettstößer (1982).

The disadvantage of formulas with form (1.1) is that the coefficients a_{kn} depend on n, and the entire set of coefficients may need to be changed whenever the number of samples is varied. This disadvantage may be avoided by reformulating (1.1) in terms of differences of the sample values. J.L. Brown (1972) considered a formula

$$\sum_{k=1}^{n} (-1)^k a^k \binom{n}{k} f(t_k) \tag{1.2}$$

for f(t), which he showed converged for $a=\cos(\pi TW)$ if $TW<1/2$. This formula extended an earlier result in Wainstein and Zubakov (1962), which did not include the a term and which converged only for $TW\leq 1/3$. In Mugler and Splettstößer (1986), the authors showed that one could rewrite (1.2) as

$$a \sum_{k=0}^{n-1} \nabla_a^k f(t_1), \tag{1.3}$$

where the modified backward differences used in this sum are defined as follows, $\nabla_a^0 f(t) = f(t)$, $\nabla_a^1 f(t) = \nabla_a f(t) = f(t) - af(t-T)$, and $\nabla_a^{k+1} f(t) = \nabla_a^k f(t) - a \nabla_a^k f(t-T)$ for $k\geq 1$. We will later use differences of a different step size, where the step size will be denoted with a superscript. For example, $\nabla_b^2 f(t) = f(t) - bf(t-2T)$ for a two-step difference, with higher differences defined similarly to those above.

The authors have shown that (1.3) converges if $a < 2\cos\pi TW$ and that one can give an estimate of the error in a similar difference form,

$$f(t) = a \sum_{k=0}^{n-1} \nabla_a^k f(t_1) + \nabla_a^n f(t) , \tag{1.4}$$

which will be important for the discussion in section 2. The expression (1.4) may also be extended to provide a prediction at an arbitrary point.

In Section 2, we present a means to approximate $\nabla_a^n f(t)$ in (1.4), and thereby improve the accuracy of the prediction. This work was inspired by a paper of Brown and Morean (1986), who determined a set of coefficients c_{kn} so that

$$f(t) - \sum_{k=0}^{2n} c_{kn} f(t_k) = \frac{1}{\sqrt{2\pi}T} \int_{-\pi W}^{\pi W} F(\omega) e^{i\omega t} (1-ae^{-iT\omega}-be^{-i2T\omega})^n d\omega.$$

The results in section 2 provide a difference method for the algorithm presented in that paper. The methods in section 3 approach formula (1.4) in a different way, and show that it can be extended to provide an optimal prediction formula for a wide class of functions.

2. A PREDICTOR-CORRECTOR METHOD

The difference formula used for predicting the value of a signal using its past samples is only a part of equation (1.4) above, which also includes a term that would be the error in an application of that formula. From that equation, we refer to $a\Sigma_{k=0}^{n-1} \nabla_a^k f(t_1)$ as the Predictor, and to $\nabla_a^n f(t)$ as the error. The fundamental idea of this section is to provide a means to approximate the value of the error, and then combine that estimate of the error, hereafter called the Corrector, to obtain a more accurate prediction of f(t). Although truncation error estimates for the Predictor are known which show that this error eventually vanishes for any t, the use of a large number of sample values may generate a large roundoff error. One cannot continually improve accuracy by increasing the number of samples. We will show that a new level of accuracy may be obtained by the use of a predictor-corrector combination.

In this section, we first describe the steps used to compute the Corrector value and give an example. Theorem 2.1 then provides a formal approach for the method, and the section is closed with a discussion of the optimal values of the difference modifiers needed for the various TW values.

The computation of the Corrector value proceeds similarly to the computation of the Predictor itself.

Define values of a new function $f_1(t)$ at sample points and at $t = t_0$ by applying the Predictor formula, but using the twostep difference with modifier b. That is,

$$f_1(t_j) = b \sum_{k=0}^{n-1-j} {}^2\nabla_b^k f(t_{j+2}), \tag{2.1}$$

for $0 \leq j \leq n-1$, and $f_1(t_n) = 0$. Note that the number of samples must be increased from n to 2n for this calculation. From (2.1), it follows that $f_1(t)$ is based on sample values at $t_2, t_4, \ldots, t_{2n}$, $f_1(t_1)$ is based on samples at $t_3, t_5, \ldots, t_{2n-1}$, etc. Finally, the values $f_1(t)$, $f_1(t_1), \ldots, f_1(t_n)$ are used in a difference table using the original modifier a, and the value of $\nabla_a^n f_1(t)$ is used as the value of the Corrector.

Example. The following is an example to show how the calculation would proceed to predict and correct the value of $si(\pi t) := \sin(\pi t)/(\pi t)$ at $t = 0$, using $n = 3$ samples.

Predictor (with a = 1)

t	$si(\pi t)$	∇_a	∇_a^2
-.1	.98363		
		.04814	
-.2	.93549		-.02896
		.07710	
-.3	.85839		

Predictor = 1.00281 (Error = $0.28 \cdot 10^{-2}$)

Corrector (with b=-0.2)

t	$si(\pi t)$	${}^2\nabla_b$	${}^2\nabla_b^2$
-.2	.93549		
-.3	.85839	1.08686	
-.4	.75683	.98571	1.25841
-.5	.63662	.85774	
-.6	.50455		

t	$f_1(t)$	∇_a	∇_a^2	∇_a^3
0	-.65615			
		-.28733		
-.1	-.36882		-.06988	
		-.21745		-.00380
-.2	-.15137		-.06608	
		-.15137		
-.3	0			

Table 1 Predictor + Corrector = 1.00281-.00380 = .99901 (Error = $-.99 \cdot 10^{-3}$).

In the theorem and proof below, we assume that the limits on the integrals go from $-\pi TW$ to πTW, and that integrals include a multiplication factor of $(\sqrt{2\pi}\ T)^{-1}$.

Theorem 2.1.

With $f_1(t)$ values derived from the samples of f(t) as described above,

$$f(t) - \text{Predictor} - \nabla_a^n f_1(t) = \int F_T(\omega)(1-ae^{-i\omega}-be^{-i2\omega})^n d\omega \tag{2.2}$$

where n is the number of samples, and $F_T(\omega) = F(\omega)e^{i\omega t/T}$.

Proof. Write

$$(1 - ae^{-i\omega}-be^{-i2\omega})^n = ([1-ae^{-i\omega}] - be^{i2\omega})^n$$

$$= (1-ae^{-i\omega})^n + \sum_{j=1}^{n} \binom{n}{j}(1-ae^{-i\omega})^{n-j}(-be^{-i2\omega})^j.$$

Thus

$$\int F_T(\omega)(1-ae^{-i\omega}-be^{-i2\omega})^n\, d\omega$$

$$= \int F_T(\omega)(1-ae^{-i\omega})^n d\omega + \sum_{j=1}^{n} \binom{n}{j}\int F_T(\omega)(1-ae^{-i\omega})^{n-j}(-be^{-i2\omega})^j d\omega.$$

It is not difficult to show that the first integral on the right side of this equation equals $f(t) - a\sum_{k=0}^{n-1} \nabla_a^k f(t_1)$, see Mugler and Splettstößer (1985). Thus the above equation equals $f(t) - a\sum_{k=0}^{n-1} \nabla_a^k f(t_1)$ -Corrector, where the Corrector is $-\sum_{j=1}^{n}\binom{n}{j}\int F_T(\omega)\sum_{k=0}^{n-j}\binom{n-j}{k}(-ae^{-i\omega})^k(-be^{-i2\omega})^j\, d\omega.$

But $\sum_{j=1}^{n}\binom{n}{j}\sum_{k=0}^{n-j}\binom{n-j}{k} = \sum_{k=0}^{n-1}\sum_{j=1}^{n-k}\binom{n}{k}\binom{n-k}{j}$, and with the Corrector as given by the above equation, this yields

$$\text{Corrector} = -\sum_{k=0}^{n-1}\binom{n}{k}(-a)^k \int F_T(\omega)e^{ik\omega}\sum_{j=1}^{n-k}\binom{n-k}{j}(-be^{-i2\omega})^j d\omega.$$

But the sum inside the integral equals $(1-be^{-i2\omega})^{n-k}-1$, which makes the integral equal to $f(t-kT) - b\,\Sigma_{j=0}^{n-k-1} \nabla_b^{2j} f(t_{k+2}) - f(t-kT)$. Cancelling the common values in that expression leaves

$$\text{Corrector} = \sum_{k=0}^{n-1} \binom{n}{k} (-a)^k \Big[b \sum_{j=0}^{n-k-1} \nabla_b^{2j} f(t_{k+2})\Big].$$

But with values of $f_1(t)$ as defined in (2.1), the last expression equals $\nabla_a^n f_1(t)$. This completes the proof.

From Theorem 2.1, it follows that

$$|f(t)\text{-Predictor} - \nabla_a^n f_1(t)| \leq \|f\|_2 \left(\frac{1}{\sqrt{2\pi T}} \int |1-ae^{-i\omega}-be^{-i2\omega}|^{2n} d\omega\right)^{1/2}.$$

The predictor-corrector method above is actually a rearrangement of the Brown and Morean (1986) method mentioned earlier, and they considered values for the a,b modifiers that would minimize the maximal value of

$$A(\omega) := |1-ae^{-i\omega}-be^{-i2\omega}|^2 \qquad (2.3)$$

on the interval $[-\pi TW, \pi TW]$ for $TW = 1/2$. They found optimal values as $a = 2 - \sqrt{2}$, $b = 1 - \sqrt{2}$, for which the corresponding maximum of $A(\omega)$ is 0.6863, which leads to convergence of the prediction sums for each sample spacing T with $TW \leq 1/2$. In general, an error reduction is affected more by decreasing T than by increasing n (see Mugler and Splettstößer (1985)), particularly because of round-off error effects. Our own numerical trials have shown that it is necessary to determine the optimal values of a and b when TW is small, and that a and b should be quite different from the values above. For example, when $TW \leq 1/4$, another choice of modifiers leads to a value for max $A(\omega)$ that is much less than 0.6863.

For small values of TW, there is a relationship between the modifiers a and b that aids in the calculation of the maximum of $A(\omega)$. Starting with the limit case as $T \to 0$, which implies $\omega \to 0$ in (2.3), note that $A(0) = [1-(a+b)]^2$. We accordingly choose modifiers such that $a + b = 1$, so as to have a small error for small T. Setting $b = 1 - a$ in (2.3), one has

$$A(\omega) = 2[(1-\cos 2\omega)(1-a) + a^2(1-\cos \omega)] .$$

To determine the maximum over $[-\pi TW, \pi TW]$, note that

$$dA/d\omega = 2[a^2 \sin\omega - 2(a-1)\sin 2\omega] = 2 \sin\omega[a^2+4(1-a) \cos\omega]$$

is positive for each $a \in [0,1]$. For $a \in (1,2)$, there are no zero-crossings of the derivative, since in that case there would have to be a value of ω such that $\cos \omega = a^2/[4(a-1)]$. But this latter expression is always larger than 1, and is decreasing and equal to 1 only at a = 2. This shows that A is an increasing function of ω, and thus takes its maximum A_o at the right end of the interval.

To find the optimal modifier a, note that

$$\frac{dA_o}{da} = 2\{\cos 2\omega - 1 + 2a(1-\cos\omega)\} = 0 \Longleftrightarrow a = \frac{1-\cos 2\omega}{2(1-\cos\omega)} .$$

With other conditions as above, the maximum of $A(\omega)$ would thus be given as

$$\max A(\omega) = (1 - \cos 2\omega)[2 - \frac{1-\cos 2\omega}{2(1-\cos\omega)}] , \qquad (2.4)$$

and the maximum over the interval would occur at $\omega = \pi TW$.

In the following table, we list the optimal values of the modifiers a, b along with the corresponding value of max $A(\omega)$ as described by (2.4) and the discussion above, for different sample spacings.

TW	a	b	max A
0.10	1.91	-0.91	0.0172
0.20	1.76	-0.76	0.1328
0.30	1.59	-0.59	0.5396
1/3	1.50	-0.50	0.75
0.50	0.586	-0.414	0.6863

Table 1.

As can be seen from the above table, there is a region between TW = 0.30 and TW = 1/3 where the optimal value for a changes rapidly, and the theory given by Brown and Morean applies for the angles greater than $\pi/3$. Note that our modifier values are optimal under the assumption that a+b = 1, which was the assumption used above to determine the optimal values for small TW.

Finally, we list some results of numerical trials, performed to test the predictor-corrector algorithm as applied to predict si(πWt) for various t and W values. For each value of TW, the number n listed is the lowest number of samples (for the predictor) to achieve the accuracy of the combination method (using 2n samples). The modifiers a,b listed are those that allowed that accuracy.

TW	n	a	b	Order of the Error
.05	9	1.99	-.99	10^{-14}
.10	10	1.95	-.95	10^{-13}
.15	15	1.88	-.89	10^{-11}
.20	18	1.80	-.81	10^{-10}
.25	21	1.69	-.75	10^{-9}
.30	30	.9	-.4	10^{-9}
.40	30	.8	-.4	10^{-6}
	50	.5	-.2	10^{-7}
.50	30	.6	-.4	10^{-4}
	60	.5	-.3	10^{-5}

Table 2.

3. PREDICTION WITH A DIFFERENCE FORMULA AS AN OPTIMAL METHOD

As previously stated, the difference formulas above are suboptimal in the sense of producing the least truncation error for a given number of samples, particularly if the exact value of TW is known. In this section we present a formula which is known to be optimal for a wide class of functions, and we show that this formula is an extension of (1.4).

We first extend the definition of modified differences a little. In the predictor-corrector method of section 2, the differences are used with a modifier (i.e. a or b) that is fixed in the difference table. We now define a set of differences depending on a function h(t) which allows the modifier to vary between different diagonals in the table. Let $\nabla_{R(1,2)}g(t_1) = g(t_1) - R(1,2)g(t_2)$, where $R(j,k) = h(t_j)/h(t_k)$,

$\nabla_{R(2,3)} g(t_2) = g(t_2) - R(2,3)g(t_3)$, and with similar definitions at other t_k values. Further define

$$\nabla^{k+1}_{R(1,2)} g(t_1) = \nabla^{k}_{R(1,2)} g(t_1) - R(1,2)\ \nabla^{k}_{R(2,3)} g(t_2) \tag{3.1}$$

for $k \geq 1$, and with similar definitions for differences of form $\nabla^{k}_{R(j,j+1)} g(t_j)$. For example,

$$\nabla^{2}_{R(1,2)} g(t_1) = (g(t_1) - R(1,2)g(t_2)) - R(1,2)(g(t_2) - R(2,3)g(t_3))$$

$$= h(t_1)[(g(t_1)/h(t_1) - g(t_2)/h(t_2)) - (g(t_2)/h(t_2) - g(t_3)/h(t_3))].$$

Note that if $h(t) = a^{t/T}$, then the ratios above will be identical throughout the calculation. That is, $R(j,j+1) = a$ for every integer j, since $t_j = t - jT$. In this case, the differences in (3.1) are precisely those of the previous section.

Using the above differences, the difference formula that we consider in the following is an application of an interpolation formula given by A.A. Melkman (1985), who provided an existence theorem and a general method to produce an optimal formula. This method may be applied in the prediction case to provide an optimal formula, in the sense of a formula which minimizes error for a given number of samples.

Melkman considered a class of functions that can be related to duration - limited as well as bandlimited functions. Let $B(\sigma, T_M)$ be the class of entire functions of exponential type σ which are bounded by one for $|t| > T_M$. For fixed t in $(-T_M, T_M)$ and a set of points $(t_i)_1^n$ in that interval, Melkman's optimal formula is a linear formula with real coefficients $(a_i)_1^n$, which gives the minimum over all complex coefficients of

$$\max_{f \in B(\sigma, T_M)} \left| f(t) - \sum_{i=1}^{n} a_i f(t_i) \right|. \tag{3.2}$$

He also showed that the same formula is optimal over the interval above, with absolute values in (3.2) replaced by the maximum norm over $[-T_M, T_M]$.

In the following, the concept of an optimal formula is as described above.

Theorem 3.1 Let $f \in B(\sigma, T_M)$ and $t_j = t-jT, j\geq 1$. Suppose that all of the t_j, $1\leq j\leq n$, lie in the interval $(-T_M, T_M)$ and that p is any real number so that t + pT is also in that interval. Then

$$f(t+pT) = R(-p,1) \sum_{k=0}^{n-1} \binom{p+k}{k} \nabla^k_{R(1,2)} f(t_1) \tag{3.3}$$

is an optimal formula for prediction based on these samples, where $R(-p,1) = h(t+pT)/h(t_1)$.

Remark. This production formula is very similar in form to the extension of (1.4) and would reduce to the extension of that formula if $h(t) = a^{t/T}$. The proof of Theorem 3.1 is not too difficult, and can be done by an induction. The proof will be omitted on account of a lack of space.

Some of the properties of the function h(t), which is used for the differences, were given in Melkman (1985). For example, h(t) is an entire function of exponential type σ, it is real on the real axis, and it has zeros only outside of the interval $(-T_M, T_M)$. However, the constructions of h(t) and the particular values of h(t) needed to form the ratios R for the above modified differences seem difficult to provide. In closing, we note that the form of these ratios and the corresponding differences for a given set of sample times is a topic which still merits further investigation.

4. ACKNOWLEDGEMENTS

The present research was initiated while D.H. Mugler was on leave at Lehrstuhl A für Mathematik, RWTH Aachen, Oct. 83 - June 84, which was supported by the Alexander von Humboldt - Stiftung. The authors like to thank Professor Dr. P.L. Butzer for his valuable advice during that time. The work of W. Splettstößer is supported by the Stiftung Volkswagenwerk.

5. REFERENCES

Brown, J.L., Jr., (1972) "Uniform Linear Prediction of Bandlimited Processes from Past Samples", *IEEE Trans. Inform. Theory IT* - **18**, 67-75.

Brown, J.L. Jr. and Morean, O., (1986) "Robust Prediction of Bandlimited Signals from Past Samples", *IEEE Trans. Inform Theo y IT-32*, 410-412.

Melkman, A.A., (1985) "n-widths and Optimal Interpolation of Time- and Band-Limited Functions II", *SIAM J. Math. Anal.* **16**, 803-813.

Mugler, D.H. and Splettstößer, W., (1986) "Difference Methods for the Prediction of Bandlimited Signals", *SIAM J. Appl. Math.* **46**, 930-941.

Mugler, D.H. and Splettstößer, W., (1985) "Difference Methods and Round-off Error Bounds for the Prediction of Bandlimited Functions from Past Samples". *Frequenz* 39, 182-188.

Splettstößer, W., (1982) "On the Prediction of Bandlimited Signals from Past Samples". *Information Sciences* 28, 115-130.

Wainstein, L.A. and Zubakov, V.D., (1962) "Extraction of Signals from Noise". Prentice-Hall, New York.

A CLASS OF INTEGER SEQUENCES WITH IDEAL CORRELATION PROPERTIES IN THE MÖBIUS DOMAIN

R.N. Gorgui-Naguib and R.A. King
(Department of Electrical Engineering, Imperial College of Science and Technology, London)

ABSTRACT

The concept of nonlinear mapping of the autocorrelation and crosscorrelation functions is introduced. By using the Möbius μ-function which is a Number Theoretic function, integer sequences (which will be referred to as μ-sequences) can be generated and which have the property that the corresponding μ-values of the elements of the autocorrelation function of each will be zero except for the zero-shift term which has a μ-value of ± 1.

A μ-sequence is said to be a mate of another if they both are of the same length and the mapping of their crosscorrelation values is zero everywhere in the μ-domain. Theorems allowing the synthesis of such integer sequences of length $\geqslant 2$ are given and their proofs demonstrated in detail.

1. INTRODUCTION

Generally, signal design approaches, for application of signal extraction from noise, depend on the properties of the signal autocorrelation function, the design objective being that of achieving an autocorrelation function with a high main lobe and low side lobes. The problem of compromising the width of the main lobe with the degree of the side lobes is well known in linear signal design. In this paper, the concept of nonlinear mapping of the autocorrelation function is introduced. When the mapping domain is the set of positive integers and the range of the mapping contains three elements (in this case, 0,+1 and -1), the resulting mapped autocorrelation function can have the desired spike property with no side lobes.

In this paper, the Möbius μ-function is used to effect this nonlinear mapping of the autocorrelation function, the original sequence being an integer sequence and the nonlinear mapping operation function results in a sequence which is zero for all shifts except the zero-shift. Also, two integer sequences can be designed such that the nonlinear mapping of their crosscorrelation function is zero at every shift. This proves useful in implementing non-interacting sequences for certain applications such as data transmission.

First in the paper, we introduce some general definitions relating to the correlation functions and we define the Number Theoretic Möbius μ-function. Then, algorithms for constructing integer sequences of length 2 are presented and which are extended to synthesize new longer sequences. Finally, the subject of the mate of a sequence is also discussed and rules for generating mates of a sequence are presented.

2. DEFINITIONS AND TERMINOLOGY

We define A as the sequence of integer elements of length N given by:

$$A = \{\alpha_1, \alpha_2, \ldots, \alpha_N\}$$

Define $R^k(A)$ as the sequence of length N that results from shifting the elements of A by k positions to the right and filling the empty positions with zeros, as in Habbab (1983):

$$R^k(A) = \{0, 0, \ldots, \alpha_1, \ldots, \alpha_{N-k}\}$$

Similarly we define the left-shift operator $L^k(A)$ given by:

$$L^k(A) = \{\alpha_{k+1}, \ldots, \alpha_N, 0, \ldots, 0\}$$

Note that

$$L^k(A) = R^{-k}(A)$$

Let Ω_{AA} denote the autocorrelation function of the sequence A and let $\Omega_{AA}(k)$ denote the kth element of the autocorrelation. Hence:

$$\Omega_{AA}(k) = \sum_{i=1}^{N-k} \alpha_i \, \alpha_{i+k} \qquad , \; k = -N+1, \ldots, N-1$$

$\Omega_{AA}(k)$ can also be considered as the inner product of the two sequences A and $R^k(A)$:

$$\Omega_{AA}(k) = (A, R^k(A))$$

and it is worthwhile noting here that the zero-shift element is given by:

$$\Omega_{AA}(0) = \| A \|^2$$

Now we introduce the definition of a μ-sequence: an integer sequence A of length N is termed a μ-sequence if the Möbius function of the elements of its autocorrelation function yields zero at every shift except the zero-shift as given by:

$$\mu(\Omega_{AA}(k)) = \delta_{0,k} \quad , k = -N+1, \ldots, N-1$$

where $\delta_{i,j}$ is defined as:

$$\delta_{i,j} = \pm 1 \quad i=j$$
$$= 0 \quad i \neq j$$

3. DEFINITION OF THE MÖBIUS FUNCTION

The Möbius μ-function of any integer, d, is defined as:

$$\mu(d) = \begin{cases} 1 & \text{if } d = 1 \\ 0 & \text{if } p^2 | d \text{ where p is any prime number} \\ (-1)^t & \text{if } d = p_1 p_2 \ldots p_t \text{ where all } p_i \text{ are distinct primes.} \end{cases}$$

Hence, if P is a prime, $\mu(P) = -1$. If Q is also a prime, $\mu(PQ) = +1$, while $\mu(P^2) = 0$.

The Möbius function is commonly used in Number Theory and has many attractive properties. One such property, which will be of major interest in the following analysis, is the multiplicative property of the function, when gcd (A,B) = 1, thus,

$$\mu(A.B) = \mu(A) \, . \, \mu(B)$$

The function is used to map obtained integer values of the autocorrelation function of a sequence into its corresponding μ-values of 0, +1 or -1. This concept of nonlinear mapping will allow the synthesis of integer sequences having the property discussed in the previous section; and whereas, in the case of complementary sets of sequences, two or more sequences are required to achieve the delta configuration, by using the Möbius function and the mapping concept, only one sequence is needed.

Throughout our analysis the following notation will be adopted for the three different sets of μ-values:

μ_0 for the set of integers, d, having $\mu(d) = 0$

μ_1 for the set of integers, d, having $\mu(d) = +1$

μ_{-1} for the set of integers, d, having $\mu(d) = -1$

Table 1 shows the classification of the integers 1 to 30 into their corresponding μ-values.

μ_0	μ_1	μ_{-1}
	1	
		2
		3
4		
		5
	6	
		7
8		
9		
	10	
		11
12		
		13
	14	
	15	
16		
		17
18		
		19
20		
	21	
	22	
		23
24		
25		
	26	
27		
28		
		29
		30

Table 1 Mapping of the Möbius μ-function for the integers 1 to 30.

4. GENERATION OF μ-SEQUENCES OF LENGTH 2

Theorem 1: For the sequence $A = \{\alpha_i, \alpha_j\}$, if α_i and $\alpha_j \in \mu_0$ and $\gcd(\alpha_i, \alpha_j) = 1$, then the sequence $A = \{\alpha_i, \alpha_j\}$ is a μ-sequence, if and only if:

$$p^2 t' \neq n_j^2 + p_i^2 t$$

for any prime p and any integers t and t' and where n_j and p_i are defined by the relations:

$$\alpha_i \equiv 0 \pmod{p_i^2}$$
$$\alpha_j \equiv n_j \pmod{p_i^2}$$

(Notice that the subscripts i and j above are to be interchanged if the case where $\alpha_j \equiv 0 \pmod{p_j^2}$ is considered.)

Proof: For the sequence $A = \{\alpha_i, \alpha_j\}$, Ω_{AA} is given by:

$$\Omega_{AA} = \{\alpha_i\alpha_j,\ \alpha_i^2+\alpha_j^2,\ \alpha_i\alpha_j\}$$

(a) For the side lobe terms, since α_i, $\alpha_j \in \mu_0$, then by using the multiplicative property of the μ-function

$$\mu(\alpha_i\alpha_j) = \mu(\alpha_i)\ \mu(\alpha_j) = 0$$

(b) For the zero-shift term, if $\gcd(\alpha_i, \alpha_j) > 1$, say any prime p (or a multiple of p), then $\exists\ p^2 \mid (\alpha_i^2+\alpha_j^2)$ which would lead to $\mu(\alpha_i^2+\alpha_j^2) = 0$. But for $\mu(\alpha_i^2+\alpha_j^2) = \pm 1$, then as a necessary but insufficient condition, $\gcd(\alpha_i, \alpha_j)$ must equal 1. Now, since the μ-function is multiplicative then, under addition (in the case of the zero-shift element where $\Omega_{AA}(0) = \alpha_i^2+\alpha_j^2$), an extra constraint must be added, preventing the case where $\exists\ p^2 \mid \Omega_{AA}(0)$ for any p; since α_i, $\alpha_j \in \mu_0$, then $p_i^2 \mid \alpha_i$ and $p_j^2 \mid \alpha_j$ and either of the following pairs of congruences is true:

$$\alpha_i \equiv 0 \pmod{p_i^2} \quad \text{and} \quad \alpha_j \equiv n_j \pmod{p_i^2}$$

or

$$\alpha_i \equiv n_i \pmod{p_j^2} \quad \text{and} \quad \alpha_j \equiv 0 \pmod{p_j^2}$$

We will only consider the first case, the second being exactly similar but with a corresponding change in subscripts. Thus:

$$\alpha_i^2 + \alpha_j^2 \equiv n_j^2 \pmod{p_i^2}$$

But, on the other hand, it is required that

$$\alpha_i^2 + \alpha_j^2 \not\equiv 0 \pmod{p^2} \text{ for any prime } p.$$

The above congruences may be written in the equation forms:

$$\alpha_i^2 + \alpha_j^2 = n_j^2 + p_i^2 t \text{ for any } t$$

and

$$\alpha_i^2 + \alpha_j^2 \neq p^2 t' \qquad \text{for any } t'$$

Hence:

$$p^2 t' \neq n_j^2 + p_i^2 t \quad \text{for particular values of } p, t \text{ and } t'$$

in order that the sequence $A = \{\alpha_i, \alpha_j\}$ be a μ-sequence.

Table 2 shows a tabulation of the μ-sequences of length 2 where α_i and $\alpha_j \in \mu_0$.

Theorem 2: For the sequence $A = \{\alpha_i, \alpha_j\}$, if $\alpha_i \in \mu_0$ and $\alpha_j \in \mu_1$ or μ_{-1} (or vice-versa), and fulfilling the conditions of Theorem 1, then the sequence $A = \{\alpha_i, \alpha_j\}$ is a μ-sequence if either of the two conditions below is satisfied:

1. α_i and α_j are both odd, or,

2. if r and s are any integers such that $\alpha_i = r^2 - s^2$ and $\alpha_j \neq 2rs$ (or vice-versa)

$\alpha_j \in \mu_0$ → / $\alpha_i \in \mu_0$ ↓	4	8	9	12	16	18	20
4			*				
8			*				
9	*	*			*		*
12							
16			*				
18							
20			*				

Table 2 Valid combinations of μ-sequences of length 2 where $\alpha_i, \alpha_j \in \mu_0$ (in the range $1 \leqslant \alpha_i, \alpha_j \leqslant 20$)

<u>Proof</u>: Again, for the sequence $A = \{\alpha_i, \alpha_j\}$, Ω_{AA} is given by:

$$\Omega_{AA} = \{\alpha_i\alpha_j,\ \alpha_i^2+\alpha_j^2,\ \alpha_i\alpha_j\}$$

Three possibilities arise for the selection of α_i and α_j

(a) α_i and α_j both even:

Although the out-of-phase coefficients will always have $\mu(\alpha_i\alpha_j) = 0$, for the zero-shift element on the other hand:

$$\Omega_{AA}(0) = \alpha_i^2+\alpha_j^2 \equiv 0 \pmod 4$$

Consequently, $\mu(\alpha_i^2+\alpha_j^2) = 0$ which is a prohibited case. Hence, the initial condition:

$$\gcd(\alpha_i,\alpha_j) = 1$$

(b) α_i and α_j both odd:

Again, for the side lobe terms, $\mu(\alpha_i\alpha_j) = 0$. For the zero-shift element in this case and after satisfying the condition of Theorem 1:

$$p^2t' \neq n_j^2 + p_i^2t \text{ for specific values of } p, t \text{ and } t'$$

But

$$\alpha_i^2 = \alpha_j^2 \equiv 1 \pmod 4$$

Thus

$$\Omega_{AA}(0) = \alpha_i^2+\alpha_j^2 \equiv 2 \pmod 4$$

and since any square is only congruent with 0 or 1 (mod 4), hence $\nexists$ p such that $p^2 \mid (\alpha_i^2+\alpha_j^2)$ and $\mu(\alpha_i^2+\alpha_j^2) = \underline{+}1$

(c) α_i even and α_j odd:

The side lobe terms have $\mu(\alpha_i\alpha_j) = 0$. But, in this case we are also concerned with the occurrence of Pythagorean triples in the zero-shift element where if α_i and α_j are in the form:

$$\alpha_i = r^2 - s^2$$

and $\qquad \alpha_j = 2rs, \qquad$ for any integers r and s,

then the zero-shift element will be of the form:

$$\Omega_{AA}(0) = \alpha_i^2+\alpha_j^2 = (r^2+s^2)^2$$

which will have a μ-value of 0, and hence condition 2 satisfies the requirement for $A = \{\alpha_i,\alpha_j\}$ to be a μ-sequence.

In Table 3 such μ-sequences of length 2 constructed from $\alpha_i \in \mu_0$ and $\alpha_j \in \mu_{\underline{+}1}$ are shown.

$\alpha_j \in \mu_{\pm 1}$ → / $\alpha_i \in \mu_0$ ↓	1	2	3	5	6	7	10	11	13	14	15	17	19
4	*			*		*		*	*		*	*	*
8	*		*	*		*		*	*			*	
9	*	*			*	*	*	*		*		*	*
12	*					*		*	*			*	*
16	*		*	*		*		*			*	*	*
18				*		*		*	*			*	*
20	*		*			*		*	*			*	*

Table 3 Valid combinations of μ-sequences of length 2 where $\alpha_i \in \mu_0$ and $\alpha_j \in \mu_{\pm 1}$ (in the range $1 \leqslant \alpha_i, \alpha_j \leqslant 20$)

5. SYNTHESIS OF μ-SEQUENCES OF LONGER LENGTHS

Given a μ-sequence of length 2, other sequences of lengths > 2 can be obtained according to the following theorem:

Theorem 3: If the sequence $A = \{\alpha_i, \alpha_j\}$ is a μ-sequence where $\alpha_i, \alpha_j \in \mu_0$, then the L+M+2 length sequence

$$A' = \{\underbrace{\alpha_i, \ldots, \alpha_i}_{\leftarrow L \rightarrow}, \alpha_i, \alpha_j, \underbrace{\alpha_i, \ldots, \alpha_i}_{\leftarrow M \rightarrow}\}$$

is also a μ-sequence if and only if

$$L+M+1 \not\equiv 0 \pmod{\alpha_j}$$

Proof: Consider the autocorrelation function of the sequence

$$A' = \{\alpha_i, \ldots, \alpha_i, \alpha_i, \alpha_j, \alpha_i, \ldots, \alpha_i\}$$

$$\leftarrow \quad L \quad \rightarrow \qquad\qquad \leftarrow \quad M \quad \rightarrow$$

of length L+M+2. The values of the side lobes will always be multiples of α_i, and since $\alpha_i \in \mu_0$, so will all the out-of-phase terms.

The zero-shift element, on the other hand, will be in the form:

$$\Omega_{A'A'}(0) = (L+M+1)\ \alpha_i^2 + \alpha_j^2$$

Now, for this term to have a μ-function = ±1, the following two conditions must be satisfied:

(a) α_i and α_j are relatively prime

(b) L+M+1 and α_j are relatively prime

Since, initially $A = \{\alpha_i, \alpha_j\}$ is a μ-sequence, then this implicitly implies that $\gcd(\alpha_i, \alpha_j) = 1$ and hence condition (a) is fulfilled and only (b) remains, hence the theorem.

As a corollary to Theorem 3, a μ-sequence of specific length L can be obtained from a μ-sequence of length 2:

Corollary 1: If $A = \{\alpha_i, \alpha_j\}$ is a μ-sequence where $\alpha_i, \alpha_j \in \mu_0$, then the sequence

$$A_{L'} = \{\alpha_i, \ldots, \alpha_i, \alpha_j, \alpha_i, \ldots, \alpha_i\}$$

of specified overall length L' > 2 is also a μ-sequence if and only if

$$L'-1 \not\equiv 0 \pmod{\alpha_j}$$

The proof of this corollary follows immediately from the proof of Theorem 3.

6. MATE OF A μ-SEQUENCE

The mate of a μ-sequence, A, is defined as another μ-sequence, $B = \{\beta_1, \beta_2, \ldots, \beta_N\}$, of the same length as A such that the crosscorrelation of A and B yields integer values, the μ-function of which is zero for every shift. The kth element in the crosscorrelation of A and B, both of length N, is given by

$$\Omega_{AB}(k) = \sum_{i=1}^{N-k} \alpha_i \beta_{i+k}$$

$$k = -N+1, \ldots, N-1$$

<u>Theorem 4</u>: For a μ-sequence $A = \{\alpha_i, \alpha_j, \ldots, \alpha_j\}$ of length N, and where $\alpha_i, \alpha_j \in \mu_O$, there exists a corresponding set of μ-sequences each of length N and of the form

$$B = \{\beta_i, \beta_j, \ldots, \beta_j\}, \quad \beta_i, \beta_j \in \mu_O$$

satisfying

$$\mu(\Omega_{AB}) = O$$

if and only if

$$\beta_i \equiv O \pmod{\alpha_j}$$

or $$\beta_j \equiv O \pmod{\alpha_i}$$

or $$\alpha_i \equiv O \pmod{\beta_j}$$

or $$\alpha_j \equiv O \pmod{\beta_i}$$

<u>Proof</u>: To prove this theorem we will only consider one possibility where $\beta_i \equiv O \pmod{\alpha_j}$. The other possibilities follow the same pattern.

Consider the two μ-sequences A and B of length N given by:

$$A = \{\alpha_i, \alpha_j, \ldots, \alpha_j\}$$

and $$B = \{\beta_i, \beta_j, \ldots, \beta_j\}$$

where $\alpha_j | \beta_i$. Then the crosscorrelation of these two sequences will result in their side lobes being multiples of either

α_i, α_j, β_i or β_j and since they all belong to μ_O, then the corresponding side lobes μ-values = O. The zero-shift element can be written in the form

$$\Omega_{AB}(O) = \alpha_i\beta_i + (N-1)\ \alpha_j\beta_j$$

and since $\alpha_j | \beta_i$, then the zero-shift element will also have a μ-value = O.

7. CONCLUSION

We have introduced in this paper the concept of nonlinear mapping of the correlation functions. This led to the design of sequences such that their nonlinearly mapped autocorrelation function has a spike at zero-shift and is zero everywhere else. By using the Möbius function, different rules for the generation and synthesis of such sequences are given. These rules greatly involve proofs from the field of number theory.

Also, we have shown that sequences of length 2 can be tabulated for further use and can be extended to longer sequences according to Theorem 3 (and Corollary 1).

Although Golay's complementary sequences, Golay (1949), (1961) and (1982), also achieve the above objective (i.e. an autocorrelation function with zero side lobes), this is done at the expense of using at least two sequences and adding their autocorrelation together. On the other hand, Golay sequences are developed in linear arithmetic while the sequences developed here use nonlinear arithmetic and, although the advantage of using one sequence in our case instead of two or more sequences in Golay's case is desirable, one must not forget the fundamental difference in which the objective is being achieved. This fundamental difference will not automatically lend these sequences to the same applications of Golay sequences but, by virtue of their nonlinear structure, they have other applications which are currently being investigated.

ACKNOWLEDGEMENTS

The authors wish to thank Dr. I. Habbab of the Department of Electrical Engineering for his valuable contribution to this work and his various comments on the manuscript, Professor R.C. Vaughan, Dr. W.W. Chen and Dr. G. Stephenson of the Department of Mathematics, Imperial College, for their interest and the many important discussions and advice during the formulation of this paper.

REFERENCES

Habbab, I., (1983) Improving the Reliability of Data Transmission in the Presence of Impulsive Noise, PhD Thesis, Imperial College of Science and Technology, University of London.

Golay, M.J.E., (1949) Multislit Spectrometry, J. Opt. Soc. Amer., vol. 39, pp. 437.

Golay, M.J.E., (1961) Complementary Series, IRE Trans. Inform. Theory, vol. IT-7, pp. 82-87.

Golay, M.J.E., (1982) The Merit Factor of Long Low Autocorrelation Binary Sequences, IEEE Trans. Inform. Theory, vol. IT-28, No. 3, pp. 543-549.

GENERALISED SAMPLING EXPANSIONS USING TAYLOR COEFFICIENTS

R.F. Hoskins
(Department of Applied Computing and Mathematics, Cranfield Institute of Technology, Bedford)

and

J. de Sousa Pinto
(Department of Mathematics, University of Aveiro, Portugal)

ABSTRACT

The classical sampling expansion for band-limited functions, usually attributed to C.E. Shannon, is well-established in the communications theory literature, and rigorous proofs exist when the functions concerned have Fourier transforms defined in the usual Fourier Integral sense. For functions whose Fourier transforms are distributions the W.K.S. expansion may or may not be valid. A class of generally valid expansions has been developed in which a band-limited function f(t) is represented as the sum of a (truncated) Taylor series and a sampling expansion. If there is relatively complete local knowledge (in the neighbourhood of t = 0) of the behaviour of f(t), then this type of expansion can be used to provide an estimate of the Taylor remainder. When the initial information is limited to uniformly sampled values of f(t) only, an algorithm can be established which computes the required Taylor coefficients.

1. INTRODUCTION

The classical Whittaker-Kotel'nikov-Shannon (W.K.S.) sampling theorem is usually stated in the form,

If f(t) has a Fourier transform $\tilde{f}(\omega)$, which vanishes outside the interval $(-\pi\Omega,\pi\Omega)$, then it admits the representation

$$f(t) = \sum_{n=-\infty}^{+\infty} f\left(\frac{n}{\Omega}\right) \text{sinc}\,(\Omega t-n) \qquad (1.1)$$

Rigorous proofs of (1.1) are given in the literature when f(t) is a function whose Fourier transform, $f(\omega)$, is defined in terms of a Fourier integral in either the L^1 or L^2 sense. Typically, Butzer and Splettstösser [1977] give the result:-

The W.K.S. expansion (1.1) is valid for any function $f(t)$ which is continuous, belongs to $L^1(\mathbb{R})$, and is such that $\tilde{f}(\omega) = 0$ for all ω such that $|\omega| > \pi\Omega$.

For a function $f(t)$ which has a Fourier transform $\tilde{\mu}(\omega)$ in the distributional sense, where $\tilde{\mu}$ has support contained in $[-\pi\Omega, \pi\Omega]$, it may or may not be the case that the classical expansion (1.1) is valid. Campbell [1968] gave simple examples showing the validity of (1.1) for a case when $\tilde{\mu}$ was a distribution of order 0, and failure of the expansion in a case when $\tilde{\mu}$ was a distribution of order 1. He went on to establish a form of sampling expansion which is generally valid for functions which are inverse Fourier transforms of distributions of compact support. In subsequent work, such as that of Lee [1977], attention has generally been directed towards the rate of growth of $f(t)$ rather than to the order of the distributional spectrum $\tilde{\mu}(\omega)$. The main purpose of this note is to draw attention to a family of modified sampling expansions which are again generally valid and which are related more directly with the structure of the spectrum $\tilde{\mu}(\omega)$.

2. DISTRIBUTIONAL BAND-LIMITED FUNCTIONS

We define $B(\Omega)$ to be the space of all functions f such that, (i) $f(z) : z \in \mathbb{C}$, is an entire function of exponential type $\leqslant \pi\Omega$, and (ii) $|f(t)| : t \in \mathbb{R}$, is bounded by a polynomial as $|t| \to \infty$. The Paley-Wiener-Schwartz Theorem shows immediately that if f is the inverse Fourier transform of a distribution $\tilde{\mu}$ of compact support inside the interval $[-\pi\Omega, \pi\Omega]$, then $f \in B(\Omega)$.

Again, for $1 \leqslant k < \infty$, we define $L_k^2(\mathbb{R})$ to be the space of all functions $f(t)$ which are square-integrable with respect to the measure

$$dm_k(t) = (1 + t^2)^{-k} dt.$$

Then, for $0 \leqslant r < \infty$, we define $B_r(\Omega)$ to be intersection of $B(\Omega)$ and $L_{r+1}^2(\mathbb{R})$:-

$$B_r(\Omega) = B(\Omega) \cap L^2_{r+1}(\mathbb{R})\,;\ 0 \leqslant r < \infty.$$

Thus $B_r(\Omega)$ is the subspace of $B(\Omega)$ comprising all functions $f(t)$ in $B(\Omega)$ which grow no faster than $|t|^r$ as $|t| \to \infty$. Appealing once more to the Paley-Wiener-Schwartz Theorem it is easy to show that if f is any member of $B_r(\Omega)$ then it is necessarily the inverse Fourier transform of a distribution $\tilde{\mu}(\omega)$ of compact support inside $[-\pi\Omega,\pi\Omega]$, and of order $\leqslant r + 1$.

We now use the fact that if $\tilde{\mu}$ is any distribution of order r whose support is contained in $[-\pi\Omega,\pi\Omega]$ then we can always find representations of $\tilde{\mu}$ of the form

$$\tilde{\mu} = \sum_{k=o}^{r} D^k \tilde{\mu}_k \tag{2.1}$$

where the $\tilde{\mu}_k$ are all distributions of order 0 (measures) with supports contained in $[-\pi\Omega,\pi\Omega]$. Taking inverse Fourier transforms of both sides of (2.1) and differentiating ν times gives

$$f^{(\nu)}(t) = \sum_{k=o}^{\nu} \binom{\nu}{k} (-i)^k \left\{ \sum_{\sigma=k}^{r} \frac{\sigma!}{(\sigma-k)!} (-it)^{\sigma-k} f_\sigma^{(\nu-k)}(t) \right\}.$$

It follows that for a given $\tilde{\mu}$ we can choose the measures $\tilde{\mu}_k$ arbitrarily in (2.1) for k=0, 1, ..., r-1, provided only that the following r equations are satisfied:

$$\sum_{k=o}^{\nu} \binom{\nu}{k} k! \frac{(-i)^k}{2\pi} < \tilde{\mu}_k(\omega),\ (i\omega)^{\nu-k} > = f^{(\nu)}(0) \tag{2.2}$$

where $\nu = 0, 1, \ldots, r-1$.

Finally, suppose that f is any member of the space $B_{r-1}(\Omega)$ for some r such that $1 \leqslant r < \infty$. Then $f(t)$ is the inverse Fourier transform of a distribution $\tilde{\mu}(\omega)$ of compact support contained in $[-\pi\Omega,\pi\Omega]$ and of order $\leqslant r$. Let $\{\mu_k\}$, k=0, 1, ..., r-1, be any set of measures with supports contained in $[-\pi\Omega,\pi\Omega]$ which are such that the system of equations (2.2) is satisfied and write

$$f_r(t) = (-it)^{-r}\left\{f(t) - \sum_{k=o}^{r-1} (-it)^k f_k(t)\right\}$$

where $f_k(t) = F^{-1}\{\tilde{\mu}_k(\omega)\}$, k=O, 1, ..., r-1. Then f, and each of the f_k, extend to the complex plane as entire functions of exponential type $\leqslant \pi\Omega$; also, as $|t| \to \infty$ we have that f is $O(|t|^{r-1})$ and each f_k is O(1). The equations (2.2) ensure that $f_r(t)$ is bounded at t=O and so f_r itself extends as an entire function of exponential type $\leqslant \pi\Omega$. Therefore we have a decomposition of $\tilde{\mu}$ of the form

$$\tilde{\mu} = \sum_{k=o}^{r-1} D^k\tilde{\mu}_k + D^r\tilde{\Phi} \tag{2.3}$$

where $\tilde{\Phi}$ is a regular distribution, with support contained in $[-\pi\Omega,\pi\Omega]$ which is defined by an L^2 function.

3. GENERAL SAMPLING EXPANSIONS OF MIXED TYPE

In Hoskins and Pinto [1984] the decomposition (2.3) is used to generate a family of modified sampling expansions by adopting a particular strategy for choosing the $\tilde{\mu}_k$. Given a specific fixed distribution $\tilde{\eta}(\omega)$ of order O, with support contained in $[-\pi\Omega,\pi\Omega]$ we take $\mu_k = a_k\tilde{\eta}$ for k=O, 1, 2, ..., r-1, where the coefficients a_k are chosen so that the equations (2.2) are satisfied. The function $\Phi(t) = F^{-1}\{\tilde{\Phi}(\omega)\}$ is an L^2 function band-limited in the classical sense to $[-\pi\Omega,\pi\Omega]$, and therefore admits the ordinary W.K.S. sampling expansion,

$$\Phi(t) = \sum_{n=-\infty}^{+\infty} \Phi\left(\frac{n}{\Omega}\right) \operatorname{sinc}(\Omega t-n)$$

Expressing the $\Phi(^n/\Omega)$ in terms of the $f\,(^n/\Omega)$, and taking account of the constraints on the coefficients a_k we obtain the following result (cf. Hoskins and Pinto, op.cit.):-

Let $f \in B_{r-1}(\Omega)$, $1 \leqslant r < \infty$, and let $\tilde{\eta}$ be an arbitrary distribution of order 0 with support contained in $[-\pi\Omega, \pi\Omega]$, and such that $<\tilde{\eta}(\omega), 1> \neq 0$. Then, uniformly on compacts,

$$f(t) = P_r(t)\eta(t) + \sum_{n=-\infty}^{+\infty} \left(\frac{\Omega t}{n}\right)^r \left\{ f\left(\frac{n}{\Omega}\right) - P_r\left(\frac{n}{\Omega}\right)\eta\left(\frac{n}{\Omega}\right)\right\} \operatorname{sinc}(\Omega t - n) \tag{3.1}$$

where $\eta(t) = F^{-1}\{\tilde{\eta}(\omega)\}$, and $P_r(t)$ is the Taylor polynomial

$$P_r(t) = \sum_{k=1}^{r-1} \frac{t^k}{k!} \left\{ \frac{d^k}{d\tau^k} \left[\frac{f(\tau)}{\eta(\tau)}\right]_{\tau=0} \right\}$$

The term corresponding to n=0 in (3.1) is to be interpreted in the sense of

$$\lim_{\tau \to 0} \left(\frac{t}{\tau}\right)^r \left[f(\tau) - P_r(\tau)\eta(\tau)\right] \operatorname{sinc}\{\Omega(t-\tau)\}$$

$$= \frac{t^r}{r!} \operatorname{sinc}(\Omega t) \left\{ \frac{d^r}{d\tau^r} \left[\frac{f(\tau)}{\eta(\tau)}\right]_{\tau=0} \right\} \tag{3.2}$$

Taking $\tilde{\eta}(\omega) = \pi[\delta(\omega + \pi\Omega) + \delta(\omega - \pi\Omega)]$ in (3.1) gives the sampling expansion suggested by Pfaffelhuber [1971]; for $\tilde{\eta}(\omega) = 2\pi\delta(\omega - a)$, $-\pi\Omega < a < \pi\Omega$, we obtain the form of expansion derived by Lee [1976]. Other choices of $\tilde{\eta}(\omega)$ may be made to generate new expansions; for example, with $\tilde{\eta}(\omega)$ taken to be $\Omega^{-1}H(\pi^2\Omega^2 - \omega^2)$ we get

$$f(t) = P_{r+1}(t)\operatorname{sinc}(\Omega t) + \sum_{n\neq 0} \left(\frac{\Omega t}{n}\right)^r f\left(\frac{n}{\Omega}\right) \operatorname{sinc}(\Omega t - n) \tag{3.3}$$

where $$P_{r+1}(t) = \sum_{k=0}^{r} \frac{t^k}{k!} \left\{ \frac{d^k}{d\tau^k} \left[\frac{f(\tau)}{\operatorname{sinc}(\Omega\tau)}\right]_{\tau=0} \right\}$$

4. APPLICATIONS

4.1 The W.K.S. expansion

$B_o(\Omega_o)$ is the space of all functions f(t) which are bounded on the real line and which extend in the complex plane to entire functions of exponential type $\leqslant \pi\Omega_o$. It is readily confirmed that $B_o(\Omega_o)$ contains every function of the form $f(t) = F^{-1}\{\tilde{\mu}(\omega)\}$ where $\tilde{\mu}$ is a distribution of order 0 with support contained in $[-\pi\Omega_o, \pi\Omega_o]$. On the other hand $B_o(\Omega_o)$ also contains functions like

$$Si(\pi\Omega_o t) = \frac{1}{2\pi} \text{ P.V.} \int_{-\pi\Omega_o}^{\pi\Omega_o} \frac{\pi.\exp(i\omega t)}{i\omega} d\omega$$

whose Fourier transform is the pseudo-function,

$$Pf\left(\frac{\pi}{i\omega}\right) . H(\pi^2\Omega_o^2 - \omega^2).$$

This is a distribution of order 1 but not a distribution of order 0.

Now for any function $f \in B_o(\Omega_o)$ we can choose $\tilde{\eta}(\omega) = \delta(\omega)$ and for any $\Omega > \Omega_o$ obtain the expansion

$$f(t) = f(0) + \sum_{n=-\infty}^{+\infty} \frac{\Omega t}{n} \left\{ f\left(\frac{n}{\Omega}\right) - f(0) \right\} \text{sinc}(\Omega t - n)$$

which can be re-arranged to give

$$f(t) = f(0) + \frac{\sin(\pi\Omega t)}{\pi\Omega} \left[f'(0) + \sum_{n\neq o} (-1)^n \frac{f(n/\Omega) - f(o)}{(^n/\Omega)} \right]$$
$$+ \sum_{n=o} \left[f\left(\frac{n}{\Omega}\right) - f(0) \right] \text{sinc}(\Omega t - n) \qquad (4.1)$$

Following the argument used in Zakai [1965] we can now derive the W.K.S. expansion for functions in $B_o(\Omega_o)$ in a very straightforward manner:- a simple application of the Riemann - Lebesgue Lemma shows that for any $g \in L^2(\mathbb{R})$ such that

$\tilde{g}(\omega) = 0$ for $|\omega| > \pi\Omega_o$ we must have

$$\sum_{n=-\infty}^{+\infty} (-1)^n g\left(\frac{n}{\Omega}\right) = 0.$$

Since $\sum_{-\infty}^{+\infty} \text{sinc}(\Omega t-n) = 1$ this means that equation (4.1) can be reduced to

$$f(t) = \sum_{n=-\infty}^{+\infty} f\left(\frac{n}{\Omega}\right) \text{sinc}(\Omega t-n)$$

4.2 Mixed Taylor/Cardinal Expansions

With $\tilde{\eta}(\omega) = \delta(\omega)$ the general form of the expansion (7) becomes

$$f(t) = \sum_{k=0}^{r-1} \frac{f^{(k)}(0)}{k!} t^k + \sum_{n=-\infty}^{+\infty} \left(\frac{\Omega t}{n}\right)^r \left[f\left(\frac{n}{\Omega}\right) - P_r\left(\frac{n}{\Omega}\right)\right] \text{sinc}(\Omega t-n) \tag{4.2}$$

The first sum on the R.H.S. of (4.2) is, of course, a finite Taylor series for f(t) about the origin, and this suggests that the sampling series represented by the second sum could be used, suitably truncated, as an estimate for the Taylor remainder. From another point of view we could regard the Taylor terms, $P_r(t)$, as a means of improving the reconstruction from sampled data when the sampling rate is low and when only a relatively small number of terms are available. The point can be illustrated by taking a somewhat extreme example. The function cost has the (distributional) Fourier transform $\pi[\delta(\omega+1) + \delta(\omega-1)]$. Sampling at just above the Nyquist rate inevitably yields a reconstruction formula from the W.K.S. series which is slow to converge. The table below shows the results obtained when the W.K.S. expansion is truncated at $n = 9$. The improvement when "mixed" Taylor/Cardinal expansions are used is very marked. (T2 and T4 denote Taylor expansions of 2 and 4 terms respectively; S_n denotes the appropriate truncated sampling expansion of (2n+1) terms).

t	f(t)	S9 (W.K.S.)	Taylor (T2)	T2 +S4	Taylor (T4)	T4 +S4
-1.309	0.258819	0.285782	0.143264	0.253749	0.265597	0.259532
-0.995	0.544639	0.562427	0.505149	0.542713	0.545962	0.544796
-0.681	0.777146	0.786276	0.768339	0.776684	0.777283	0.777164
-0.367	0.933581	0.936380	0.932832	0.933539	0.933584	0.933581
-0.052	0.998630	0.998688	0.998629	0.998630	0.998630	0.998630
0.262	0.965926	0.967369	0.965731	0.965915	0.965926	0.965926
0.576	0.838670	0.845356	0.834136	0.838428	0.838721	0.838677
0.890	0.629320	0.644067	0.603845	0.628043	0.630002	0.629403
1.204	0.358368	0.382341	0.274858	0.354557	0.362497	0.358822
1.518	0.052335	0.084679	-0.15284	0.044131	0.068676	0.053888

Note that it is a straightforward matter to implement the mixed expansion even when the initial data is limited to a finite set of uniformly sampled data values $f\left(\frac{n}{\Omega}\right)$, for $n = 0, \pm 1, \pm 2, \ldots, \pm N$. We can develop from this an algorithm which generates a "corrected" Taylor expansion as follows:-

Step 0.

$$f_o(t) = \sum_{n=-N}^{N} f\left(\frac{n}{\Omega}\right) \operatorname{sinc}(\Omega t-n)$$

$$= f(o)\operatorname{sinc}(\Omega t) + \sum_{n=-N}^{N}{}' f\left(\frac{n}{\Omega}\right) \operatorname{sinc}(\Omega t-n)$$

where we use Σ' to indicate omission of the term corresponding to n=0. This gives $f_o(0) = f(0)$, and we write

$$f_1'(0) \equiv \Big[f_o'(t)\Big]_{t=o} = \sum_{n=-N}^{N}{}' (-1)^{n+1} \frac{\Omega}{n} f\left(\frac{n}{\Omega}\right).$$

Step j. Assuming $f_i(t)$ defined for $i = 0, \ldots, j-1$, write

$$f_j(t) = f(0) + tf_1'(0) + \ldots + \frac{t^{j-1}}{(j-1)!} f_{j-1}^{(j-1)}(0)$$

$$+ \sum_{n=-N}^{N} \left(\frac{\Omega t}{n}\right)^j \left\{ f\left(\frac{n}{\Omega}\right) - \sum_{i=o}^{j-1} \frac{f_i^{(i)}(o)}{i!} \left(\frac{n}{\Omega}\right)^i \right\} \operatorname{sinc}(\Omega t-n)$$

$$\text{i.e. } f_j(t) = \sum_{i=o}^{j-1} \frac{f_i^{(i)}(o)}{i!} t^i + \frac{t^j}{j!} f_j^{(j)}(o).\operatorname{sinc}(\Omega t)$$

$$+ \sum_{n=-N}^{N}{}' \left(\frac{\Omega t}{n}\right)^j \left\{ f\left(\frac{n}{\Omega}\right) - \sum_{i=o}^{j-1} \frac{f_i^{(i)}(o)}{i!} \left(\frac{n}{\Omega}\right)^i \right\} \operatorname{sinc}(\Omega t-n)$$

which gives

$$f_{j+1}^{(j+1)}(0) \equiv \left[f_j^{(j+1)}(t) \right]_{t=0}$$

$$= (j+1)! \sum_{n=-N}^{N}{}' (-1)^{n+1} \left(\frac{\Omega}{n}\right)^{j+1} \left\{ f\left(\frac{n}{\Omega}\right) - \sum_{i=0}^{j-1} \frac{f^{(i)}(0)}{i!} \left(\frac{n}{\Omega}\right)^{i} \right\}$$

The result is therefore to produce the reconstruction formula,

$$f_r(t) = \sum_{i=0}^{r-1} \frac{f_i^{(i)}(0)}{i!} t^i$$

$$+ \sum_{n=-N}^{N} \left(\frac{\Omega t}{n}\right)^{r} \left\{ f\left(\frac{n}{\Omega}\right) - \sum_{i=0}^{r-1} \frac{f_i^{(i)}(0)}{i!} \left(\frac{n}{\Omega}\right)^{i} \right\} \operatorname{sinc}(\Omega t - n) \tag{4.3}$$

5. CONCLUSIONS

For functions which are inverse Fourier transforms of distributions of compact support various types of sampling expansions may be developed. These may either take the form of a modified W.K.S. expansion, as exemplified by Campbell's expansion, or else that of a "mixed" expansion involving a Taylor polynomial and a true sampling series. Expansions of this latter type deserve further investigation to assess the possible advantages which their form suggests.

6. ACKNOWLEDGMENT

The authors would like to acknowledge the work of Miss E. de Dias, of the University of Aveiro, who was responsible for the numerical calculations.

7. REFERENCES

Butzer, P.L. and Splettstosser, W. (1977), "A Sampling Theorem for Duration-Limited Functions with Error Estimates", *Inform. and Control*, 34, 55-63.

Campbell, L.L. (1968), "Sampling Theorem for the Fourier Transform of a Distribution with Bounded Support", *SIAM J. Appl. Math.*, 16, 626-636.

Hoskins, R.F. and Sousa Pinto, J.R. (1984), "Sampling Expansions for Functions Band-Limited in the Distributional Sense", *SIAM J. Appl. Math.*, **44**, 605-610.

Lee, A.J. (1976), "Characterization of band-limited functions and processes", *Inform. and Control*, **31**, 258-271.

Pfaffelhuber, E. (1971), "Sampling Series for Band-Limited Generalized Functions", *IEEE Trans. Inf. Th.*, IT-**17**, 650-654.

Zakai, M. (1965), "Band-Limited functions and the Sampling Theorem", *Inform. and Control*, **8**, 143-158.

ANALYSIS OF BINARY QUANTISATION EFFECTS IN THE PROCESSING OF CHIRP AND SYNTHETIC APERTURE RADAR SIGNALS

B.C. Barber
(Space Department, RAE, Farnborough, Hants)

ABSTRACT

Some coherent radars employing pulse compression techniques - especially synthetic aperture radars - are digitally processed. Usually, signals reflected from targets are coherently digitised via an analogue to digital converter with many bits of resolution. In the case of synthetic aperture radars the resulting digitised signals are range compressed and then azimuth compressed to produce a radar image. It has been observed by many experimenters that images with wide dynamic range can be produced from signals with relatively coarse digitisation. Indeed subjectively good images can be produced from data which has only one bit digitisation.

This paper presents an analysis of this phenomenon. The signal from a single point target is constructed analytically with one bit digitisation for the real and imaginary components of the chirp. The signal is then compressed. It is shown that the resulting point spread function has a simple asymptotic form for large time-bandwidth product and has a modulus which is a triangular function up to the first zeroes. The two-dimensional point spread function is then analysed and is shown to consist of a sinc^2 function.

The effects of noise and clutter are considered. It is shown that the range compression process in this case is a random walk on a two-dimensional square lattice with non-isotropic directional probability distribution. The expected point spread function is worked out for the asymptotic case of large time-bandwidth product assuming non-correlated noise and clutter. It is shown that one effect of the noise and clutter is to modulate the zero crossings of the one bit data in such a way as to encode amplitude information. Finally, for a given time-bandwidth product, hard clipping raw radar signal data introduces a loss of about 5.9 dB in signal-to-clutter ratio in the compressed signal for a point target in uncorrelated clutter for small signal-to-noise ratio in the raw, uncompressed signal.

1. INTRODUCTION

Some coherent radars employing pulse compression techniques - especially synthetic aperture radars - are digitally processed. Usually, signals reflected from targets are coherently digitised via an analogue to digital converter with many bits of resolution. In the case of imaging synthetic aperture radars the resulting digitised signals are range compressed and then azimuth compressed to produce a radar image. It has been observed by many experimenters that images with wide dynamic range can be produced from signals with relatively coarse digitisation. Indeed, subjectively good images can be produced from data which has only one bit digitisation.

This effect is interesting from several points of view. The use of one bit data implies a very considerable reduction in data rate over 'many bit' data. This may help with future imaging radar systems where improvements in resolution and area of coverage lead to greatly increased data rates. In addition the one bit correlations needed to compress the signals in range are relatively simple, easy and fast with simple hardware. This is an advantage when real time processing is required.

There are other interesting implications for the binary encoding of signals in general but these aspects are not discussed in this paper.

The use of one bit correlators has been widespread for many years in signal processing, especially for signals which can be considered 'random' in some sense, see for example Davies (1969). The performance of signal processors using such one bit or 'hard-clipped' data is invariably analysed using numerical methods, see Blythe et al (1970, 1973), or by simulation.

In addition there has been continuing interest over a number of years in the reconstruction of signals from their zero crossing and from the zero crossings of their Fourier transforms. See, for example, Logan (1977), Oppenheimer et al (1981, 1983).

One result of the analysis presented here is that it is shown how a stochastic signal may be encoded so that, in a certain asymptotic sense, the error in reconstructing it from the one bit encoding may be made as small as one pleases. This specific problem will be addressed in a future paper.

The analysis in this paper, unlike previous analyses is not numerical. It may be of interest to note that it uses ideas from analytical number theory.

The one-bit problem is analysed in this paper for a point target. In sections 2 and 3 the point target is taken to be on an absorbing, black background. In the remaining sections the point is taken to be an a cluttered background, the image clutter resulting from other uncorrelated scatterers and noise.

Considering just a single point has limitations. There are, for example, possible effects caused by the interactions between close targets such as capture effects which can occur when highly non-linear operations like hard clipping are performed on signals. An approximate analysis of several close targets is possible on the basis of the analysis given here for a single target but is not carried out in this paper.

2. THE POINT SPREAD FUNCTION IN RANGE

Consider a sequence of samples of a phase modulated analytic signal of unit modulus:

$$A(n) = \exp i\phi(n) , \quad 0 \leqslant n \leqslant N , \qquad (1)$$

where n is an integer.

Let this sequence be digitised with one bit for the real and imaginary parts so that the phase of the signal vector is now quantised so that it has the values $\pm\pi/4$ and $\pm 3\pi/4$ only. The sequence (1) then becomes

$$B(n) = \exp i\,(2k + 1)\pi/4, \qquad (2)$$

where k is the integer part of $2\phi(n)/\pi$, ie $k = [\,2\phi(n)/\pi\,]$ and square brackets are used to denote the 'integer part of'. We consider here the correlation of B(n) against a signal sequence C(n) :

$$C(n) = \exp i(2\ell + 1)\pi/4 , \qquad (3)$$

where $\ell = [\,2\psi(n)/\pi\,]$ and $\psi(n) = \phi(n - \gamma)$, that is $\psi(n)$ is the delayed or 'lagged' version of ϕ. The lag, γ, is a real number. In the correlation process considered here each sample of the one bit lagged signal C(n) is multiplied by the conjugate of the one bit (transmitted) undelayed signal B(n) and added. This correlation is here called the point spread function, $g(\gamma)$.

Hence

$$g(\gamma) = \sum_{n=0}^{N} C(n - \gamma)B^*(n) \qquad (4)$$

$$= \sum_{n=0}^{N} \exp i(\ell - k)\pi/2 \ , \qquad (5)$$

where $k = [\,2\phi(n)/\pi\,]$ and $\ell = [\,2\psi(n - \gamma)/\pi\,]$.

In general $g(\gamma)$ can be a complicated and somewhat erratic function. It is self-evident that $|g(\gamma)| \leqslant N+1$, and that if $\psi(n) = \phi(n)$ for all n $|g(0)| = N+1$.

To proceed further an analytic representation of the digitisation process is required and this is given in Appendix A.

2.1 Analytic representation of the correlation

In Appendix A it is demonstrated that if $\exp i\eta(\chi)$ is the 'phase part' of an analytic signal then the one bit digitised version with unit modulus can be written as a Fourier series

$$z(\chi) = \frac{2\sqrt{2}}{\pi} \sum_{a=-\infty}^{\infty} \frac{\exp\{i(4a + 1)\eta(\chi)\}}{4a + 1} \ , \qquad (6)$$

except on a set of measure zero. When $\eta(\chi)$ is an integer multiple of $\pi/2$ then $z(\chi)$ is understood to be $\frac{1}{2}z(\chi-) + \frac{1}{2}z(\chi+)$. Using this representation $B^*(n)$ becomes

$$B^*(n) = \frac{2\sqrt{2}}{\pi} \sum_{a=-\infty}^{\infty} \frac{\exp\{-i(4a + 1)\phi(n)\}}{4a + 1} \qquad (7)$$

and

$$C(n - \gamma) = \frac{2\sqrt{2}}{\pi} \sum_{b=-\infty}^{\infty} \frac{\exp\{i(4b + 1)\psi(n - \gamma)\}}{4b + 1} \ . \qquad (8)$$

Hence the point spread function is

$$g(\gamma) = \frac{8}{\pi^2} \sum_{n=0}^{N} \sum_{a=-\infty}^{\infty} \sum_{b=-\infty}^{\infty} \frac{\exp i\{(4b+1)\psi(n-\gamma) - (4a+1)\phi(n)\}}{(4a+1)(4b+1)}. \tag{9}$$

There are no difficulties here in multiplying the Fourier series together and neither this nor reversing the order of the sums affects the summability of the series. So

$$g(\gamma) = \frac{8}{\pi^2} \sum_{a=-\infty}^{\infty} \sum_{b=-\infty}^{\infty} \frac{1}{(4a+1)(4b+1)} \sum_{n=0}^{N} \exp i\{(4b+1)\psi(n-\gamma) - (4a+1)\phi(n)\}. \tag{10}$$

2.2 The point spread function for quadratic phase modulation

In Appendix B $\phi(n)$ is worked out for a typical pulse compression radar and is shown to be

$$\phi(n) = \pi\lambda n\left(\frac{n}{N} - 1\right), \tag{11}$$

where $0 \leqslant n \leqslant N$ and $0 < \lambda \leqslant 1$. λ is restricted to a rational number, ie $\lambda = c/d$, $(c,d) = 1$ in the usual notation of number theory.

$\psi(n - \gamma)$ is taken to be a delayed replica of $\phi(n)$ but with the addition of a constant phase offset θ (which varies from pulse-to-pulse and forms the basis for a synthetic aperture, see Appendix B). In this section any random phase terms are ignored and a purely noise-free signal is considered.

Then

$$\psi(n-\gamma) = \pi\lambda(n-\gamma)\left\{\frac{n-\gamma}{N} - 1\right\} + \theta. \tag{12}$$

Substituting for ϕ and ψ into (10) then gives

$$g(\gamma) = \frac{8}{\pi^2} \sum_{a=-\infty}^{\infty} \sum_{b=-\infty}^{\infty} \frac{\exp i\{\pi\lambda\gamma(4b+1)(\frac{\gamma}{N}+1) + \theta(4b+1)\}}{(4a+1)(4b+1)} \times \sum_{n=0}^{N} \exp i\left\{\frac{4\lambda\pi n^2(b-a)}{N} - 2\pi\lambda n\{2(b-a) + \frac{\gamma}{N}(4b+1)\}\right\}. \tag{13}$$

This expression for $g(\gamma)$ is obviously quite complicated, but it does have a simple asymptotic form for N large (which is the practical case of interest) and this is now examined.

2.3 *The asymptotic form of the point spread function*

The major part of the sum in (13) comes from the diagonal where $a = b$ and this is $O(N)$. The remainder is $O(N^{\frac{1}{2}+\varepsilon})$. Let

$$\xi(a,b,\gamma) = \sum_{n=0}^{N} \exp i\left\{\frac{4\lambda\pi n^2 (b-a)}{N} - 2\pi\lambda n\left\{2(b-a) + \frac{\gamma}{N}(4b+1)\right\}\right\} \tag{14}$$

Now separate the sum for $g(\gamma)$ in (13) into the sum over the diagonal $a = b$ and the rest so that

$$g(\gamma) = h(\gamma) + f(\gamma) , \tag{15}$$

where

$$h(\gamma) = \frac{8}{\pi^2} \sum_{b=-\infty}^{\infty} \frac{\exp i\left\{\pi\lambda\gamma(4b+1)(\frac{\gamma}{N}+1) + \theta(4b+1)\right\}}{(4b+1)^2} \xi(b,b,\gamma) \tag{16}$$

$$= \frac{8(N+1)}{\pi^2} \sum_{b=-\infty}^{\infty} \frac{\exp i(4b+1)\theta}{(4b+1)^2} \frac{\sin \pi\lambda\gamma(4b+1)}{\pi\lambda\gamma(4b+1)} + O(\gamma) + O(1).$$

Thus $h(\gamma) = O(N)$ and it remains to be shown that $f(\gamma) = O(N^{\frac{1}{2}+\varepsilon})$.

The sum $\xi(a,b,\gamma)$ may be estimated by Weyl's inequality (Vaughan (1981), Lemma 2.4, p. 11, see also Landau (1969), Vol. 1, pp 250-260).

If

$$\phi(x) = \alpha x^k + \alpha_1 x^{k-1} + \ldots + \alpha_k \tag{18}$$

and

$$T(\phi) = \sum_{x=1}^{N} \exp 2\pi i\phi(x) \tag{19}$$

and α has the rational approximation $|\alpha - a/q| \leqslant 1/q^2$, which it always has by Dirichlet's lemma (Vaughan 1981 lemma 2.1, p. 9).

Then

$$T(\phi) \ll N^{1+\varepsilon}\left(\frac{1}{q} + \frac{1}{N} + \frac{q}{N^k}\right)^{1/K}, \tag{20}$$

where $K = 2^{k-1}$. The notation $f \ll g$ means $|f| \leqslant cg$, c constant. And so

$$\xi(a,b,\gamma) = O\left(\sum_{n=0}^{N} \exp i\, 2\pi \frac{2cn^2(b - a)}{Nd}\right) \tag{21}$$

and in this case $K = k = 2$.

In Appendix C it is shown that

$$\sum_{n=1}^{N} \exp i\, 2\pi\alpha n^2 = O(N^{\frac{1}{2}+\varepsilon}) \tag{22}$$

where if α is rational (as has been assumed here) ie. $\alpha = r/s$, say, then it is necessary that $s \geqslant N$.

In the case of eqn. 21 $r = 2c(b-a)$
and $s = Nd$ (23)

When r and s have no common factors it immediately follows that $\xi(a,b,\gamma) = O(N^{\frac{1}{2}+\varepsilon})$. In addition, $c/d = \lambda \leqslant 1$, $(c,d) = 1$ and so $c/Nd < {}^1\!/_N$ so that whatever common factors c and N have, $s > N$ and again $\xi(a,b,\gamma) = O(N^{\frac{1}{2}+\varepsilon})$.

If $\lambda = 1$ then $d = 1$ and if N is even a factor of 2 cancels. However in this case it is easily demonstrated that again $\xi(a,b,\gamma) = O(N^{\frac{1}{2}+\varepsilon})$ and the constant in the order term increases by 2. Obviously in all other cases (ie. $\lambda < 1$) $d \geqslant 2$ and then $s \geqslant N$ so cancelling the factor of 2 makes no difference.

The last case occurs when (a-b) and N have a common factor which is greater than d. Let (a-b) and N be divisible by p ie. $(a-b, N) = p$

where $$d < p \leqslant N \tag{24}$$

The sum in (21) is then

$$\xi(a,b,\gamma) = O\left(\sum_{n=o}^{N} \exp i\, 2\pi \frac{2c(b-a)n^2/p}{Nd/p}\right) \tag{25}$$

where $d/p < 1$.

In this case there is a result due to Hua Loo-Keng (1957):

$$\sum_{x=1}^{P} \exp i\, 2\pi \frac{hx^k}{q} = \frac{P}{q} \sum_{x=1}^{q} \exp i \frac{2\pi hx^k}{q} + O(q^{\frac{1}{2}+\varepsilon}) \tag{26}$$

where $k \geqslant 2$, $(h,q) = 1$ and applying this gives

$$\xi(a,b,\gamma) = O\left(\left(\frac{Np}{d}\right)^{\frac{1}{2}+\varepsilon}\right) \tag{27}$$

i.e.

$$= O\left((Np)^{\frac{1}{2}+\varepsilon}\right) \tag{28}$$

Now consider the factor $^1/(4a+1)$. $^1/(4b+1)$ in the sum for $f(\gamma)$ and let $b-a = p.t$ say

$$\text{ie.} \quad b = a + pt \tag{29}$$

Then

$$\frac{1}{4a+1} \cdot \frac{1}{4b+1} = \frac{1}{4pt} \left\{ \frac{1}{4a+1} - \frac{1}{4a+4pt+1} \right\} \tag{30}$$

$$= O\left((pt)^{-1} \right) \tag{31}$$

Combining this with (25) gives the order of this term in the sum for $f(\gamma)$, $f_p(\gamma)$ as

$$f_p(\gamma) = O\left(\frac{N^{\frac{1}{2}+\varepsilon}}{p^{\frac{1}{2}+\varepsilon}} \cdot \frac{1}{t} \right) \tag{32}$$

$$\text{ie.} \quad f_p(\gamma) = O(N^{\frac{1}{2}+\varepsilon}) \tag{33}$$

The net result from all of the cases considered above is that

$$f(\gamma) = O(N^{\frac{1}{2}+\varepsilon}). \tag{34}$$

Combining (34) with (15) and (17) then gives

$$g(\gamma) = \frac{8(N+1)}{\pi^2} \sum_{b=-\infty}^{\infty} \frac{\exp i(4b+1)\theta}{(4b+1)^2} \frac{\sin \pi \lambda\gamma(4b+1)}{\pi\lambda\gamma(4b+1)} + O(N^{\frac{1}{2}}) \tag{35}$$

This expression has been proved for λ rational and N large

$|g(\gamma)|$ is independent of θ. We have then

$$|g(\gamma)| = \frac{8(N+1)}{\pi^2} \sum_{b=-\infty}^{\infty} \frac{\sin \pi\lambda\gamma(4b+1)}{\pi\lambda\gamma(4b+1)^3} + O(N^{\frac{1}{2}}). \quad (36)$$

The sum of the series in (36) is given by Bromwich (1955) (p. 362, equation (2.7)) and this gives for $|g(\gamma)|$:

$$|g(\gamma)| = (N+1)(1-|\lambda\gamma|)\ , \quad -1 \leqslant \lambda\gamma \leqslant 1\ . \quad (37)$$

So that up to the first zeroes the asymptotic form of the point spread function is a triangular function which is a very simple result for such an apparently complicated expression as (13).

$g(\gamma)$ has zeroes wherever $\gamma\lambda$ is an integer, it is therefore an interpolating function as $N \to \infty$.

The 'exact' $|g(\gamma)|$ given by (5) is plotted for $N = 100$ $\lambda = 1$ in fig. 1 and $N = 10000$, $\lambda = 1$ in fig. 2. Note the triangular nature of the centre lobe up to the first zeroes.

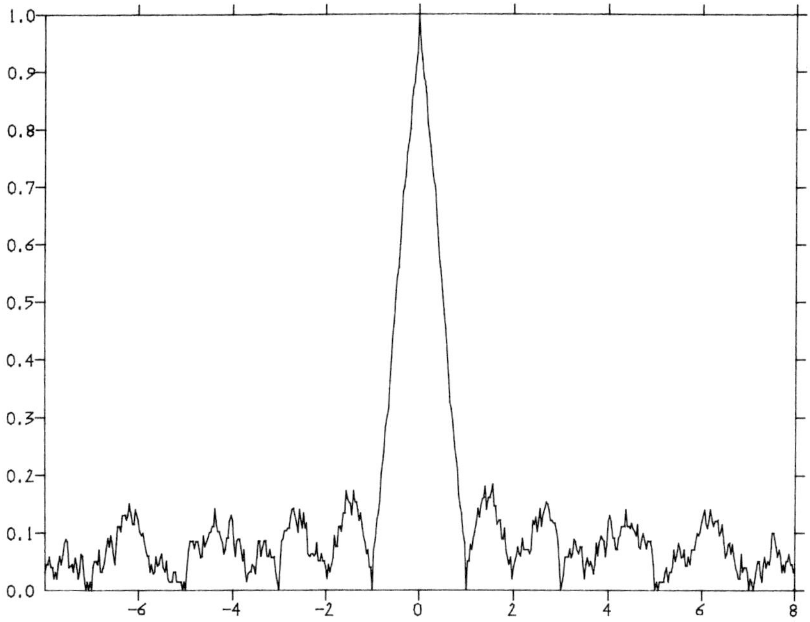

Fig. 1 $g(\gamma)$ for $N = 100$

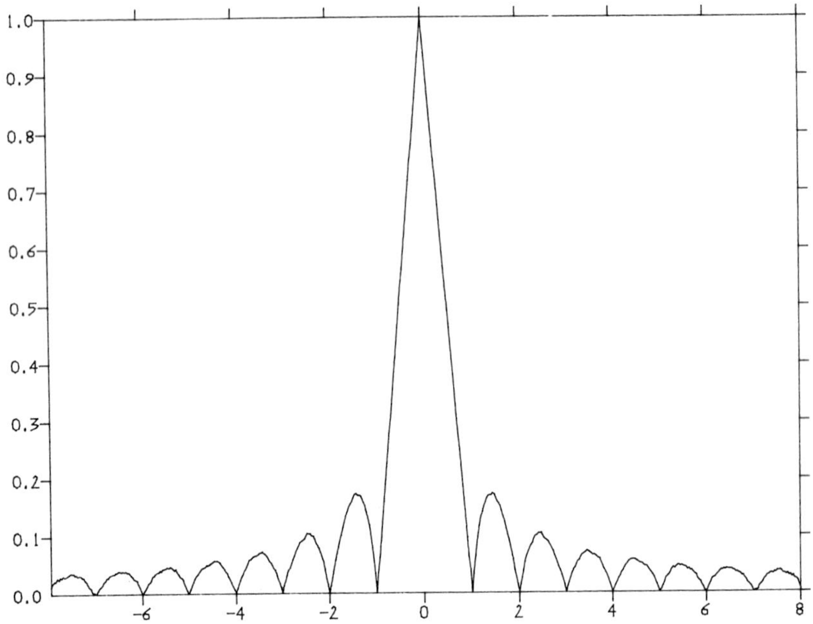

Fig. 2 $g(\gamma)$ for N = 10000

2.4 The range spectrum: aliassing

The range point spread function given in (35) is 'nominal' in the sense that it is assumed implicitly that γ is continuous. The parameter γ is essentially the difference between the range coordinate of the target point and the correlation variable or 'lag'. If the correlation is performed as a one bit correlation γ is sampled at the same rate as the original data. The lag γ could be sampled at a higher rate either by choosing a small value for λ or by complicating the correlation. These measures would negate the reasons for performing one bit correlations (i.e. lower data rate, simple processing etc) and anyway are not necessary as will now be demonstrated.

Consider the spectrum corresponding to the function in (35). The spectrum of $\sin \pi\lambda\gamma(4b + 1)/\pi\lambda\gamma(4b + 1)$ defined by

$$G(\omega) = \frac{1}{2\pi} \int_{-\infty}^{\infty} \frac{\sin \pi\lambda\gamma(4b + 1)}{\pi\lambda\gamma(4b + 1)} \exp - i\omega\gamma \, d\gamma \qquad (38)$$

is a rectangular function of height $1/2\pi\lambda(4b + 1)$ extending from $-\pi\lambda(4b + 1)$ to $+\pi\lambda(4b + 1)$ of unit area.

Suppose that the original signal is sampled at the Nyquist rate, i.e. $\lambda = 1$, in this case all of the spectrum outside the interval from $-\pi$ to $+\pi$ (corresponding to $b = 0$) is aliassed back into this interval. All the spectra corresponding to the various $\text{sinc}\{\pi\lambda\gamma(4b + 1)\}$ terms in (35) have the same area (equal to unity). Hence the result of sampling γ at the Nyquist rate is to remove the b dependence of the sinc $\{\pi\lambda\gamma(4b + 1)\}$ terms in (35) so that they all become $\sin \pi\gamma/\pi\gamma$. The point spread function then becomes

$$g(\gamma) = \frac{8(N + 1)}{\pi^2} \frac{\sin \pi\gamma}{\pi\gamma} \sum_{b=-\infty}^{\infty} \frac{\exp i(4b + 1)\theta}{(4b + 1)^2} . \qquad (39)$$

3. THE AZIMUTH POINT SPREAD FUNCTION

The formation of a synthetic aperture for a linear system is outlined in Appendix B.

In the case of one bit data the range point spread function for each pulse is given by (35). The phase corresponding to the time delay from radar to target is θ and in Appendix B equation (102) it is demonstrated that

$$\theta = - \frac{\pi\mu m^2}{M} , \qquad (40)$$

where $0 < \mu \leqslant 1$, $0 \leqslant m \leqslant M$ and there are $M + 1$ pulses in the set used for azimuth compression. The phase function against which the correlation is performed is $\exp (i\pi\mu/M)(m - \nu)^2$ where ν is the azimuth 'lag' in the correlation.

The two-dimensional point spread function in range and azimuth is then

$$g_R(\gamma)g_A(\nu) = \sum_{m=0}^{M} \frac{8(N + 1)}{\pi^2} \sum_{b=-\infty}^{\infty} \frac{\sin \pi\lambda\gamma(4b + 1)}{(4b + 1)^2 \pi\lambda\gamma(4b + 1)} \exp \frac{i\pi\mu}{M}(m - \nu)^2$$

$$\exp - i\pi \frac{\mu m^2}{M} (4b + 1) \qquad (41)$$

$$= \frac{8(N+1)(M+1)}{\pi^2}\,\frac{\sin \pi\lambda\gamma}{\pi\lambda\gamma}\,\frac{\sin \pi\mu\nu}{\pi\mu\nu} + \frac{8(N+1)}{\pi^2}\sum_{b=1}^{\infty}\frac{\sin \pi\lambda\gamma(2b+1)}{(2b+1)^2\,\pi\lambda\gamma(2b+1}$$

$$\times \exp\frac{i\pi\mu\nu^2}{M}\sum_{m=0}^{M}\exp -\frac{i2\pi\mu m\nu}{M}\cos\frac{4\pi\mu b m^2}{M}\,. \qquad (42)$$

From Weyl's inequality (20) the sum over m in (42) is $O(M^{\frac{1}{2}})$ and contributes to the background clutter against which the image of the point target appears.

Hence we have

$$g_R(\gamma)g_A(\nu) = \frac{8(N+1)(M+1)}{\pi^2}\,\frac{\sin \pi\lambda\gamma}{\pi\lambda\gamma}\,\frac{\sin \pi\mu\nu}{\pi\mu\nu} + O(N^{\frac{1}{2}}M) + O(M^{\frac{1}{2}}N) \qquad (43)$$

Of course, there is no change in amplitude of the image with changes in target cross-section due to the loss of all amplitude information in the hard clipping operation. In the next sections it is shown that amplitude information is encoded into the hard clipped signals (via the zero crossings, since there is no other way) when the signal from the target point is added to signals from a clutter background and noise.

4. RANDOM PROCESS IN RANGE

In reality the signal digitised by the one bit process consists of the resultant of the noise and clutter and the signal scattered from the target point under discussion.

For example, let R be the magnitude of the resultant vector, ζ the phase of R, P the magnitude of the signal vector from the target point and σ the rms noise amplitude. The joint probability density function for R and ζ is given by Rice (1954), p. 238:

$$p(R,\zeta) = \exp\left\{-\frac{1}{2\sigma^2}(R^2 + P^2 - 2RP\cos\zeta)\right\}\frac{RdRd\zeta}{2\pi\sigma^2} \qquad (44)$$

and this gives the pdf for one sample. For N + 1 samples we have a joint density involving N + 1 R variates and N + 1 ζ variates. After the one bit digitisation process the magnitude of the resultant vector is constant and thus the density function which appears in the evaluation of the expected point spread function in this case is the joint marginal phase density for the N + 1 samples $p(\zeta_0 \ldots \zeta_N)$. It is self-evident that $p(\zeta_0 \ldots \zeta_N)$ is periodic in the N + 1 phase variates and each has a period of 2π. That is $p(\zeta_0 \ldots \zeta_N)$ can be considered to be the surface density of an N + 2 dimensional unit sphere (and incidentally can be related to an 'equivalent' unwrapped density via an N + 1 dimensional Poisson sum).

Writing $p(\zeta_0 \ldots \zeta_N)$ as a Fourier series

$$p(\zeta_0\ldots\zeta_N) = \left(\frac{1}{2\pi}\right)^{N+1} \sum_{k_0=-\infty}^{\infty} \sum_{k_N=-\infty}^{\infty} \Gamma(k_0\ldots k_N) \exp i(k_0\zeta_0 + \ldots + k_N\zeta_N), \tag{45}$$

where

$$\Gamma(k_0..k_N) = \int_{-\pi}^{\pi} \ldots \int_{-\pi}^{\pi} p(\zeta_0\ldots\zeta_N) \exp - i(k_0\zeta_0 + \ldots + k_N\zeta_N) d\zeta_0\ldots d\zeta_N. \tag{46}$$

In the analysis which follows in the next section we shall require the joint marginal density of ζ_n and ζ_m:

$$p(\zeta_n,\zeta_m) = \frac{1}{4\pi^2} \sum_{k_n=-\infty}^{\infty} \sum_{k_m=-\infty}^{\infty} \Gamma(k_n,k_m) \exp i(k_n\zeta_n + k_m\zeta_m). \tag{47}$$

One of the difficulties which arises is that as matters stand the mean of each of the phase samples varies from sample to sample. The signal vector P is added to a random walk and the phase density of the random walk from the clutter and noise is uniform over $[-\pi,\pi]$ and hence the mean phase of the resultant R is equal to the phase of the signal P. This rotates from sample to sample and is the phase $\psi(n - \gamma)$ of section 1.

This problem can, however, be easily overcome by rotating the angles ζ_n and ζ_m in (47) so that $\langle\zeta_n\rangle = \langle\zeta_m\rangle = 0$. Define $p^0(\zeta_n,\zeta_m) = p(\zeta_n,\zeta_m)$ when $\langle\zeta_n\rangle = \langle\zeta_m\rangle = 0$; let the corresponding Fourier coefficient be $\Gamma^0(k_n k_m)$.

Then

$$p(\zeta_n,\zeta_m) = \frac{1}{4\pi^2} \sum_{k_n=-\infty}^{\infty} \sum_{k_m=-\infty}^{\infty} \Gamma^0(k_n,k_m) \exp - i(k_n\psi_n + k_m\psi_m) \exp i(k_n\zeta_n + k_m\zeta_m), \tag{48}$$

where $\psi_n = \psi(n - \gamma) = \langle\zeta_n\rangle$, $\psi_m = \psi(m - \gamma) = \langle\zeta_m\rangle$.

$\Gamma^o(k_n k_m)$ now relates to a stationary zero mean process. It is now necessary to examine the correlation properties of the process. The properties of phase correlations are still not well understood except in a few special cases - see for example the discussion in Mardia (1975), and some care is necessary. The joint density of two phase measurements in the case of a constant vector (with constant phase) plus a band-limited noise process has been given by Middleton (1960) (p. 419, example 9.6(c) equations (6), (7) and (8)), see also Middleton (1948) equation (5.20), pp. 472-473. The correlation is determined by the noise process and evidently if the process is sampled at the Nyquist rate, ie $\lambda = 1$ where λ is defined in Appendix B, and the noise process is strictly band limited (and has the same bandwidth as the signal vector) then there will be no correlation between the phase samples. However in practical cases the process is always over-sampled by 10-20%. i.e. λ is usually somewhat less than 1 and there will be a small correlation between the phase samples. Indeed if the noise spectrum is rectangular then the correlation coefficient for two consecutive phase samples is $\rho = \sin \pi\lambda/\pi\lambda$, i.e. about 0.2 for $\lambda = 0.8$. It is therefore evident that it is really necessary to consider first-order terms in ρ in any approximation of $p^o(\zeta_n,\zeta_m)$.

In addition the case of most interest here is that of a weak signal in strong noise. This will, in practice, always be the real situation to which this analysis relates. The reason is that the signal from the target point is added to signals from all points within a distance equivalent to one half pulse width and for a chirp with a long pulse width this random clutter signal will completely swamp the signal from the target point.

Define

$$a_o = \frac{P}{\sigma\sqrt{2}}, \quad \text{see equation (44).} \tag{49}$$

Then $a_o < 1$.

Hence an approximation for $p^o(\zeta_n,\zeta_m)$ is required including terms $O(a_o^2)$ and $O(\rho)$ but excluding higher order terms. Expanding equation (6), example 9.6(c), p. 419 of Middleton (1960) then gives

$$p^{o}(\zeta_n,\zeta_m) = \frac{1}{4\pi^2}\left\{1 + \frac{\rho\pi}{2}\cos(\zeta_n - \zeta_m) + \pi a_o^2 \cos\zeta_n \cos\zeta_m + a_o^2 \cos 2\zeta_n + a_o^2 \cos 2\zeta_m + \frac{\rho a_o}{\pi^2}\cos(\zeta_n - \zeta_M) \times (\cos\zeta_n + \cos\zeta_m)\right\}$$

$$+ O(\rho^2) + O\left(\rho^2 a_o\right) + O\left(a_o^2\rho\right) + O\left(a_o^4\right). \qquad (50)$$

In the next section we shall require $\Gamma^{o}(k_n = -1, k_m = +1) \equiv \Gamma^{o}_{nm}(-1,+1)$ and this is easily shown from (46) and (50) to be

$$\Gamma^{o}_{nm}(-1,+1) = \frac{\pi}{4}\left(a_o^2 + \rho\right) + O(\rho^2) + O\left(\rho a_o^2\right) + O\left(a_o^4\right). \qquad (51)$$

The correlation ρ is a function of the difference between n and m since the noise process is wide sense stationary, ie $\rho = \rho(|n - m|)$. In this instance

$$\rho = \frac{\sin \pi\lambda(n - m)}{\pi\lambda(n - m)} \quad . \qquad (52)$$

A further parameter which will be required in the next section is the marginal probability density $p(\zeta_n)$ where from equation (48)

$$p(\zeta_n) = \frac{1}{2\pi} \sum_{k_n=-\infty}^{\infty} \Gamma^{o}(k_n) \exp - ik_n\psi_n \exp ik_n\zeta_n \qquad (53)$$

and from equation (44)

$$\Gamma^{o}(k_n = -1) \equiv \Gamma^{o}_n(-1) = \frac{a_o\sqrt{\pi}}{2} + O\left(a_o^3\right). \qquad (54)$$

5. THE EXPECTED RANGE POINT SPREAD FUNCTION

5.1 The mean square point spread function

The psf is given by equation (5) for the noise-free case. In the case of a signal plus noise the phase ψ is a random variable and the accompanying random process was discussed in the last section. In this section the expected intensity $\langle g(\gamma)g^*(\gamma)\rangle$ is calculated for a chirp radar (i.e. in the range direction). The averaging operation is carried out over the random phases introduced by the noise.

Then

$$\langle g(\gamma)g^*(\gamma)\rangle = \sum_{n=0}^{N} \sum_{m=0}^{N} \exp i\,\frac{\pi}{2}(u - k)\ \langle \exp i\,\frac{\pi}{2}(\ell - v)\rangle\ , \tag{55}$$

where $u = [2\phi(m)/\pi]$, $k = [2\phi(n)/\pi]$, $\ell = [2\zeta_n/\pi]$, $v = [2\zeta_m/\pi]$ and ζ_n, ζ_m are the phase angles defined in the previous section. Note that $\langle g(\gamma)g^*(\gamma)\rangle$ is the mean square resultant of a non-isotropic random walk on the unit square lattice.

Let

$$\left.\begin{aligned} \langle \exp i\,\frac{\pi}{2}(\ell - v)\rangle &= G(n,m) \quad & n \neq m \\ &= 1 \quad & n = m\,. \end{aligned}\right\} \tag{56}$$

Substituting the one bit process (6) gives, for G(n,m)

$$G(n,m) = \frac{8}{\pi^2} \sum_{b=-\infty}^{\infty} \sum_{c=-\infty}^{\infty} \frac{1}{4b+1}\,\frac{1}{4c+1} \int_{-\pi}^{\pi}\int_{-\pi}^{\pi} \exp\left\{i(4b+1)\zeta_n\right\} \times$$

$$\exp\left\{-i(4c+1)\zeta_m\right\} p(\zeta_n,\zeta_m)\,d\zeta_n\,d\zeta,\quad n \neq m\,. \tag{57}$$

Substituting the Fourier series (45) for $p(\zeta_n,\zeta_m)$ and integrating over ζ_n and ζ_m gives

$$G(n,m) = \frac{8}{\pi^2} \sum_{b=-\infty}^{\infty} \sum_{c=-\infty}^{\infty} \frac{1}{4b+1}\,\frac{1}{4c+1}\,\Gamma^{o}_{nm}\{-(4b+1),(4c+1)\} \times$$

$$\exp\{-i(4b+1)\psi_n\} \times \exp\{i(4c+1)\psi_m\},\ n \neq m, \tag{58}$$

$$= 1 \quad n = m\,.$$

Substituting for G(n,m) into (55) and also the one bit process for $\exp i\frac{\pi}{2}(u - k)$ gives

$$\langle g(\gamma)g^*(\gamma)\rangle = N + 1 + \frac{64}{\pi^4}\sum_{a=-\infty}^{\infty}\sum_{b=-\infty}^{\infty}\sum_{c=-\infty}^{\infty}\sum_{d=-\infty}^{\infty}\frac{1}{4a+1}\frac{1}{4b+1}\times$$

$$\frac{1}{4c+1}\frac{1}{4d+1}\sum_{n=0}^{N}\sum_{\substack{m=0\\ \neq n}}^{N}\times$$

$$\Gamma^{0}_{nm}\{-(4b+1),(4c+1\}\exp i\{-(4b+1)\psi(n) +$$

$$(4c+1)\psi(m)\}\times\exp i\{(4a+1)\phi(n) - (4d+1)\phi(m)\}, \quad (59)$$

where $\phi(n) = \pi\lambda n\left(\frac{n}{N} - 1\right)$,

$$\psi(n) = \pi\lambda(n - \gamma)\left\{\frac{n-\gamma}{N} - 1\right\} + \theta,$$

with similar expressions for $\phi(m)$ and $\psi(m)$. The same approximation may be made in the calculation of the sums over n and m as was made in the evaluation of the deterministic point spread function in section 1 by applying Weyl's inequality. That is for large N the terms for which a = b and c = d are $O(N^2)$ and the remainder are O(N). The constant implied in the order depends upon a_0 the signal-to-noise ratio, which is small. Hence the constant N + 1 corresponding to n = m in G(n,m) is retained. Hence

$$\langle g(\gamma)g^*(\gamma)\rangle = N + 1 + \frac{64}{\pi^4}\sum_{a=-\infty}^{\infty}\sum_{c=-\infty}^{\infty}\left(\frac{1}{4a+1}\right)^2\left(\frac{1}{4c+1}\right)^2$$

$$\exp - i(4a+1)\left\{\pi\lambda\gamma\left(\frac{\gamma}{N}+1\right)+\theta\right\}\times\exp i(4c+1)\left\{\pi\lambda\gamma\left(\frac{\gamma}{N}+1\right)+\theta\right\}\times$$

$$\sum_{n=0}^{N}\sum_{\substack{m=0\\ \neq n}}^{N}\Gamma^{0}_{nm}\left\{-(4a+1),(4c+1)\right\}\exp i\frac{2\pi\lambda\gamma}{N}\left\{n(4a+1)-m(4c+1)\right\}+O(N). \quad (60)$$

Now, the case of small signal-to-noise ratio and small correlation was discussed in the last section and this is the case of interest here. It is evident from (50) that if terms $O(\rho^2)$, $O\left(\rho a_o^2\right)$ and $O\left(a_o^4\right)$ are discarded then $-(4a + 1)$ and $(4c + 1)$ can only be -1 and $+1$, ie $a = c = 0$ in (60).

Hence in this case

$$\langle g(\gamma)g^*(\gamma)\rangle = N + 1 + \frac{64}{\pi^4} \sum_{n=0}^{N} \sum_{\substack{n=0 \\ \neq m}}^{N} \Gamma^{o}_{nm}\{-1,+1\} \exp i \frac{2\pi\lambda\gamma}{N}(n - m)$$

$$+ O(N) + O(\rho^2) + O\left(\rho a_o^2\right) + O\left(a_o^4\right). \tag{61}$$

Substituting for $\Gamma^{o}_{nm}(-1,+1)$ from (51) then gives

$$\langle g(\gamma)g^*(\gamma)\rangle = N + 1 + \frac{16a_o^2(N + 1)^2}{\pi^3}\left(\frac{\sin \pi\lambda\gamma}{\pi\lambda\gamma}\right)^2 + \frac{16}{\pi^3} \times$$

$$\sum_{n=0}^{N} \sum_{\substack{m=o \\ \neq n}}^{N} \rho(|n - m|) \exp i \frac{2\pi\lambda\gamma}{N}(n - m) + O\left(a_o^2 N\right) + O(\rho^2) + O\left(\rho a_o^2\right) + O\left(a_o^4\right)$$

$$\tag{62}$$

The double sum can be written

$$\frac{16}{\pi^3} \sum_{n=0}^{N} \sum_{\substack{m=0 \\ \neq n}}^{N} \frac{\sin \pi\lambda(n - m)}{\pi\lambda(n - m)} \exp i \frac{2\pi\lambda\gamma}{N}(n - m) = \frac{32}{\pi^3} \sum_{t=1}^{N} \left\{(N + 1) - t\right\} \times$$

$$\frac{\sin \pi\lambda t}{\pi\lambda t} \cos \frac{2\pi\lambda\gamma t}{N}, \tag{63}$$

where $t = n - m$ and it is straightforward to show that this sum is $O(N)$ and the 'constant' in the order term depends on the correlation.

Hence the final result for $\langle g(\gamma)g^*(\gamma)\rangle$ is

$$\langle g(\gamma)g^*(\gamma)\rangle = (N + 1) + \frac{16a_o^2(N + 1)^2}{\pi^3}\left(\frac{\sin \pi\lambda\gamma}{\pi\lambda\gamma}\right)^2 + O\left(a_o^2 N\right) + O(\rho N). \tag{64}$$

Notice that when $a_o = 0$ and $\lambda = 1$, ie no signal and no correlation then $\langle g(\gamma)g^*(\gamma)\rangle = N + 1$ and this is then the mean clutter intensity in the image against which the image of the point is observed.

It is also evident that for small a_o and small ρ the effects of the correlation from sample to sample are $O(N)$ and can be discounted. In other words the samples can be treated as if they are uncorrelated.

4.2 The mean amplitude point spread function

The expected amplitude point spread function is from equation (4)

$$\langle g(\gamma)\rangle = \sum_{n=0}^{N} \exp - \frac{ik\pi}{2} \langle \exp i \frac{\ell\pi}{2} \rangle , \qquad (65)$$

where $\ell = [\,2\psi(n - \gamma)/\pi\,]$, $k = [\,2\phi(n)/\pi\,]$.

Applying the one bit operation (6) and also (53) for the marginal phase density and integrating over the ζ_n gives

$$\langle g(\gamma)\rangle = \frac{8}{\pi^2} \sum_{a=-\infty}^{\infty} \sum_{b=-\infty}^{\infty} \frac{1}{4a+1} \frac{1}{4b+1} \sum_{n=0}^{N} \Gamma_n^o\{-(4a+1)\}$$

$$\exp - i\,(4b+1)\phi(n) \exp i(4a+1)\psi(n) , \qquad (66)$$

where as before $\phi(n) = \pi\lambda n\left(\frac{n}{N} - 1\right)$,

$$\psi(n) = \pi\lambda(n - \gamma)\left\{\frac{n-\gamma}{N} - 1\right\} + \theta .$$

The same approximation is made as in the case of the deterministic point spread function and then for large N the terms for which $a = b$ are $O(N)$ and the remainder are $O(N^{\frac{1}{2}})$. Then

$$\langle g(\gamma)\rangle = \frac{8}{\pi^2} \sum_{a=-\infty}^{\infty} \frac{\exp i(4a+1)\theta}{(4a+1)^2} \sum_{n=0}^{N} \Gamma_n^o\left\{-(4a+1)\right\} \exp - i(4a+1) x$$

$$\frac{2\pi n\lambda\gamma}{N} + O(N^{\frac{1}{2}}) + O(\gamma) + O(1) . \qquad (67)$$

For small a_0 only $\Gamma_n^0(-1)$ is of significance and is given by (54). Hence

$$\langle g(\gamma)\rangle = \frac{4a_0(N+1)}{\pi^{3/2}} \exp(i\theta)\frac{\sin \pi\lambda\gamma}{\pi\lambda\gamma} + O(\sqrt{a_0 N}) + O(\gamma) + O\left(a_0^3\right). \tag{68}$$

Note that the mean phase in the psf in (68) is θ which is the Doppler phase upon which the formation of a synthetic aperture depends (Appendix B). Hence range compressed one bit data can be used to construct a synthetic aperture. There are, however, phase fluctuations and also amplitude fluctuations and these must be taken into account.

6. THE AZIMUTH CORRELATION PROCESS

In this section the azimuth correlation is carried out for the range image function $g_R(\gamma)$ generated by the one bit data discussed in the previous sections. After range compression the data effectively has many bits of resolution and any sample can have any value between $-(N + 1)$ and $(N + 1)$. For large N therefore one can consider that the signal is continuous (although still sampled, of course) and that assumption is now made. Thus the azimuth correlation is just a simple linear correlation.

However, as pointed out in the last section, the range point spread function $g_R(\gamma)$, ie the range image function in this case, has fluctuation in phase and amplitude. Hence, in principle, the calculation of the azimuth image function for the target point under consideration is just a simple partial coherence problem.

Appendix B equation (102) gives the azimuth correlation process in the case of a linear deterministic signal. If the amplitude of the m th azimuth sample is r_m and the random part of the phase of this sample is η_m then equation (102) gives after some simplification

$$g_A(\nu) = \exp i\,\frac{4\pi f_0 a_0}{c}\;\exp\frac{i\pi\mu\nu^2}{M}\sum_{m=0}^{M} r_m \exp - i\left\{\frac{2\pi\mu\nu m}{M} + \eta_m\right\}, \tag{69}$$

where in equation (102) $\zeta - \xi$ has been replaced by ν.

Hence the expected intensity image function is

$$\langle g_A(\nu) g_A^*(\nu)\rangle = \sum_{n=0}^{M} \sum_{m=0}^{M} \exp i \frac{2\pi\mu\nu(n-m)}{M} \langle r_n r_m \exp i\,(\eta_n - \eta_m)\rangle \tag{70}$$

$$= \sum_{n=0}^{M} \langle r_n^2\rangle + \sum_{n=0}^{M} \sum_{\substack{m=0\\ \neq n}}^{M} \exp i \frac{2\pi\mu\nu(n-m)}{M} \langle r_n r_m \exp i(\eta_n - \eta_m)\rangle. \tag{71}$$

It is assumed that the random process due to noise and clutter is wide sense stationary and also that the correlation between any two range samples is insignificant.

Hence

$$\langle r_n^2\rangle = \langle r^2\rangle = \langle g_R(\gamma) g_R^*(\gamma)\rangle \tag{72}$$

and so

$$\sum_{n=0}^{M} \langle r_n^2\rangle = (N+1)(M+1) + \frac{16a_O^2}{\pi^3}(N+1)^2(M+1)\left(\frac{\sin \pi\lambda\gamma}{\pi\lambda\gamma}\right)^2 + O\left(a_O^2 NM\right) + O(\rho NM) \tag{73}$$

from (64).

Also

$$\langle r_n r_m \exp i\,(\eta_n - \eta_m)\rangle = \langle r \exp i\eta\rangle \langle r \exp - i\eta\rangle \tag{74}$$

$$= \langle g_R(\gamma)\rangle\langle g_R^*(\gamma)\rangle \tag{75}$$

$$= \frac{16a_O^2(N+1)^2}{\pi^3}\left(\frac{\sin \pi\lambda\gamma}{\pi\lambda\gamma}\right)^2 + O\left(a_O^2 N\right) \quad \text{from (68).}$$

Hence

$$\langle g_A(\nu) g_n^*(\nu)\rangle = (N+1)(M+1) + \frac{16a_O^2}{\pi^3}(N+1)^2(M+1)^2 \times \left(\frac{\sin \pi\lambda\gamma}{\pi\lambda\gamma}\right)^2\left(\frac{\sin \pi\mu\nu}{\pi\mu\nu}\right)^2 + C\left(a_O^2 NM\right) + O(\rho NM) \tag{76}$$

which gives the two-dimensional image in range and azimuth of a single point on a speckled background. The first term $(N + 1)(M + 1)$ represents the mean intensity of the speckle and the second term gives the intensity point spread function.

The corresponding point spread function for the case of a linear system (ie very many bits) is given in Appendix B, equation (105).

Define α as the ratio of the intensity point spread function for one bit data to the psf for continuous data for a single point target for a given target cross-section and clutter level. Then from (76), (105) and (49) we have

$$\alpha = \frac{8}{\pi^3} . \tag{77}$$

That is, hard clipping the raw radar data to one bit introduces a loss of about 5.9 dB in signal-to-clutter ratio for a single point target for large time-bandwidth product in range (ie large N). This situation can be rectified by suitably increasing the range time-bandwidth product.

7. CONCLUSIONS AND COMMENTS

(1) For a given (large) time-bandwidth product, hard clipping raw radar signal data introduces a loss of about 5.9 dB in signal-to-clutter ratio in the compressed signal.

(2) In the limit of very large time-bandwidth product the radar and processing system behaves, in a sense, as if it is linear since the mean intensity psf is proportional to signal power. In addition the variance of the point spread function is $O(N)$ and the mean intensity is $O(N^2)$ so the ratio of variance to mean intensity $\rightarrow 0$ as $N \rightarrow \infty$. Hence the fluctuations of the point spread function in relation to its magnitude disappear.

(3) The radar and processing system differs from a linear system in one important respect however. Consider what happens when the target cross-section is kept constant but the clutter and noise level is increased. For a completely linear system the image of the point stays constant and the image clutter increases. For a hard clipped system, however, the mean background clutter in the image stays constant (the term $(N + 1)(M + 1)$ in equation (76) represents the mean clutter level). Tis is to be expected because of the hard clipping. The intensity of the target point in the image, however, decreases (in equation (76) the psf is proportional to the

ratio of signal power to mean noise/clutter power). Hence in the image all intensity levels are relative and all 'absolute' levels are lost.

However, the absolute levels could be restored if the mean received signal power is measured independently and if the system is calibrated. Absolute levels only have significance in a calibrated system anyway.

APPENDIX A

ANALYTIC REPRESENTATION OF THE ONE-BIT FUNCTION

Consider a phase $\eta(\chi)$. We require a representation of the function:

$$\begin{aligned} z(\chi) &= \exp i\,3\pi/4 \quad \text{if} \quad +\pi + 2n\pi \leqslant \eta(\chi) < +\pi/2 + 2n\pi \\ &= \exp i\,\pi/4 \quad \text{if} \quad +\pi/2 + 2n\pi \leqslant \eta(\chi) < 2n\pi \\ &= \exp -i\pi/4 \quad \text{if} \quad 2n\pi \leqslant \eta(\chi) < -\pi/2 + 2n\pi \\ &= \exp -i3\pi/4 \quad \text{if} \quad -\pi/2 + 2n\pi \leqslant \eta(\chi) < -\pi + 2n\pi, \end{aligned} \tag{78}$$

and the easiest way to do this is to expand the real and imaginary parts as Fourier series. The real part is easily shown to be

$$\mathcal{R}\{z(\chi)\} = \frac{2\sqrt{2}}{\pi} \sum_{n=0}^{\infty} (-)^n \frac{\cos(2n+1)\eta(\chi)}{2n+1} \tag{79}$$

and the imaginary part is

$$\mathcal{I}\{z(\chi)\} = \frac{2\sqrt{2}}{\pi} \sum_{n=0}^{\infty} \frac{\sin(2n+1)\eta(\chi)}{2n+1} \tag{80}$$

and so

$$z(\chi) = \frac{2\sqrt{2}}{\pi} \sum_{n=0}^{\infty} \frac{(-)^n \cos(2n+1)\,\eta(\chi)}{2n+1} + i\,\frac{\sin(2n+1)\,\eta(\chi)}{2n+1}\,. \tag{81}$$

Hence

$$z(\chi) = \frac{2\sqrt{2}}{\pi} \sum_{n=-\infty}^{\infty} \frac{\exp i\,(4n+1)\eta(\chi)}{4n+1}\,. \tag{82}$$

At points of discontinuity where $\eta(\chi) = m\pi/2$ the series sums to give $\frac{1}{2}z(\chi)$ at $\eta(\chi)+$ plus $\frac{1}{2}z(\chi)$ at $\eta(\chi)-$. The set of such points of discontinuity has zero measure and as such has no significance for the analysis in the main paper.

APPENDIX B

THE POINT SPREAD FUNCTION FOR A LINEAR SYSTEM

The treatment given here follows that given in Barber (1985).

The transmitted chirp is

$$A(t) = \exp i\, 2\pi t\left\{f_o + \frac{\Delta F_R}{2}\left(\frac{t}{T_R} - 1\right)\right\}, \qquad (83)$$

t is the local time coordinate measured from the start of the pulse. The centre frequency is f_o, the bandwidth ΔF_R and the pulse length is T_R. The pulse extends from $t = 0$ to $t = T_R$. Expression (83) therefore represents a linear frequency sweep from $f_o - \Delta F_R/2$ at $t = 0$ to $f_o + \Delta F_R/2$ at $t = T_R$.

Let the chirp be mixed down to zero baseband (it is an analytic signal). Then setting $f_o = 0$ in (83) gives

$$B(t) = \exp i\, \pi \Delta F_R t\left(\frac{t}{T_R} - 1\right), \qquad (84)$$

let B(t) be sampled (but not quantised) at intervals of Δt. Let there be $N + 1$ samples and hence N intervals in the pulse interval T_R.

Then

$$N\Delta t = T_R, \qquad (85)$$

also, let

$$\Delta t \Delta F_R = \lambda, \qquad (86)$$

where $0 < \lambda \leqslant 1$.

The case $\lambda = 1$ corresponds to sampling at the Nyquist rate. Then for the n th sample, since $t = n\Delta t$,

$$c(n) = \exp i\, \pi\lambda n\left(\frac{n}{N} - 1\right), \qquad (87)$$

let the time delay sufffered by the chirp in being scattered by a target point be t_D. The received chirp is then

$$P(t) = \exp i\ 2\pi(t - t_D) \left\{ f_O + \frac{\Delta F_R}{2} \left(\frac{t - t_D}{T_R} - 1 \right) \right\}. \quad (88)$$

It is supposed that t_D is constant over a pulse interval of T_R (see Barber (1985) for an examination of this approximation - this is the stop-start approximation). Mixing P(t) down to zero baseband gives

$$Q(t) = \exp i\ \pi\Delta F_R(t - t_D) \left[\frac{t - t_D}{T_R} - 1 \right] \exp - i\ 2\pi t_D f_O \ . \quad (89)$$

Now let Q(t) be sampled in the same way as B(t).

Then

$$R(n) = \exp i\ \pi\lambda(n - \alpha) \left[\frac{n - \alpha}{N}\right] - 1 \quad \exp - i2\pi f_O \alpha\Delta t \ , \quad (90)$$

where α is the time delay measured in units of Δt and is not necessarily an integer.

In performing pulse compression (90) is correlated against a replica of the transmitted chirp (87) and then

$$g_R(\beta - \alpha) = \sum_{n=0}^{N} \exp i\ \pi\lambda\ [n + (\beta - \alpha)] \left[\frac{n + (\beta - \alpha)}{N} - 1\right] \times$$

$$\exp - i\ \pi\lambda n \left[\frac{n}{N} - 1\right] \exp i\ 2\pi\alpha f_O \Delta t \ , \quad (91)$$

where β is the correlation variable (the 'lag') in units of Δt,
g_R is the point spread function or correlation.

Hence

$$g_R(\beta - \alpha) = \exp i\ 2\pi\alpha f_O \Delta t \exp i\ \pi\lambda(\beta - \alpha)^2/N \ \frac{\sin \pi\lambda(N + 1)(\beta - \alpha)/N}{\sin \pi\lambda\ (\beta - \alpha)/N} \quad (92)$$

which for N large gives

$$g_R(\beta - \alpha) \sim N \exp i\ 2\pi\alpha f_O \Delta t \ \frac{\sin \pi\lambda\ (\beta - \alpha)}{\pi\lambda(\beta - \alpha)} \ . \quad (93)$$

B.1 Pulse compression in azimuth

The phase factor in (91) $\exp i\, 2\pi\alpha f_O \Delta t = \exp i\, 2\pi f_O t_D$ gives a rotating vector from pulse-to-pulse if the radar is perfectly coherent. The rate of rotation represents the Doppler frequency. Let τ be the azimuth time coordinate. Let the time coordinate of the target point under consideration be τ_A in the azimuth direction. This coordinate τ_A is chosen in relation to the set of pulses which form the synthetic aperture so that the first pulse of the set occurs at $\tau = 0$ and the last at $\tau = T_A$. There are M + 1 pulses and hence M interpulse periods each of the time increment $\Delta\tau$. Note that the set of pulses (which may be a contiguous subset of all possible pulses which illuminates the point) may cover the target asymmetrically. The azimuth time coordimate is assumed constant during a pulse - called the stop-start approximation in Barber (1985).

Let the range from the radar to the target point be

$$a(t) = a_O + a_1(\tau - \tau_A) + a_2(\tau - \tau_A)^2 + \ldots , \qquad (94)$$

where the range is expanded as a Taylor series about $\tau = \tau_A$, the azimuth time coordinate of the target point.

Then

$$t_D = \frac{2a(t)}{c} = \frac{2}{c}(a_O + a_1(\tau - \tau_A) + a_2(\tau - \tau_A)^2 + \ldots) . \qquad (95)$$

We assume that the cubic and higher powers are negligibly small. The phase term in (9) is then

$$\exp i\, 2\pi\alpha f_O \Delta t = \exp i\, 4\pi f_O \left[a_O + a_1(\tau - \tau_A) + a_2(\tau - \tau_A)^2\right]$$

$$= \exp i\, \psi . \qquad (96)$$

$$\text{The Doppler frequency} = \frac{\partial\psi}{2\pi\partial\tau} = \frac{2f_O}{c}(a_1 + 2a_2(\tau - \tau_A)) , \qquad (97)$$

given that $0 \leqslant \tau \leqslant T_A$ and $0 \leqslant \tau_A \leqslant T_A$ the Doppler bandwidth is then

$$\Delta F_A = 4a_2 f_o T_A / c \ . \tag{98}$$

The Doppler frequency corresponding to the azimuth time coordinate τ_A is thus

$$f_A = 2f_o a_1 / c \ . \tag{99}$$

Substituting for a_1 and a_2 from (99) and (98) into (96) then gives

$$= \exp i \frac{4\pi f_o a_o}{c} \exp i\, 2\pi(\tau - \tau_A) \left[f_A + \frac{\Delta F_A}{2} \, \frac{\tau - \tau_A}{T_A} \right] \tag{100}$$

Compare with (88) for the range process.

We shall suppose next that the offset Doppler frequency f_A is mixed to zero. This simplifies the expression without loss of generality.

Also, let $\tau = m\Delta\tau$ and $\Delta\tau\Delta F_A = \mu$ where $0 < \mu \leqslant 1$. The case $\mu = 1$ corresponds to the Nyquist rate. In addition $T_A = M\Delta\tau$ and putting $\tau_A = \zeta\Delta\tau$ gives

$$V(\tau) = \exp i\, 4\pi \frac{f_o a_o}{c} \exp i \frac{\pi\mu}{M} (m - \zeta)^2 \ . \tag{101}$$

Correlating this against the expected phase function for the target point (mixed to zero Doppler offset) with a correlation variable (or 'lag') ξ expressed in units of $\Delta\tau$ gives

$$g_A(\zeta - \xi) = \sum_{m=0}^{M} \exp i\, 4\pi \frac{f_o a_o}{c} \exp i \frac{\pi\mu}{M} \left\{ m - (\zeta - \xi) \right\}^2 \exp - i \frac{\pi\mu m^2}{M} \tag{102}$$

$$= \exp i\, 4\pi \frac{f_o a_o}{c} \exp i \frac{\pi\mu}{M} (\zeta - \xi)^2 \exp - i\pi\mu(\zeta - \xi) \frac{\sin \pi\mu(M+1)(\zeta-\xi)/M}{\sin \pi\mu(\zeta-\xi)/M} \tag{103}$$

which for M large gives

$$g_A(\zeta - \xi) \sim M \exp i\, 4\pi \frac{f_o a_o}{c} \exp - i\, \pi\mu(\zeta - \xi) \frac{\sin \pi\mu\,(\zeta-\xi)}{\pi\mu(\zeta-\xi)} . \tag{104}$$

The two-dimensional point spread function in range and azimuth for large N and M is then

$$|g_R(\beta - \alpha)\|g_A(\zeta - \xi)| \sim NM \left|\frac{\sin \pi\lambda(\beta - \alpha)}{\pi\lambda(\beta - \alpha)}\right|\left|\frac{\sin \pi\mu(\zeta - \xi)}{\pi\mu(\zeta - \xi)}\right| . \tag{105}$$

APPENDIX C

WEYL's INEQUALITY

As pointed out in the main text the sum $\xi(a,b,\gamma)$ may be estimated by Weyl's inequality.

If $$\phi(x) = \alpha x^k + \alpha_1 x^{k-1} + \ldots + \alpha_k \tag{106}$$

and $$T(\phi) = \sum_{x=1}^{N} \exp i\, 2\pi\phi(x) \tag{107}$$

and α has the rational approximation $|\alpha - a/q| \leq {}^1/_{q^2}$

then $$T(\phi) << N^{1+\varepsilon} \left(\frac{1}{q} + \frac{1}{N} + \frac{q}{Nk}\right)^{1/K} \tag{108}$$

where $K = 2^{k-1}$ and $k \geq 2$.

This inequality can be further refined in special circumstances and for the present purposes the following is required: For the inequality $T(\phi) << N^{1-1/K+\varepsilon}$ to hold it is sufficient that

(i) the quotients a_n in the continued fraction expansion of α increase no more rapidly than N^{k-2} $(k \geq 2)$

(ii) in the event of α being rational so that $\alpha = r/s$, say, then $s \geq N$.

Proof.

Put $$N^{1+\varepsilon}\left(\frac{1}{q} + \frac{1}{N} + \frac{q}{Nk}\right)^{1/k} = N^{1-1/K+\varepsilon}\ \gamma^{1/K} \tag{109}$$

where
$$\gamma = \frac{N}{q} + 1 + \frac{q}{N^{k-1}} \tag{110}$$

and it is now necessary to show that γ is bounded under conditions (i) and (ii) above.

Let the n th convergent in the continued fraction expansion of α be p_n/q_n:

$$p_n/q_n = a_o + \frac{1}{a_1+}\frac{1}{a_2+} \cdots \frac{1}{a_n} \tag{111}$$

then $|\alpha - a/q| \leqslant 1/q^2$ is satisfied by all the convergents p_n/q_n in the continued fraction expansion see Hardy and Wright (1954) chapter 10. Now any q_n from the set $\{q_n\}$ which satisfies $|\alpha - a/q_n| \leqslant 1/q_n^2$ may be chosen.

Let
$$a/q = p_n/q_n \tag{112}$$

Now $q_n < q_{n+1}$ so q_n can always be chosen such that $q_{n-1} < N \leqslant q_n$ provided condition (ii) is satisfied.

Hence
$$1 \leqslant q_n N^{-1} < q_n/q_{n-1} \tag{113}$$

Now $q_n = a_n q_{n-1} + q_{n-2}$ see Hardy and Wright (1954) or Khinchin (1964).

And so $1 \leqslant q_n N^{-1} < a_n + 1$ since $q_{n-2} < q_{n-1}$

Let
$$\gamma_n = \frac{N}{q_n} + 1 + \frac{q_n}{N^{k-1}} \tag{114}$$

then
$$\frac{q_n}{N} = N^{k-2} \left(\gamma_n - \frac{N}{q_n} -1\right) \tag{115}$$

which gives

$$N^{2-k} + \frac{N}{q_n} + 1 \leqslant \gamma_n < N^{2-k}(a_{n+1}) + \frac{N}{q_n} + 1 \tag{116}$$

and since
$$N \lesssim q_n \text{ and } k \gtrsim 2$$

$$1 < \gamma_n < N^{2-k} (a_n + 1) + 2 \tag{117}$$

which means that if a_n grows no more rapidly than N^{k-2} then γ_n will be bounded.

In the case of the analysis presented in the paper K = k = 2 and so $T(\phi) = O(N^{\frac{1}{2}+\varepsilon})$ provided conditions (i) and (ii) of the above proof are satisfied.

Among the class of numbers which have bounded continued fraction quotients are the rationals (because they have finite expansions) and the quadratic surds (because their expansions are periodic). Beyond this, however, little seems to be known except that Khinchin showed that the set of numbers which have bounded quotients has zero measure ie almost all numbers have unbounded quotients and further that almost everywhere $a_n > n \ln n$ see Khinchin (1964) p. 93. Notice that it has not been proven that it is necessary that the a_n be bounded only that it is sufficient. Indeed it is probably not necessary but to prove this seems to require completely different methods. This whole area of number theory is somewhat difficult and the situation is saved by the observation that all numbers which exist inside a digital computer are rational or rational approximations to irrationals so that in the practical situations to which this analysis relates α is always rational and the analysis presented here is all that is required for present purposes. Of course π itself is represented by a rational (decimal) approximation in the computer so that $\pi - e/f = \delta$, say, where e/f is the rational approximation. This is just a simple decimal approximation, in no sense a best approximation. It is very easy to show that the error in $T(\phi)$ is of order δ and thus not significant:

$$\text{Error} = \sum_{x=1}^{N} \exp i\, 2\pi\phi(x) - \sum_{x=1}^{N} \exp i\, 2\frac{e}{f}\,\phi(x) \tag{118}$$

$$= \sum_{x=1}^{N} \exp i\, 2\pi\phi(x) \left\{1 - \exp i\, 2\delta\phi(x)\right\} \tag{119}$$

$$= i\, 2\delta\phi(x).T(\phi) + O(\delta^2.\phi^2(x)) \tag{120}$$

$$\text{Hence Error} / T(\phi) = O(\phi(x).\delta) \tag{121}$$

8. REFERENCES

Barber, B.C., (1985) *Int. J. Remote Sensing,* **6**, 1009-1057.

Blythe, J.H., Edgcombe, K., Ryley, A., Worthy, W.D., (1970), *Proc. IEE., London,* **117**, 1767-1776.

Blythe, J.H., Edgcombe, K., Ryley, A., Worthy, W.D., (1973), *Proc. IEE, London,* **120**, 825-832.

Bromwich, T.J.I'A., (1955) Theory of infinite series, Macmillan, London.

Davies, R.D., Ponsonby, J.E.B., Pointon, L., De Jager, G., (1969), *Nature, London,* **222**, 933-937.

Hardy, G.H. and Wright, E.M., (1954) An introduction to the theory of numbers, Oxford, Clarendon Press 3rd Ed.

Hua Loo-Keng, (1957) Science Record 1 1-4, Also in Loo-Keng Hua, Selected Papers, Ed. Halberstam, H., 1983, Springer-Verlag, New York, 277-280.

Khinchin, A.Ya., (1964) Continued Fractions, English translation of 3rd Ed. University of Chicago Press.

Landau, E., (1969) Vorlesungen über Zahlentheorie, Chelsea, New York.

Logan, B.F., (1977) *Bell System Technical Jnl.,* **56**, 487-510.

Mardia, K.V., (1975) *J. Royal Statistical Soc.,* **37**, 349-393.

Middleton, D., (1948) *Quart. App. Math.,* **5**, 445-498.

Middleton, D., (1960) Statistical Communication Theory, McGraw-Hill, New York.

Oppenheim, A.V., Lim, J.S., (1981), *Proc. IEEE.,* **69**, 529-541.

Oppenheim, A.V., Lim, J.S., Curtis, (1983), *J. Optical Soc. of America,* **73**, 1413-1420.

Rice, S.O., (1954) In: Noise and Stochastic Processes, Ed. Wax, N., Dover, New York.

Vaughan, R.C., (1981) The Hardy-Littlewood Method - (Cambridge tracts in mathematics; Vol. 80), CUP Cambridge.

II SPECTRAL ANALYSIS

ITERATIVE HARMONIC DECOMPOSITION OF NONSTATIONARY RANDOM PROCESSES: THEORY AND APPLICATION TO SPECTRAL LINE TRACKING

M.H. Hayes, D.M. Wilkes and M.A. Clements
*(School of Electrical Engineering,
Georgia Institute of Technology, Atlanta, Georgia, USA)*

ABSTRACT

The harmonic decomposition of a random process into a sum of sinusoids in white noise is an important problem with applications in a number of different areas. As a result of the work of V.F. Pisarenko, it has been shown that the sinusoidal frequencies and the white noise power are determined by the minimum eigenvalue and the corresponding eigenvector of the autocorrelation matrix. In this paper, a new iterative approach for performing the harmonic decomposition is presented. This solution is based on the Levinson recursion and uses the reflection coefficients to determine convergence. The iterative harmonic decomposition is then applied to the spectral line tracking problem. Unlike other adaptive algorithms, this method gives good tracking performance in only 2 or 3 iterations with computational requirements comparable to those of the adaptive algorithms which makes it viable for implementation in real time. Simulation examples using stationary and nonstationary data are given and the performance of the iterative solution is compared to several adaptive methods.

1. INTRODUCTION

The Pisarenko Harmonic Decomposition (PHD) is an effective tool for estimating the frequencies of sinusoids in an additive white noise background. For p real sinusoids in white noise it is well known that the white noise power corresponds to the minimum eigenvalue of the $(2p+1)\times(2p+1)$ autocorrelation matrix, $\mathbf{R}$, and that the frequencies of the sinusoids may be found from the corresponding eigenvector, $\mathbf{a}$, see Pisarenko (1973). Stated in matrix notation we have

$$\mathbf{Ra} = \lambda_{MIN}\mathbf{a} \tag{1.1}$$

where $\sigma^2=\lambda_{MIN}$ is the Pisarenko estimate of the white noise power. The frequencies of the sinusoids correspond to the roots of the eigenfilter A(z) which is the polynomial formed from the vector $\mathbf{a}=[\,a(0),\ a(1),\ \ldots\ ,\ a(2p)\,]^T$, i.e.,

$$A(z) = \sum_{i=0}^{2p} a(i)z^{-i} \tag{1.2}$$

The minimum eigenvalue and the corresponding eigenvector may be found using any one of a number of algorithms such as the inverse power method, the Rayleigh quotient iteration, or a gradient algorithm, see Parlett (1980). Alternatively, the minimum eigenvalue and the eigenvector may be computed efficiently using Levinson's recursion, Hayes and Clements. Specifically, note that the minimum eigenvalue is the smallest non-negative value of $\hat{\sigma}^2$ for which the matrix

$$\mathbf{R}_s = \mathbf{R} - \hat{\sigma}^2\mathbf{I} \tag{1.3}$$

is singular. Since $\mathbf{R}_s$ is a symmetric Toeplitz matrix, Levinson's recursion may be used to generate a set of reflection coefficients, k_i , i=1,2,...,2p. Based on these reflection coefficients we have the following properties, from Hayes and Clements:

1. If $\mathbf{R}_s$ is positive definite then all the reflection coefficients, k_i i=1,2,...,2p, will fall in the range $-1<k_i<1$, and all of the roots of the eigenfilter, A(z), will lie inside the unit circle.

2. If $\mathbf{R}_s$ is positive semidefinite and singular then $|k_i|<1$ for i=1,2,...,2p-1 and $|k_{2p}|=1$, and all of the roots of the eigenfilter, A(z) will lie on the unit circle.

3. If $\mathbf{R}_s$ is indefinite (having both positive and negative eigenvalues) then at least one k_i will be such that $|k_i|>1$, and the roots of the eigenfilter may lie both inside and outside the unit circle.

The Pisarenko problem reduces to finding the value of $\hat{\sigma}^2$ for which $\mathbf{R}_s$ is positive semidefinite and singular. Using the properties of Toeplitz symmetric matrices listed above, an iterative technique for finding the white noise power, $\hat{\sigma}=\lambda_{MIN}$, has been developed for off-line signal processing. This technique, proposed by Hayes and Clements, involves successive bisection of an interval containing $\hat{\sigma}^2$ until convergence is obtained.

What is interesting to observe is the behaviour of the last reflection coefficient as a function of $\hat{\sigma}^2$. Shown in Figure 1 is a typical plot of the final reflection coefficient, k_{2p}, as $\hat{\sigma}^2$ is varied. In this example, the value of the noise variance is 0.1 so $|k_{2p}|<1$ for $\hat{\sigma}^2 < 0.1$. When $\hat{\sigma}^2 = 0.1$, however, we have $k_{2p}=1$ which corresponds to case 2 above where $\mathbf{R}_s$ is singular. Finally, as $\hat{\sigma}^2$ is increased beyond 0.1, k_{2p} approaches infinity as $\hat{\sigma}^2$ approaches its first asymptote. The value of $\hat{\sigma}^2$ at this asymptote defines where the next to the last reflection coefficient, k_{2p-1}, has a magnitude of 1. Beyond this asymptote $|k_i|>1$ for at least one value of i. Obviously, the steeper the slope of the curve at $k_{2p} = 1$, the greater the precision required on $\hat{\sigma}^2$ to produce a value of k_{2p} close to 1. In fact, this slope strongly affects how difficult it is to converge to the correct value of $\hat{\sigma}^2$.

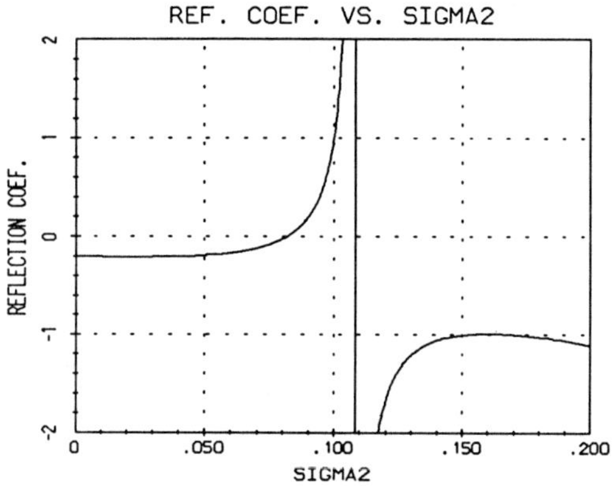

Fig. 1 Plot of the last reflection coefficient vs. the noise power estimate.

2. RECURSIVE UPDATE OF THE AUTOCORRELATION FUNCTION

Before considering the problem of spectral line tracking, we look at two methods for recursively updating the autocorrelations. The first method uses a sliding rectangular data window whereas the second method uses an exponential data window.

With a sliding rectangular window of length N, the estimated autocorrelation, $\hat{\phi}_n(k)$, is given by

$$\hat{\phi}_n(k) = \frac{1}{N-k} \sum_{i=0}^{N-1} x(n-i)x(n-k-i) \tag{2.1}$$

where n corresponds to the most recent point. Note that since

$$\hat{\phi}_{n+1}(k) = \frac{1}{N-k} \sum_{i=0}^{N-1} x(n+1-i)x(n+1-k-i) \tag{2.2}$$

then

$$\hat{\phi}_{n+1}(k) = \hat{\phi}_n(k) + \frac{1}{N-k}\left[x(n+1)x(n+1-k) - x(n+k-N+1)x(n-N+1)\right] \tag{2.3}$$

Therefore, with the arrival of a new data point the autocorrelation may be updated with only 3 multiplies and 2 adds. In addition to being computationally efficient, this estimator is an unbiased estimate of the autocorrelation function for data which is stationary over intervals longer than the window length. The frequency estimates generated from the windowed data are taken as estimates of the instantaneous frequencies at the sample number in the centre of the window.

With an exponential window the computational requirements are even smaller. Specifically, the recursive update equation is given by

$$\hat{\phi}_n(k) = x(n)x(n-k) + \alpha\hat{\phi}_{n-1}(k) \tag{2.4}$$

where n is the position of the newest data point and α is an exponential damping factor for which $0<\alpha<1$. The estimator reduces the effects of past values on the data in a more gradual fashion than the sliding window.

3. THE ITERATIVE SPECTRAL LINE TRACKER

Now we turn our attention to two iterative interpolation methods for updating the noise power. Both methods require that an interval containing the white noise power, $\sigma^2=\lambda_{MIN}$, be specified. Fortunately, such an interval is easy to define. In particular, given the previous estimate $\mathbf{a}(n-1)$ of the eigenvector let

$$\rho[\mathbf{a}(n-1),\mathbf{R}_n] = \mathbf{a}^T(n-1)\mathbf{R}_n\mathbf{a}(n-1)/\mathbf{a}^T(n-1)\mathbf{a}(n-1) \qquad (3.1)$$

be the Rayleigh Quotient where $\mathbf{R}_n$ is the Toeplitz autocorrelation matrix based on the autocorrelation values computed from the data up to time n. Since $\lambda_{MIN} \leq \rho$, we may choose the Rayleigh Quotient as the upper bound on $\hat{\sigma}^2$ and zero as the lower bound, see Hayes and Clements also Hu and Kung (1983).

Both interpolation methods use an iterative method to choose successively better values of $\hat{\sigma}^2$. When a value of $\hat{\sigma}^2$ is chosen, it is subtracted from $\hat{\phi}_n(0)$ and the Levinson recursion is performed on the modified autocorrelations to produce a set of 2p reflection coefficients and a 2p+1 long feedback coefficient vector, $\mathbf{a}$. The updating of $\hat{\sigma}^2$ is then based on the following observations. If $|k_i|>1$ for any $i=1,2,\ldots,2p-1$ then this $\hat{\sigma}^2$ lies beyond the first asymptote in Figure 1 so the corresponding $\mathbf{a}$ will not yield very accurate frequency estimates. Therefore, the interval for $\hat{\sigma}^2$ is updated by making $\hat{\sigma}^2$ the new upper bound and the next value of $\hat{\sigma}^2$ is chosen as the midpoint of the new interval. If, on the other hand, $|k_i|<1$ for all $i=1,2,\ldots,2p-1$ with $k_{2p} < 0$ then two of the roots lie on the real axis. In this case, the value of $\hat{\sigma}^2$ is too small, so $\hat{\sigma}^2$ is used as the new lower bound interval and the midpoint is used as next value of $\hat{\sigma}^2$. Finally, if $|k_i| < 1$ for all $i=1,2,\ldots,2p-1$ and $k_{2p} > 0$ then the corresponding $\mathbf{a}$ may produce good estimates of the sinusoidal frequencies. Therefore, this value of $\mathbf{a}$ is saved along with the corresponding values of $\hat{\sigma}^2$ and k_{2p}. If $k_{2p} > 1$ then the upper bound of the interval is set equal to $\hat{\sigma}^2$, whereas if $k_{2p} < 1$ then the lower bound is set equal to $\hat{\sigma}^2$.

After two estimates are found which exhibit the property that $|k_i| < 1$ for $i=,1,2,...,2p-1$ and $k_{2p} > 0$, then interpolation techniques can be used to find the next $\hat{\sigma}^2$ estimate. The first technique is linear interpolation which involves two previous estimates to define a line in the $(\hat{\sigma}^2, k_{2p})$ plane to find the value of $\hat{\sigma}^2$ for which $k_{2p} = 1$, see also Hu and Kung (1981). With the previous estimates denoted by $\hat{\sigma}^2(1)$, $\hat{\sigma}^2(2)$, $k_{2p}(1)$, and $k_{2p}(2)$, it is easily shown that the next $\hat{\sigma}^2$ estimate is given by

$$\hat{\sigma}^2 = \frac{\hat{\sigma}^2(2)[1-k_{2p}(1)] - \hat{\sigma}^2(1)[1-k_{2p}(2)]}{k_{2p}(2) - k_{2p}(1)} \qquad (3.2)$$

The second interpolation technique is single pole interpolation. Again, two previously saved estimates are used, but this time the model assumed is a single pole model of the form

$$k_{2p} = a/(b-\hat{\sigma}^2)$$

where $a>0$ and $b>0$. The next value of $\hat{\sigma}^2$ is chosen to make $k_{2p} = 1$. This occurs for

$$\hat{\sigma}^2 = \frac{k_{2p}(2)[1-k_{2p}(1)]\hat{\sigma}^2(2) - k_{2p}(1)[1-k_{2p}(2)]\hat{\sigma}^2(1)}{k_{2p}(2) - k_{2p}(1)} \qquad (3.3)$$

In performing the iterative line tracking method, it is necessary to choose the number of local iterations to be performed at each update (often 3 to 5 will be sufficient). After performing this number of iterations the best of the results from the Levinson recursion is chosen as the estimate. The best estimate is the one for which $|k_i|<1$ $i=1,2,...,2p-1$ and k_{2p} is positive and closest to 1. If none of the results qualify, then the last estimate of the minimum eigenvector is used.

4. ADAPTIVE PISARENKO ALGORITHMS

Recently, much attention has been given to adaptive solutions to the Pisarenko problem. Some of the methods proposed include the inverse power method, in Hu (1985) and Vacarro (1984), a least squares type algorithm, Reddy et al.

(1982), and several stochastic gradient algorithms including Thompson's LMS algorithm, in Thompson (1979), a steepest descent method, Hu (1985), and a conjugate gradient algorithm, in Hu (1985) and Sarkar et al. (1983). So far the papers on this subject have emphasized the convergence properties of the various techniques for stationary data. In practice, however, the data will typically be nonstationary (e.g. phase modulated sinusoids) sc that the performance of these algorithms in tracking the sinusoidal frequencies becomes important. In particular, the ability to track the instantaneous frequencies may be very important. We intend to compare the performance of these adaptive algorithms on non-stationary data to the performance of the Iterative Spectral Line Tracker described in Section 3. First, however, the adaptive algorithms mentioned above are briefly described.

The adaptive inverse power method utilizes the property that repeated multiplication of an initial vector by the inverse of the covariance matrix will converge to the minimum eigenvector, of the covariance matrix, is the initial vector is not orthogonal to the minimum eigenvector see Hu (1985). The recursion formula for the inverse covariance matrix is

$$C^{-1}(n) = \frac{1}{\alpha}\left[C^{-1}(n-1) - \frac{C^{-1}(n-1)\,\mathbf{x}(n)\,\mathbf{x}^T(n)\,C^{-1}(n-1)}{\alpha + \mathbf{x}^T(n)\,C^{-1}(n-1)\,\mathbf{x}(n)}\right] \tag{4.1}$$

where

α = the exponential window factor, $0<\alpha<1$

$\mathbf{x}(n) = [\,x(n)\;\; x(n-1)\;\; \ldots\;\; x(n-2p)\,]^T$

An exponential window is typically used to track nonstationary signals whereas, for stationary data, a growing rectangular window is used which is obtained by simply setting $\alpha = 1$. The update of the coefficient vector is given by

$$\mathbf{a}(n) = C^{-1}(n)\mathbf{a}(n-1) \tag{4.2}$$

where $\mathbf{a}(n)$ is normalized to unit magnitude after every update.

The least squares type algorithm is rather lengthy to describe here. It is explained in detail in Reddy et al. (1982). A deterministic error functional is defined and an approximate unconstrained least squares solution is formulated using an approximate Hessian. The algorithm is given by

$$e(n) = \mathbf{x}^T(n)\mathbf{a}(n-1)$$

$$\psi(n) = \mathbf{x}(n) - e(n)\mathbf{a}(n-1)$$

$$P(n) = \frac{1}{\alpha}\left[P(n-1) - \frac{P(n-1)\psi(n)\psi^T(n)P(n-1)}{\alpha + \psi^T(n)P(n-1)\psi(n)}\right] \tag{4.3}$$

$$\mathbf{a}(n) = \mathbf{a}(n-1) - P(n)\psi(n)e(n)$$

and $\mathbf{a}(n)$ is normalized to unity magnitude after every update.

Next we describe the gradient algorithms. The Thompson algorithm, Thompson (1979), uses a simple instantaneous estimate of the gradient. This algorithm minimizes $\mathbf{a}^T\mathbf{C}\mathbf{a}$ subject to the constraint that $\mathbf{a}^T\mathbf{a} = 1$. The gradient is thus $\mathbf{C}\mathbf{a}$. The covariance matrix, C, is approximated by $\mathbf{x}(n)\ \mathbf{x}^T(n)$, so the gradient estimate becomes $\mu\varepsilon\mathbf{x}(n)$ where

$$\varepsilon = \mathbf{x}^T(n)\mathbf{a}(n-1) \quad \mu = \text{step size} \tag{4.4}$$

and $\mathbf{x}(n)$ is the same as above. The coefficient update is then reduced to

$$\mathbf{a}(n) = \mathbf{a}(n-1) - \mu\varepsilon\mathbf{x}(n) \tag{4.5}$$

The coefficient vector is normalized to unity magnitude after each iteration.

The steepest descent algorithm minimizes the Rayleigh Quotient, $\rho[\mathbf{a}(n-1), C(n)]$, Hu (1985). The resulting gradient is

$$\mathbf{g}(n) = C(n)\mathbf{a}(n-1) - \rho[\mathbf{a}(n-1), C(n)]\mathbf{a}(n-1) \tag{4.6}$$

Updating the covariance matrix is accomplished by

$$C(n) = \alpha C(n-1) + \mathbf{x}(n)\mathbf{x}^T(n), \quad 0<\alpha<1 \tag{4.7}$$

Again an exponential window is used on the covariance matrix in order to track nonstationary signals. The coefficients are updated by

$$\mathbf{a}(n) = \mathbf{a}(n-1) - \mu\mathbf{g}(n) \tag{4.8}$$

and $\mathbf{a}(n)$ is normalized after every update.

The conjugate gradient algorithm uses the same covariance matrix recursion and gradient computation as the steepest descent algorithm above, see Hu (1985). The coefficient update, however, is given by

$$\mathbf{a}(n) = \mathbf{a}(n-1) - \mu\mathbf{f}(n) \tag{4.9}$$

where

$$\mathbf{f}(n) = \mathbf{g}(n) + t\mathbf{f}(n-1) \tag{4.10}$$

The scalar, t, is chosen so that $\mathbf{f}(n)$ and $\mathbf{f}(n-1)$ are C(n) conjugate, that is

$$\mathbf{f}^T(n-1)C(n)\mathbf{f}(n) = 0 \tag{4.11}$$

The resulting formula for the scalar is

$$t = -\frac{\mathbf{f}^T(n-1)C(n)g(n)}{\mathbf{f}^T(n-1)C(n)f(n-1)} \tag{4.12}$$

As a result of the conjugate search direction, the conjugate gradient algorithm generally converges faster than the steepest descent algorithm when the estimate is close to the correct answer, Sarkar et al. (1983).

A new gradient algorithm is the Toeplitz gradient algorithm which uses a covariance matrix, $\mathbf{R}_n$, constrained to have a Toeplitz structure. This matrix is formed by estimating the autocorrelations, which are updated using either of the two methods described above. The coefficient update equation is the same as the steepest descent algorithm except that $\mathbf{R}_n$ is used in place of C(n).

5. SIMULATIONS

Now we will examine some simulations to compare the performance of the different methods. First we will consider stationary sinusoidal data.

Example 1: This example consists of 3 unit amplitude sinusoids having constant frequencies of 0.0556, 0.1667, and 0.2778 Hz where the sampling frequency is 1 second. These frequencies correspond to angles of 20, 60 and 100 degrees around the unit circle. The noise is white with a variance of 0.1; thus the SNR is 7dB per sinusoid.

The simulation results are shown in Figure 2, and illustrate the results which typically have been observed for the examples we considered. Since the algorithms generally had the most difficulty tracking the lowest frequency sinusoid, it was chosen to illustrate the properties of the algorithms. All of the methods converged to the correct frequencies within 100 iterations except for the Thompson algorithm which needed over 400 iterations. Notice that the iterative line tracker immediately produces a good estimate of the frequency. Only 3 iterations of the Levinson recursion were used at each new data point.

Example 2: In this example the three sinusoids in example 1 are phase modulated. They have the same centre frequencies as before, but now the frequencies vary by .0278 Hz (10 degrees) in a sinusoidal manner about their respective centre frequencies. Again, only the results for the lowest frequency sinusoid are provided. An accurate initial vector was used so that we could study the tracking behaviour.

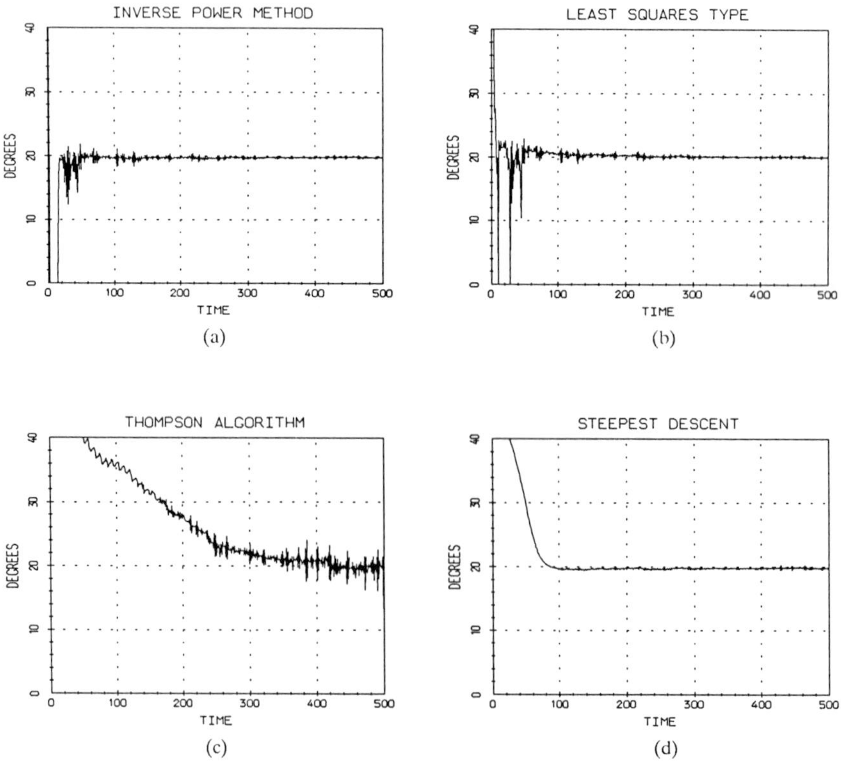

Figure 2: Results from Example 1
(a) The inverse power method. (b) The least squares type method.
(c) The Thompson algorithm. (d) The steepest descent algorithm.

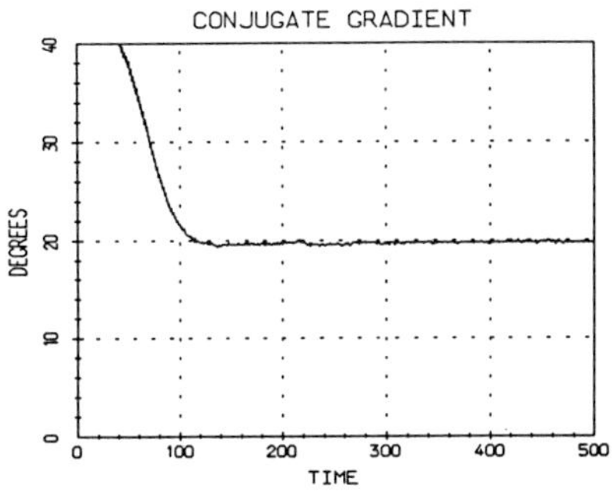

(e)

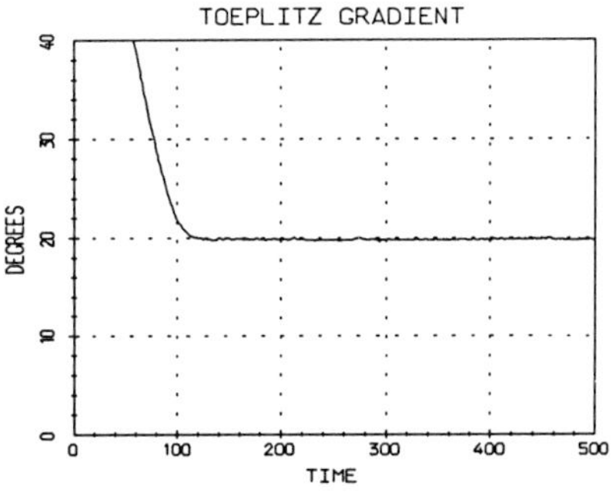

(f)

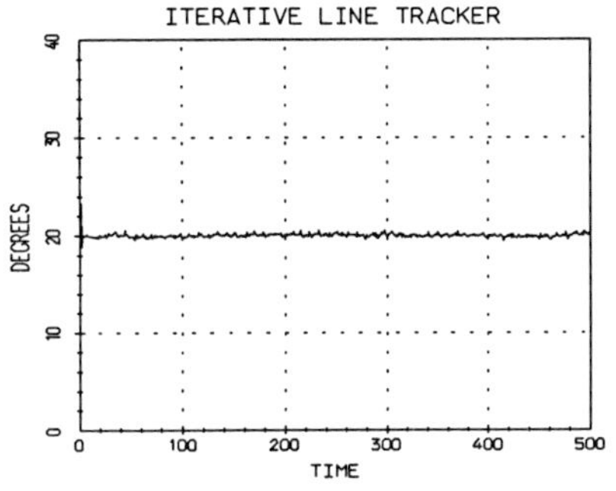

(g)

Figure 2: Results from Example 1
(e) The conjugate gradient algorithm. (f) The Toeplitz gradient algorithm. (g) The iterative line tracker.

Figure 3 illustrates the performance of the techniques. The inverse power method generally exhibits good performance once the convergence transients have died out. However, even after the transients have disappeared, spurious estimates were occasionally produced. The least squares type algorithm performed poorly since it was not able to track the full range of frequencies. The Thompson algorithm was very sensitive to the choice of the step size, μ. The estimate exhibits a large variance and was not able to track all the way up to the 30 degree estimates. The steepest descent and conjugate gradient algorithms performed similarly. They both had rather good tracking performance. The Toeplitz gradient algorithm did a very good job of tracking the frequencies. Notice that it produces a very smooth estimate. It was not very sensitive to the choice of the step size. The performance of the iterative spectral line tracker was also quite good. This result is very close to what is obtained by performing the Pisarenko harmonic decomposition at every point. The variance in the observed estimate is the result of sliding a rectangular window along the data.

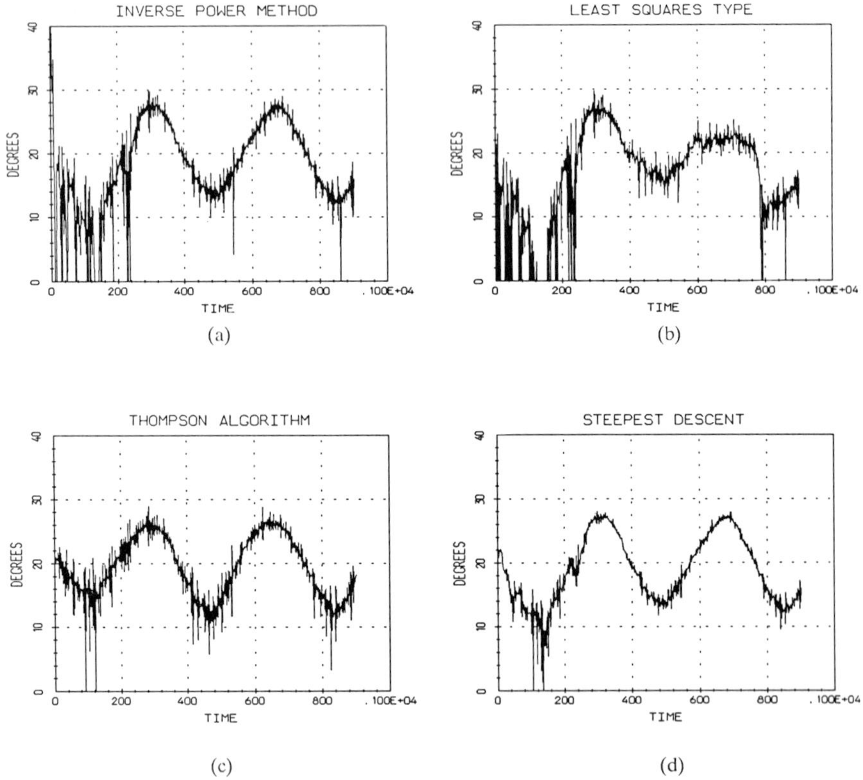

Figure 3: Results from Example 2
(a) The inverse power method. (b) The least squares type method.
(c) The Thompson algorithm. (d) The steepest descent algorithm.

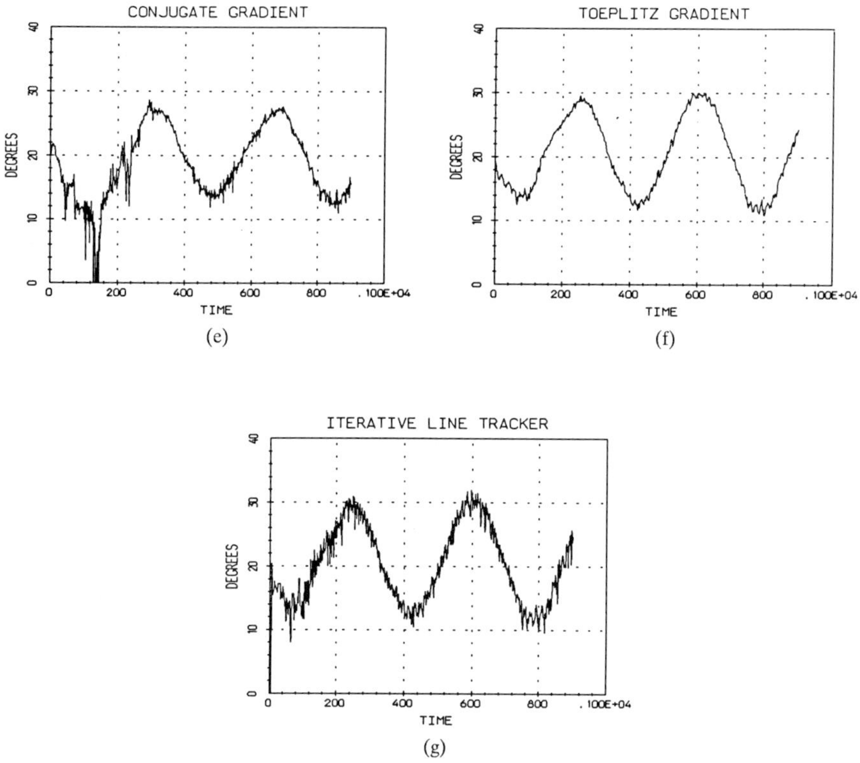

Figure 3: Results from Example 2
(e) The conjugate gradient algorithm. (f) The Toeplitz gradient algorithm. (g) The iterative line tracker.

Visual comparison of these methods is not the only factor in determining their relative merits. The computational burden for each new update is another important consideration. For simplicity we let N=2p+1 where p is the number of real sinusoids to track. Then the inverse power method and the least squares type algorithm require approximately $3.5N^2$ multiplications per update. The Thompson algorithm is the simplest of the methods considered with only 4N multiplications needed for every new estimate. The steepest descent, conjugate gradient, and Toeplitz gradient require $2N^2$, $3N^2$, and N^2 respectively. Notice that the Toeplitz gradient showed good tracking performance and had the lowest computational burden except for the Thompson algorithm, which did not track as well. The amount of computation required by the iterative line tracking algorithm depends on the number of times that Levinson's recursion is used at each update. In general, for k recursions per update, $(k+1)N^2$ multiplications are needed. For the three recursions per update used in the examples, this means $4N^2$ multiplies. Notice that this is clearly at the same order of complexity as the other adaptive algorithms. Since the Toeplitz gradient algorithm and the iterative line tracker use the same covariance updating, further savings can be obtained by using the iterative method only once every 50 points or so and using the Toeplitz gradient method on all the other updates. This is an appealing hybrid tracking system, since the iterative line tracker has the advantage of being able to tell if it is tracking the minimum eigenvalue by examining the reflection coefficients of Levinson's recursion. This kind of error checking is very desirable for ensuring accurate tracking.

6. CONCLUSION

We have discussed spectral line tracking with the intention of addressing the problem of nonstationary sinusoids in noise. A new iterative spectral line tracker which is based on Levinson's recursion was described in the paper. Since this recursion uses autocorrelation estimates as its input, two recursive ways of updating the autocorrelations were discussed. Other adaptive Pisarenko methods were described as well as a new Toeplitz gradient algorithm which uses the same covariance updating technique as the iterative line tracker. Finally, we conducted simulations which compared the iterative line tracker and the adaptive methods. The iterative line tracker was found to converge very quickly to the correct frequencies and to perform at least as well as the other adaptive techniques in tracking time-varying frequencies. In addition, the computational complexity of the various algorithms was compared and the iterative line tracker had the same level of computation as the best of the other algorithms considered.

REFERENCES

Hayes, M.H. and Clements, M.A., "An efficient algorithm for computing Pisarenko's harmonic decomposition using Levinson's recursion," *IEEE Trans. ASSP,* Vol. ASSP- 3T, pp. 485-491, 1986.

Hu, Y.H., (1985) "Adaptive methods for real time Pisarenko spectrum estimate," *Proc. ICASSP,* Tampa, FL.

Hu, Y.H. and Kung, S.Y., (1981) "Computation of minimum eigenvalue of Toeplitz matrix by Levinson algorithm," *Proc. SPIE Real-Time Signal Processing IV,* pp. 40-45.

Hu, Y.H. and Kung, S.Y., (1983) "Highly concurrent Toeplitz eigen-system solver for high resolution spectral estimation," *Proc. ICASSP,* pp. 1422-1425.

Parlett, B.N., (1980) The Symmetric Eigenvalue Problem, Englewood Cliffs, NJ: Prentice-Hall.

Pisarenko, V.F., (1973) "The retrieval of harmonics from a covariance function," *Geophysical J. Royal Astronomical Soc.,* vol. 33, pp. 347-366.

Reddy, V.U., Egardt, B., and Kailath, T., (1982) "Least-squares type algorithm for adaptive implementation of Pisarenko's harmonic retrieval method," *IEEE Trans. ASSP,* vol. 30, no. 3, pp. 399-405.

Sarkar, T.K., Dianat, S.A. and Rao, S.M., (1983) "A finite step adaptive implementation of the Pisarenko harmonic retrieval method for colored noise," *Proc. ICASSP,* vol. 3, pp. 1102-1105.

Thompson, P., (1979) "An adaptive spectral analysis technique for unbiased frequency estimation in the presence of white noise," *Proc. 13th Asilomar Conf. Circuits, Syst., Computers,* Pacific Grove, CA, pp. 529-533.

Vaccaro, R.J., (1984) "On adaptive implementations of Pisarenko's harmonic retrieval method," *Proc. ICASSP,* vol. 1, pp. 6.1.1-6.1.4.

USING A PRIORI INFORMATION AND L_1-NORM MINIMIZATION IN CADZOW'S AND PRONY'S LINE SPECTRAL ANALYSIS METHODS

A.R. Figueiras-Vidal, J.M. Páez-Borrallo*,
J. Alcázar-Fernández and B. Ruiz-Mezcua

(ETSI Telecommunicacion-UPM, Cdad. Universitaria, Madrid, Spain)

ABSTRACT

In this paper, we start from two high-resolution methods to detect sinusoids in noise, Cadzow's second all-pass model and Prony-Hildebrand's procedures, and explore the advantages of applying L_1-norm minimization algorithms in finding their parameters. Additionally, we demonstrate the possibility of extending the first method to coloured noise cases, and the improvements and drawbacks originated by using bandwidth knowledge and imposing symmetrical coefficients in this case.

1. INTRODUCTION

Many methods to estimate sinusoids buried in noise have been presented in the literature; two of the more effective are Cadzow's procedure to solve overdetermined autocorrelation equations associated with an all-pass ARMA model for sinusoids in white noise, Cadzow (1982), and Prony-Hildebrand's method to locate sinusoids in noise from equivalent overdetermined time equations, Kay and Marple (1981), Hildebrand (1956).

Both methods are based on the fact that, if s(n) is a sum of P sinusoids, it is easy to prove that

$$\sum_{i=0}^{2P} a_i \, s(n-i) = 0 \qquad (1.1)$$

where the a_i ($a_0=1$, $a_i=a_{2P-i}$) are the coefficients of the polynomial that has roots exp $(\pm j\omega_k)$, ω_k being the frequencies of the sinusoids. If we add a white noise, that can be modelled as the all-pass ARMA filter

$$H(z) = \sum_{i=0}^{2P} a_i z^{-i} / \sum_{i=0}^{2P} a_i z^{-i} \tag{1.2}$$

driven by (the same) white noise, the resulting sinusoids-plus-noise signal is mathematically equivalent to the same all-pass ARMA model, because (1.1) holds. This is equivalent to saying that (1.2) is undefined at $z^{-1} = \exp(\pm j\omega_k)$, and the output can be assumed to be a sinusoidal component at this frequency.

In Prony-Hildebrand's case, a (generally coloured) noise is added to s(n) and the expression

$$\sum_{i=0}^{2P} a_i x(m-i) = \varepsilon(n) \quad (= \sum_{i=0}^{2P} a_i r(n-i)) \tag{1.3}$$

where x(n) is the signal plus noise and r(n) the noise, is used as the starting point.

Both methods offer high-resolution performances, due to their overdetermined formulations. Any possibility of improving their effectiveness, at least in particular situations, is, then, an interesting topic.

In fact, Cadzow proposes two alternatives of using over-determined autocorrelation equations in his paper: the first excludes center equations, theoretically affected by the variance of the white noise; the second accepts this error as a part of the overall error to be minimized. This second option is preferable when the number of data points is low, since high lag autocorrelation estimates are poor in this case.

But, accepting this second version, the effect of noise variance is similar, as we will see, to the presence of outliers in the centre equations. Thus, to essay an L_1-norm minimization of the overall error can be advantageous. This will be the first subject we will discuss here.

Additionally, this second version is equivalent to considering a Prony-Hildebrand's model in the autocorrelation domain. Then, to obtain polynomial zeroes on the unit circle, we can try to force symmetrical coefficients. This approach will be also considered.

According to this interpretation, we can accept the presence of coloured noise when applying this second version of Cadzow's method. We demonstrate with some simulation examples the feasibility of this application.

If we obtain a first spectral preestimator $\hat{S}(\omega)$, we can (with some risk of error) obtain an estimate B of the band corresponding to the present sinusoids: that in which the estimated spectral density is higher. This band estimate can be obtained from other prior knowledge too (physical constraints, etc.). Using as autocorrelation estimates

$$\hat{R}_B(m) = F^{-1}[\, S(\omega)\; I_B(\omega)\,] \qquad (1.4)$$

where

$$I_B(\omega) = \begin{cases} 1, \omega \in B \\ 0, \omega \notin B \end{cases} \qquad (1.5)$$

serves to improve the performance of Cadzow's method(s), as we will show in this paper.

Finally, we consider that some of the possibilities we have discussed above can be extended to Prony-Hildebrand's method: specifically, we will illustrate why an L_1-norm minimization is better than an L_2-norm when the model order is low and, simultaneously, there are outliers in the data register.

2. CADZOW'S SECOND FORMULATION AND L_1-NORM MINIMIZATION

We have seen that P sinusoids in white noise can be modelled as the output of an all-pass filter driven by white noise. Assuming a uniform random phase for the sinusoids, the associated autocorrelation equations have the theoretical form:

$$\sum_{i=0}^{2P} a_i R(m-i) = \sigma^2 \sum_{i=0}^{2P} a_i\; \delta(m-i) \qquad (a_0 = 1) \qquad (2.1)$$

where R(.) is the data autocorrelation sequence, $\sum_{i=0}^{2P} a_i z^{-i}$ the filter polynomial, and σ^2 the noise power. (2.1) can be easily obtained from the equality of the left and right terms of (1.3), by multiplying by $x(n') = s(n') + r(n')$, keeping in mind that s and r are uncorrelated, putting $m = n-n'$, and forcing a white character for r. If we consider $m > 2P$ in (2.1), we have no right terms; using more than 2P equations, we can improve the estimates of the a_i selecting the minimum L_2-norm solution of the corresponding overdetermined system

$$\sum_{i=0}^{2P} a_i \hat{R}(m-i)=\varepsilon(m), \quad m=2P+1,\ldots,M>4P+1 \quad (a_0=1) \qquad (2.2)$$

The oversized character of the equations seems to compensate for the errors originated by leakage; in fact, this procedure and the one indicated below frequently offer better performance than Prony-Hildebrand's method, that works directly in the time domain, under high noise level conditions.

Another possibility is to accept the right term of (2.1) as a part of $\varepsilon(m)$, and to apply the same approach. Taking into account the non-symmetrical character of the a_i, Cadzow proposes this second version:

$$\min \|\underline{\varepsilon}\|^2 \qquad (2.3a)$$

under

$$\underline{\underline{\hat{R}}}\,\underline{a} = \underline{\varepsilon} \qquad (2.3b)$$

where $\|.\|$ is the usual (L_2) norm of a vector, $\underline{a} = [a_0 \ldots a_{2P}]^T$, and

$$\underline{\underline{\hat{R}}} = \begin{bmatrix} \hat{R}(-N/2+2P) \ldots \hat{R}(-N/2) \\ \cdot \qquad\qquad \cdot \\ \cdot \qquad\qquad \cdot \\ \cdot \qquad\qquad \cdot \\ \hat{R}(N/2) \quad \ldots \hat{R}(N/2-2P) \end{bmatrix} \qquad (2.3c)$$

(a_0 is normalized to 1 in the applied computational procedure).

This approach is better when the data sequence is short and the noise comparatively low.

Note that, theoretically, the centre values of ε (those offering $\delta(m-i)\neq 0$) have the form $\sigma^2 a_i$, according to (2.1). This is similar to having outliers in the centre equations. Thus, using an L_1-norm minimization can be useful here. In fact, to apply Simplex provides the alternative of maintaining approximately these values of ε for the centre equations: since this algorithm provides a basic solution, we can have 2P nonzero a_i and 2P zero ε, since the number of constraints is enough.

Since $a_i = 0$ ($i \neq 0$) forces theoretically $\varepsilon(m) = \hat{R}(m)$ in the noncentre equations, to have low $\hat{R}(m)$ values there (to have zero crossings and few peaks) will favour the appearance of adequate solutions (as, clearly, to have a relatively high σ^2). Then, intermediate frequencies will be preferable, a high noise is acceptable, and to use a moderate N is helpful, Figueiras-Vidal et al. (1985).

To apply Simplex, it is necessary to have non-negative unknowns. Keybl and Eichmann (1983) have proposed, in a similar case, to use bounds (derived from Cardano's relationships) for the a_i in order to introduce non-negative representations of these coefficients. This approach requires a lot of calculus when P is high; but we have repeatedly verified that these bounds are inactive in most cases, and, in the rest, useless, since they do not guarantee unit modulus roots nor preestablished frequency bands for these roots. A simpler possibility is to split the a_i in their positive and negative parts. Thus, we have:

$$\min \sum_{-N/2+2P}^{N/2} [\varepsilon^+(m) + \varepsilon^-(m)] \qquad (2.4a)$$

under

$$\underline{\underline{\hat{R}}}[\underline{a}^+ - \underline{a}^-] = \underline{\varepsilon}^+ - \underline{\varepsilon}^- \qquad (2.4b)$$

where $\underline{a}^+$, $\underline{a}^-$, $\underline{\varepsilon}^+$ and $\underline{\varepsilon}^-$ have obvious expressions ($a_0^+ = 1$, $a_0^- = 0$).

Figs. 1a , b , show a confirmation of our comments on the possible advantage of this new approach with respect to that proposed by Cadzow. The computational cost is, certainly, increased.

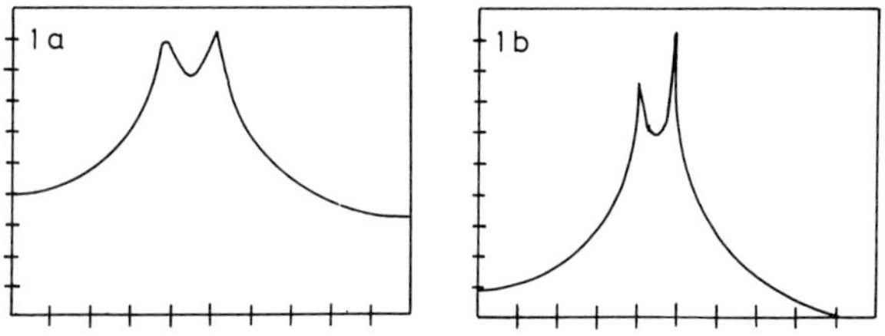

Fig. 1: $1/|A(\omega)|^2$ for: a) minimum L_2-norm, and
b) minimum L_1-norm solutions.

Simulated data: cos. 4π n+cos .5π n+w(n), $0 \le n \le 59$; P=2, N=20
Gaussian, white w(n), 3dB under the sinusoid powers
$\hat{R}$: classical biased estimator.

3. USING A PRIORI BANDWIDTH KNOWLEDGE. SYMMETRY AND COLOURED NOISE

It is well established today that the incorporation of previous knowledge about the signal characteristics into a spectral analysis model or its parameter estimation will improve the performance. One of the most effective approaches is to use bandlimitation of the solution, Levy et al. (1982), Jain and Ranganath (1981).

Here, we can obtain clearly better performance if $\hat{R}(.)$ is computed as the inverse transform of a bandlimited version of a first spectral estimate, as we can see in the examples shown in this Section. This colours the corresponding noise; but this is acceptable, since, on the other hand, to apply Cadzow's second version equations is equivalent to accepting a model for the autocorrelation sequence

$$\hat{R}(m) = \sum_{i=1}^{P} P(i) \cos \omega_i m + R_w(m) \tag{3.1}$$

where $P(i)$ is the power of the i-th sinusoid, and $R_w(.)$ the noise autocorrelation sequence. Then, we can consider $\sum_{i=1} P(i) \cos \omega_i m$ as a sinusoidal signal and $R_w(m)$ as noise. From this point of view, Cadzow's second version is equivalent to the Prony-Hildebrand method applied in the autocorrelation domain; and

a) it becomes evident that we can apply it to coloured noise case;

b) we can introduce immediately a symmetrical version of the coefficients (this reduces the number of constraints to N/2+1).

Remark a) increases the interest in considering an L_1-norm minimization, since the comparative performances of this and an L_2-norm based method will depend on the noise colour.

Some examples will clarify the effectiveness of all these possibilities. We will show results of simulations taking as data:

$$\cos .44\pi n + \cos .48\pi n + \cos .58\pi n + w(n), \quad 0 \le n \le 63$$

where w(.) is white or coloured noise having spectral shape as Fig. 2 shows. We use P=3 as the model order, and N=20.

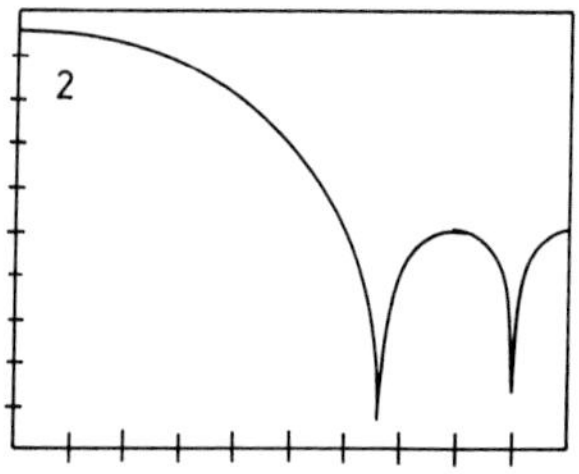

Fig. 2 Coloured noise spectral shape

Figs. 3a , b , c , show the periodogram (and assumed bandlimitation), and asymmetrical Cadzow's minimum L_1-norm estimates without and with bandlimitation, respectively, under white noise; SNR=3dB. The advantage of using bandlimitation is obvious, although the periodogram does not discriminate two lines. Fig. 4a , b , c , are the equivalent curves under coloured noise, SNR=-8dB. The comments are similar, and we can see how these approaches offer good results under coloured noise conditions.

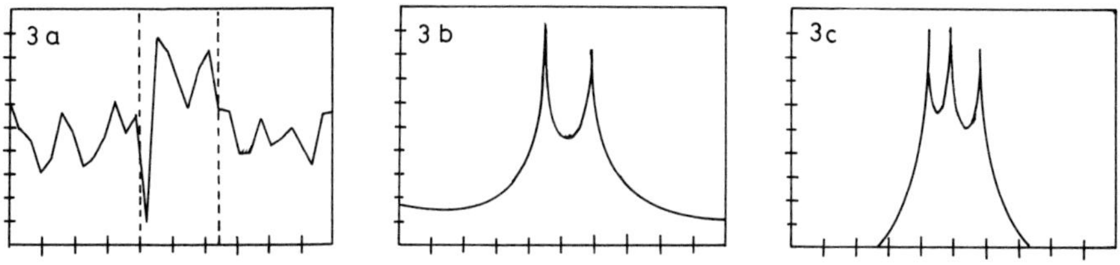

Fig. 3 a) Periodogram and band used in inverse $\hat{R}$;
b) Symmetrical Cadzow's minimum L_1-norm estimate $1/|A(\omega)|^2$ without bandlimitation;
c) Same as b) using $\hat{R}$ as the inverse of .2-.32 band of the periodogram.

Signal and parameters as given in the text; white noise, 3 dB. under sinusoid power.

Associated frequency estimates: b) .224, .295 (there is a double root)

c) .211, .245, .289.

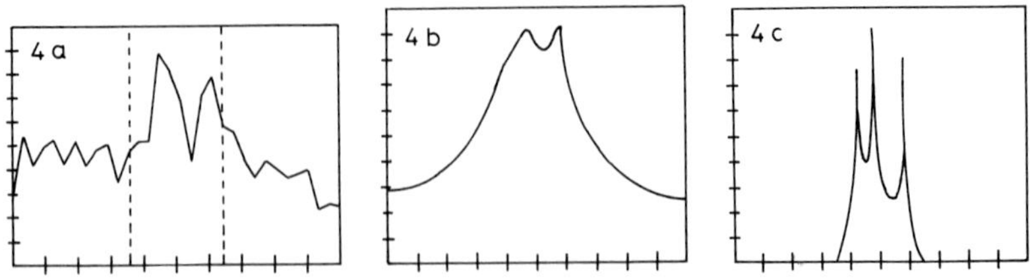

Fig. 4 a), b), c), as in Fig. 2; coloured noise, 8dB. over sinusoid power.

Bandlimitation is from .18 to .32.

Associated frequency estimates: b) .193, .238, .290
c) .211, .238, .290

Equivalent results forcing a symmetrical polynomial are shown in Figs. 5 a, b, c (white noise) and 6 a, b, c (coloured noise), with different noise realizations, but equal conditions.

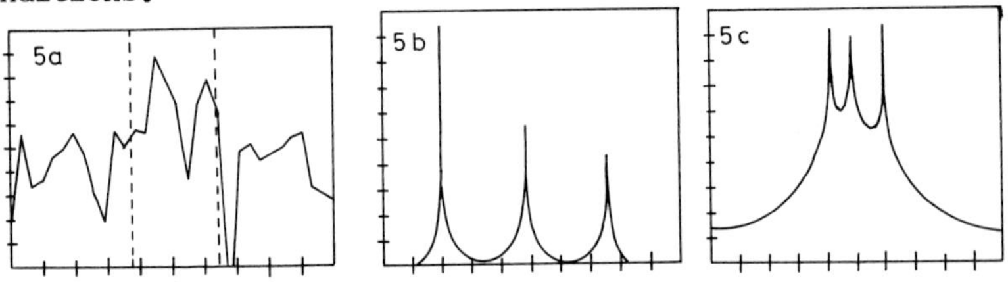

Fig. 5 a), b), c), as in Fig. 3, but under symmetrical model.

Bandlimitation from .19 to .32; white noise, SNR=3 dB.

Frequency estimates: b) .099, .241, .377
c) .205, .240, .297

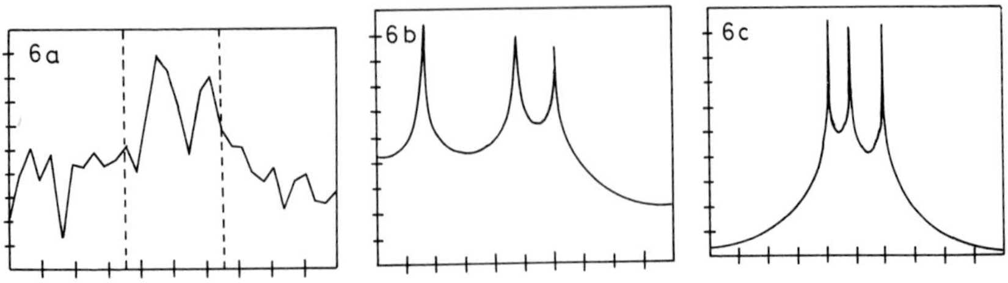

Fig. 6 a), b), c), as in Fig. 3, but under symmetrical model.

Bandlimitation from .18 to .32; coloured noise, SNR=-8dB.

Frequency estimates: b) .089, .238, .310
c) .204, .240, .295

Note that, here, the estimates obtained without bandlimitation show three sharp peaks, originated by the symmetry; but the frequency estimates are clearly worse. When using bandlimitation, the observed tendency to disperse frequencies is maintained, although the side information reduces it. In any case, these symmetrical procedures offer sharper peaks but worse frequency estimates: the last seems to be a consequence of reducing the degrees of freedom in the model.

4. COMMENTS ON PRONY-HILDEBRAND's METHOD

Since this method is equivalent to the symmetrical version of Cadzow's second procedure, but applied in the time domain, we can extrapolate most of the results of the previous discussions: for example, the interest of using a previous bandpass filtering (at least with long data registers) when we know sinusoid band (or we can estimate it from an FFT), or the differences between a symmetrical and a non-symmetrical (overdetermined linear prediction analysis) formulations.

Clearly, there is not a direct equivalent of the effect of centre equations here; but, as it is well-known, the use of an L_1-norm minimization will be advantageous if the errors offer long tail distribution or outliers.

Since the errors in the minimum L_1-norm equations

$$\min \sum_{n=P+1}^{N-P} [\varepsilon^{+}(n) + \varepsilon^{-}(n)] \tag{4.1a}$$

under

$$x(n) + \sum_{i=0}^{P-1} [a_i^{+} - a_i^{-}]\,[x(n-i) + x(n+i)] = \varepsilon^{+}(n) + \varepsilon^{-}(n) \tag{4.1b}$$

$$P+1 \leq n \leq N-P$$

(normalizing $a_P = 1/2$; this has not any particular effect), are all-zero filtered versions of the present additive noise and outliers, only when the tails are high or the outliers important and the model order not very high is Simplex interesting. But, under these conditions, the use of the L_1-norm minimization could be very convenient, as Table I demonstrates.

	L_2		L_1		
#	$\hat{f}_1$	$\hat{f}_2$	$\hat{f}_1$	$\hat{f}_2$	0
1	.229	.229	.200	.251	38
2	.200	.250	.201	.251	–
3	.229	.246	.205	.255	7
4	.200	.250	.200	.249	–
5	.200	.251	.200	.248	–
6	.199	.215	.202	.252	–
7	.229	.242	.205	.252	6
8	.201	.250	.201	.251	–
9	.232	.242	.203	.254	12
10	.232	.245	.204	.259	10

Table I L_1 and L_2-norm frequency estimates for ten experiments. Register cos $.4\pi n$+cos $(.5\pi n+\pi/4)$, $0 \le n \le 39$.

w(n): white noise 30 dB. under sinusoid power plus fixed level random outliers 8 dB. over sinusoid peak with probability of appearance 0.1.

0 indicates position of the outlier.

5. CONCLUSIONS

We have shown the potential interest of using L_1-norm minimization in Cadzow's second procedure in order to compensate white noise effects. Additionally, we have considered the possibility of applying this procedure under coloured noise conditions, and we have shown that:

- to use bandwidth knowledge improves the high resolution properties;
- to use symmetrical coefficients offers sharper peaks but worse frequency estimates.

These conclusions can be extended, with the appropriate equivalences, to Prony-Hildebrand's method; when using this, to apply L_1-norm minimization will be advantageous if there are outliers or long tail noise distributions and the model order is low.

6. REFERENCES

Cadzow, J.A., (1982) "Spectral Estimation: An Overdetermined Rational Model Equation Approach", *Proc. IEEE* **70**, 907-939.

Kay, S.M. and Marple, Jr., S.L., (1981) "Spectrum Analysis-A Modern Perspective", *Proc. IEEE* **69**, 1380-1419.

Hildebrand, F.B., (1956) "Introduction to Numerical Analysis". McGraw-Hill, New York.

Figueiras-Vidal, A.R., et al., (1985) "L_1-Norm Versus L_2-Norm Minimization in Parametric Spectral Analysis: A General Discussion", Proc. ICASSP'85, Tampa, FL, 304-307.

Keybl, J. and Eichmann, G., (1983) "Estimation of Two Closely Spaced Frequencies Buried in White Noise Using Linear Programming", Proc. Topical Meeting on Sig. Recovery and Synth., Incline Village, Nevada, WA 18-1/WA 18-4.

Levy, S., et al., (1982) "A Linear Programming Approach to the Estimation of the Power Spectra of Harmonic Processes", *IEEE Trans. ASSP,* **30**, 675-679.

Jain, A.K. and Ranganath, S., (1981) "Extrapolation Algorithms for Discrete Signals with Application in Spectral Estimation", *IEEE Trans. ASSP,* **29**, 830-845.

APPROXIMATE EVALUATION OF BOUNDS FOR PEAK ESTIMATES

R. Knötsch and J.F. Böhme
*(Lehrstuhl für Signaltheorie, Ruhr-Universität,
Federal Republic of Germany)*

ABSTRACT

Peak estimates are certain high-resolution methods for estimating the spectra of a wave field received by an array of sensors. The problem is to predict the behaviour of such methods when only an estimate of the spectral matrix of the sensor outputs is known. The properties of the eigensystem of the spectral matrix in the case of weak ambient noise motivate the methods of orthogonal beamforming, especially the projection estimates of Owsley. In applications asymptotic expressions for the peak estimates, e.g. for the first and second moment, cannot be calculated because the eigensystem of the exact spectral matrix is required. Therefore we developed bounds for the deviation of the peak estimates which only use weak knowledge about the matrix. We applied some results on pertubation of Hermitian operators. The behaviour of the Owsley-estimates is investigated and discussed. Approximate expressions to evaluate the bounds, when only an estimate of the spectral matrix is known, are introduced. These expressions are motivated by the asymptotic distribution of the eigenvalues of the estimated spectral matrix. Extensive simulations with random matrices were performed to evaluate the bounds and the approximate expression. As a result we have found that if the eigenvalues of interest are sufficiently separated from the rest, the considered estimators behave stably. Furthermore approximate expressions can be used to predict the behaviour of the peak estimates.

1. INTRODUCTION

High-resolution spectral estimation methods are used for passive array processing in seismology and sonar applications. Signals received by sensor arrays are processed to obtain estimates of the directions of arrival. The methods use the special propagation models of the wavefield, for example plane waves for farfield sources. The sensor outputs may consist of sensor noise, ambient noise, and noise from a finite set of discrete sources. The spectral matrix of the sensor outputs, which are modelled by a stationary stochastic process as the response to the waves, has a simple structure.

The methods of orthogonal beamforming estimate the propagation model by an analysis of the eigenvalues and orthogonal eigenvectors of the spectral matrix, which is assumed to be well estimated. The propagation model may be known except for some parameters, for example directions or wavenumbers, and we are interested in the behaviour of diagrams over these parameters. The diagrams are interpreted like power spectra and can be compared with the corresponding classical beamforming spectra. We pick up well known methods, which we call Owsley estimates and try to predict their behaviour. The asymptotic statistical behaviour of the Owsley estimates are shown in [1]. The asymptotic expressions, for example for the first and second moments, cannot be calculated in application since the eigensystem of the exact spectral matrix is required. Therefore we developed bounds for the deviation of the peak estimates computed from the estimated spectral matrix to the best possible computed from the exact matrix, which only use limited knowledge about the exact matrix. The behaviour of the bounds is discussed, and approximate evaluation of the bounds are performed. The evaluation of the bounds only use the eigenvalues of the estimated spectral matrix.

Numerical experiments with random hermitian matrices were executed to show how good in the mean the bounds predict the deviation and how good the bounds can be approximately evaluated.

2. DATA MODEL

The array may consist of N identical omnidirectional sensors, which have low-pass character, located at points x_n $(n = 0,\ldots,N-1)$.

The model for the sensor outputs which are sampled in parallel with period T is a discrete stationary stochastic vector process with N components. The expectation of the stochastic process is assumed to be zero. The spectral matrix of the process is $F(\lambda) = (f_{nj}(\lambda))$, where f_{nj} is the cross spectrum of the components from sensors at location x_n and x_j, $\lambda = \omega T$ and ω is a frequency in cycles/s. Similar to [2] we suppose that the process consists of three independent parts: nearfield noise, farfield or ambient noise and signals of K discrete sources, where K is assumed to be less than N. Hence $F(\lambda)$ is the sum of three nonnegative definite hermitian matrices:

$$F(\lambda) = F_n(\lambda) + F_a(\lambda) + F_s(\lambda). \tag{1}$$

$F_n(\lambda)$ describes the nearfield noise. We assume either that we have some a priori knowledge about the structure of the noise and transform the data, or that we can remove the nearfield component as in [2]. Hence, we assume that $F_n = f_n I$, where I is the identity matrix. The ambient noise can be represented as

$$F_a(\lambda) = \int f_a(\lambda,\theta) D_\theta(\lambda) \overline{D_\theta(\lambda)} d\theta, \tag{2}$$

where $f_a(\lambda,\theta)$ is assumed to be a smooth spectrum, D_θ is the steering column vector for waves described by a parameter θ, and the bar denotes transposition and conjugation. If we assume plane waves of a known velocity ν of propagation, ambient noise can be interpreted as a homogeneous random field. D_θ is the direction vector with elements $\exp(i\omega\overline{u(\theta)}x_n/\nu)$ $(n=0,\ldots,N-1)$, and $-u(\theta)$ is a unit vector pointing in the direction of propagation of the wave.

F_s describes the signals from K sources with parameters $\theta_0,\ldots,\theta_{K-1}$. Let P be the matrix with column vectors D_{θ_k} $(k=0,\ldots,K-1)$ and G be the KxK matrix of cross spectra of signals received at the origin. Thus

$$F_s(\lambda) = P(\lambda)G(\lambda)\overline{P(\lambda)}. \tag{3}$$

The rank of F_s is not greater than K. Since F_s has rank (F_s) positive eigenvalues the corresponding orthogonal eigenvectors are analysed to determine the steering vectors D_{θ_k} and the matrix G, cf. [2,3,4].

In applications the exact spectral matrix F is unknown. The matrix can be estimated if the sample is long and if stationarity is assumed. Then the estimate $\hat{F}$ has to be analysed to determine the signal parameters $\theta_0,\ldots,\theta_{K-1}$. We assume that $\hat{F}$ is estimated by one of the classical methods like smoothing the periodogram matrix of the data over frequencies or by the mean of periodogram matrices of succeeding stretches of data. Then it is well known [5] that the estimated spectral matrix $\hat{F}$ is asymptotically distributed as $W_N^c(L, \frac{1}{L}F(\lambda))$ if λ is not a multiple of π, where $W_N^c(L,\Sigma)$ denotes the complex Wishart distribution with L degrees of freedom and parameter matrix Σ. The number of degrees of freedom is determined by the numbers of frequencies over which the periodogram matrix is smoothed or by the numbers of stretches. In our investigation we suppose that we have chosen a suitable frequency λ and therefore we shall omit the notation of λ.

3. PEAK ESTIMATES

If the signals are uncorrelated, the spectral matrix F_s can be described similar to (2) with a discrete spectrum $f_s(\theta)$ which has δ-contributions for $\theta=\theta_k$ and weights proportional to the magnitude of the spectrum of the signal k. The diagram

$$g(\hat{\theta}) = D_\theta \hat{F} \overline{D_\theta} \tag{4}$$

is used for estimating the shape of the sum $f_a(\theta) + f_s(\theta)$. Sometimes the elements of D_θ are modified by tapering. The equation (4) corresponds to classical beamforming. If the steering vectors D_{θ_k} are well separated and there is no strong coherence between signals, the diagram will show a peak for $\theta=\theta_k$. The resolution of the method is limited by the geometry of the sensorfield. Therefore peak estimates of orthogonal beamforming are introduced to improve the resolution. They are functions of quadratic forms similar to (4), where $\hat{F}$ is replaced by a matrix which is a function of the eigensystem of $\hat{F}$. Let $\nu_0 \geq .. \geq \nu_{N-1}$ be the eigenvalues and $V_0,..,V_{N-1}$ the corresponding orthonormal eigenvectors of $\hat{F}$. In this paper we consider estimates which we call <u>Owsley estimates</u>,

$$\hat{g}_1(\theta) = \overline{D_\theta}\left(\sum_{j=r}^{s} V_j \overline{V_j}\right) D_\theta = \sum_{j=r}^{s} |\overline{D_\theta} V_j|^2 \tag{5}$$

$$\hat{g}_2(\theta) = \overline{D_\theta}\left(\sum_{j=r}^{s} \nu_j V_j \overline{V_j}\right) D_\theta = \sum_{j=r}^{s} \nu_j |\overline{D_\theta} V_j|^2 . \tag{6}$$

These estimates are determined by a real interval, which contains the eigenvalues $\nu_r,\ldots,\nu_s$. Owsley [6] used the estimate $\hat{g}_1(\theta)$ for r=s=0, which means the largest eigenvalue. Other authors use the smallest ones. The use of the smallest eigenvalues has the effect, that the diagrams show valleys instead of peaks for $\theta=\theta_k$. Therefore an application of $1/\hat{g}_1(\theta)$ seems to be useful.

4. BOUNDS

The asymptotic behaviour of the peak estimates $\hat{g}_1(\theta)$, $\hat{g}_2(\theta)$ are shown in [1]. Unfortunately the asymptotic expressions require the knowledge of the total eigensystem of the exact spectral matrix F. Estimates of these expressions by replacing the eigensystem of F by the eigensystem of $\hat{F}$ would not lead to good results, because the estimated eigenvectors are unstable estimators for the exact eigenvectors of F. Therefore bounds were developed to predict the behaviour of the peak estimates in a concrete situation. Let

$$d_i(\theta) = |\hat{g}_i(\theta) - g_i(\theta)| \quad (i=1,2).$$

We are interested in a bound for $\max_\theta d_i(\theta)$. In [1] it was shown that

$$d_i(\theta) \le \|D_\theta\|^2 \|A_i\|_1, \tag{7}$$

where

$$A_1 = Q_{rs} - P_{rs}, \quad Q_{rs} = \sum_{j=r}^{s} V_j\bar{V}_j, \quad P_{rs} = \sum_{j=r}^{s} U_j\bar{U}_j,$$

$$A_2 = \sum_{j=r}^{s} (\nu_j Q_{jj} - \mu_j P_{jj}),$$

$\mu_0 \ge \ldots \ge \mu_{N-1}$ are the eigenvalues of F and $U_0,\ldots,U_{N-1}$ the corresponding eigenvectors. The norm $\|A_i\|_1$ is defined by $\|A_i\|_1 = (\text{max. eigenvalue of } (\bar{A}_i\ A_i))^{1/2}$.

The bound (7) cannot be improved in general. Bounds for $\|A_i\|$ (i=1,2) were found by applications of results on weak perturbation of hermitian operators [7,8]:

$$\|A_1\|_1 = \|Q_{rs} - P_{rs}\|_1$$

$$\leq \begin{cases} (.5(1-(1-4\|H\|_1^2/\Delta^2)^{1/2}))^{1/2} & \text{if } \|H\|_1 \leq \Delta/2 \\ 1 & \text{if } \|H\|_1 > \Delta/2 \end{cases}, \quad (8)$$

where $H = \hat{F} - F$ is called the perturbation and Δ is the minimum distance of the eigenvalues $(\nu_j, j=r,\ldots,s)$ to the eigenvalues $(\nu_j, j\neq r,..,s)$. We denote the right hand side of (8) by $S(r,s)$. Bounds for $\|A_2\|_1$ were shown in [8]. We have:

$$\|A_2\|_1 \leq \|H\|_1 + \min_{c\geq 0}(cS(r,s) + \sum_{j=r}^{s}|\nu_j - c|S(j,j)) \quad (9a)$$

$$\|A_2\|_1 \leq 2\|H\|_1 + \min_{c\geq 0}(cS(r,s) + \sum_{j\neq r}^{s}|\nu_j - c|S(j,j)) \quad (9b)$$

where $\sum_{j\neq r}^{s} x_j$ means the sum over all x_j $(j\neq r,..,s)$. A bound for $\|A_2\|_1$ is now found by taking the minimum of the right sides of (9a) and (9b) and we denote it by (9). Further improvement of the bound (9a) can be made if we replace $\|H\|_1$ by $\|\text{diag } \gamma_j\|$, where

$$\gamma_j = \begin{cases} \mu_j - \nu_j & j=r,\ldots,s \\ 0 & \text{otherwise} \end{cases},$$

because $\|H\|_1 \geq \max_j|\mu_j - \nu_j|$.

Then we get

$$\|A_2\|_1 \leq \|\text{diag } \gamma_j\|_1 + \min_{c\geq 0}(cS(r,s) + \sum_{j=r}^{s}|\nu_j - c|S(j,j)) \quad (10a)$$

The bound (9b) can be improved if we replace $\|H\|_1$ by $\|\text{diag } \delta_j\|$, where

$$\delta_j = \begin{cases} 0 & j=r,\ldots,s \\ \mu_j - \nu_j & \text{otherwise} \end{cases}.$$

We can only replace $2\|H\|_1$ by $\|H\|_1 + \|\text{diag } \delta_i\|$ because only one term is determined by the difference of the theoretical and estimated eigenvalues.

Then we get

$$\|A_2\|_1 \leq \|H\|_1 + \|\text{diag } \delta_i\| + \min_{c \geq 0}(cS(r,s) + \sum_{j \neq r}^{s} |\nu_j - c| S(j,j)) . \qquad (10b)$$

We denote the minimum of the right sides of (10a) and of (10b) with (10).

Extensive simulations with random hermitian matrices were executed to check how well the bounds work. The model (2) was used for a line array with N=8,12,16 sensors with a mutual distance of a half wavelength. We used K=1,2,3,4 incoherent signals with bearings $\theta_0=10^\circ$, $\theta_1=40^\circ$, $\theta_2=-10^\circ$, $\theta_3=-30^\circ$ and powers $g_{00}=1$, $g_{11}=0.5$, $g_{22}=0.5$ and $g_{33}=1$. Various signal to noise ratios (S/N) were considered. For comparing the different situations, we used the factor by which a bound is greater than the corresponding norm. For each situation various degrees of freedom were used. Table 1 shows the typical behaviour of the bound (8) and of the composite bound of (9a), (9b) and of the composite bound of (10a), (10b). Simulation with 100 degrees of freedom were performed with an (S/N) of 3dB and 0dB respectively. Table 1 shows the mean factor, by which the bounds are greater than the norms. We remark that in different situations the bounds behave in the same manner. The values for r and s are 0 and K-1, i.e. we assume that the number of sources is known.

Table 1: Comparison of the bounds

Bound:	(8)	(9)	(10)		(8)	(9)	(10)
Sensors		1 Signal		Sensors	2	Signals	
8	2.125	2.085	1.611	8	5.074	2.789	1.956
12	2.120	2.069	1.634	12	4.858	2.940	2.139
16	2.136	2.071	1.695	16	4.925	3.020	2.229
Bound:	(8)	(9)	(10)		(8)	(9)	(10)
Sensors		3 Signals		Sensors	4	Signals	
8	7.444	2.683	1.802	8	8.518	2.496	1.571
12	6.488	2.761	1.888	12	8.575	2.567	1.651
16	6.697	2.860	1.997	16	8.310	2.636	1.731

From Table 1 we can see that the behaviour of the bounds does not depend on the number of sensors. The dependence of the number of signals is evident. Especially the bound (8) is influenced by the numbers of signals. We see from equation (8) that the bound depends on the distance between the smallest eigenvalue of the signal space to the greatest eigenvalue of the noise space. As we have chosen signals (signal 2 and 3) with a relatively low signal to noise ratio, the behaviour of the bound for more than 1 signal could be expected. The bound (8) needs either a great signal to noise ratio or a great number of degrees of freedom. Though the bounds (9a), (10a) increase very quickly if the number of signal increases, the composite bounds (9), (10) do not, because the bounds (9b), (10b) will decrease if the number of signals increases.

5. APPROXIMATE EVALUATION OF BOUNDS

The only unknowns of the bounds (8), (9a) and (9b) is the $\|H\|_1$. As it is very difficult to get the distribution of $\|H\|_1$ if $\hat{F}$ is given, we investigated estimates of $E\|H\|_1$. But even the calculation of $E\|H\|_1$ is in general very difficult because the distribution of $\|H\|_1$ is not known. Therefore we investigated how far $\|H\|_{sq}$ deviates from $\|H\|_1$ in the above situations, where $\|H\|_{sq} = \mathrm{tr}(H\bar{H})^{1/2}$. At first we have $\|H\|_1 \leq \|H\|_{sq}$, i.e. we can bound $\|H\|_1$ by $\|H\|_{sq}$, and the bounds (8), (9a), (9b) are still valid. Table 2 shows the factor by which in the mean $\|H\|_{sq}$ is greater than $\|H\|_1$ in the situation of Table 1. We see, that $\|H\|_{sq}$ is a good and tight bound for $\|H\|_1$. In other performed situations $\|H\|_{sq}$ behaves in the same manner.

Table 2: Comparison of $\|H\|_1$ and $\|H\|_{sq}$.

Sensors: Signals	8	12	16
1	1.177	1.182	1.155
2	1.224	1.223	1.221
3	1.285	1.283	1.289
4	1.345	1.370	1.369

Now we develop an estimate for $E\|H\|_{sq}$. We have $E\|H\|^2_{sq} = E\sum_{i,k}|h_{ik}|^2 = E\sum_{i,k}|\hat{f}_{ik}-f_{ik}|^2$. Now $\hat{F}$ is approximately distributed like $W_N^c(L,\frac{1}{L}F)$, [5], and we get the result

$$E\,|\hat{f}_{ik} - f_{ik}|^2 = \frac{1}{L}\, f_{ii}f_{kk}\,, \qquad (11)$$

$$E\,\|H\|^2_{sq} = \frac{1}{L}\,(\mathrm{tr}\,F)^2. \qquad (12)$$

That motivates

$$(L)^{-1/2} \text{tr } \hat{F} \tag{13}$$

as an estimate for $E\|H\|_{sq}$, because $E \text{ tr } \hat{F} = \text{tr } F$ and $E\|H\|_{sq} \leq \sqrt{E\|H\|^2_{sq}}$.

The properties of the estimate (13) can be calculated from the characteristic function of tr $\hat{F}$ given by Goodman [9]. We get

$$E(L)^{-1/2} \text{tr } \hat{F} = (L)^{-1/2} \text{tr } F = (E\|H\|^2_{sq})1/2, \tag{14}$$

$$\text{Var}(L)^{-1/2} \text{tr } \hat{F} = \frac{1}{L} \|F\|^2_{sq} \tag{15}$$

Table 3 shows how good the estimate (13) approximates $\|H\|_{sq}$. Indicated is the bias of $(\|H\|_{sq} - (L)^{-1/2} \text{tr } \hat{F})$ and the mse $(\|H\|_{sq} - (L)^{-1/2} \text{tr } \hat{F})$.

Table 3: Behaviour of estimate (13)

Sensors:	8		12		16	
Signals	bias	mse	bias	mse	bias	mse
1	-0.018	0.163	-0.020	0.290	0.073	0.444
2	0.027	0.134	-0.025	0.305	0.078	0.484
3	0.025	0.140	0.034	0.290	0.106	0.490
4	0.018	0.213	0.021	0.376	-0.001	0.581

A better approximation for $\|H\|_1$ is the expression $\sum_{j=0}^{N-1} |\nu_j - \mu_j|$,

which was proved by numerical simulations. These approximations work very well in the situations considered. We can estimate this expression by

$$(2/(\pi L))^{1/2} \text{tr } \hat{F}. \tag{16}$$

Table 4 compares the estimates (13) and (16). Indicated is the bias of $(\|H\|_1 - (L)^{-1/2} \text{tr } \hat{F})$, of $(\|H\|_1 - (2/(\pi L))^{1/2} \text{tr } \hat{F})$, and the mse, respectively.

Table 4: Comparison of the estimates (13) and (16)

Signals	(13) bias	(13) mse	(16) bias	(16) mse
1	-0.159	0.231	-0.081	0.212
2	-0.239	0.223	0.082	0.174
3	-0.391	0.321	0.011	0.170
4	-0.699	0.694	-0.137	0.224
12 sensors				
Signals	(13) bias	(13) mse	(16) bias	(16) mse
1	-0.228	0.452	0.139	0.421
2	-0.408	0.538	0.080	0.378
3	-0.588	0.695	0.019	0.352
4	-1.080	1.525	-0.231	0.408
16 sensors				
Signals	(13) bias	(13) mse	(16) bias	(16) mse
1	-0.181	0.636	0.297	0.684
2	-0.455	0.791	0.183	0.610
3	-0.754	1.123	0.046	0.556
4	-1.451	2.622	-0.333	0.629

Estimates for $\|\text{diag}\,\gamma_i\|$ and $\|\text{diag}\,\delta_i\|$ can be found by using the asymptotic properties of the eigenvalues of $\hat{F}$. Brillinger [5] has shown that, if the eigenvalues of F are different, the eigenvalues ν_j are approximately normal distributed with $\overline{\text{ave}}\ \nu_j = \mu_j$ and $\overline{\text{var}}\ \nu_j = \frac{1}{L}\mu_j^2$, where the notations $\overline{\text{ave}}$, $\overline{\text{var}}$ mean expected values derived in a term by term manner from a Taylor expansion. Then we choose as an estimate for $\|\text{diag}\,\gamma_i\|$

$$\left(\frac{1}{L}\sum_{j=r}^{s}\nu_j^2\right)^{1/2} \tag{17}$$

and as an estimate for $\|\text{diag}\,\delta_i\|$

$$\left(\frac{1}{L}\sum_{j\neq r}^{s}\nu_j^2\right)^{1/2} . \tag{18}$$

The estimates (17) and (18) have the following asymptotic properties:

$$E\left(\frac{1}{L}\sum_{j=r}^{s}\nu_j^2\right) = E\ \|\text{diag}\,\gamma_i\|_{sq}^2 \tag{19}$$

$$E\left(\frac{1}{L}\sum_{j\neq r}^{s}\nu_j^2\right) = E\ \|\text{diag}\,\delta_i\|_{sq}^2 \tag{20}$$

Table 5 shows the bias and the mse for the estimates (17) and (18) with respect to $\|\text{diag } \gamma_i\|$ and $\|\text{diag } \delta_i\|$ in the situation of Table 1.

Table 5:

Estimate (17):						
Sensors:	8		12		16	
Signals	bias	mse	bias	mse	bias	mse
1	-0.159	0.349	-0.183	0.693	-0.089	0.967
2	-0.075	0.248	-0.139	0.559	-0.081	0.811
3	-0.044	0.252	0.106	0.487	-0.059	0.662
4	-0.078	0.390	0.529	0.788	-0.057	0.703
Estimate (18):						
Sensors:	8		12		16	
Signals	bias	mse	bias	mse	bias	mse
1	0.073	0.008	0.160	0.027	0.259	0.069
2	0.052	0.005	0.129	0.017	0.224	0.052
3	0.034	0.003	0.101	0.011	0.193	0.039
4	0.021	0.002	0.077	0.007	0.163	0.028

The results of Table 4 and Table 5 show that the estimates have relatively low bias and that it is possible to use the estimates in the bounds (8), (9) and (10). Table 6 shows that the estimated bounds behave like the bounds (8), (9), (10). Indicated is the mean factor by which the estimated bounds are greater than the corresponding norm, the bias between the estimate and the bound and the corresponding mse. We see that the estimates using (16) approximate the bounds better than those using (13). For 12 and 16 sensors and in other different situations the estimates behave in the same manner. The estimates using (16) have relatively low bias and a relatively low mse, whereas the estimates using (13) tend to overestimate the bounds. In the mean the estimates using (16) are good estimates for the corresponding bounds.

Table 6: Comparison of estimates (8 sensors)

Estimates for (8):						
Estimates using	(16)			(13)		
Signals	factor	bias	mse	factor	bias	mse
1	1.915	-0.014	0.005	2.413	0.020	0.005
2	4.289	-0.072	0.052	6.491	0.138	0.077
3	8.328	0.100	0.059	8.685	0.139	0.071
4	8.556	0.002	0.002	8.556	0.004	0.002
Estimates for (9):						
Estimates using	(16)			(13)		
Signals	factor	bias	mse	factor	bias	mse
1	2.195	-0.187	1.041	2.758	0.332	1.117
2	2.814	-0.190	0.843	3.387	0.541	1.089
3	2.825	0.024	0.785	3.360	0.846	1.497
4	2.718	0.276	0.902	3.276	1.401	2.783
Estimates for (10):						
Estimates using	(16)			(13)		
Signals	factor	bias	mse	factor	bias	mse
1	2.070	0.167	0.754	2.373	0.445	0.929
2	1.924	-0.162	0.270	2.245	0.248	0.303
3	1.872	0.001	0.235	2.145	0.421	0.412
4	1.673	0.118	0.227	1.952	0.681	0.676

6. DISCUSSION

We are interested in bounds for peak estimates. Therefore we developed bounds for the norm $\|A_i\|_1$ (i=1,2). The bound for $\|A_1\|_1$ behaves well, if the signal to noise ratio is high or if $\|H\|_1$ is low, which means that the number of degrees of freedom is high. If the number of degrees of freedom is too small or the signal to noise ratio is too low, then $\|A_1\|_1$ can only be bounded by 1 and the bound gives no information. The bound for $\|A_2\|_1$ works well even for low signal to noise ratios. Estimates of the bounds were developed which behave like the bounds. So we can predict the behaviour of the peak estimates by means of the estimated spectral density matrix $\hat{F}$. We remark that in other situations like circular arrays etc. the bounds and the estimates behave similarly. The bounds are still valid, if we consider wideband plane waves and construct estimates in the same way as described in [10], i.e. the estimates rely on an approximately coherent combination of the spatial signal spaces of the temporally narrowband decomposition of the received signal vector. Thus, the properties of the eigensystem of the estimated matrix has the same structure as in the narrowband case and the above bounds can be applied. To estimate these bounds only slight modifications of the estimates above have to be made.

REFERENCES

1. Böhme, J.F., "Remarks on the statistical behaviour of orthogonal beamforming", *IEE Proc.* Vol. 130, Pt.F, 1983, pp. 256-260.

2. Liggett, W.S., Hr., "Passive sonar processing for noise with unknown covariance structure", *J. Acoust. Soc. Am.*, 1972, 51, pp. 24-30.

3. Owsley, N.L., "Spectral signal set extraction" in Tacconi, G., (Ed.), "Aspects of signal processing", D. Reidel, 1977, pp. 469-475.

4. Reddi, S.S., "Multiple source location: a digital approach", *IEEE Trans.* 1979, AES - 15, pp. 95-105.

5. Brillinger, D.R., "Time series: Data Analysis and theory", Holt, Rinehart and Winston, 1975.

6. Owsley, N.L., "A recent trend in adaptive spectral processing for sensor arrays: constrained adaption", in Griffith, J.W.R, (Ed.): "Signal processing", Academic Press, 1973, pp. 591-604.

7. Davis, C. amd Kahan, W.M., "The rotation of eigenvectors by a perturbation III", *SIAM J. Num. Anal.* **7**, 1970, pp. 1-46.

8. Knötsch, R., "Eine neue Schranke für Peak- schätzer" Proc. 5. Aachener Kolloquium, 1984, pp. 212-215.

9. Goodman, N.R., "Statistical analysis based upon a certain multivariate complex Gaussian distribution (an introduction)", *Ann. Math. Stat.* **34**, 1963, pp. 152-177.

10. Wang, H. and Kaveh, M., "Coherent signal-subspace processing for the detection and estimation of angles of arrival of multiple wideband sources", *IEEE Trans. ASSP 33*, pp. 823-831.

PRINCIPAL COMPONENT ANALYSIS OF THE SPECTRA OF POINT PROCESSES - AN APPLICATION IN ELECTROMYOGRAPHY

N.B. Jones
(University of Leicester)

P.J. Lago
(University of Porto)

and

A. Parekh
(University of Sussex)

ABSTRACT

An objective method of classifying the spectrum of a point process is by approximating it by the formula $M(\omega) + a_1 P_1(\omega) + a_2 P_2(\omega)$ where M is the mean spectrum and p_1 and p_2 the first two principal component spectra derived from a set of standard spectra of processes with known characteristics.

Any new spectrum is classified according to the values assumed by a_1 and a_2 after a least squares fitting procedure.

Examples are given of the use of the technique in electromyography. In this case the source of the information is a continuous signal and a point process is generated from this as an intermediate step. The continuous signal is known to consist of a limited number of component waves which occur stochastically distributed in time. The objective is to gain information on the number of peaks in the component waves. The point process is generated by marking all significant peaks in the signal.

It is shown, by simulation, that different types of spectra derived from signals typical of normal, myopathic and neurogenic subjects can be classified successfully using the method of principal component analysis. It is shown that care needs to be exercised in the procedure used to identify peaks. The progress of disorder can also be charted using this approach.

Finally, it is shown that a three dimensional space of these coefficients can be used to improve the discrimination when necessary.

1. INTRODUCTION AND BACKGROUND

1.1 Point Processes and Their Spectra

A point process is one which is defined in terms of a sequence of occurrence times only. Such processes are important in science and engineering, arising, for example, from consideration of nerve cell discharges, radio-active decay and of engineering component failure. It is important to be able to analyse point processes and to make deductions about the nature of the system producing them.

The example shown here refers to point processes derived from the electrical waves (the electromyogram or EMG) generated as a human muscle contracts. However, the method may have application in other areas where the system produces signals consisting of component waves of a few, fixed, but unknown shapes which are obscured by mutual interference.

Spectral analysis of the point process is attractive because it is likely to reveal time locking between some of the events in the train which is not discernible by visual inspection. Spectral analysis of the point process requires that the process be regarded as stationary for the period of the analysis as defined in Cox and Miller (1965).

The most appropriate spectral description of a point process is its spectrum of intervals, or Bartlett spectrum, defined as the Fourier Transform of the covariance density function (Bartlett (1963)). It has been shown by Lago & Jones (1983) that for a sufficiently small sampling interval, the spectrum of a train of delta functions located at the nearest sample time to the occurrence times of the events approaches this Bartlett spectrum and, furthermore, that this is, for all practical terms, the same as the spectral estimate of the equivalent discrete 0-1 signal. In addition it has been shown (Lago & Jones (1983)) that, in practical cases, a significant decimation of the 0-1 data before the Fourier transformation takes place generates very little error.

A recommended procedure for estimating the spectrum of a point process has therefore been derived and is shown in Fig. 1.

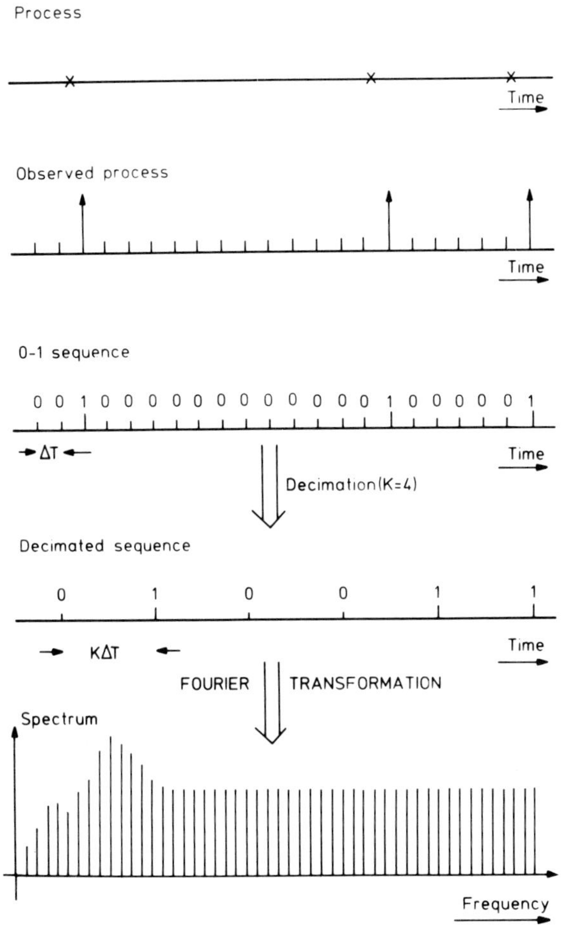

Fig. 1 Formation of the spectrum of a point process

The spectrum of any of these types of process tends to a constant at high frequency. The behaviour in the base and mid-band depends on the statistics of the point process and the degree of time dependence between the events.

Various methods have been proposed for extracting information on the time relationship between the events in the point process from this mid and base band region of the spectrum which depend on splitting or thinning the process into several new processes until the individual spectra are essentially flat (Jones & Lago (1983)).

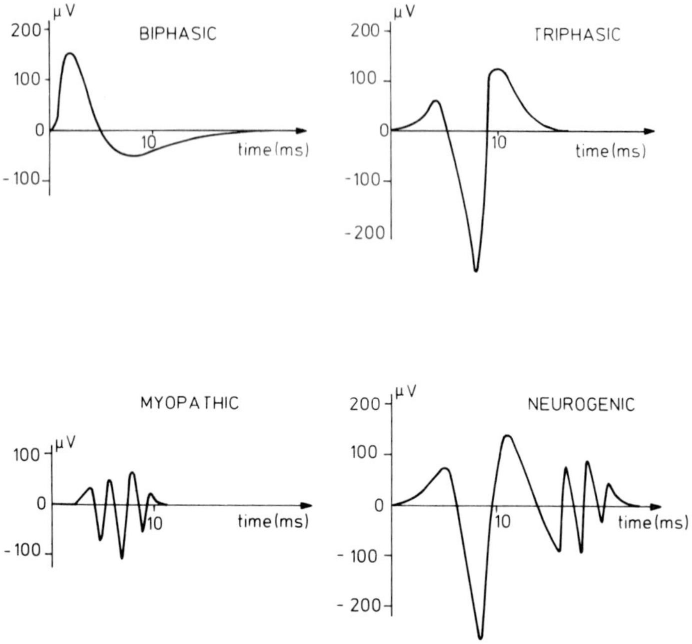

Fig. 2 Action potentials used in simulation

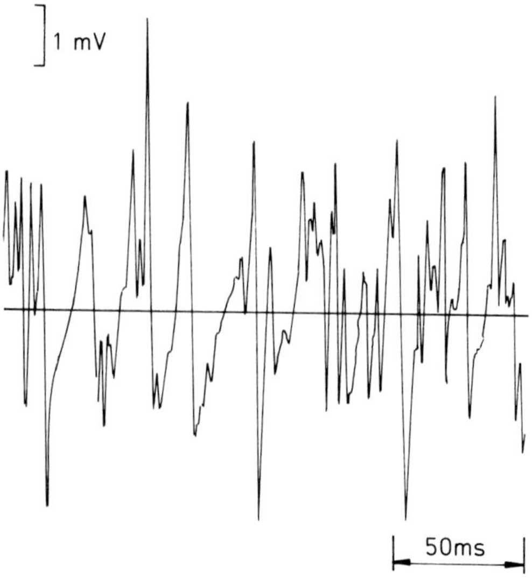

Fig. 3 Interference EMG

The effectiveness of these methods in practical situations is as yet not fully evaluated. However it is necessary to estimate when a spectrum is sufficiently flat, and also, in the present application, mixtures of component waves will produce point processes which will only partially whiten by these methods and complex component waves may require record lengths which will begin to weaken the assumption of stationarity.

The method of principal component analysis of the spectrum of the point process, proposed here as an alternative, can reliably discriminate between point processes derived from signals containing different component waves, and is more objective and perhaps more reliable than thinning or splitting.

1.2 The Electromyogram (EMG)

A muscle is made up of a large number of fibres which contract when producing force. The contraction of these fibres is elicited by the arrival of nerve impulses via the motoneurons. As a fibre contracts a wave of depolarisation travels along the membrane surrounding it. A group of fibres fed by a single motoneuron should contract at the same time and as a single unit. Under these circumstances a concentric needle electrode placed near to these fibres will detect the depolarisation currents and will usually record, in normals, a sequence of voltage waves of the biphasic or triphasic types as shown in fig. 2. These waves (known as motor unit action potentials) can be seen as separate waves when the contraction levels are low but they overlap and produce an interference pattern of the type shown in fig. 3 as the contraction level increases.

In certain myopathic and neurogenic disease states, for example, those fibres in a single unit which should all contract at the same time suffer from weakened and/or delayed contractions and, as a result, produce waves with multiple peaks, also shown in fig. 2. During strong contractions these waves will again also interfere and again produce signals similar to fig. 3 thus making it difficult to differentiate between cases.

The shape of the motor unit action potentials such as shown in fig. 6, are important indicators of disease and clinicians will try to isolate these action potentials for study [7] as described, for example, in Swash and Schwartz (1981). However, it is not always either possible or desirable to study the isolated action potentials alone and the problem is then to extract information from the interference EMG (fig. 7). The main reasons for studying the interference EMG are, firstly because it is easier to obtain, particularly in children, mentally handicapped and other classes of patient, and secondly because sometimes the diseased motor units only come into action at high force levels.

The property which distinguishes pathological motor unit action potentials, such as the myopathic or neurogenic types, from normal action potentials such as the biphasic or triphasic types, are the number of slope reversals or "phases" in the wave. The analysis described here not only attempts to distinguish between interference EMG signals arising from polyphasic as opposed to bi- or triphasic action potentials but also attempts to discriminate between interference signals arising in different disease states. In order to proceed with this analysis the interference EMG is first turned into a point process by marking all the significant slope reversals. The point process is then analysed according to the method described and justified by Lago & Jones (1983), to obtain a spectrum. This spectrum is then classified by principal component analysis.

Direct spectral analysis of the interference EMG is not very successful in discriminating between EMGs arising from different constituent action potentials because there are so many factors influencing the shape of the spectrum in the frequency band of interest as shown by Jones & Lago (1977) and Lago & Jones (1977). Turning point spectral analysis avoids these problems to some extent and is, moreover, less expensive in computer time.

1.3 Principal Component Analysis

If we have a set of random variables $x_1 \ldots\ldots x_p$ arising from spectrum measurements taken from some signal of interest, then N samples of these variables from N such signals can be expressed as an N x p data matrix. We can, from this, estimate a mean vector M and a covariance matrix C.

A set of new variables $y_1, y_2 \ldots\ldots$ can be defined made up of linear combinations of the measured variables $x_1 \ldots\ldots x_p$, such that y_1 contains the greatest sample variance, y_2 the next greatest and so on. These new variables are known as the principal components of the observations. It can be shown that the weights which should be assigned for the linear combination to be as required are the elements of the characteristic vectors of C. The vector associated with the largest root being allocated to y_1, those associated with the next largest to y_2 etc. as described by Morrison (1978). The characteristic vectors P_1, P_2 of the covariance matrix have the same dimensionality (p) as the original data and are known, in this context, as the principal component spectra. In effect the data vectors are being expressed as orthogonal series made up of the mean vector plus weighted sums of these principal component spectra. Elimination of the higher order principal component terms produce errors which are minimum in the mean square sense.

Any set of variables $x_1 \ldots\ldots x_p$ can now be represented by a combination $M + a_1 P_1 + a_2 P_2 \ldots$ where $a_1\ a_2 \ldots$ are principal component coefficients.

In this way it is possible to combine measurements in such a way that a large number of data points (say 256) could be reduced to a smaller number (say 2 or 3) of principal component coefficients which will still account for most of the variance in the data (50-70% in the cases studied here). This reduction in dimensionality has significant potential for extracting the features of interest from a signal. In this example the signal in question is the Turning Point Spectrum of an EMG.

Methods for computing the principal components from measured data are described by Morrison (1967). Applications to signal processing are described, for example, by Stevens & Fulton (1982).

In an example drawn from the present study, a group of 3 spectra are used as the data base. These produce the mean spectrum M and the first three principal component spectra P_1, P_2 and P_3. The appropriate principal component coefficients, a_1 and a_2, which minimised the mean square error in the representation of any particular spectrum by the expression $M + a_1 P_1 + a_2 P_2$ are plotted on orthogonal axes. It will be seen that, in this example, good separation is achieved between types which are basically normal and types which are pathological.

1.4 Simulated Interference EMG

Simulation of the interference EMG is very useful as a means of providing a controlled study of this method of analysis. The simulated EMG is obtained by generating a series of up to 20 gaussian renewal point processes and allocating to each point of each progress one of the motor unit action potentials shown in fig. 2. These are derived from models by Parker & Scott (1973) and by Marranza Figini et al (1981). Each of the resulting trains of waves has action potentials of the same amplitude and shape but they both may be different from those of other trains. The interval between action potentials in each train is a gaussian random variable with a standard deviation (in ms) equal to 3.5 plus the coefficient of variation multiplied by the mean (in ms). The trains of action potentials so produced are then summed to give the synthetic interference EMG. This procedure provides a good and simple fit to experimental data derived by Clamman (1969).

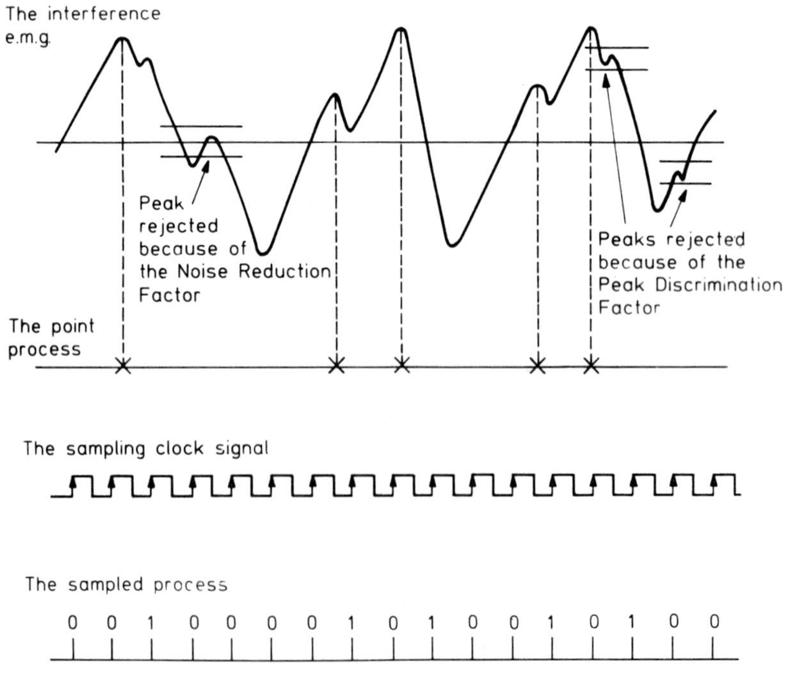

Fig. 4 Marking the peaks of the interference EMG.

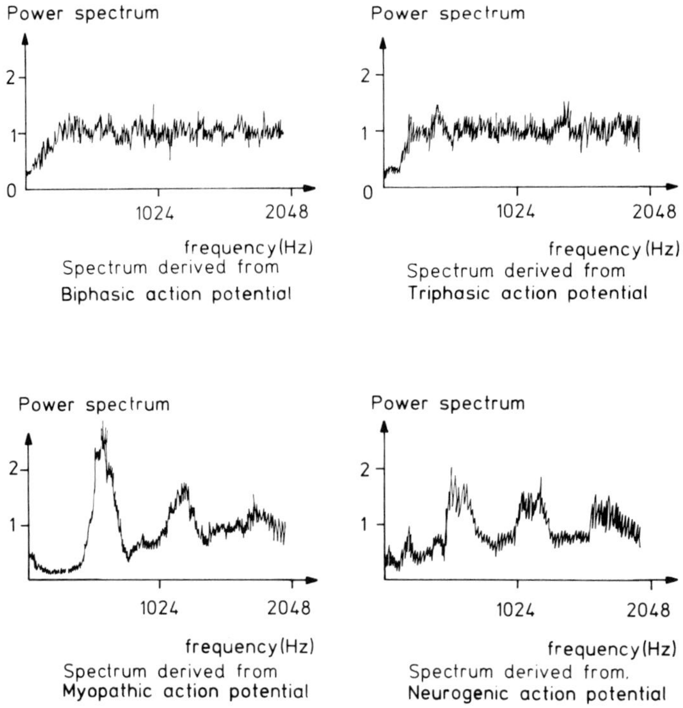

Fig. 5 Spectra of the turning point processes derived from EMGs made up of 100% of each of the action potentials shown in fig. 2.

1.5 Deriving a Point Process from the EMG

An interference EMG of the type shown in fig 3, on an expanded time scale, is similar to that shown in fig 4. A point process is derived from this by setting a marker (binary 1) at the next sampling point after the occurrence of a significant maximum. A significant maximum is defined as one which occurs beyond a certain distance (the Noise Reduction Factor) of the base line and is also more than another distance (the Peak Discrimination Factor) from either of the adjacent minima. These concepts are illustrated in fig 4.

2. EXPERIMENTS AND RESULTS

2.1 Standard Spectra and the training set

EMGs generated entirely of one or other of the action potentials shown in fig 2, when processed as described provide spectra as shown in fig 5. The notable feature of these spectra is the relative flatness and simplicity of those arising from normal (biphasic or triphasic) EMGs compared with those derived from the two pathological cases. This provides the user with a simple standard by which he may judge the need for further tests and/or processing.

One such example of further processing could be the method of principal component analysis proposed. For this a training set containing a variety of spectra ranging from totally normal through to totally neurogenic or myopathic is required. A noise reduction factor of 10μV and a peak discrimination factor of 50μV were used.

The training set used to generate these results was derived from 36 EMGs, each containing 20 action potential trains representing 2 or 3 different action potentials. The resulting mean spectrum and the first three principal component spectra are shown in fig 6. Three examples of each of the four pure cases were fitted, as described, with the mean and first two components shown in fig 6. The coefficients are plotted in fig 7. It can be seen that there is good clustering within classes and, importantly, good separation of the classes except for the near coincidence of two classes regarded as normal (the biphasic and triphasic).

2.2 The peak discrimination factor

Tests were then carried out to discover the effect of changing the peak discrimination factor on the classification of 100% normal and 100% pathological cases (fig 8) and on 50% normal and 50% myopathic cases (fig 9). For the remaining tests a peak discrimination factor of 50μV was used. The noise reduction factor was fixed at 10μV (the simulated EMGs are virtually noise free).

2.3 The number of motor units

Figs 10 and 11 show the effect of the number of motor units on the classification.

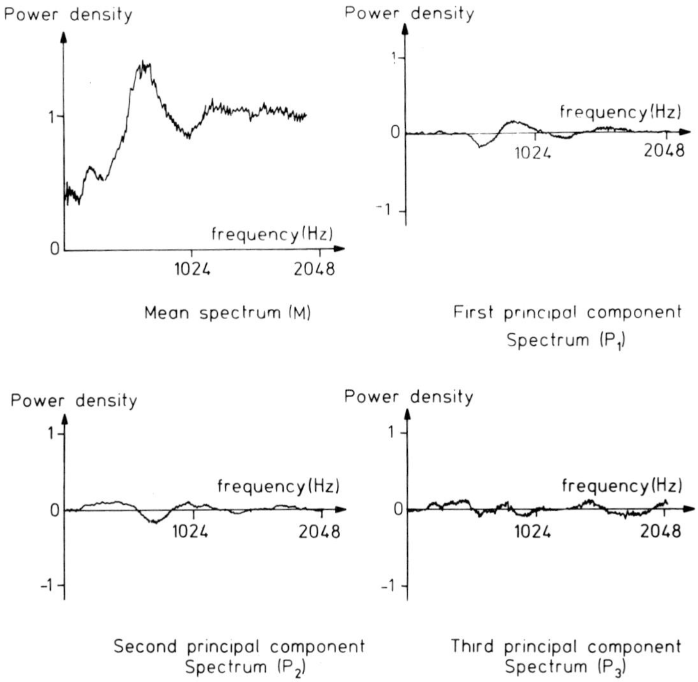

Fig. 6 Mean spectrum and first three principal component spectra of data base.

2.4 The progress of disorder

Figs 12 and 13 show the progress of a developing disorder from 100% normal to 100% pathological. It can be seen from these figs that a progressing myopathy can, in this two dimensional analysis, be mistaken for a neurogenic disorder.

Figs 14 and 15 indicate that the inclusion of a third principal component coefficient helps to resolve these ambiguities.

3. CONCLUSIONS

3.1 Conclusions based on Simulation Study

General

1. Spectral analysis is a powerful method for examining point processes.

2. Principal Component Analysis is an effective way of classifying the spectra of the turning point process of interference EMGs.

3. Principal component analysis of simulated interference EMGs in the cases tried allows successful separation of normals and cases of neurogenic and of myopathic disorders.

4. The use of two principal component coefficients can give adequate discrimination if extra evidence is available to separate neurogenic disorders from early myopathic disorders.

5. The use of three coefficients provides useful enhancement to the discrimination in difficult cases.

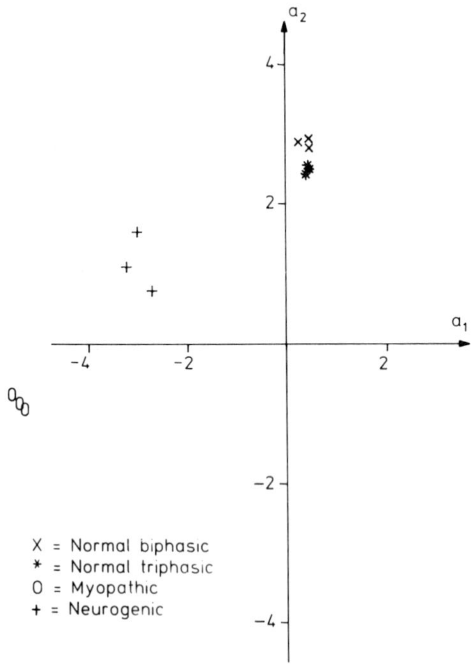

Fig. 7 Principal component coefficients of 100% of each standard type (three examples of each illustrated). These are repeated in the remaining figs.

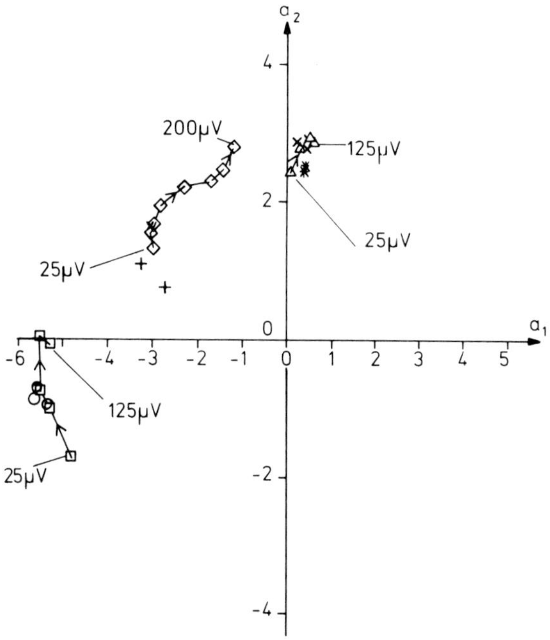

Fig. 8 Effect of peak discrimination factor. 100% of each standard type Δ = Normal □ = 50% Normal and 50% Myopathic.

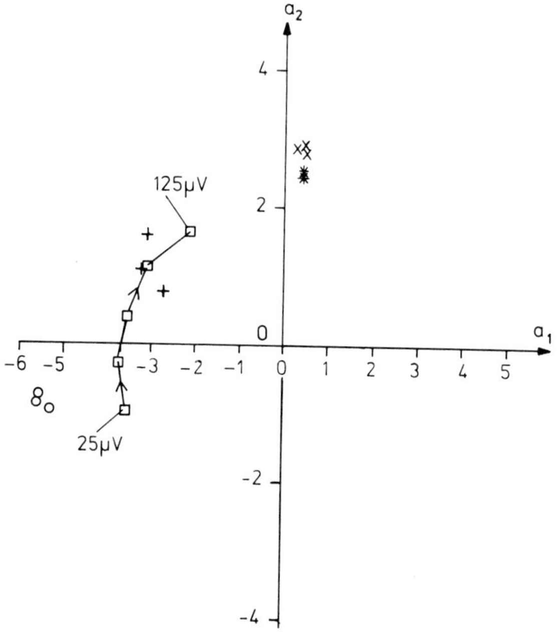

Fig. 9 Effect of peak discrimination factor 50% Myopathic units and 50% Normal units □ = 50% Normal and 50% Myopathic.

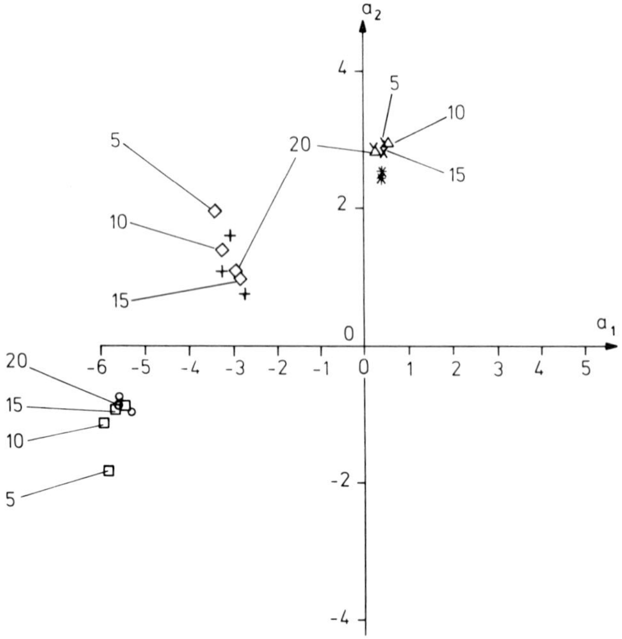

Fig. 10 Effect of number of motor units 100% of each standard type Δ = Normal, □ = Myopathic, ◇ = Neurogenic.

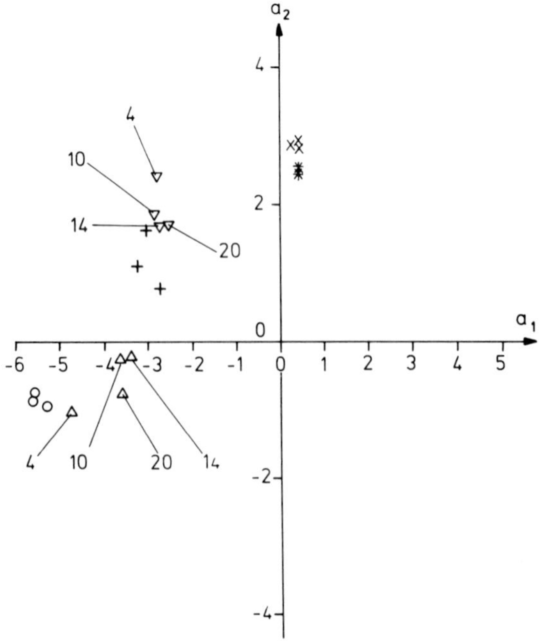

Fig. 11 Effect of number of motor units. 50% Normal and 50% Pathological Δ = 50% Normal, 50% Myopathic ∇ = 50% Normal, 50% Neurogenic.

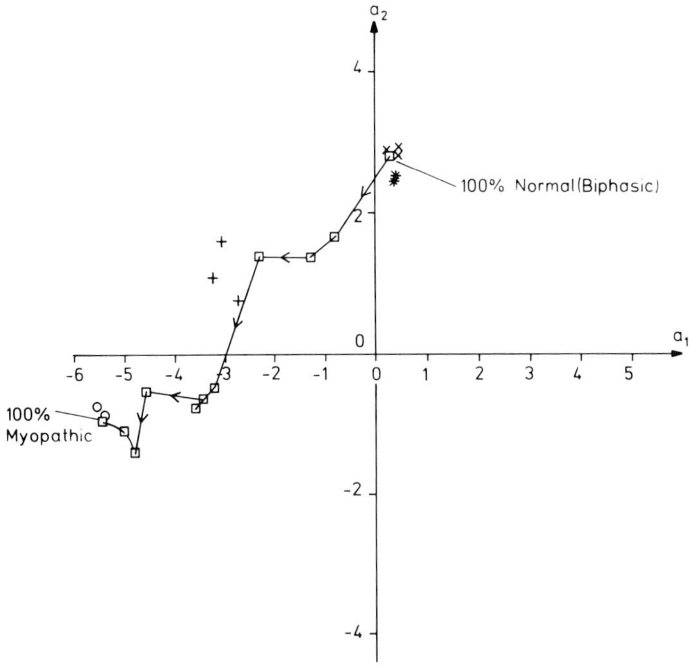

Fig. 12 Progress of myopathic disorder (in steps of 10%)

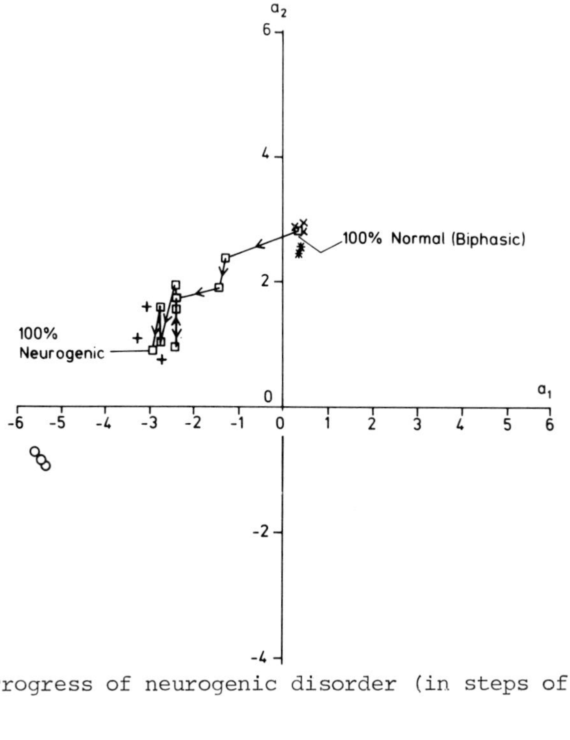

Fig. 13 Progress of neurogenic disorder (in steps of 10%).

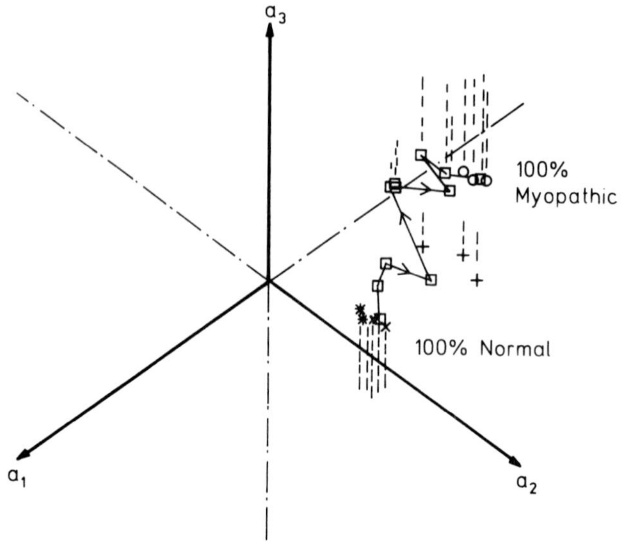

Fig. 14 Progress of myopathic disorder in three dimensions.

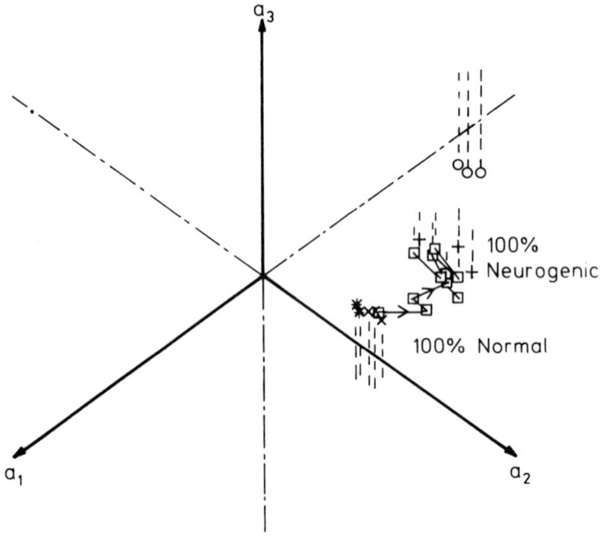

Fig. 15 Progress of neurogenic disorder in three dimensions.

Specific

1. The number of motor units active does not effect classification when the case is totally normal, totally neurogenic or totally myopathic.

2. Changing the coefficient of variation of the underlying gaussian renewal processes does not effect the classification of totally normal, totally neurogenic or totally myopathic EMGs (not illustrated here).

3. The choice of Peak Discrimination Factor has no effect on the classification of normal or of myopathic EMGs.

4. If the Peak Discrimination Factor is too large (greater than 50μV) then errors in classification can occur.

5. Mixtures of normal and myopathic EMGs with more than 10% and fewer than 40% abnormal units are usually classified as neurogenic in two dimensional analysis. EMGs with fewer than 10% abnormal units are classified as normal and those with more than 40% myopathic units are classified as myopathic.

6. Mixtures of normal and neurogenic EMGs with more than 10% abnormal units are correctly classified as neurogenic in two dimensional analysis. EMGs with fewer than 10% abnormal units are classified as normal.

7. The addition of a third principal component coefficient into the procedure allows the correct classification of all mixtures of normal and myopathic units when there are more than 20% abnormal units present.

3.2 Preliminary Conclusions Based on Real Data
(Not illustrated here)

1. The method has successfully separated one sample each from normals and cases of polymyositis, myotonia, dermatomyositis and limb girdle muscular dystrophy.

2. One case of motor neurone disease was classified as normal using two dimensional analysis but was separated by using three dimensional analysis.

3. Collection of the data from different muscles does not affect the results in the examples tested so far.

REFERENCES

Bartlett, M.S., (1963) "The Spectral Analysis of Point-Processes" *J. Roy. Statist. Soc.* B.25, 264-280.

Cox, D.R. and Mill r, H.D., (1965) "The Theory of Stochastic Processes" 356, Chapman and Hall, London.

Jones, N.B. and Lago, P.J., (1977) "Factors Affecting the Interpretation of EMG Spectra" IEE Conf. Pub. 159, 65-81.

Jones, N.B. and Lago, P.J., (1983) "Characterising the Interference EMG" Proc. Vth Annual Conference IEEE Engineering in Med. & Biol. Soc. 477-482.

Lago, P.J. and Jones, N.B., (1977) "The Effect of Motor Unit Firing Time Statistics on EMG Spectra" Med. & Biol. Eng. & Comput. 15, 643-655.

Lago, P.J. and Jones, N.B., (1982) "A Note on the Spectral Analysis of Neural Spike Trains". Med. & Biol. Eng. & Comput. 2o, 44-48.

Lago, P.J. and Jones, N.B., (1983) "Turning Points Spectral Analysis of the Interference Myoelectric Activity" Med. & Biol. Eng. & Comput. 21, 333-342.

Marranza Figini, M. and Fabbro, M., (1981) "A Simulation Model for the Study of EMG Signals in Normal & Pathological Conditions" Electroenceph. & Clin. Neurophysiol. 52, 778-781.

Morrison, D.F., (1978) "Multivariate Statistical Methods" McGraw-Hill. 2nd Edition 266-299.

Parker, P.A. & Scott, R.N., (1973) "Statistics of the Myoelectric Signal from Monopolar and Bipolar Electrodes" Med. & Biol. Eng. & Comput. 11, 591-596.

Stevens, A.L. and Fulton, T.J., (1982) "Digital Processing of Doppler Signals in Digital Signal Processing" (N.B. Jones Ed) 447-466. Peter Pereginus, London.

Swash, M. and Schwartz, M., (1981) "Neuro-muscular Diseases - a Practical Approach to Diagnosis & Management" Verlag Springer, Berlin.

III INVERSE PROBLEMS

THE USE OF SINGULAR SYSTEMS AND OTHER INVERSION METHODS IN WEIGHTED L^2 - SPACES

E.R. Pike
(Royal Signals and Radar Establishment, Malvern, UK)

M. Bertero
(Dipartimento di Fisica dell'Universita and Istituto Nazionale di Fisica Nucleare, Genova, Italy)

M. Defrise
(Theoretische Natuurkunde, Vrije Universiteit Brussel, Belgium)

C. De Mol
(Departement de Mathematique, Universite Libre de Bruxelles, Belgium)

ABSTRACT

We consider the object reconstruction problem from sampled data. The solution is supposed to be square integrable with respect to some weight function which reflects prior knowledge about the shape and localization of the unknown object. As shown by numerical examples, the use of a suitable weight function allows us to improve the reconstructions and to avoid undesirable edge effects. We describe various algorithms for computing regularized solutions in weighted L^2 spaces, such as the use of singular systems, successive approximations and the conjugate-gradient method. We then compare their performances in a particular case.

1. CONTINUOUS-DISCRETE INVERSE PROBLEMS IN WEIGHTED L^2-SPACES

Let us consider an inverse problem modelled by a Fredholm integral equation of the first kind:

$$g(t) = \int_{-\infty}^{+\infty} K(t,s)\ f(s)\ ds \qquad (1)$$

As usual in optical inverse problems, we will call "object" the function f(s) which is to be determined from its "image" g(t). We will assume that the image g(t) is measured in N given sampling points $t=t_n$, n=1,2,...,N, so that the data of the inverse problem are the components

$$g_n = g(t_n) \qquad n=1,2,\ldots,N \tag{2}$$

of a vector g. Hence we can take as data space a N-dimensional vector space Y_N with the following weighted scalar product

$$(g,h) = \sum_{m,n=1}^{N} w_{nm}\, g_n\, h_m^* \tag{3}$$

where the w_{nm}'s are the elements of a positive definite hermitian matrix $[W]$. The choice of that matrix depends on the particular application one has in mind. For instance, the choice $w_{nm} = \delta_{nm}$ corresponds to the usual l^2-norm representing the mean square error. Another choice used in linear regression theory is to take $[W] = [C]^{-1}$, where $[C]$ is the covariance matrix of the errors δg_n on the components of the data vector g. In other cases, it could also be convenient to choose the weights w_{nm} according to some quadrature formula, in order to approximate by (3) the scalar product in some functional space of data.

Although the data are discrete, we still assume that the unknown object f(s) is an element of a function space, i.e. we consider a "continuous-discrete" inverse problem. As also done in other reconstruction problems (e.g. in [1,2,3]), we will assume that the object space is a weighted L^2-space X, with scalar product

$$(f,h) = \int_{-\infty}^{+\infty} \frac{f(s)\, h^*(s)}{|P(s)|^2}\, ds \tag{4}$$

where P(s) is a given "profile" function which allows us to incorporate in the norm of the space X some prior knowledge about the localization and shape of the object. Indeed, if the norm of the solution f(s) is controlled in some way, as done in regularization theory, then the unknown function is forced to vanish outside the support of the profile function and to be small where P(s) is small.

We will now call A the linear operator from X into Y_N which maps the solution f onto the data vector g, i.e.

$$(Af)_n = \int_{-\infty}^{+\infty} K(t_n, s)\, f(s)\, ds \tag{5}$$

The adjoint operator A^*, acting from the space Y_N with the scalar product (3) into the space X with scalar product (4), is then given by

$$(A^* g)\ (s) = |P(s)|^2 \sum_{m,n=1}^{N} w_{mn}\ K^*(t_n,s)\ g_m \qquad (6)$$

Let us denote by $\{u_k, v_k; \alpha_k\}_{k=1}^{N'}$ $(N' \leqslant N)$ the singular system of operator A. We recall that the singular vectors v_k are the normalised eigenvectors of the self-adjoint operator AA^* which can be represented here by the NxM matrix $[AA^*] = [G]^t [W]^t$, where $[G]^t$ is the transposed of the so-called Gram matrix of elements

$$[G]_{nm} = \int_{-\infty}^{+\infty} K^*(t_n,s)\ |P(s)|^2 K(t_m,s)\ ds \qquad (7)$$

The corresponding N' non-zero eigenvalues of $[AA^*]$ are denoted by α_k^2 and the real and positive numbers α_k are called the singular values of A. They are supposed to be ordered in decreasing magnitude $(\alpha_1 \geqslant \alpha_2 \geqslant \ldots \geqslant \alpha_{N'} > 0)$ and each value is counted as many times as required by its multiplicity. The functions

$$u_k = \frac{1}{\alpha_k} A^*\ v_k \qquad k = 1,2,\ldots,N' \qquad (8)$$

called the singular functions, are eigenfunctions of the finite-rank integral operator A^*A and form an orthonormal basis in the orthogonal complement $(\mathrm{Ker}\ A)^{\perp}$ of the null-space of A. The properties of the singular system can be summarised by the following formulae $(k=1,2,\ldots,N')$

$$\begin{aligned} Au_k &= \alpha_k v_k & \qquad A^*Au_k &= \alpha_k^2\ u_k \\ A^*v_k &= \alpha_k u_k & \qquad AA^*v_k &= \alpha_k^2\ v_k \end{aligned} \qquad (9)$$

The inverse problem we consider consists in finding a solution to the equation

$$Af = g \qquad (10)$$

where $A:X \to Y_N$ is the operator defined hereabove. But solving equation (10) is not a well-posed problem: a solution does not necessarily exist for any data vector g. Furthermore, when a solution exists, in general it is not unique. In the absence of further information about the solution, it is then usual to look for the so-called generalised or pseudo-normal solution f^+ [4, 5]. Let us recall that a pseudo-solution (or least-squares solution) is a function f minimizing the quantity $\|Af - g\|_Y$; the pseudo-normal solution is then the pseudo-solution which has minimal norm. It is of course unique and can be expressed in the following way by means of the singular system of operator A. Since f^+ belongs to $(\mathrm{Ker}\ A)^{\perp}$, it can be expanded into the basis $\{u_k\}_{k=1}^{N'}$:

$$f^+ = \sum_{k=1}^{N'} f_k^+ u_k \tag{11}$$

where $f_k^+ = (f^+, u_k)$ are the corresponding "Fourier" coefficients. Putting (11) into (10) and identifying the coefficients, we find by using (9) that $f_k^+ = (g, v_k)/\alpha_k$. Hence we have

$$f^+ = \sum_{k=1}^{N'} \frac{(g, v_k)}{\alpha_k} u_k \tag{12}$$

which is the singular function expansion of the pseudo-normal solution.

2. THE USE OF SINGULAR SYSTEMS

Formula (12) provides a way of computing numerically an approximation of the pseudo-normal solution f^+. We first have to compute the singular system $\{u_k, v_k; \alpha_k\}_{k=1}^{N'}$ which can be stored once for all in a given practical situation (characterised by a given operator A). Then the solution f^+ corresponding to a given data vector is easily and quickly computed; indeed we just have to compute the scalar products (g, v_k) and to perform the summation on index k. However in doing this we must always keep in mind the problem of stability. When the norm $\|g\|$ of the data vector tends to zero, the norm $\|f^+\|$ also tends to zero. Therefore, strictly speaking, the solution f^+ is continuous in the data and its computation is a well-posed problem. Nevertheless, as it often happens in inverse problems, this can be an ill-conditioned problem, in which case the solution f^+ will be affected by numerical instabilities in the presence of noise on the data. It is not difficult to show that error propagation from the data into the pseudo-normal solution is controlled by the ratio $\alpha_{N'}/\alpha_1$ called the condition number. When this ratio is very large, the problem has to be regularized, i.e. the components in (12) corresponding to the smallest singular values should be damped or removed by means of some appropriate filtering procedure. For instance, as is well-known in regularization theory [4,6] , a stable solution is obtained simply by truncating the expansion (12)

$$f^+ = \sum_{k=1}^{N_o} \frac{(g\ ,v_k)}{\alpha_k} u_k \tag{13}$$

where the cut-off is chosen in order that, in the class of truncated solutions, the following functional

$$\Phi = \| Af^+ - g\|^2 + \mu \| f^+\|^2 \tag{14}$$

is minimum with respect to N_o. A possible choice for the regularization parameter μ is

$$\mu = (\varepsilon/E)^2 \tag{15}$$

where ε is an upper bound for the norm of the error (in the Y_N-space) whereas E is an upper bound for the norm of f^+ (in the X-space) [6,9] . The quantity $(E/\varepsilon)^2$ then appears as a "signal-to-noise ratio". One can easily show that N_o is the largest integer k such that $\alpha_k \geqslant \sqrt{\mu}$. The quantity N_o may be called the "number of degrees of freedom" since it represents the number of components in the solution which can be retrieved from the data with sufficient accuracy.

The computation of (13) only requires the determination of the eigenvectors v_k corresponding to the largest eigenvalues α_k^2 $(k=1,2,\ldots,N_o)$ of the matrix [AA*] . This can be done using standard numerical algorithms, provided that they are appropriate for computing the largest eigenvalues of ill-conditioned matrices, e.g. the power method combined with deflation [7,8] or Sturm sequence methods [8] . The corresponding eigenvectors v_k can be computed by means of the inverse iteration method [7,8] and then the singular functions u_k by means of formula (8).

3. ITERATIVE METHODS

An alternative to the computation of the singular system is provided by iterative methods. Indeed we may express the pseudo-normal solution as follows

$$f^+ = A^* h \qquad h \in Y_N \tag{16}$$

so that we have to solve for h the equation

$$AA^* h = g \tag{17}$$

reducing by means of the matrix representation of AA^* to a NxN algebraic system which can be solved iteratively. We will assume in this section that the operation AA^* has no zero eigenvalue, so that (17) admits a solution for any data vector g.

A first very simple iterative procedure is the method of successive approximations [7]:

$$h_{i+1} = h_i - \tau r_i \tag{18}$$

where τ is a relaxation parameter used for improving convergence and r_i is the residue at the ith iteration step

$$r_i = [AA^*] \, h_i - g \tag{19}$$

When the initial guess is $h_o = \tau g$, the ith iteration for the vector h is also given by

$$h_i = R_i \, g \tag{20}$$

with

$$R_i = \tau \sum_{k=0}^{i} (1 - \tau [AA^*])^k \tag{21}$$

The method converges when the norm $\| [AA^*] \|$ is strictly inferior to $2/\tau$.

Another iterative method which can be used when the matrix $[AA^*]$ is hermitian is the steepest-descent algorithm [7]

$$h_{i+1} = h_i - \tau_i \, r_i \tag{22}$$

where

$$\tau_i = \frac{(r_i, r_i)}{(r_i, [AA^*] r_i)} \tag{23}$$

Also in this case, a third possibility is the conjugate-gradient method [7], the first step of which is identical with the first step in steepest descent:

$$h_1 = h_0 - \tau_0 p_0 \tag{24}$$

with $p_0 = r_0$ and $\tau_0 = (r_0, r_0)/(p_0, [AA^*] p_0)$, whereas the next steps are given by the following formulae:

$$h_{i+1} = h_i - \tau_i p_i \qquad (i=1,2,\ldots)$$

$$r_i = r_{i-1} - \tau_{i-1} [AA^*] p_{i-1}$$

$$\tau_{i-1} = \frac{(r_{i-1}, p_{i-1})}{(p_{i-1}, [AA^*] p_{i-1})} \tag{25}$$

$$p_i = r_i + \beta_{i-1} p_{i-1}$$

$$\beta_{i-1} = - \frac{(r_i, [AA^*] p_{i-1})}{(p_{i-1}, [AA^*] p_{i-1})}$$

When the system (17) is ill-conditioned, the iterative process will lead to unstable solutions with respect to noise on the data. Some regularizing procedure has then to be applied. Instead of modifying equation (17), we can just stop the iterative process at a suitably chosen number of iterations I, before instabilities would occur. The number 1/I can indeed be shown to play the role of a regularization parameter [10,4]. For choosing I, one can apply the so-called "residual principle" [11], i.e. one can stop the iterative process when the norm of the residue r_i, given by (19), becomes smaller than ε, the norm of the error on the data. This criterion appeared to be quite satisfactory for the practical examples we considered.

4. NUMERICAL EXAMPLES

We have tested the preceding methods on a particular problem, for which the kernel of the integral equation (1) is given by

$$K(t,s) = \frac{\sin \Omega\,(t-s)}{\Pi(t-s)} \quad \text{with } \Omega = \Pi \tag{26}$$

The function g(t) represents then the image of an object f(s) through an ideal diffraction-limited system (its transfer function being the characteristic function of the frequency interval (-1/2,+1/2)). The corresponding classical resolution distance R, i.e. the Rayleigh limit, is then given by R = 1. The inverse problem of object reconstruction is also called the superresolution problem if one pretends to resolve characteristic features of the object on a smaller scale than the Rayleigh distance R [6,12] .

Figure 1 shows reconstructions of the following object (solid curve)

$$f(s) = \exp\left(-\frac{(s+s_0)^2}{(0.3)^2}\right) + \frac{1}{2}\exp\left(-\frac{(s-s_0)^2}{(0.4)^2}\right) \tag{27}$$

where $2s_0=0.75$, i.e. two gaussian peaks separated by $\frac{3}{4}$ R. The corresponding image (dotted curve) has been computed numerically and sampled, according to the Shannon theorem, in N equidistant points, symmetrically placed around the origin and separated by the Nyquist or Rayleigh distance R. In order to approximate by (3) the scalar product of two band-limited image functions, we have put in this section $w_{nm} = \delta_{nm}$ (see [12]). The two reconstructions we present were obtained by means of formula (13) and the singular system was computed along the lines described in section 2. We started from N=21 data points and the cut-off $N_0=6$ was chosen in order to take into account all singular values greater than 10^{-3}; no noise was added to the data. We assumed a priori that the support of the object was of the same order of magnitude as the Rayleigh distance and the profile function was chosen accordingly. The curve with crosses represents the reconstructed object when the profile function is the characteristic function of the interval [-1,+1] :

$$P(s) \begin{cases} = 1 & \text{for } |s| \leq 1 \\ = 0 & \text{for } |s| > 1 \end{cases} \tag{28}$$

Let us remark that in this case, the singular function u_k corresponding to the largest singular values are close approximations to the so-called prolate spheroidal wave functions which occur in the case of continuous data (see [12]). On the other hand, the dashed curve represents the reconstruction obtained with a gaussian profile of width equal to the Rayleigh distance:

$$P(s) = \exp\left(-\frac{s^2}{2\sigma^2}\right) \tag{29}$$

with $\sigma = 0.5$. Figure 1 clearly demonstrates the improvement obtained by the use of a smooth profile, which allows in particular to avoid edge effects.

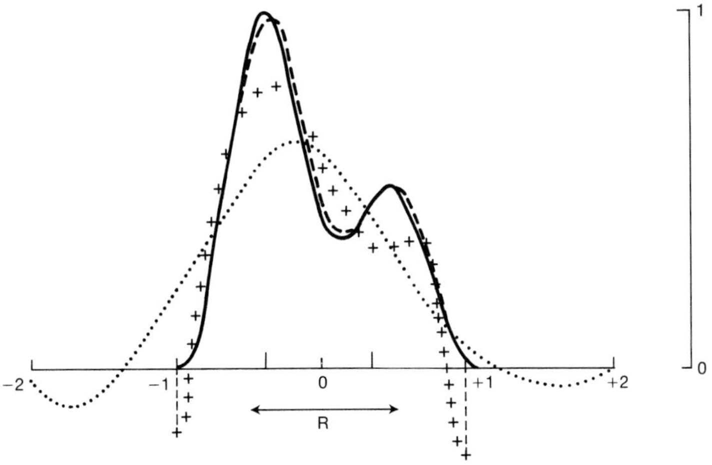

Fig. 1

In order to investigate the possibilities of superresolution, we have also considered the reconstruction of two delta peaks.* Their image, given by

$$g(t) = \frac{\sin \Pi(t-t_0)}{\Pi(t-t_0)} + \frac{\sin \Pi(t+t_0)}{\Pi(t+t_0)} \tag{30}$$

---

* Similar reconstructions for the case of Laplace transform inversion are presented in [13] .

is represented by the dotted curve in Figure 2, for $2t_o$=0.75, i.e. for a separation distance between the peaks of $\frac{3}{4}$ R. The dashed curve shows the reconstruction obtained when using the gaussian profile (29), again with σ = 0.5. This reconstruction was obtained by means of formula (13), with N_o=6 (i.e. summing on all singular values greater than 10^{-3}) and was not affected in a significant way when adding gaussian noise on the data up to 5% and choosing N_o accordingly. We should now stress the fact that in order to obtain superresolution, the choice of the width of the profile function, based on prior knowledge, is essential. Indeed, with a broader gaussian profile it was impossible to resolve two delta peaks separated by 3/4 of the Rayleigh distance. With σ = 0.5, we were able to resolve the two peaks as long as they were separated by more than $\frac{2}{3}$ R. In order to supperresolve beyond that limit, we would have to take a sharper profile, assuming that we know a priori that the unknown object is really concentrated on a smaller support. For instance, two delta peaks separated by 0.5 R were resolved with σ = 0.25. Although the width of the profile is a critical parameter, we have checked that its particular shape, provided that it is smooth enough, is not very important.

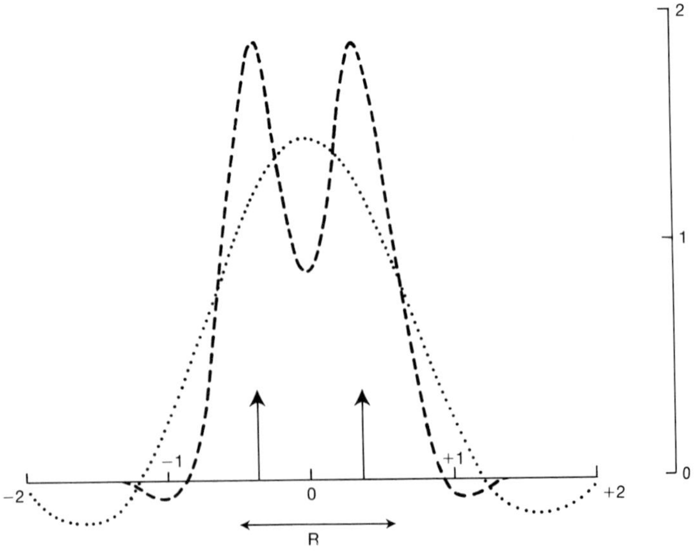

Fig. 2

The figures 1 and 2 were obtained using 21 data points and this is in fact a situation which is very close to a continuous-data inverse problem. One expects however that a much smaller number of data points, roughly of the order of the number of degrees of freedom N_o, would be sufficient to obtain equivalent reconstructions. As discussed in [12, 13], this is indeed true and with only 7 data points, we could reproduce the reconstruction of Figure 2. We were still able to resolve the peaks with N=5.

The iterative methods described in section 3 also allowed us to obtain very similar reconstructions. For example, in exactly the same situation, we got with the successive approximation method (18) ($\tau = 1.9$) a reconstruction which on the graphical scale used cannot be distinguished from the dashed curve of Figure 2. However 10 000 iterations were needed in order to recuperate the same number of singular values. An almost identical reconstruction was also obtained with the conjugate-gradient algorithm (25), but only in 3 steps! The quality of the reconstructions being of the same order, we should then compare the different methods in terms of computing time. We give here a rough idea of the computing times needed on a fast and big machine. The computation of the singular system, up to the singular values equal to 10^{-3}, required about 50 seconds when using a 21 x 21 matrix and about 6 seconds for a 7 x 7 matrix. Once the singular system was pre-stored for a given application, one reconstruction was obtained in less than 1 second, a time which is comparable to the time needed for the conjugate-gradient algorithm (3 iterations, with N=21). The successive approximations methods is however much slower because of its poor rate of convergence, so that about 45 seconds were needed for the 10 000 iterations with a 21 x 21 matrix. We also tried the Gerchberg algorithm where the FFT algorithm is used back and forth between the object space and its conjugate space, at each iteration step [14, 15]. Again when modifying the original algorithm in order to incorporate a profile function, the reconstructions obtained were of similar quality but the method is still more time consuming (about 100 times). An advantage of the latter method is that a positivity constraint could be easily incorporated at each iteration step. In the examples considered here however, such a constraint did not appear to be very helpful since the reconstructions without positivity constraint have already very small negative lobes. From the previous examples we can conclude that among the iterative methods considered here, only the conjugate-gradient algorithm can compete with singular function expansions. However this conclusion concerns only problems characterised by a small number of degrees freedom and should not be extrapolated to inverse problems involving very large systems and numbers of degrees of freedom.

ACKNOWLEDGEMENTS

The support of the NATO Grant for international collaboration in research N.463/84 is gratefully acknowledged. M. Defrise is in receipt of a Fellowship of the Interuniversity Institute for Nuclear Sciences and C. De Mol is Research Associate with the National Fund for Scientific Research (Belgium).

REFERENCES

[1] Davison, M., and Grünbaum, F.A., (1979) *IEEE Trans. Sci.* NS-26, 2670-73.

[2] Byrne, C.L., and Fitzgerald, R.M., (1982) *SIAM J. Appl. Math.* **42**, 933-940.

[3] Hall, T.J., Darling, A.M., and Fiddy, M., (1982) *Opt. Letters* 7, 467-468.

[4] Nashed, M.Z., (1981) *IEEE Trans. Prop. AP-29,* 220-231.

[5] Groetsch, C.W., (1977) "Generalized Inverses of Linear Operators", Dekker, M.

[6] Bertero, M., De Mol, C., and Viano, G.A., (1980) The Stability of Inverse Problems" in "Inverse Scattering Problems in Optics" (H.P. Baltes, ed.), Springer.

[7] Faddeev, D.K., and Faddeeva, V.N., (1963) "Computational Methods of Linear Algebra", Freeman, W.H.

[8] Jennings, A., (1977) "Matrix Computations for Engineers and Scientists" Wiley.

[9] Miller, K., (1970) *SIAM J. Math. Anal.* **1**, 52-74.

[10] Bakushinskii, A.B., (1967) *USSR Comput. Math. and Phys.* **7**, 279-287.

[11] Morozov, V.A., (1984) "Methods for Solving Incorrectly Posed Problems" Springer.

[12] Bertero, M., Brianzi, P., Parker, P., and Pike, E.R., (1984) *Optica Acta* **31**, 181-201.

[13] Bertero, M., Brianzi, P., and Pike, E.R., (1985) Inverse Problems **1**, 1-15.

[14] Gerchburg, R.W., (1974) *Optica Acta* **30**, 709-20.

[15] Abbiss, J.B., Defrise, M., De Mol, C. and Dhadwal, H.S., (1983) *JOSA* **73**, 1470-75.

J-INNER SYSTEMS AND STATE SPACE DESCRIPTIONS

C.V.K. Prabhakara Rao
(Department of ECE, Drexel University, Philadelphia, USA)

and

P. Dewilde
*(Department of Electrical Engineering,
Delft University of Technology, The Netherlands)*

ABSTRACT

The study of J-lossless transfer functions forms the starting point for the lossless modelling (or synthesis) theory referred to as the LIS (Lossless Inverse Scattering) Theory. Recursive Schur algorithm leads to practical schemes to solve the LIS problem. This unifies a number of engineering problems such as, Orthogonal Digital Filter synthesis, Linear Prediction and Stochastic Modelling of processes, and Seismic and Acoustic Inversion. On the mathematical side of the problem, the LIS solution can be viewed as solving NP (Nevanlinna Pick) interpolation problem or generating certain generalized orthogonal functions.

Encouraged by this varied use of J-inner systems theory, we began to examine the structure of the state space of J-inner systems and have arrived at some interesting characterizations. As a consequence, we are provided with a powerful approach (using Reproducing Kernel Hilbert Space concepts) to solve the multiport LIS problem.

We considered a strictly stable J-inner system $\theta(z)$ and characterized the null space and the input/output state spaces in the classical Nerode sense. It turns out that, although inner product with respect to the J-signature matrix is an indefinite metric, functions in the state space will have nonnegative J-norms. Further, the Reproducing Kernel has a striking form, which is later used to solve the inverse synthesis problem.

Moreover, we can show that the state space is completely specified by the transmission zeros and a related scattering function of $\theta(z)$ only. Using now properties of NP interpolation, we can construct the state space directly from the given data to be interpolated. With the help of the formulae for the Reproducing Kernel or by going through an abstract realization procedure, we can obtain a global description of the LIS solution $\theta(z)$.

Also, recursive orthonormalization procedures are derived (imitating Schur algorithm) for the LIS solution.

1. INTRODUCTION AND HISTORY

Consider the system shown below with inputs $a_1(t)$ and $b_2(t)$, called incident waves, and outputs $a_2(t)$ and $b_1(t)$, called reflected waves.

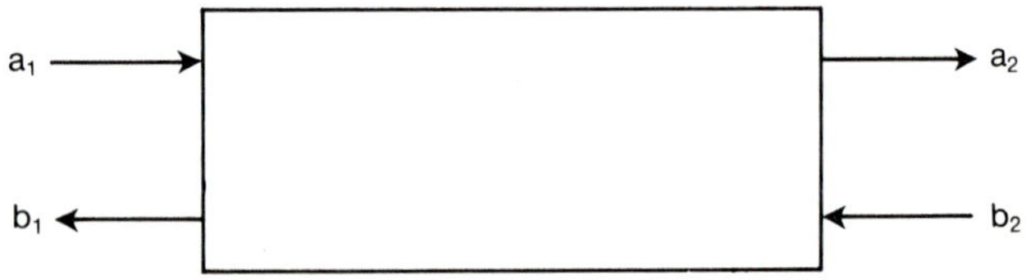

Fig. 1 Incident and Reflected Waves from the System

In general, the right going waves (a_1,a_2) and the left going waves (b_1,b_2) are pxl and qxl vector valued time series. The system is said to be lossless if the l_2-norms (energies) of the incident and reflected waves are preserved; ie,

$$\sum_{-\infty}^{\infty} (a_1^*(t)a_1(t)+b_2^*(t)b_2(t)) \tag{1}$$

$$= \sum_{-\infty}^{\infty} (a_2^*(t)a_2(t)+b_1^*(t)b_1(t))$$

where * indicates conjugate transpose. Moreover, if the system is causal,

$$\sum_{-\infty}^{\tau} (a_1^*(t)a_1(t)+b_2^*(t)b_2(t)) \geq \tag{2}$$

$$\sum_{-\infty}^{\tau} (a_2^*(t)a_2(t)+b_1^*(t)b_1(t)) \quad \text{for all } \tau$$

Define the z-transform of a time series x(t) formally as,

$$X(z) = \sum_t x(t) z^t \tag{3}$$

and let the system in fig. 1 be described by the so-called chain scattering matrix, $\theta(z)$, as:

$$\begin{bmatrix} A_2(z) \\ \\ B_2(z) \end{bmatrix} = \theta(z) \begin{bmatrix} A_1(z) \\ \\ B_1(z) \end{bmatrix} \tag{4}$$

The losslessness property (1) becomes the J-losslessness of $\theta(z)$:

$$J - \theta(z) J \theta^*(z) \geq 0 \quad z \in D = (z : |z| < 1) \tag{5}$$

$$= 0 \quad z \in T = (z : |z| = 1)$$

with $J = I_p \oplus I_{-q}$. We shall call such systems J-lossless systems.

J-lossless systems are being studied with increasing interest, because of their ubiquitous role in a number of applications. For instance, their role in classical network theory can be seen in the works of Youla [1] while a summarizing description can be found in Efimov and Potapov [2]. More recently, the theory and synthesis of J-lossless systems have been used in the design of low sensitivity digital filters, now called orthogonal filters [3,19]. Further, there is an important connection between such systems and stochastic estimation, as explored by a number of researchers [4,5,6]. (Perhaps not) Lastly, lossless modelling is being applied in seismic and acoustic inversion problems, since this leads to robust numerical algorithms [7,8].

The first comprehensive work dealing with the J-lossless transfer functions and their properties is due to Potapov [9] who described the multiplicative structure (or factorisation) of these systems. Further work extended the results to J-lossless operator functions, eg [10]. Since state space theories give more insight into the system dynamics, research effort is now being put in this direction. Questions regarding realization of J-lossless systems, their connection to Lyapunov theory, and to orthogonal polynomials have been studied in [11,12,13]. In a more mathematical setting the results in [14,15,16] lie in this direction too.

2. OUTLINE OF PAPER

In Section 3, the state is defined in the Nerode sense with the novelty being that we split the input space into null and state spaces with respect to J-metric, instead of the usual Euclidean metric. This enables the extension of the various results of the lossless systems to the J-lossless case, as demonstrated in Section 4. A noteworthy feature is that the state space turns out to be a Reproducing Kernel Hilbert Space (RKHS) with respect to the indefinite J-metric, with a special form for the reproducing kernel. Further, it follows that the state space is completely determined by the zeros of $\theta(z)$ and the values of the associated scattering function $S(z)$ at these points. See Section 5. This suggests a close connection to the interpolation theory. Indeed, the Lossless Inverse Scattering problem (which essentially solves the Nevanlinna - Pick interpolation problem) can be stated and solved in the state space language. See Section 6.

3. STATE SPACE IN THE NERODE SENSE

In out study of the state space of a 2x2 rational J-lossless system $\theta(z)$, we follow the Nerode approach [16]. Let K_2 denote the space of 2x1 vector valued power series in $\{z^{-1},z^{-2},\ldots\}$ with finite l_2-norm, and H_2 denote the space of 2x1 vector valued power series in $\{1,z,z^2,\ldots\}$ with finite l_2-norm. In time domain K_2 and H_2 denote spaces of finite energy signals in $t<0$ and $t\geq 0$ resp. Consider the following Hankel operator built on $\theta(z)$ and which maps K_2 into H_2 as:

$$\Gamma = P\theta|_{K_2} \tag{6}$$

where P denotes orthogonal projection onto H_2 with respect to the usual $L_2(d\theta)$ metric.

The kernel of Γ is called the nullspace N of $\theta(z)$, and has the meaning that inputs $u(z)$ in $N(\subset K_2)$, bring the system to the zero state at $t=0$, and produce zero output for $t\geq 0$. The range of $\Gamma(\subset H_2)$ is called the "output state space" H^o of $\theta(z)$ (note! this usage is not standard).

It has the meaning that every element in H^o represents a possible state of the system $\theta(z)$. We can conveniently define an "input state space" H to be the J-orthogonal complement of

N in K_2, ie.

$$H = K_2 \ominus JN \tag{7}$$

The unique representative of a state of $\theta(z)$ in H is now a (Nerode) equivalence class, defined as $f + N$ where $f \in H$. The meaning is that all inputs in such an equivalent class bring the system to the same state at t=0, and hence produce the same output for $t \geq 0$.

4. STATE SPACE of $\theta(z)$

Lemma 1: The null space N, the input state space H, and the output state space H^o of $\theta(z)$ are given by,

$$N = \theta^{-1} K_2 \tag{8}$$

$$H = K_2 \ominus JN \tag{9}$$

$$H^o = \theta H = H_2 \ominus J\theta H_2 \tag{10}$$

Note that if J=1, $\theta(z)$ is simply a scalar inner function for which these properties are easily checked.

Remark 1: From (10, we observe that if an input $u(z) \in H$, the output $y(z) = \theta(z).u(z)$ lies completely in H_2. This property is specially due to the J-losslessness of $\theta(z)$.

Define now an indefinite (J-) inner product on H (H^o) as follows:

for $f,q \in H$ (or H^o)

$$[f,g] := \langle Jf,g \rangle = \int_{-\pi}^{\pi} g^*(e^{i\theta}) Jf(e^{i\theta})\, d\theta/2\pi \tag{11}$$

Lemma 2: The state space H (or H^o) is J-positive, ie. for $f \in H$ (or H^o),

$$[f,f] \geq 0 \tag{12}$$

This property follows from the J-losslessness condition (1), causality (2) and remark (1). Thus the state space H (or H^o) is a Hilbert space with respect to the J-innerproduct and further (certainly for finite dimensional systems) is a Reproducing Kernel Hilbert Space (RKHS).

Lemma 3: The reproducing kernels of H and H^o are given by

$$H: \quad K_\omega(z) = (J-\theta^{-1}(z)J\theta^{-1}(\omega)^*)/(-1+\omega^* z) \quad \omega\in D \tag{13}$$

$$H^o: \quad K^o_\omega(z) = (J-\theta(z)J\theta(\omega))/(1-\omega^* z) \quad \omega\in D \tag{14}$$

We mean that the columns of K_ω and K^o_ω belong to H, H^o resp. and that for $f \in H$ (or H^o), the reproducing property holds:

$$[f, K_\omega \xi] = \xi^* f(\omega) \tag{15}$$

$$[f, K^o_\omega \xi] = \xi^* f(\omega)$$

where ξ is any 2x1 vector. Observe that the reproducing kernel completely determines $\theta(z)$ up to a normalisation (say $\theta(\alpha)=I$, for some $\alpha\in T$), as

$$\theta^{-1}(z) = I+(1-\alpha^* z)K_\alpha(z)J \tag{16.a}$$

$$\theta(z) = I-(1-\alpha^* z)K^o_\alpha(z)J \tag{16.b}$$

5. A SIMPLE DESCRIPTION OF THE STATE SPACE:

It is known [5] that every J-lossless system $\theta(z)$ defines a scattering function $S(z)$ and a complementary scattering function $S_c(z)$ as,

$$S(z) = -\theta_{21}\theta_{22}^{-1} \tag{17.a}$$

$$S_c(z) = -\theta_{22}^{-1}\theta_{12} \tag{17.b}$$

ie, S and S_c are analytic and bounded in magnitude by 1 in D. We will now obtain a simple description of the state space H and H^o in terms of the zeros $\{\omega_j\}$ and S, S_c associated with $\theta(z)$.

Consider the input state space H and let the zeros $\{\omega_j\}$ be distinct for the time being. Let

$$f_j(z) = \frac{\begin{bmatrix} 1 \\ S(\omega_j) \end{bmatrix}}{(z-\omega_j)} \tag{18}$$

It is seen that $f_j \in K_2$ and direct evaluation reveals that

$$\langle J\, f_j, \theta^{-1} g\rangle = 0, \quad \text{for } g \in K_2 \tag{19}$$

Thus we have

$$H = \text{span}\{f_j, j=1,2,\ldots\} \tag{20}$$

In general, when the points are not necessarily distinct, we can show that H can be expressed compactly as follows:

Define the space M based on the zeros of $\theta(z)$ as:

$$M = \text{span}\left\{\frac{1}{(z-\omega_j)^i} : j=1,2,\ldots,\ i=1,2,\ldots v_j\right\} \tag{21}$$

Note that higher powers appear when zeros are not distinct. Then H can be written as:

$$H = \left\{\begin{bmatrix} f \\ QSf \end{bmatrix} : f \in M\right\} \tag{22}$$

Q denotes orthogonal projection on K_2. (Observe that $f \in K_2$; while $S \in H^\infty$, so that this projection is not trivial). This formula can be easily verified to give (20) for the distinct points case. As an example of the coalescent point case, if all the zeros are at z=0, we obtain from (22),

$$f_j = \frac{1}{z^{(j+1)}},\ QS.f_j = \frac{s_0+s_1 z+\ldots+s_j z^j}{z^{(j+1)}}\ j=1,2\ldots \tag{23}$$

In a similar way H^O can be expressed compactly in terms of the zeros $\{\omega_j\}$ and the complementary scattering function $S_c(z)$.

6. APPLICATION TO LIS-PROBLEM

The so-called Lossless Inverse Scattering (LIS) problem, we are given S(z) and are required to find a J-lossless $\theta(z)$ and a scattering function $S_L(z)$ such that

$$S = -(\theta_{21} - \theta_{11}S_L)/(\theta_{22} - \theta_{12}S_L) \tag{24}$$

This means that with respect to fig. 1

$$B_1(z) = S(z)A_1(z) \quad \text{when } B_2(z) = S_L(z)A_2(z)$$

Let us require the LIS solution to have zeros (in the Smith - McMillan sense) at $\{\omega_j, j=1\ldots n\} \in D$.

Define the space

$$H = \text{span}\left\{\frac{u_j}{(z-\omega_j)}\right\}, \quad u_j = \begin{bmatrix} 1 \\ S(\omega_j) \end{bmatrix} \quad j=1\ldots n \tag{25}$$

with respect to the J-inner product defined in (11). The Grammian of H with respect to the basis vectors $\{f_j\}$ is nothing but the Pick interpolation matrix, which by assumption (that $S(z)$ is a scattering function) is nonnegative definite.

So H is a genuine RKHS with respect to the J-metric. We can construct the LIS-solution $\theta(z)$, up to a normalization, in 3 different ways.

6.1 Algorithm 1

This consists in solving for a realization $\{A,B,C,D\}$ such that

$$\theta(z) = D + zC(I-zA)^{-1}B \tag{26}$$

whose state space is H. Let the operators A and B be:

$$A = \Pi_H z^{-1}|_H, \quad B = \Pi_H 1 \tag{27}$$

Evaluate A,B as matrices choosing a set of J-orthonormal basis, which gives

$$A\,A^* + B\,J\,B^* = I \tag{28}$$

Compute C,D as:

$$C = J.B^*.(I-A^*)^{-1}.(I-A) \tag{29.a}$$

$$D = I - C.(I-A)^{-1}.B \tag{29.b}$$

$\{A,B,C,D\}$ form a J-orthogonal realization for a J-lossless $\theta(z)$ such that $\theta(1)=I$, ie

$$\begin{bmatrix} A & B \\ C & D \end{bmatrix} \begin{bmatrix} I & O \\ O & J \end{bmatrix} \begin{bmatrix} A & B \\ C & D \end{bmatrix}^* = \begin{bmatrix} I & O \\ O & J \end{bmatrix} \qquad (30)$$

See (12) related discussion on J-orthogonal realizations.

6.2 *Algorithm 2*

Compute the reproducing kernel of H as:

$$K_\omega(z) = \sum_k \psi_k(z)\psi_k(\omega)^* \qquad (31)$$

where ψ_k are J-orthonormal basis of H.

It is known [15] that $K_\omega(z)$ has the form

$$K_\omega(z) = \frac{J-\theta^{-1}(z)J\theta^{-1}(\omega)^*}{-1+\omega^* z} \qquad (32)$$

from which $\theta(z)$ can be solved up to a normalization $\theta(1) = I$, as

$$\theta^{-1}(z) = I + (1-z)K_1(z)J \qquad (33)$$

6.3 *Algorithm 3*

This is a recursive algorithm and consists of J-orthogonalizing the basis vectors $f_j(z)$ successively into $\psi_j(z)$. Essentially, this is a Schur procedure.

Start with $\psi_1 = f_1$.

ψ_2 is obtained as

$$\psi_2 = \theta_1^{-1}\frac{V_2}{(z-\omega_2} \qquad (34)$$

where

$$\theta_1 = I + (\xi_{\omega_1}-1)\ V_1V_1^*J \qquad \xi_{\omega_1} = \frac{z-\omega_1}{1-\omega_1^* z} \qquad (35)$$

with $V_1 = u_1$ defined in (25), and V_2 is obtained from u_2 as:

$$V_2 = \theta_1(\omega_2)u_2 \qquad (36)$$

This procedure can be continued to give

$$\psi_j = \theta_1^{-1} \ldots \theta_{j-1}^{-1} \frac{V_j}{(z-\omega_j)} \tag{37}$$

where θ_i has the form given in (35) with

$$v_j = \theta_{j-1}(\omega_j)\ldots\theta_1(\omega_j)u_j \tag{38}$$

Also the role played by the Pick interpolation matrices is seen as:

$$V_j^* J V_j = \frac{\det(P_j).(1-|\omega_j|^2)}{\det(P_{j-1}).(1-|\omega_{j-1}|^2)} \tag{39}$$

where det denotes the determinant and P_i is the Pick matrix built on the ordered set of points $\omega_1 \ldots \omega_i$.

REMARK: For a more detailed discussion of these results see [17].

7. ACKNOWLEDGEMENT

The first author wishes to acknowledge gratefully the guidance and suggestions offered by Professor H. Dym of the Weizmann Institute of Sciences, Israel.

8. REFERENCES

[1] Youla, D.C., Castriota, L.J. and Carlin, H.J., (1959) "Bounded Real Scattering Matrices and the Foundations of Linear Passive Network Theory" *IRE Trans. CT-4* (1) pp. 102 - 124.

[2] Efimov, A.V. and Potapov, V.P., (1973) "J-Expanding Matrix Functions and their role in Analytical Theory of Electric Circuits", *Russ. Math. Surveys, Uspeki,* vol. 21, no. 1, pp. 69-140.

[3] Deprettere, E.F., (1981) "Orthogonal Filters", Ph.D. Dissertation, Delft Univ. of Technology, Delft.

[4] Dewilde, P., Vieira, A. and Kailath, T., (1978) "on the generalized Szego-Levinson Algorithm for Optimal Linear Predictors based on a Network Synthesis Approach", *IEEE Trans. CAS,* vol. 25, no. 9.

[5] Dewilde, P. and Dym, H., (1984) "Lossless Inverse Scattering, Estimation Theory and Digital Filters", *IEEE Trans. Infor. Thy.*, vol. IT-30, no. 4, pp. 644-662.

[6] Lev-Ari, H. and Kailath, T., (1984) "Lattice Filter Parametrization and Modelling of Nonstationary Processes", *IEEE Trans. Info. Thy.*, vol. IT-30.

[7] Dewilde, P., Fokkema, J. and Widya, I., (1981) "Inverse Scattering and Linear Prediction: the time continuous case", Stochastic Systems: The Mathematics of Filtering and Identification and Applications, D. Riedel Publ. Co., pp. 351-382.

[8] Levy, B. and Tsitsiklis, J., (1983) "An Inverse Scattering Procedure for the Linear Estimation of Two-Dimensional Isotropic Random Fields", Rep. no. LIDS-P-1192, MIT.

[9] Potapov, V.P., (1960) "The Multiplicative Structure of J-Contractive Matrix Functions", *Transl. AMS,* Series 2, vol. 15, pp. 131-243.

[10] Ginzburg, Yu. P., (1967) "On Multiplicative Representations of J-contractive Operator Functions" *Mat. Issled.,* 2, no. 2, pp. 52-83, no. 3, pp. 20-51.

[11] Dewilde, P., (1982) Lecture Notes on Modern Signal Processing, Stanford Univ.

[12] Genin, Y., van Dooren, P., Kailath, T., Delosme, J-M. and Morf, M., (1983) "On Σ-Lossless Transfer Functions and Related Questions", Linear Algebra and Appln., vol. 50, pp. 251-275.

[13] Kailath, T., and Porat, B., (1982) "State Space Generators for Orthogonal Polynomials", Harmonic Analysis and Prediction Theory: Essays in honor of P. Masani, ed. by H. Saleh and V. Madrekar, pp. 131-163, North Holland.

[14] de Branges, L., (1963) "Some Hilbert Spaces of Analytic Functions I", *Trans. Amer. Math. Soc.,* vol. 108, pp. 445-468.

[15] Alpay, D. and Dym, H., (1984) "Hilbert Spaces of Analytic Functions, Inverse Scattering and Operator Models", Integral Equations and Operator Theory, Vol. 7, pp. 589-641.

[16] Kailath, T., (1980) "Linear Systems", Prentice Hall, New Jersey.

[17] Prabhakara Rao, C.V.K. and Dewilde, P., (1985) "A State Space Approach to the Lossless Inverse Scattering Problem", *Proc. ISCAS-85*, Kyoto, Japan.

[18] Dym, H., (1984) "J-inner Matrices and Their Application" to appear in CBMS Lecture Notes, Cleveland.

[19] Rao, S.K. and Kailath, T., (1985) *IEEE Trans. on CAS*, pp. 1105-1117, Vol. CAS-32, no. 11, and Rao, S.K., Kailath, T., (1984) *IEEE Trans.* on CAS, Vol. CAS-31, no.11, pp.933-945.

DECONVOLUTION IN SLOW MOTION AND THE EXPERIMENTAL DETERMINATION OF THE IMPULSE RESPONSE OF A LINEAR TIME INVARIANT SYSTEM*

A.S. Carasso
(Center for Applied Mathematics,
National Bureau of Standards, Gaithersburg)

ABSTRACT

The class of one sided infinitely divisible probability density functions includes a rich variety of multimodal C^∞ approximations to the Dirac δ-function. When such pulses are used as probe waveforms in linear time invariant systems, the output waveform is an approximation to the system's impulse response in which the singularities have been smoothed out. Such singularities are usually of vital interest in characterizing the linear system. The ill-posed deconvolution problem for the output waveform aims at reconstructing the singularities in the presence of noise. The infinite divisibility of the probe pulse plays a key role in such singular reconstruction problems, by allowing reformulation of the Volterra integral equation as a Cauchy problem for a linear parabolic pseudo differential equation in two independent variables. In turn, this leads to the concept of regularized partial deconvolution for which strong error bounds can be obtained, in the L^∞ norm, under L^2 a- priori constraints on the data noise and on the unknown system response. In addition, the Cauchy problem formulation allows for deconvolution in slow motion, and enables the monitoring of the development of sharp features as well as suspected artifacts. The theory considerably extends previous work on the Inverse Gaussian pulse, where the relevant partial differential equation is the sideways heat equation. A computational procedure is developed, based on the Poisson Summation Formula and FFT algorithms, using equispaced data on sufficient long but finite time intervals. The ideas are illustrated with a computational experiment on reconstructing the singular dynamic Green's

*Research supported in part by the U.S. Army Research Office under MIPR No. ARO 63-82.

function for an elastic plate, using smooth synthetic noisy data resulting from an infinitely divisible multimodal input pulse.

1. INTRODUCTION

Let S be a linear time invariant system, and let g(t) be the impulse response of S, i.e. g(t) is the causal time signal representing the system's response to a Dirac δ-function input at t=0. Knowledge of g(t) characterizes the dynamic behaviour of S, in that the output z(t) for any given input v(t) can be obtained by convolution of v with g,

$$z(t) = \int_0^t v(t-\tau)\; g(\tau)\; d\tau = v * g. \tag{1.1}$$

While the experimental determination of g(t) is an important task in diverse areas of measurement science, the "band-limited" nature of signals which can be synthesized in the laboratory precludes the use of such infinitely sharp probe waveforms as the Dirac δ-function or the Heaviside unit step function, H(t). Instead, one must use approximations p(t) in the form of narrow pulses, which vanish for $t < 0$ but are of class C^∞ on the whole real line. As a result, the sharp features in g(t) are blurred or smoothed out, generally beyond recognition. In order to reconstruct the sharp signal g(t) from smooth data w(t), an ill-posed deconvolution problem must be solved,

$$w(t) = \int_0^t p(t - \tau)\; g(\tau)\; d\tau, \quad t \geqslant 0, \tag{1.2}$$

where p(t) is the probe pulse. The difficulty stems from the fact that one must solve (1.2) with noisy data $w_n(t)$ in lieu of the (unknown) exact data w(t). Thus, in order to stabilize the inversion procedure in (1.2) and obtain useful bounds for the error, (as a function of the noise level as the latter tends to zero) strong regularizing constraints must be imposed on g(t). Typically, see e.g. [1], [2], a priori bounds are placed on the first m derivatives of g(t) and/or g(t) is required to be nonnegative. However, such constraints are not applicable in many signal reconstruction problems where g(t) is not positive and has non-differentiable singularities such as jumps, spikes, cusps, and the like. The ability to reconstruct such features is important in several areas of nondestructive testing such as ultrasonic flaw detection, [3], acoustic emission, [4], [5], and optical fibre characterization, [6], [7].

A way of circumventing this difficulty is developed in [8], [9] and [10] and the discussion is illustrated with several examples of successful computational reconstructions of singular elastic Green's functions, from smooth synthetic noisy data. We advocate the use of a specifically designed probe in the form of an "inverse Gaussian" pulse,

$$p(\sigma;t) = \frac{\sigma}{\sqrt{4\pi t^3}} \exp(-\sigma^2/4t), \quad t \geq 0. \qquad (1.3)$$

Here, σ is a fixed positive parameter which controls the pulse width, and $p(\sigma;t)$ approaches $\delta(t)$ as $\sigma \downarrow 0$. The inverse Gaussian pulse is the causal fundamental solution for the one dimensional heat equation in the quarter plane, with zero initial data. As a consequence, the deconvolution problem (1.2) with (1.3) as kernel is equivalent to the "inverse heat conduction" problem in a homogeneous semi-infinite rod, [11], whereby the temperature history $g(t)$ at the boundary $x = 0$, is reconstructed from the observed history $w(t)$, at the interior location $x = \sigma$. This interpretation of (1.2) as a heat flow problem leads naturally to the notion of partial deconvolution, namely the reconstruction of the temperature history at an interior point y, intermediate between $x = 0$ and $x = \sigma$. For small values of y, the temperature history at y is a faithful approximation to the singular boundary data $g(t)$. On the other hand, even if $g(t)$ is singular, the temperature history at any positive x along the rod is a C^∞ function of t, which is bounded together with any of its derivatives, in the L^∞ norm, in terms of the L^2 norm of $g(t)$. This is a reflection of the smoothing properties of the heat flow problem associated with (1.2). In turn, this leads to a useful strategy for regularizing (1.2) using weak a-priori constraints. We assume only an a-priori L^2 bound, M, for the unknown impulse response $g(t)$, together with an a-priori estimate, ε, for the L^2 norm of the noise in $w_n(t)$. In [10], we show this leads to L^∞ error bounds for the partial deconvolution at any positive x, in terms of the L^2 bounds ε and M. Thus, the pointwise structure of the solution near $x = 0$ can be reconstructed from a sufficiently close L^2 approximation to the data at $x = \sigma$. These error estimates degenerate at $x=0$, corresponding to total deconvolution of (1.2).

The link between the inverse Gaussian pulse (1.3) and heat conduction, provides another useful feature. First, as discussed in detail in [9] and [11], regularization of (1.2) is equivalent to appropriately filtering the given noisy data

$w_n(t)$ using the L^2 "noise to signal" ratio $\omega = (\varepsilon/M)$ as the only a- priori information. Next, the deconvolution problem (1.2) can be implemented as a Cauchy problem for the sideways heat equation. Using the filtered noisy data $w_n(t)$ as initial values on the line $x = \sigma$, one may use a step-by-step marching finite difference scheme and solve the heat equation as an initial value problem in the x-variable, from $x = \sigma$ to $x = 0$. By displaying the solution at selected intermediate values of x as x tends to zero, the evolution of the deconvolution process can be monitored. This represents continuous deconvolution of (1.2). Such an evolutionary display enables the analyst to visually distinguish systematic trends from noise artifacts, provided the L^2 noise level ε is reasonably small. See Figure 4 below and other examples in [13]. In effect, by using the fact that the partial devoncolutions at positive values of x are pointwise reliable as $\varepsilon \downarrow 0$, and by performing the deconvolution in slow motion, one can compensate for the lack of an error bound for the desired total deconvolution.

2. ONE SIDED C^∞ INFINITELY DIVISIBLE PROBABILITY DENSITY FUNCTIONS

In a laboratory environment, pulses of the general character of (1.3) are commonly synthesized either by using an analog circuit or by using a digital to analog (D/A) converter. In applications to ultrasonic testing, [3], a high fidelity transducer is used to convert the synthesized voltage into a mechanical pulse with a prescribed time dependence. Interfacing devices such as power amplifiers and transducers generally introduce distortions, superimposed small oscillations and other perturbations, on any given pulse shape. In order to more closely model such perturbations, a more comprehensive pulse theory is needed allowing a rich variety of pulse shapes, including multimodal pulses. Extending the inverse Gaussian theory leads naturally to the question of identifying the class of one-sided non-negative pulses p(t), of unit mass, for which the notions of partial and continuous deconvolution can be made meaningful, i.e., those pulses which can be linked to a Cauchy problem. It turns out that this question is effectively answered by the work of Paul Lévy on the Central Limit Theorem. See [12, Chapter 9]. The class of pulses we seek form a subclass of the class of one-sided, infinitely divisible, probability density functions. In the present paper we shall sketch these ideas while directing the reader to [13] for a detailed discussion, including proofs and appropriate references to the literature.

We restrict attention to non-negative probe waveforms p(t), defined on $0 \leqslant t \leqslant \infty$ and such that

$$\int_0^\infty p(t)\, dt = 1. \tag{2.1}$$

In addition, if H(t) is the Heaviside function, we require H(t)p(t) to be of class C^∞ on the whole real line. Let

$$\{p(t)\}^{*n}, \quad n \text{ a positive integer,} \tag{2.2}$$

denote the n-fold convolution of p(t) with itself. Let L and F be, respectively, the Laplace and Fourier transforms of p(t),

$$L\{p(t)\} = P(s) = \int_0^\infty e^{-st} p(t)\, dt, \text{ Re } s \geqslant 0, \tag{2.3}$$

$$F\{p(t)\} = \hat{p}(\xi) = \frac{1}{\sqrt{2\pi}} \int_0^\infty e^{-i\xi t}\, p(t)\, dt, \ \xi \text{ real.} \tag{2.4}$$

Clearly,

$$\sqrt{2\pi}\ \hat{p}(\xi) = P(i\xi), \ \xi \text{ real,} \tag{2.5}$$

$$L\{p(t)\}^{*n} = \{P(s)\}^n, \text{ Re } s \geqslant 0, \tag{2.6}$$

$$F\{p(t)\}^{*n} = (2\pi)^{\frac{n-1}{2}} \{\hat{p}(\xi)\}^n, \ \xi \text{ real.} \tag{2.7}$$

Definition 1

A one-sided probability density p(t) is infinitely divisible if and only if, given any positive integer n, there exists a one-sided probability density $q_n(t)$ such that,

$$p(t) = \{q_n(t)\}^{*n}. \tag{2.8}$$

Definition 2

A one-sided infinitely divisible density p(t) belongs to the class $\mathcal{D}$ if and only if there exist positive constants A, c, and β with $A \geqslant 1$ and $\beta \leqslant 1$, such that,

$$\sqrt{2\pi}\,|\hat{p}(\xi)| \leqslant A e^{-c|\xi|^\beta}. \tag{2.9}$$

It can be shown, see [13], that the class $\mathcal{D}$ includes the one-sided Lévy stable density functions as well as a rich variety of multimodal pulses. Next, we consider the convolution equation

$$w(t) = \int_0^t p(t-\tau)\, g(\tau)\, d\tau, \quad t \geq 0 \tag{2.10}$$

where p(t) is in class $\mathcal{D}$. As shown in [13], one can associate a generalized heat conduction problem with (2.10) as follows. Given p(t), define the linear pseudo-differential operator $\mathcal{P}$ acting on the t variable, by means of

$$(\mathcal{P}f)(t) = \mathcal{F}^{-1}\left[\hat{f}(\xi)\log(\sqrt{2\pi}\,\hat{p}(\xi))\right], \tag{2.11}$$

where the principal branch of the logarithm is understood. Consider the Cauchy problem

$$\begin{cases} \dfrac{\partial u}{\partial x} = \mathcal{P}u, & x > 0,\ t > 0, \\ u(x,0) \equiv 0, & x \geq 0, \\ u(0,t) = g(t), & t \geq 0. \end{cases} \tag{2.12}$$

Suppose that in (2.10) the left-hand side w(t) is such that a (unique) $L^2(0,\infty)$ solution, g(t), exists. Then w(t) = u(1,t) where u(x,t) is the unique solution of (2.12). Using w(t) as initial data on the line x = 1 in (2.12), and integrating the differential equation in the direction of decreasing x, from x = 1 to x = 0, we obtain

$$u(0,t) = g(t). \tag{2.13}$$

For any fixed x such that $0 < x < 1$, u(x,t) is a partial deconvolution of (2.10) while u(0,t) is the total deconvolution. Even with boundary data g(t) of class $L^2(0,\infty)$, the solution u(x,t) in (2.12) is a C^∞ function of t for each fixed positive x, whenever p(t) lies in $\mathcal{D}$. In this sense, (2.12) is a heat conduction problem. Note that when p(t) is the inverse Gaussian pulse (1.3), the pseudo-differential equation in (2.12) becomes

$$\frac{\partial u}{\partial x} = -\sigma\left(\frac{\partial}{\partial t}\right)^{\frac{1}{2}} u, \quad x > 0,\ t > 0. \tag{2.14}$$

On differentiating both sides of (2.14) with respect to x, we recover the heat equation. The connection between (2.10) and (2.12) when p(t) is a C^∞ infinitely divisible pulse in class $\mathcal{D}$, provides the framework for a far reaching extension of the inverse Gaussian theory discussed in [8], [9] and [10]. The wide variety of pulse shapes included in the class $\mathcal{D}$ frees the experimentalist from the limitation of having to achieve a perfect, unimodal, inverse Gaussian probe pulse. On the other hand, in the presence of noise, it will be advantageous to synthesize probe waveforms in the form of narrow pulses, concentrated near t = 0, and with relatively small deviations from unimodality.

3. L^2 REGULARIZATION; L^∞ ERROR BOUNDS

We now consider the case when the left hand side of (2.10) is replaced by noisy data $w_n(t)$. Let $\varepsilon > 0$ be an a-priori estimate for the L^2 norm of the noise in $w_n(t)$,

$$\|w-w_n\|^2 = \int_{-\infty}^{\infty} |\hat{w}(\xi) - \hat{w}_n(\xi)|^2 \, d\xi \leq \varepsilon^2. \tag{3.1}$$

Let $M > 0$ be an a-priori bound for the L^2 norm of the unique solution of (2.10),

$$\frac{1}{2\pi} \int_{-\infty}^{\infty} |\hat{w}(\xi)|^2 |\hat{p}(\xi)|^{-2} d\xi = \int_{-\infty}^{\infty} |\hat{g}(\xi)|^2 \, d\xi \leq M^2. \tag{3.2}$$

We assume ε and M to be known and compatible. Let ω be the L^2 "noise to signal" ratio,

$$\omega = (\varepsilon/M) \ll 1. \tag{3.3}$$

Given the recorded data $w_n(t)$ and the probe pulse p(t) in (2.10), we define the regularized problem as follows in Laplace transform space,*

$$P(s)H(s) = \frac{|P(s)|^2 W_n(s)}{|P(s)|^2 + \omega^2}, \quad \text{Re } s > 0, \tag{3.4}$$

*From a practical computational standpoint, for small values of ω, q(t) in (3.6) is a one-sided function of t if $w_n(t)$ is one-sided. This is so despite the fact that the filter in (3.4) or (3.5) is not holomorphic.

or equivalently, in Fourier space,

$$\sqrt{2\pi}\ \hat{p}(\xi)\hat{h}(\xi) = \frac{|\hat{p}(\xi)|^2\ \hat{w}_n(\xi)}{|\hat{p}(\xi)|^2 + (\omega^2/2\pi)}, \quad \xi \text{ real.} \tag{3.5}$$

The unique solution h(t) of (3.4) or (3.5) corresponds to the regularised total deconvolution of (2.10) in the presence of noise. The regularization procedure consists of replacing the given noisy data $w_n(t)$ by new filtered data q(t), where

$$q(t) = \mathcal{F}^{-1}\left\{\frac{|\hat{p}(\xi)|^2\ \hat{w}_n(\xi)}{|\hat{p}(\xi)|^2 + (\omega^2/2\pi)}\right\}, \quad t \geq 0. \tag{3.6}$$

Moreover, the value of ω is the only a-priori information used in constructing q(t). We now consider the related regularized Cauchy problem associated with (2.10). Using (3.6) and (2.11), this is,

$$\begin{cases} \dfrac{\partial v}{\partial x} = \mathcal{P}v, & 0 < x < 1,\ t > 0, \\ v(x,0) \equiv 0, & 0 \leq x \leq 1, \\ v(1,t) = q(t) & t \geq 0. \end{cases} \tag{3.7}$$

In (3.7), the differential equation is to be integrated in the direction of decreasing x, from x=1 to x=0, using the filtered data q(t) as initial values on x = 1. For each fixed x with $0 < x < 1$, the unique solution v(x,t) of (3.7) is a regularized partial deconvolution of (2.10). Let u(x,t) be the unique solution of (2.12) and consider the error (v(x,t) - u(x,t)). The essential point is that no useful error bound is possible at x = 0, i.e., for the function (h(t) - g(t)), under the weak a- priori constraints (3.1), (3.2). However, if p(t) lies in $\mathcal{D}$, one can estimate the error for each fixed x, $0 < x < 1$, in arbitrarily high Sobolev norms. See [13, Theorem 3]. By way of Sobolev's Lemma, this leads to error bounds in the L^∞ norm for the regularized partial deconvolution and any of its derivatives.

4. NUMERICAL PARTIAL DECONVOLUTION

A means of numerically solving (3.7), using sufficiently long but finite records of the probe pulse p(t) and the output waveform $w_n(t)$, may be based on the following formula,

$$f(t) = \frac{e^{at}}{2T} \sum_{k=-\infty}^{\infty} F(a + \frac{ik\pi}{T}) e^{\frac{ik\pi t}{T}} + O(e^{-2aT}), \quad 0 \leq t \leq 2T. \tag{4.1}$$

In (4.1), $f(t)$ is a function in $L^2(0,\infty)$, F is the Laplace transform of f, a is any positive number, and the approximation is valid in the $L^2[0, 2T]$ norm. See [13]. This formula is a consequence of the Poisson Summation Formula and is sometimes used as a method for numerically inverting the Laplace transform, given a closed-form analytical expression for $F(s)$ valid in some right half-plane, [14], [15]. In the present framework, (4.1) is viewed as a vehicle for performing sophisticated deconvolution operations on time domain data $p(t)$ and $w_n(t)$, given at $2N$ equispaced points on the finite interval $[0,2T]$, with N and T sufficiently large. The algorithm proceeds as follows:

We choose $a > 0$ so that the $O(e^{-2aT})$ term is negligible in (4.1) and we form

$$e^{-at} p(t), \quad e^{-at} w_n(t), \quad 0 \leq t \leq 2T. \tag{4.2}$$

Using FFT algorithms together with complex arithmetic, we find approximations to the first $(2N + 1)$ Fourier coefficients in the complex Fourier series for each of the functions in (4.2). From (4.1), we may interpret these coefficients respectively as,

$$(\frac{1}{2T}) P(a + \frac{ik\pi}{T}), \quad (\frac{1}{2T}) W_n (a + \frac{ik\pi}{T}), \quad |k| \leq N. \tag{4.3}$$

Next, with ω as in (3.3) we filter the noisy data using the prescription on the right hand side of (3.4). We form the $(2N+1)$ coefficients $Q(a + ik\pi/T)$, where,

$$Q(a + \frac{ik\pi}{T}) = \frac{|P(a + ik\pi/T)|^2 W_n(a + ik\pi/T)}{|P(a + ik\pi/T)|^2 + \omega^2}, \quad |k| \leq N. \tag{4.4}$$

Since (3.4) is valid for $\mathrm{Re}\ s > 0$, we have in particular,

$$P(a + \frac{ik\pi}{T}) H(a + \frac{ik\pi}{T}) = Q(a + \frac{ik\pi}{T}), \quad |k| \leq N. \tag{4.5}$$

A consequence of the assumed infinite divisibility of p(t) is that P(s) never vanishes, [12, Chapter 13]. This guarantees uniqueness in (2.10) and enables us to divide both sides of (4.5) by $P(a + {}^{ik\pi}/T)$, to get

$$\left(\frac{1}{2T}\right) H\left(a + \frac{ik\pi}{T}\right) = \left(\frac{1}{2T}\right) \frac{Q(a + {}^{ik\pi}/_T)}{P(a + {}^{ik\pi}/_T)}, \quad |k| \leq N. \tag{4.6}$$

From (3.4) and (4.1), it follows that the left hand side of (4.6) consists of the first 2N+1 complex Fourier coefficients for e^{-at} h(t) on [0,2T], where h(t) is the regularized total deconvolution of (2.10). An inverse FFT followed by multiplication by e^{at}, yields h(t). This is equivalent to obtaining the solution of the Cauchy problem (3.7) at the boundary x = 0.

Consider now partial deconvolution of (2.10) corresponding to the solution of (3.7) at a value of x such that $0 < x < 1$. Using (3.4), it is easy to see that the partial deconvolution at x has the Laplace transform,

$$V(s) = [P(s)]^x H(s), \quad \text{Re } s > 0, \tag{4.7}$$

with the principal branch of the root understood in (4.7). Thus, from (4.6), we get

$$\left(\frac{1}{2T}\right) V\left(a + \frac{ik\pi}{T}\right) = \left(\frac{1}{2T}\right) \frac{Q(a + ik\pi/T)}{[P(a + ik\pi/T)]^{1-x}}, \quad |k| \leq N. \tag{4.8}$$

We then obtain v(x,t) by an inverse FFT applied to (4.8), followed by multiplication by e^{at}. One may also use (4.8) to simultaneously form and store the partial deconvolution Fourier coefficients at preselected values of x, $(x_1, x_2, x_3, \ldots, x_n)$ where,

$$0 \leq x_n < x_{n-1} < \ldots < x_1 < 1. \tag{4.9}$$

When the corresponding time domain signals $(v(x_1,t), v(x_2,t), v(x_3,t) \ldots, v(x_n,t))$ are viewed in the same order, the gradual unfolding of the deconvolution process may be observed. This represents continuous deconvolution of (2.10). See e.g. Figure 4 below.

In practice, the finite word length of computing equipment precludes the use of large values of a in the approximation (4.1). Too large a value of a emphasizes the early parts of the signals p(t) and $w_n(t)$, while effectively destroying the information content of the late parts, prior to processing by the deconvolution algorithm. After processing, post multiplication by e^{at} merely amplifies noise in the late parts of the signals, as manifested by an oscillatory behaviour, with amplitude growing exponentially with t. This limitation may be circumvented by making T large instead in (4.1), i.e., by recording the probe pulse p(t) and the output waveform $w_n(t)$ in (2.10), on an interval [0,2T] substantially larger than is required to identify the features of physical interest in the system's impulse response g(t). Another procedure which has been successfully used, consists in smoothly extending the given signals to a larger interval [0,2T']. The deconvolution procedure is then applied on 2T' with a small value of a, with the final results retained on [0,2T]. This extension procedure rests on the easily proved observation that in a convolution equation such as (1.2), where the kernel has a non-vanishing Laplace transform, knowledge of w(t) on any finite interval [0,2T] uniquely determines g(t) on the same interval.

5. A COMPUTATIONAL EXPERIMENT

The preceding developments will now be illustrated with a numerical experiment where we reconstruct a singular elastic Green's function from synthetic noisy data. These data were created by numerical convolution of a known exact Green's function with a multimodal C^∞ infinitely divisible pulse in class $\mathcal{D}$.

The dynamic Green's function for an infinite elastic plate plays a central role in several theoretical and experimental studies involving transient wave propagation in elastic structures, in such fields as Acoustic Emission and Ultrasonic Flaw Detection, [3], [4], [5]. This function provides the normal displacement as a function of time, at a point ξ on one face of the plate, due to a normal point force with Dirac δ-function or Heaviside time dependence, applied at some other point η, on the same side or on the opposite side of the plate. By summarizing the wave propagation process within the plate, the dynamic Green's function characterizes the plate as a linear time invariant system for each test configuration (ξ,η). A comprehensive treatment of the Green's tensor based on generalized rays and Cagniard's method is given in [5]. Recently, [16], computational software has been developed which enables explicit calculation of the Green's function given the test configuration (ξ,η). The availability of such exact solutions is crucial in the development and calibration of refined measurement techniques, in a broad experimental research program conducted by the Ultrasonics Standards Group at the National Bureau of Standards, [4] .

The theory described in the present paper was developed in connection with one facet of the above research program, dealing with the experimental determination of the dynamic Green's function for a plate. The method is based on deconvolution of the recorded displacement at the point ξ on the plate, due to a predesigned mechanical pulse approximating the Dirac δ-function, applied at some other point η on the plate. A detailed discussion of the experimental technique will appear in forthcoming papers in the Journal of the Acoustical Society of America. See also [3] . The ultimate aim of this experimental technique is to obtain the impulse response for more complex elastic structures for which exact solutions are not known.

The figures shown below display normalized displacement versus normalized time where,

Normalized Displacement = π* shear modulus* plate thickness* actual displacement/force. (5.1)

Normalized Time = actual time * shear wave speed/plate thickness. (5.2)

In Figure 1, the top drawing is an example of a plate Green's function for a point source with Heaviside time dependence. The source and receiver are on opposite sides of the plate, with the receiver located two plate thicknesses away from the epicentre. This exact solution was calculated using the software discussed in [16] . Evidently, this solution is neither positive nor differentiable. Indeed, the singularities are of vital interest as they serve to identify the times of arrival of various wave components of the source signal, as well as the subsequent multiple reflections from the plate faces. The corresponding response to a Dirac δ-function time dependence is given by the distribution derivative of the top drawing in Figure 1, and is even more singular. The middle drawing in Figure 1 is an example of a multimodal infinitely divisible pulse in class $\mathcal{D}$. This pulse p(t) was chosen primarily to illustrate the mathematical theory in Section 2 and is not otherwise particularly significant. (Much more complicated pulses are included in $\mathcal{D}$; see [13]). The waveform for p(t) is obtained by putting $\sigma=0.5$ and $\lambda=0.2$ in the expression:

$$p(t) = e^{-\lambda} \sum_{m=0}^{\infty} \frac{\lambda^m}{m!} q(\sigma; t-m), \quad t \geq 0, \qquad (5.3)$$

where,

$$q(\sigma;\ t) = \frac{H(t)\, e^{\sigma-t}\, e^{-\sigma^2/4t}}{\sqrt{\pi t}} . \qquad (5.4)$$

The pseudo-differential equation in (3.7) determined by (5.3) has a right hand side given by,

$$(pv)(x,t) = F^{-1}[\hat{v}(x,\xi)\{(\sigma-\lambda)+ \lambda e^{-i\xi} - \sigma\sqrt{1+i\xi} - \frac{1}{2}\log(1+i\xi)\}] .$$

The bottom drawing in Figure 1 was obtained by convolution of the top two drawings, using adaptive quadrature routines. Physically, this simulates the response to a waveform which approximates the Heaviside function, and which is given by the indefinite integral of (5.3). Evidently, the sharp features in the top drawing have been smoothed out well beyond recognition. Each of the input pulse and output waveform in Figure 1 consists of 500 equispaced values on the normalized time interval $[0,5]$. Moreover, in the output waveform, each synthetic data value y was perturbed by adding to it a random number sampled from a uniform distribution in the range $\pm.001$ y. Finally, in order to permit the use of a small value of $\bar{a}$ in the approximation (4.1), these signals were artificially extended to the interval $[0,24]$ as smooth functions of t, ultimately decaying rapidly to zero.

The use of an interactive graphics language such as DATAPLOT, [17], [18], is highly desirable in reconstructing singular signals. Thus, the parameter ω in (3.3) is best found interactively, starting from a plausible guess for the ratio (ε/M). Too large a value of ω leads to a visually pleasing total deconvolution, devoid of noise artifacts but also devoid of sharp features. Too small a value of ω brings out more noise in the total deconvolution without sharpening the signal any further. A range of ω values is quickly found for which little change is apparent in the total deconvolution. At this stage, the deconvolution is performed in slow motion so as to monitor the development of sharp features in the presence of noise.

The computational algorithm discussed in Section 4 was used to generate the sequence of partial deconvolutions shown in Figures 2, 3 and 4, with $\omega = .0001$. While the total deconvolution is obscured by noise for $t \geqslant 3$, a significant amount of detail has been reconstructed on the interval $0 \leqslant t \leqslant 3$. Especially noteworthy is the development of the sharp cusp near $t = 2.2$, whose genesis can be traced back to the 75% partial deconvolution. Also, in Figure 3, the sequence appears to indicate a jump in the function at $t = 3.6$.

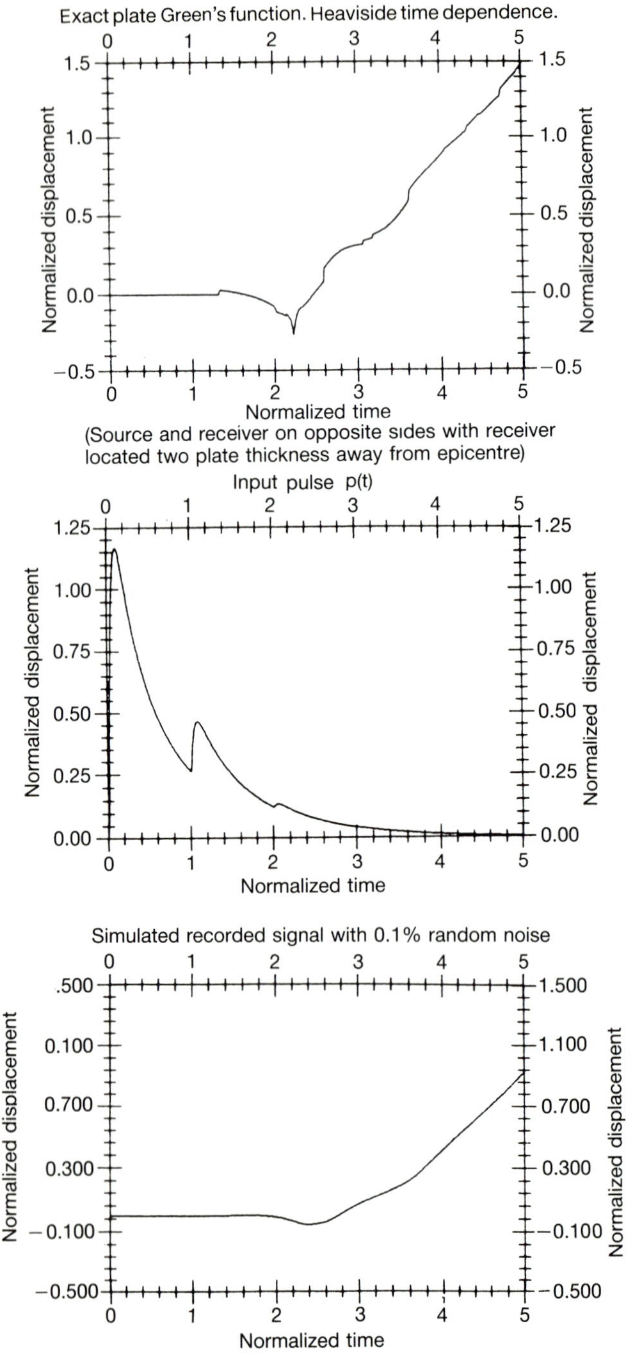
Exact plate Green's function. Heaviside time dependence.
Normalized displacement
Normalized time
(Source and receiver on opposite sides with receiver located two plate thickness away from epicentre)
Input pulse p(t)
Normalized displacement
Normalized time
Simulated recorded signal with 0.1% random noise
Normalized displacement
Normalized time

Fig. 1

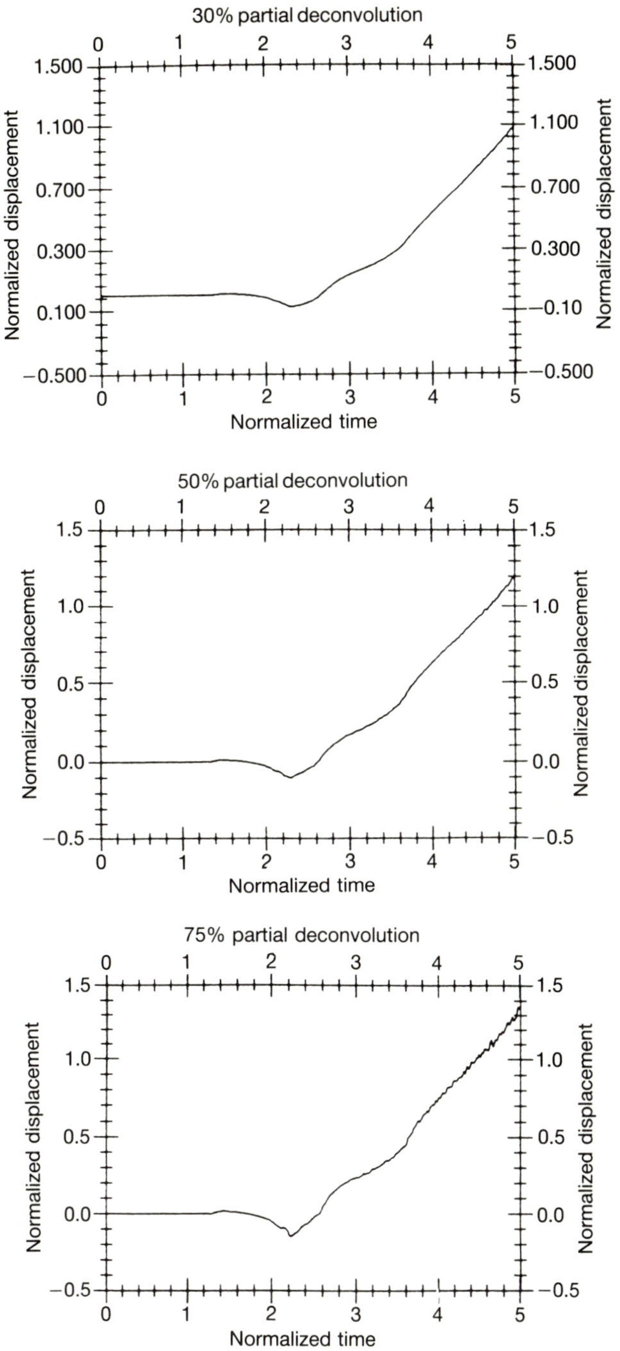
30% partial deconvolution
0
1
2
3
4
5
1.500
1.100
0.700
0.300
0.100
−0.500
−0.10
Normalized displacement
Normalized time
50% partial deconvolution
1.5
1.0
0.5
0.0
−0.5
Normalized displacement
Normalized time
75% partial deconvolution
1.5
1.0
0.5
0.0
−0.5
Normalized displacement
Normalized time

Fig. 2

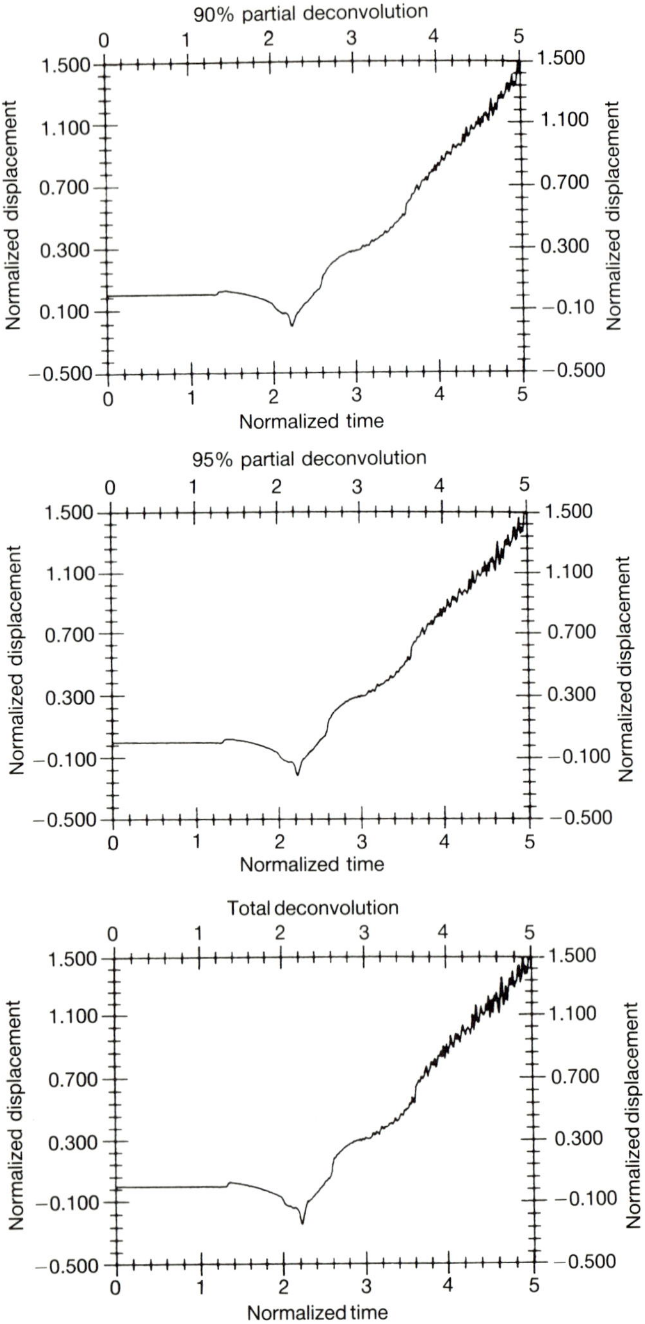
90% partial deconvolution
0 1 2 3 4 5
1.500
1.100
0.700
0.300
0.100
−0.500
1.500
1.100
0.700
0.300
−0.10
−0.500
Normalized displacement
Normalized displacement
0 1 2 3 4 5
Normalized time
95% partial deconvolution
0 1 2 3 4 5
1.500
1.100
0.700
0.300
−0.100
−0.500
1.500
1.100
0.700
0.300
−0.100
−0.500
Normalized displacement
Normalized displacement
0 1 2 3 4 5
Normalized time
Total deconvolution
0 1 2 3 4 5
1.500
1.100
0.700
0.300
−0.100
−0.500
1.500
1.100
0.700
0.300
−0.100
−0.500
Normalized displacement
Normalized displacement
0 1 2 3 4 5
Normalized time

Fig. 3

Fig. 4 Evolution of 20 successive partial deconvolutions viewed from three different angles.

Displays begin with blurred signal in foreground and end with total deconvolution in background.

We remark that under the weak a- priori constraints imposed in (3.1), (3.2), an error bound for the total deconvolution shown in the bottom drawing in Figure 3 is not possible. On the other hand, the partial deconvolutions are pointwise accurate as the L^2 noise level ε in (3.1) tends to zero. In Figure 4, each drawing is composed of twenty successive equispaced partial reconstructions, corresponding to setting $x_j = 1 - .05j$, $j = 1,20$, in equation (4.9). Such evolutionary displays may be viewed from various angles and are often helpful in evaluating the genuineness of sharp features in the desired total deconvolution. Further examples are given in [13]. In this way, the solution of the ill-posed integral equation as an evolutionary problem can provide valuable additional information.

6. CONCLUSION

A theory has been presented for pulse probing and identification of linear time invariant systems. The method is based on the use of smooth pulses, approximating the Dirac δ-function and possessing additional semi-group properties. The underlying semi-group is used to associate a fictitious heat conduction problem with the original ill-posed Volterra convolution integral equation. The heat flow problem is then exploited to perform the deconvolution continuously with respect to a parameter which reflects the pulse width. In addition, the smoothing properties of the heat flow can be used to obtain error bounds in the L^∞ norm for partial deconvolution, under L^2 a- priori constraints on the data noise and the unknown system response. Other deconvolution problems can also be linked to heat conduction problems. In [19], an important class of two-dimensional image restoration problems was likewise analyzed and solved by reformulation in terms of parabolic equations with the time direction reversed. In future work, we hope to analyze a wider class of image blurs via two-dimensional infinitely divisible probability distributions.

REFERENCES

[1] Franklin, J.N., (1974) "On Tikhonov's Method for Ill-Posed Problems," *Math. Comp.*, **28**, pp. 889-907.

[2] Franklin, J.N., (1978) "Minimum Principles for Ill-Posed Problems," *SIAM J. Math. Anal.*, **9**, pp. 638-650.

[3] Eitzen, D., Hsu, N., Carasso, A. and Proctor, T., (1985) "Deconvolution by Design - An Approach to the Inverse Problem of Ultrasonic Testing," In Review of Progress in Quantitative NDE, University of California-San Diego, LaJolla, California, July 1984. Thompson, D.O. and Chimenti, D.E., Editors, Plenum Publishing Corporation, New York, pp. 179-188.

[4] Eitzen, D.G., et al.,(1981) "Fundamental Developments for Quantitative Acoustic Emission Measurements," EPRI NP-2089, Interim Report, prepared by the National Bureau of Standards, Gaithersburg, MD 20899, for the Electric Power Research Institute, Palo Alto, CA 94304.

[5] Ceranoglu, A.N., and Pao, Y.H., (1981) "Propagation of Elastic Pulses and Acoustic Emission in a Plate," *ASME Journal of Applied Mechanics,* **48**, pp. 125-147.

[6] Chamberlain, G.E., Day, G.W., Franzen, D.L., Gallawa, R.L., Kim, E.M. and Young, M., "Optical Fiber Characterization," NBS Special Publication 637, October 1983, U.S. Department of Commerce/National Bureau of Standards, Gaithersburg, MD 20899.

[7] Guillaume, M.E. and Bizeul, J.C., (1981) "Mesure de la Réponse Impulsionnelle d'un Milieu de Transmission: Application aux Fibres Optiques," *Annales des Télécommunications, Tome* **36**, Numéros 3, 4, pp. 179-186.

[8] Carasso, A. and Hsu, N.N., (1985) "The Inverse Gaussian Pulse in the Experimental Determination of Linear System Green's Functions," Transactions of the Second Army Conference on Applied Nathematics and Computing, pp. 389-404, ARO Report 85-1, U.S. Army Research Office, P.O. Box 12211, Research Triangle Park, NC 27709.

[9] Carasso, A.S. and Hsu, N.N., (1985) "Probe Waveforms and Deconvolution in the Experimental Determination of Elastic Green's Functions," *SIAM J. Appl. Math.,* **45**, pp. 369-383.

[10] Carasso, A.S. and Hsu, N.N., (1985) "L^∞ Error Bounds in Partial Deconvolution of the Inverse Gaussian Pulse," *SIAM J. Appl. Math.,* **45**, pp. 1029-1038.

[11] Carasso, A., (1982) "Determining Surface Temperatures from Interior Observations," *SIAM J. Appl. Math.,* **42**, pp. 558-574.

[12] Feller, W., (1971) "An Introduction to Probability Theory and its Applications" Vol. 2, Second Edition, John Wiley and Sons, Inc., New York.

[13] Carasso, A.S., "Infinitely Divisible Pulses, Continuous Deconvolution, and the Characterization of Linear Time Invariant Systems," To appear in *SIAM J. Appl. Math.* **47**, (1987).

[14] Crump, K.S., (1976) "Numerical Inversion of Laplace Transforms Using A Fourier Series Approximation," *J. Assoc. Comput. Mach.*, **23**, pp. 89-96.

[15] Dubner, H. and Abate, J., (1968) "Numerical Inversion of Laplace Transforms by Relating them to Finite Cosine Transform," *J. Assoc. Comput. Mach.*, **15**, pp. 115-123.

[16] Hsu, N.N., (1985) "Dynamic Green's Functions of an Infinite Plate - A Computer Program," NBS Internal Report, NBSIR 85-3234, U.S. Department of Commerce/National Bureau of Standards, Gaithersburg, MD 20899.

[17] Filliben, J.J., (1981) "DATAPLOT--An Interactive High Level Language for Graphics, Nonlinear Fitting, Data Analysis and Mathematics," *Computer Graphics*, **15**, pp. 199-213.

[18] Filliben, J.J., (1984) "DATAPLOT-- Introduction and Overview," NBS Special Publication 667, U.S. Department of Commerce/National Bureau fo Standards, Gaithersburg, MD 20899.

[19] Carasso, A.S., Sanderson, J.G. and Hyman, J.M., (1978) "Digital Removal of Random Media Image Degradations by Solving the Diffusion Equation Backwards in Time," *SIAM J. Numer. Anal.*, **15**, pp. 344-367.

REGULARISATION PROCEDURES IN SIGNAL PROCESSING AND STATISTICS

D.M. Titterington
(Department of Statistics, University of Glasgow, Scotland)

ABSTRACT

A general structure is suggested for regularisation as applied to inverse problems in signal processing or image restoration. Within this structure it is necessary to choose a prescription for the restored image and, within that, a value for a regularisation parameter that determines the degree of smoothing imposed. A class of compromise prescriptions is set up which exemplifies the trade-off between fidelity to the data and smoothness, and which includes various well-known regularisation procedures. It is then pointed out that the same structure is apparent in various fields of statistical interest: ridge regression, nonparametric regression, multinomial smoothing and nonparametric density estimation. A major aspiration of the paper is that there might be useful cross-fertilisation between experiences in the statistical and signal processing fields. Finally, methods are described for choosing the value of the regularisation parameter.

1. INTRODUCTION

A common method for dealing with ill-posed inverse problems is to impose some form of regularisation. This has been applied in many areas, such as the solution of integral equations and, more specifically, in some versions of the problem of image restoration. It turns out that the same general structure appears in certain branches of Statistics and it is of interest to take note both of the similarities and of the variety of procedures for implementation.

2. A SIMPLE MODEL FOR IMAGES, AND THE NEED FOR REGULARISATION

Consider a picture window containing N pixels and let the true image, f, a vector of N intensities, be related to the observed image, g, by the system of N equations

$$g = Af + \varepsilon, \tag{2.1}$$

in which ε denotes the noise vector. Although the general principles we shall expound apply to more complicated models, we shall restrict attention to this simple, linear, additive model. We shall also assume, unnecessarily in principle, that $E(\xi) = 0$ and $\mathrm{cov}(\varepsilon) = \sigma^2 I$, where I is the NxN identity matrix. We shall assume that σ^2 and A, the point-spread function, are both known.

The image restoration problem is to derive, from g, a useful estimator, $\hat{f}$, of f.

A naive approach is to suggest using the <u>least-squares</u> estimator $\hat{f}_0$, which satisfies

$$A^T A\hat{f}_0 = A^T g = A^T Af + A^T \varepsilon.$$

If A is nonsingular, then

$$\hat{f}_0 = A^{-1} g = f + A^{-1}\varepsilon. \tag{2.2}$$

In spite of the facility with which (2.2) can be written down, $\hat{f}_0$ is often not a useful solution. In practice N can be very large ($\sim 2^{20}$) and A, even if nonsingular, may be very poorly conditioned. As a result, numerical construction of (2.2) may be difficult and the influence of the noise typically creates instability.

The remedy for this ailment is to impose some form of <u>regularisation</u> or, to use terminology from the statistical literature, <u>shrinkage</u> or <u>smoothing</u>. To decide upon a specific procedure we must answer two questions.

I. <u>Qualitative</u>: In what <u>manner</u> to regularise?

II. <u>Quantitative</u>: By <u>how much</u> to regularise?

Question I is answered by choosing some <u>prescription</u>, $\{\hat{f}_h, 0 \leq h \leq \infty\}$, which defines a class of estimates of f. The quantity h is a <u>smoothing parameter</u> or <u>regularisation parameter</u>, and the selection of a particular value, $\hat{h}$, for h, will constitute our answer to question II. Finally, then, we shall report $\hat{f}_{\hat{h}}$ as the restored image.

Within any prescription, the "extreme" restorations tend to represent the ultra-rough, data-based, least-squares estimator $\hat{f}_0 (h \to 0)$ and some ultra-smooth $\hat{f}_\infty (h \to \infty)$. This is usually the completely grey picture with constant intensity but might be some other prior image (c.f. Cornwell and Evans, 1985, Skilling, 1986). A prescription, usually chosen to reflect the local smoothness of the image, thus represents a path from $\hat{f}_0$ to $\hat{f}_\infty$ (see Fig. 1 of Titterington, 1985a) and we then have to decide, through the choice of h, how far along the path to go, from either end; see Section 5.

One way of describing the basic objective is as that of smoothing away the noise from $\hat{f}_0$ without introducing too much systematic distortion from the true image. Although $\hat{f}_0$ is unbiased, regularisation typically destroys this property but, in compensation, greatly reduces the "variance" of the restored image.

3. A COMPROMISE PRESCRIPTION

One general form of prescription is obtained as the solution of the problem

$$\min_f \{\Delta_1(f,\hat{f}_0) + h\Delta_2(f,\hat{f}_\infty)\}, \qquad (3.1)$$

in which Δ_1 and Δ_2 are measures of "distance" between their two arguments. Note that we shall not require Δ_1 and Δ_2 to be metrics. The above criterion represents a compromise between fidelity to the data (the first term measures lack of fidelity) and smoothness, with the value of h indicating the accepted trade-off. When $\hat{f}_\infty$ is the uniform map, $\Delta_2(f,\hat{f}_\infty)$ can often be written in the form $\Phi(f)$ + constant, in which case Φ is then called a roughness penalty function.

The compromise criterion can be interpreted in various ways. In form (3.1), it is a minimum penalised distance (MPD) criterion: the distance $\Delta_1 (f,\hat{f}_0)$ is minimised subject to a penalty on the resulting roughness. Equivalently, denoting h^{-1} by γ, we may use the criterion

$$\min_{f} \{\Phi(f) + \gamma\Delta_1(f,\hat{f}_O)\}, \tag{3.2}$$

and speak of minimum penalised roughness or minimum penalised informativeness (regarding the uniform map as totally uniformative) or maximum penalised smoothness.

Titterington (1985a) lists several candidates for $\Phi(.)$. Some reflect global roughness, such as

$$\Phi_O(f) = \sum_{i}\sum_{j} (f_i - f_j)^2,$$

in which f_i is the ith component of f. Others reflect local roughness. For a one-dimensional image, one of these is the first-difference penalty function

$$\Phi_1(f) = \sum_{i=1}^{N-1} (f_{i+1} - f_i)^2.$$

Note that both Φ_O and Φ_1 can be written

$$\Phi(f) = f^T C f, \tag{3.3}$$

for some specified nonnegative definite, symmetric C.

Another global measure is the negative entropy

$$\Phi_{NE}(f) = \sum_i p_i \log p_i,$$

where $p_i = f_i/\sum_j f_j$ and we assume $f_i \geqslant 0$ for all i.

In view of the model of Section 2.1, a natural choice for Δ_1 is

$$\Delta_1(f,\hat{f}_O) = (g - Af)^T(g - Af). \tag{3.4}$$

(Remember that $g = A\hat{f}_O$.)

If Δ_1 is as given in (3.4) and Φ is of the form (3.3), then the MPD prescription is

$$\hat{f}_h = (A^T A + hC)^{-1} A^T g, \tag{3.5}$$

a linear, regularised restoration; c.f. Phillips (1962).

If Δ_1 is, again, as given in (3.4) and $\Phi = \Phi_{NE}$, we obtain, as a special case within this structure, the maximum entropy prescription of Gull and Daniell (1978); see also Skilling (1986). In this case there is no explicit formula such as (3.5) but, thanks to the presence of the logarithm in Φ_{NE}, the resulting restorations are guaranteed to be nonnegative.

The MPD prescription can be given a helpful interpretation in terms of Bayesian statistics. If, give or take additive constants, $-\Delta_1(f,\hat{f}_O)$ is interpreted as a loglikelihood and $-h\Phi(f)$ as the logarithm of a prior density for f, then the MPD prescription corresponds to the calculation of the mode of the Bayesian posterior distribution for f. If Δ_1 is as in (3.4) and if the model is as in Section 2 with ε Gaussian, then $-\Delta_1(f,\hat{f}_O)$ does indeed fit this interpretation. Whether or not $-h\Phi(f)$ can be related to a bona fide prior is quite another matter.

4. SOME OTHER MANIFESTATIONS OF REGULARISATION PRESCRIPTIONS

In this Section we briefly point out some other problems displaying the same general structure as regularisation. For a more detailed exposition, see Titterington (1985b).

4.1 Linear regression

In statistical terms, (2.1) represents a linear regression model. More generally, such a model takes the form (2.1) but with g a vector of length M and A an MxN matrix. The least-squares estimator, $\hat{f}_O$, of f still satisfies $A^TA\hat{f}_O = A^Tg$, and, if $M < N$ and A is of full rank,

$$\hat{f}_O = (A^TA)^{-1}A^Tg. \tag{4.1.1}$$

Use of the prescription (3.5) constitutes ridge-regression, the subject of a vast literature (Draper and van Nostrand, 1979). Unlike (4.1.1), (3.5) can also be used when A^TA is singular, as will certainly occur in over-parameterised problems (the case $M > N$), such as that described by Green et al.(1983).

Note that the predictors, $A\hat{f}_h$, for g, based on $\hat{f}_h$, can be written

$$\hat{g}_h = A(A^TA + hC)^{-1}A^Tg \tag{4.1.2}$$

$$= K_h g,$$

say. If A were NxN and nonsingular, then

$$K_0 = I. \tag{4.1.3}$$

4.2 Nonparametric regression

Suppose we have a general regression model

$$\mathbb{E}g(t) = f(t)$$

$0 < t < 1$ where g(t) is the observed value of f(t) and f(.) is some unknown function. Observations $\{(g_i, t_i),\ i = 1, \ldots, N\}$ are made with

$$g_i = F_i + \varepsilon_i,$$

$i = 1, \ldots, N$, where $F_i = f(t_i)$. The ε_i are uncorrelated, with zero means and constant variance. The objective is to obtain a sensible estimate of the regression function f(t), based on the data. Silverman (1985) uses a MPD prescription of the form

$$\min_f \{(g - F)^T(g - F) + h\Phi(f)\},$$

where

$$\Phi(f) = \int \{f''(t)\}^2 dt$$

and the minimisation is over a suitable class of functions. The resulting f is a member of a class of cubic splines and it is interesting to note (Silverman, 1985) that there is a finite dimensional formulation in terms of B-splines, that leads to a prescription analogous to (4.1.2), with (4.1.3), for the estimator of E(g). Titterington (1985b, Section 7) makes similar comments about the activity known as kriging.

4.3 Multinomial smoothing and probability density estimation

Suppose f represents a set of k multinomial probabilities and $\hat{f}_0$ the set of relative frequencies obtained from a sample of data. Several approaches to the smoothing of $\hat{f}_0$ are outlined in Sections 2 and 3 of Titterington (1985b). Some of them result in linear estimators of the form

$$\hat{f}_h = K_h \hat{f}_0,$$

with $K_0 = I$ ((4.1.3) again). In this context, K_h is called a kernel matrix and $\hat{f}_h$ a kernel-based estimator.

MPD estimators have also been investigated, particularly based on the choice of the Kullback-Leibler directed divergence for Δ_1, i.e.

$$\Delta_1(f,f') = \sum_{i=1}^{k} f_i' \log(f_i'/f_i).$$

Since minimising $\Delta_1(f,\hat{f}_0)$ is equivalent to maximising the multinomial loglikelihood, we obtain maximum penalised likelihood estimators.

The estimation of multinomial probabilities is the equivalent, for a finite sample space, of nonparametric density estimation. Titterington (1985b, Section 4) shows that both the kernel method and the maximum penalised likelihood method, among others, have been widely discussed for density estimation; see also Silverman (1986). For a sample of univariate data, the equivalent of $\hat{f}_0$ is the "derivative" of the empirical distribution function. Formally, this can be written

$$\hat{f}_0(x) = n^{-1} \sum_{i=1}^{n} \delta(x - x_i),$$

where $\delta(.)$ is the Dirac delta function and $x_1, \ldots, x_n$ form the sample. The general form of kernel-based estimators is

$$\hat{f}_h(x) = h^{-1} \int K\{(x-y)/h\}\, \hat{f}_0(y)dy, \qquad (4.3.1)$$

where K(.) is a probability density function and

$$\lim_{h \to 0} h^{-1} K\{./h\} = \delta(.) . \qquad (4.3.2)$$

Equations (4.3.1) and (4.3.2) are direct analogues of (4.1.2) and (4.1.3).

5. CHOICE OF REGULARISATION PARAMETER

In this section we make some brief remarks about the second choice required of us, that of h, the degree of smoothing, in the MPD prescription.

5.1 Bayesian approach

Were $-h\Phi(f)$ truly the logarithm of a prior density then h would be a parameter of the prior and therefore, ideally, prespecified. Usually this ideal does not obtain and either h has to be elicited, or a hyperprior density must be specified for h, or h has to be estimated by an empirical Bayes technique. For instance, if $\pi_h(f)$ denotes the prior for f and $p(\hat{f}_O|f)$ denotes the likelihood factor, then

$$p_h(\hat{f}_O) \equiv \int p(\hat{f}_O|f)\pi_h(f)df$$

defines a "marginal" density for $\hat{f}_O$ within which h can be estimated, by maximum likelihood, for instance; see Titterington (1985b, Section 3.1).

5.2 Minimum risk choice and cross-validation

The motivation here is to get as close as possible to the true f, f_T, in some sense. Accordingly we solve

$$\min_{h} \; \mathop{E}_{data} \; \Delta(f_T, f_h) ,$$

where Δ is a distance measure, now thought of as a loss function. If Δ is quadratic, this leads to minimum mean-squared error choices, again reflecting trade-off between bias and variance. A practical difficulty arises in that the optimal $\hat{h}$ is a function $\hat{h}(f_T)$ of the true f_T! In some contexts an initial estimate $\tilde{f}$ can be used such that $\hat{h}(\tilde{f})$ is satisfactory. Another way of obtaining a purely data-based choice of h that is "close" to $\hat{h}(f_T)$, at least asymptotically, is to use cross-validation; see, for instance, Craven and Wahba (1979), Golub et al (1979), Silverman (1985), and O'Sullivan and Wahba (1985). Roughly speaking, and returning to the context of image restoration, the observation on each pixel is separately "predicted" using the prescription obtained from all <u>other</u> pixels. The N predictions are compared with the N datum values using the measure of distance, and $h = \tilde{h}$ is chosen to achieve optimum fit. The above-mentioned papers discuss theoretical and practical aspects of various versions of cross-validation; see also Stewart and Durrani (1986).

5.3 Choice based on fit to the data

Another philosophy for choosing the degree of regularisation is to smooth as much as possible subject to not being too far from the data. This can be expressed as

$$\min_f \Phi(f) \quad \text{subject to}$$

$$\Delta_1(f,\hat{f}_0) \leqslant c,$$

for specified c. In this context, Δ_1 has the interpretation of a measure of fit, the constraint is generally satisfied with equality, and an analysis of the problem based on a Lagrange multiplier takes us back to criterion (3.2): h is determined by γ, which is determined by c. See, for instance, Burch et al. (1983).

6. DISCUSSION

It is important to investigate whether either of the choices introduced in Section 2 is crucial as far as the restoration is concerned. Folklore, experience and some hard results in the density-estimation literature have it that the choice of prescription is not particularly important but that the choice of degree of smoothing can be. I anticipate that a similar pattern will hold in other fields of application of regularisation and I do not find myself able to accept the pre-eminence, in this context, accorded by Skilling (1986) to the maximum entropy prescription.

Recent work with Dr. P. Hall (Hall and Titterington, 1986) has established that, in at least some specialised situations, the method discussed in Section 5.3, based on values of c that are used in regularisation practice, leads to a value of h that is of the order of $\hat{h}^{\frac{1}{2}}$, where $\hat{h}$ is the minimum risk value discussed in Section 5.2. Since, in these contexts, $\hat{h}$ will be small, it is clear that the method of Section 5.3.oversmooths, compared with that of Section 5.2. Whether or not this result is of practical relevance and, if so which approach, if either, provides the more useful restorations, is a matter for future report.

7. ACKNOWLEDGEMENTS

I am grateful to Dr. P. Hall, of the Australian National University, for several helpful discussions.

8. REFERENCES

Burch, S.F., Gull, S.F. and Skilling, J., (1983) "Image restoration by a powerful maximum entropy method". Comput. Vision Graphics Im. Proc. **23**, 113-128.

Cornwell, T.J. and Evans, K.F., (1985) "A simple maximum entropy deconvolution algorithm." Astron. Astrophys. **143**, 77-85.

Craven, P. and Wahba, G., (1979) "Smoothing noisy data with spline functions: estimating the correct degree of smoothing by the method of generalized cross-validation." *Numer. Math.* **31**, 377-403.

Draper, N.R. and van Nostrand, R.C., (1979) "Ridge regression and James-Stein estimation: review and comments." Technometrics **21**, 451-466.

Golub, G.H., Heath, M. and Wahba, G., (1979) "Generalized cross-validation as a method for choosing a good ridge parameter." Technometrics **21**, 215-223.

Green, P.J., Jennison, C.R. and Seheult, A.G., (1983) "Discussion of a paper by G.N. Wilkinson et al." *J.R. Statist. Soc.* **B 45**, 193-195.

Gull, S.F. and Daniell, G.J., (1978) "Image reconstruction from incomplete and noisy data" Nature **272**, 686-690.

Hall, P. and Titterington, D.M., (1986) "On some smoothing techniques used in image restoration." *J.R. Statist. Soc. B,* **48**, to appear.

O'Sullivan, F. and Wahba, G., (1985) "A cross-validated Bayesian retrieval algorithm for nonlinear remote sensing experiments." *J. Comput. Phys.* **59**, 441-455.

Phillips, D.L., (1962) "A technique for the numerical solution of certain integral equations of the first kind." *J. Assoc. Comput. Mach.* **9**, 84-97.

Silverman, B.W., (1985) "Some aspects of the spline smoothing approach to nonparametric regression curve fitting (with discussion)." *J.R. Statist. Soc.* **B 47**, 1-52.

Silverman, B.W., (1986) "Density Estimation for Statistics and Data Analysis." Chapman and Hall, London.

Skilling, J., (1986) "Maximum entropy data analysis." This volume.

Stewart, K. and Durrani, T.S., (1986) "Estimation of the optimal regularisation parameter in bandlimited signal extraction." This volume.

Titterington, D.M., (1985a). "General structure of regularization procedures in image reconstruction." *Astron. Astrophys.* **144**, 381-387.

Titterington, D.M., (1985b) "Common structure of smoothing techniques in statistics". Int. Statist. Rev. **53** . 141-170.

COMPACT OPERATOR EQUATIONS, REGULARIZATION AND SUPER-RESOLUTION

J.B. Abbiss and P.G. Earwicker
(Space Department, Royal Aircraft Establishment, Farnborough)

ABSTRACT

We consider the problem of object restoration from blurred, incomplete or noisy image data. We assume that the imaging process can be represented by a compact linear operator A. Problems of this type arise in a variety of fields, including optical, acoustical and radar imaging and in laser anemometry, as well as in band-limited signal and image extrapolation, and are now widely known to be improperly posed, in the sense that small perturbations in the data can cause gross changes in the reconstruction; in mathematical terms, the inverse of the operator A is not continuous. We discuss two closely related aspects of the problem - firstly, efficient methods of exploiting prior knowledge about the object by appropriate modification of the operator, and, secondly, the application of regularizing techniques as a means of controlling noise and stabilizing the reconstruction process.

A problem of the object-restoration type occurs naturally in space-borne synthetic aperture radar systems, for which the object-image relationship can be written in the form of a complex two-dimensional Fredholm integral equation of the first kind. An example is presented which illustrates the application of regularization techniques to the solution of this equation.

1. INTRODUCTION

The problem of object reconstruction from blurred, incomplete or noisy image data, or some equivalent task of super-resolution type, occurs in a wide variety of scientific and engineering fields, including optical, acoustical and radar imaging and in band-limited signal extrapolation. Such improperly-posed problems are characterised by the common feature that small perturbations of the data can cause gross changes in the reconstruction. This extreme sensitivity to error arises from the strongly smoothing properties of the imaging process; under the action of a compact (completely continuous) operator, which is the class considered here, widely differing types of object may be mapped into closely similar images. Thus if the data are subject to error the reconstruction process is intrinsically unstable.

2. REGULARIZED RECONSTRUCTION

We suppose some object f, an element of a Hilbert space F, is imaged by a linear operator A into an element of another Hilbert space G, and that this image is perturbed by an unknown additive noise function which may arise in either the imaging or the observation process. The effect of the noise in practice will often be to displace the image out of the range of A - i.e. there will no longer be any element of F corresponding to the observed image g. Thus the equation g = Af will have no solution.

$\|.\|_H$ will denote the norm in the appropriate Hilbert space. An approximate least-squares solution might be sought by minimising the quantity $\|g - Af\|_G^2$ to obtain the normal equation

$$A^*Af = A^*g \ ,$$

where A* is the operator adjoint to A.

It very commonly happens, however, that the inverse of A*A is discontinuous (i.e unbounded) - for example, when A is a compact operator. In such cases the reconstruction would be characterised by extreme instability with respect to perturbations of the image. The problem is then said to be ill-posed (Tikhonov (1977)).

These difficulties can be alleviated by one or other of the various techniques of regularization which have been developed in recent years - see, for example, Nashed (1976). A commonly-used procedure is to replace the normal equation above by the regularized form

$$(A^*A + \alpha I)f_\alpha = A^*g, \qquad (2.1)$$

where α, the regularization parameter, is some 'small' positive quantity. The operator $A^*A + \alpha I$ in this equation has a continuous inverse and the reconstruction f_α will now be stable with respect to small changes in the image data.

Equation (2.1) can also be obtained (Miller (1970)) as the minimiser of the functional

$$\|g - Af\|_G^2 + \alpha\|f\|_F^2.$$

From this point of view, α can sometimes be given a physical interpretation. If, for example, ε is a measure in the norm sense of the 'noise' in the image and E a known bound on the norm of the original object, one could reasonably require that, for permissible reconstructions f',

$$\|g - Af'\|_G^2 \leq \varepsilon \text{ and } \|f'\|_F \leq E .$$

The required functional follows by combining these constraints quadratically and making the identification $\alpha = \varepsilon^2/E^2$.

The eigenvalues of the symmetric operator A*A are non-negative, so that the eigenvalues of A*A + αI are strictly positive. Hence A*A + αI is a positive definite operator, its inverse exists and the solution can be expressed in either of the forms

$$f_\alpha = (A^*A + \alpha I)^{-1}A^*g \tag{2.2}$$

or

$$f_\alpha = A^*(AA^* + \alpha I)^{-1}g. \tag{2.3}$$

It should be noted that this regularized solution cannot be completely data-consistent, since if $Af_\alpha = g$, either $\alpha = 0$ or $f_\alpha = 0$. We observe also that as $\alpha \to 0$, f_α tends to the solution obtained from the generalized inverse of A (Groetsch (1984)). Since, however, the generalized inverse of a compact operator is unbounded, some form of regularization will still be necessary if this solution is to be stable in the sense described above.

A slightly different form of regularization, which has certain useful properties in applications to coherent optical and related problems (Abbiss et al (1983a, b)) can be written as

$$[(1 - \alpha)A^*A + \alpha I]f_\alpha = A^*g . \tag{2.4}$$

The reconstruction again tends to the generalized inverse solution as $\alpha \to 0$; as $\alpha \to 1$, on the other hand, $f_\alpha \to A^*g$, which corresponds to the geometrical image in the optical case.

3. PROPERTIES OF FINITE-DIMENSIONAL APPROXIMATIONS

In practice, the image g will be available as a finite array of samples. The representation within a computer of the reconstruction f_α will also be a finite array. Hence A becomes a matrix and equation (2.1), for example, is to be interpreted as a matrix equation.

The model chosen to represent the physical system under study and the specific form of discretization used will determine the approximations to the spaces F and G, and in particular their inner products. The structure of the latter will reflect the known or desired characteristics of image or object; since the adjoint A* is defined in terms of the inner products, the form of the reconstruction will also be determined by them.

Suppose the approximation to F is K^N, the N-dimensional space of possibly complex numbers, with inner product

$$(f_1, f_2)_{K^N} = \sum_{i=1}^{N} (W_F f_1)_i \overline{f_{2i}} , \tag{3.1}$$

where W_F is a positive definite weighting matrix which may be required, for example, to predispose the reconstructed object to a particular shape. Similarly the approximation to G is taken to be K^M, with inner product

$$(g_1, g_2)_{K^M} = \sum_{i=1}^{M} (W_G g_1)_i \overline{g_{2i}} . \tag{3.2}$$

Then we find for the adjoint

$$A^* = W_F^{-1} A^H W_G , \tag{3.3}$$

where A^H is the Hermitian response of A.

Hence, from equation (2.1),

$$f_\alpha = \left(A^H W_G A + \alpha W_F\right)^{-1} A^H W_G g \tag{3.4}$$

or

$$f_\alpha = W_F^{-1} A^H \left(A W_F^{-1} A^H + \alpha W_G^{-1}\right)^{-1} g. \tag{3.5}$$

For object domains of finite extent, W_F may be used to control the behaviour of the reconstruction at the boundaries - for example, by incorporating Chebyshev weighting. On the other hand, if the covariance matrix for the image noise is known, its inverse would be a natural choice for W_G (Luenberger (1969), Luttrell (1985)). Note that if W_F and W_G can be determined with sufficient precision, the parameter α may become redundant in equations (3.4) and (3.5).

4. COMPUTED RECONSTRUCTIONS

We discuss in some detail one particular method for solving equation (3.4). Consider first the special case for which $W_F = I_N$ and $W_G = I_M$. Then equation (3.4) becomes

$$f_\alpha = \left(A^H A + \alpha I_N\right)^{-1} A^H g \ . \tag{4.1}$$

We shall suppose the computation of f_α proceeds via a singular value decomposition (SVD) of A (Golub and Reinsch (1970)). The inaccuracy inherent in forming $A^H A$ directly is then avoided; in addition, a study of the solution behaviour for a range of values of α is made easier. Thus we write

$$A = U \Sigma V^H \ , \tag{4.2}$$

where $U^H U = V^H V = V V^H = I_N$

and $\Sigma = \mathrm{diag}(\sigma_1, \ldots, \sigma_N)$, $\sigma_i \geqslant 0$.

U consists of the N orthonormalized eigenvectors u_i associated with the N largest eigenvalues of AA^H, and V consists of the N orthonormalized eigenvectors v_i of $A^H A$. In the present context orthonormality is defined by the inner products

$$(u_j, u_k)_{K^N} = \sum_{i=1}^{N} u_{ji} \overline{u_{ki}} = \delta_{jk} \tag{4.3}$$

and

$$(v_j, v_k)_{K^M} = \sum_{i=1}^{M} v_{ji} \overline{v_{ki}} = \delta_{jk} \ . \tag{4.4}$$

The σ_i are the singular values of A : they are the non-negative square roots of the eigenvalues of $A^H A$.

Then, from equation (4.1),

$$f_\alpha = V \,\mathrm{diag}\left(\ldots, \frac{\sigma_i}{\sigma_i^2 + \alpha}, \ldots\right) U^H g. \tag{4.5}$$

The stabilizing rôle played by α in the reconstruction process is apparent here; the modulus of continuity of the reconstruction operator is in fact bounded by $\frac{1}{2}\alpha^{-\frac{1}{2}}$ (Deschamps and Cabayan (1972)). Note that if a series of numerical experiments is to be performed for a range of values of α, the SVD of A and the evaluation of $U^H g$ need be computed once only. By carrying out a preliminary Choleski factorization (LINPACK Users' Guide (1979)) of the weight matrices W_F and W_G, the more general form of equation (3.4) can also be made amenable to standard numerical SVD techniques.

Thus we put

$$W_G = R_G^H R_G \,, \quad W_F = R_F^H R_F \,,$$

where R_F and R_G are upper triangular matrices.

Substituting in equation (3.4) we obtain

$$\left(B^H B + \alpha I_N\right) f'_\alpha = B^H g' \,, \tag{4.6}$$

where $B = R_G A R_F^{-1}$, $f'_\alpha = R_F f_\alpha$ and $g' = R_G g$.

SVD of B will now yield f'_α and hence f_α.

The singular system (v_i, σ_i, u_i) of A with respect to the inner products of equations (3.1) and (3.2) is simply related to the singular system (v'_i, σ'_i, u'_i) of B with respect to the inner products of equations (4.3) and (4.4). We find

$$\left.\begin{aligned} \sigma_i &= \sigma'_i \\ v_i &= R_F^{-1} v'_i \\ u_i &= R_G^{-1} u'_i \end{aligned}\right\} \tag{4.7}$$

In terms of the singular system for B, the solution to equation (4.6) can be rewritten as

$$f'_\alpha = \sum_{i=1}^{N} \left(\frac{\sigma_i}{\sigma_i^2 + \alpha} \right) \langle g', u'_i \rangle_{K^M} v'_i .$$

Hence

$$f_\alpha = \sum_{i=1}^{N} \left(\frac{\sigma_i}{\sigma_i^2 + \alpha} \right) (g, u_i)_{K^M} v_i .$$

Again it can be seen that

$$\lim_{\alpha \to 0} f_\alpha = A^\dagger g ,$$

where $A^\dagger$ is the generalized inverse of A with respect to the inner products defined by equations (3.1) and (3.2).

5. APPLICATION TO SYNTHETIC APERTURE RADAR IMAGES

We discuss briefly the application of some of the ideas considered above to the specific problem of achieving enhanced feature definition (or 'super-resolution') in images obtained with space-borne synthetic aperture radar systems. These are based on the same range and Doppler tracking methods as conventional coherent radars - in the SAR arrangement, however, the signals received at each point of the satellite's track are recorded and all of the echoes arising from the ground area illuminated by the radar 'footprint' are coherently processed to yield detailed ground reflectivity information in both azimuth (along-track) and range (across track) directions. This information is assembled into the basic SAR image. By exploiting in a further stage of digital processing prior knowledge of the SAR system parameters and of the likely extent of the features of interest, enhanced resolution in this image can be achieved.

The SAR image of a single bright point on the earth's surface, embedded in an absorbing background, would have essentially the same structure as the diffraction pattern arising from a rectangular aperture in an optical arrangement, but with certain additional phase terms in both range and azimuth directions (Barber (1983)). These are however separable, and the point spread function $K(x_1,x_2)$ can be factorised into a straightforward product of range, x_1, and azimuth, x_2, components. Thus, if x_1 and x_2 are expressed in units of metres on the ground

$$K(x_1,x_2) = K_1(x_1)K_2(x_2) \ , \tag{5.1}$$

where $K_1(x_1) = \exp\left\{-\, i2\pi\left(k_{11}x_1 + k_{12}x_1^2\right)\right\} \dfrac{\sin(\pi c_1 x_1)}{\pi c_1 x_1}$,

$$K_2(x_2) = \exp\left\{-\, i2\pi\left(k_{21}x_2 - k_{22}x_2^2\right)\right\} \frac{\sin(\pi c_2 x_2)}{\pi c_2 x_2}$$

and $k_{11} = \dfrac{2f_o}{c}\sin\theta$, $k_{12} = \dfrac{2\Delta F_R}{Tc^2}\sin^2\theta$, $c_1 = \dfrac{2\Delta F_R}{c}\sin\theta$,

$$k_{21} = \frac{2f_o a_1}{cv_p} , \qquad k_{22} = \frac{2f_o a_2}{cv_p^2} , \qquad c_2 = \frac{\Delta F_A}{v_p} .$$

In these expression, c is the velocity of light, v_p is the velocity of the radar relative to the target, f_o the centre frequency of the transmitted chirped pulse, ΔF_R the chirp bandwidth, ΔF_A the Doppler bandwidth, T the chirp duration (pulse width) and θ the angle between the direction of incidence of the radiation and the local vertical. a_1 and a_2 are the coefficients of the linear and quadratic terms in the range polynomial.

Hence the image g of an extended object f is given by a two-dimensional complex Fredholm integral equation of the first kind:

$$g(x_1,x_2) = \iint_D K_1(x_1 - y_1)K_2(x_2 - y_2)f(y_1,y_2)dy_1dy_2 \ , \tag{5.2}$$

where D is the domain of f. The super-resolution problem for SAR imagery then consists in estimating or 'reconstructing' the object f from its image g. (Note that D need not be rectangular in form, or even connected.)

We conclude with some results obtained by applying the reconstruction techniques described above to a simulated satellite - SAR image. Fig. 1 shows the simulated object, consisting of three isolated pixels of relative brightness 4 (upper left), 2 (lower left) and 3. The modulus of the resulting complex image, calculated for a SAR system corresponding closely to one actually flown on an operational satellite (SEASAT, launched in 1978 to carry out ocean surface studies), is displayed in Fig. 2, to the same scale. The unstabilized reconstruction, for which the regularization parameter was set to zero, is shown in Fig. 3; here the prior knowledge was of the area of the image in which the object was expected to lie, and, of course, the parameters of the imaging operator. Although computer round-off error was the only source of noise in the image, none of the original structure can be discerned in Fig. 3. However, with the regularization parameter set to 10^{-7} (approximately the level of the single-precision round-off error) the object distribution is well recovered. The result is shown in Fig. 4. Although there is still some blurring, the peaks are in the correct positions, with relative brightnesses of 4, 2.2 and 3.0. This reconstruction is nearly the optimum. Further regularization simply increases the blurring and reduces the total energy in the object estimate.

Fig. 1

Fig. 2

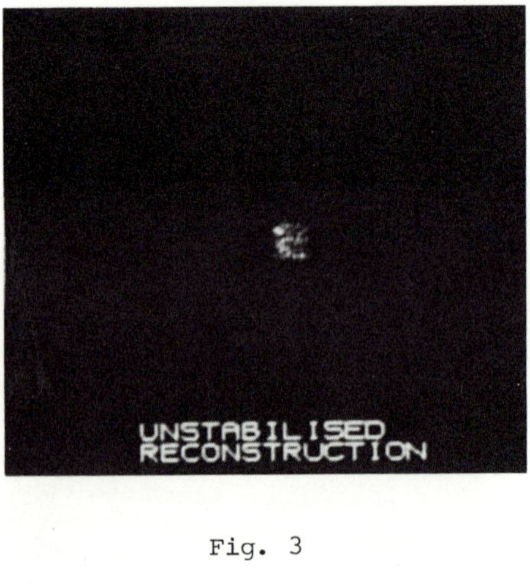

Fig. 3

Fig. 4

REFERENCES

Abbiss, J.B., De Mol, C. and Dhadwal, H.S., (1983) "Regularized iterative and non-iterative procedures for object restoration from experimental data." *Optica Acta*, **30**, 1, 107-124. See also: Abbiss, J.B., Defrise, M, De Mol, C. and Dhadwal, H.S., (1983). "Regularized iterative and noniterative procedures for object restoration in the presence of noise: an error analysis," *J. Opt. Soc., Am.*, **73**, 11, 1470-5.

Barber, B.C., (1983) "Some properties of SAR speckle." In: Satellite Microwave Remote Sensing (ed. T.D. Allan) Ellis Horwood Ltd., Chichester.

Deschamps, G.A. and Cabayan, H.S., (1972). "Antenna synthesis and solution of inverse problems by regularization methods." *IEEE Transactions on Antennas and Propagation*, AP-20, **3**, 268-74.

Dongarra, J.J., Moler, C.B., Bunch, J.R. and Stewart, G.W., (1979) LINPACK Users' Guide, SIAM, Philadelphia.

Golub, G.H. and Reinsch, C., (1970) "Singular value decomposition and least squares solutions". *Numer. Math.*, **14**, 403-20.

Groetsch, C.W., (1984) The Theory of Tikhonov Regularization for Fredholm Equations of the First Kind. Pitman Publishing Ltd., London.

Luenberger, D.G., (1969) Optimization by Vector Space Methods. John Wiley and Sons Inc, New York.

Luttrell, S.P., (1985) "Prior knowledge and object reconstruction using the best linear estimate technique." *Optica Acta,* **32**, 6, 703-16.

Miller, K., (1970) "Least squares methods for ill-posed problems with a prescribed bound." *SIAM J. Math. Anal.,* **1**, 52-74.

Nashed, M.Z., (1976) "Aspects of generalized inverses in analysis and regularization." In: Generalized Inverses and Applications (ed. M.Z. Nashed), Academic Press, New York.

Tikhonov, A.N. and Arsenin, V.Y., (1977) Solutions of Ill-Posed Problems. V.H. Winston and Sons, Washington, DC.

ESTIMATION OF THE OPTIMAL REGULARISATION PARAMETER IN BANDLIMITED SIGNAL EXTRAPOLATION

K. Stewart and T.S. Durrani
(Signal Processing Group, Department of Electronic and Electrical Engineering, University of Strathclyde, Glasgow)

1. ABSTRACT

A major difficulty encountered in the regularisation of single or multi-dimensional signal reconstruction problems when the data is noisy is that of choosing the 'best' approximation to the original problem which the regularisation scheme in question can provide.

This difficulty may be regarded in most regularisation procedures to be that of selecting the value of the regularisation or 'smoothing' parameter α which will be applied to the problem.

When the statistics of the data noise process are known (particularly the noise variance) several methods exist for choosing α. If the noise variance is unknown, α must be selected purely from a knowledge of the data, and the problem of obtaining an optimal α is made much more difficult. The method of Weighted Cross-Validation (WCV) however, provides a means of estimating an appropriate value of regularisation parameter even if the variance of the noise (assumed white, zero-mean) is unknown.

This paper discusses the application of WCV to the problem of bandlimited signal extrapolation in the presence of noise when a zero-th order, Miller-Tikhonov regularising constraint has been applied to the problem. Firstly, the general principles and justification of WCV are described. The application of WCV to the bandlimited extrapolation problem is then reviewed, and finally experimental results are presented which illustrate the performance of the technique under varying conditions of signal-noise ratio.

2. INTRODUCTION

The minimum-norm solution to the bandlimited signal extrapolation problem (or reconstruction problem in the frequency domain) from finite, discrete data is well known to be ill-conditioned with respect to the data. For this reason, it is necessary to apply regularisation techniques to the problem in order to obtain meaningful, stable solutions when the data is noisy (see Abbiss et al. (1983), and Titterington (1985) for discussions of regularised reconstruction, and regularisation methods in general.)

Unfortunately, in order for the regularisation procedure in use to be effective, the regularisation parameter α which controls the procedure must be chosen so as to give the 'best' solution. For example, if α has a small value, the reconstructed signal is in close agreement with the recorded data. Its variance is, however, very large and repetition of the same experiment gives rise to markedly different solutions. This phenomenon generally manifests itself as spurious artifacts in the F-domain reconstruction of the signal. If α is selected to have a large value, the reconstructed signal variance is small, but it also possesses little data consistency, and - in terms of its F-domain reconstruction - possesses little resolution.

The problem of choosing α is, in fact, that of deciding where on the 'trajectory' of solutions lying between $\tilde{f}_0$ and $\tilde{f}_\infty$ (the reconstructed solutions obtained using $\alpha=0$ and $\alpha=\infty$ respectively) the solution $\tilde{f}_\alpha$ should lie so as to be as close as possible to the original signal, where the trajectory itself is defined by the regularisation scheme which has been selected.

3. BANDLIMITED SIGNAL EXTRAPOLATION

Given a set of N sampled data points from a signal f whose Fourier Transform is known to be only supported over $[-B,B]$ Hz (ie. $\hat{f}(\omega) = 0 : \omega \notin [-2\pi B, 2\pi B]\ s^{-1}$), the purpose of extrapolation is to estimate f over $t\in[-\infty,\infty]$, or equivalently to reconstruct it in the frequency domain over the range $[-B,B]$ Hz.

We shall assume that the original N data samples $f(t_i) : t_i \in [t_1 \dots t_N]$ have been corrupted by a zero-mean Gaussian noise process of variance σ^2 to provide the available data as $y(t_i) : t_i \in [t_1 \dots t_N]$ where

$$y(t_i) = f(t_i) + \varepsilon(t_i) \; : \; t_i \in [t_1 \ldots t_N] \tag{3.1}$$

The original signal may be considered to lie in a subspace of functions bandlimited to $[-2\pi B,\ 2\pi B]\ s^{-1}$ in a complex Hilbert space F whilst the true data, noisy data and noise vectors defined as

$$\underline{g}^T = [f(t_1) \; \ldots . \; f(t_N)]$$

$$\underline{y}^T = [y(t_1) \; \ldots . \; y(t_N)] \tag{3.2}$$

$$\underline{\varepsilon}^T = [\varepsilon(t_1) \; \ldots . \; \varepsilon(t_N)]$$

respectively are taken to lie in an N-dimensional complex Hilbert space G. These conventions permit the operator $S:F \to G$ to be introduced to describe the basic process of sampling the N data points as

$$Sf = \underline{g} = \begin{bmatrix} (f, C_1)_F \\ \vdots \\ (f, C_N)_F \end{bmatrix} \tag{3.3}$$

$$\underline{y} = \underline{g} + \underline{\varepsilon} \tag{3.4}$$

where $(.)_F$ denotes the inner product in space F and the functions $C_i(t)$ are the sampling functions

$$C_i(t) = \frac{\sin(2\pi B(t-t_i))}{\pi(t-t_i)} \tag{3.5}$$

The minimum-norm least-squares reconstruction $\tilde{f}$ may then be expressed as

$$\tilde{f} = S^*(S.S^*)^{-1}\underline{y} \tag{3.6}$$

where S^* is the adjoint of S. This expression is, due to the nature of the operator S, highly unstable with respect to the data and the resulting high solution variance requires the application of regularisation theory. In order to restore stability, a zeroth-order Tikhonov-Miller regularising constraint (or energy constraint in physical terms) may be included and the solution $\tilde{f}_\alpha$ is now defined as

$$\tilde{f}_\alpha = \min_h \sum_{i=1}^{N} |(Sh)(t_i) - y(t_i)|^2 + \alpha||h||_F^2 \qquad (3.7)$$

although many alternative regularisation schemes could have been adopted (see Tikhonov and Arsenin (1977), Groetsch (1984).)

The solution to (3.7) is given by

$$\tilde{f}_\alpha = S^*(SS^* + \alpha I)^{-1}\underline{y} \qquad (3.8)$$

The role of α - the regularisation parameter - in controlling the trade-off between high variance, data consistent solutions ($\alpha \to 0$) and low variance, biased solutions ($\alpha \to \infty$) should now be apparent. The question of how to choose α remains.

4. METHODS OF SELECTING α

A number of techniques have been used in the past to select α. Possibly the simplest of these is to merely select several different values of α in the reconstruction process (ie. follow $\tilde{f}_\alpha$ along its trajectory) and let the observer judge when the restored signal is close to what he or she was expecting to obtain. If little or no a-priori knowledge is available as to what signal is expected however, this approach is rather arbitrary, since the selected solution merely reflects the observer's preferences, which in the absence of a-priori knowledge must themselves be arbitrary.

An alternative is to select the solution $\tilde{f}_\alpha$ which meets a regularising constraint, which in this case would be of the form $\|\tilde{f}_\alpha\|_F^2 = E^2$ where E is a constant, resulting in a Lagrange multiplier problem. This method exchanges the problem of choosing α for that of choosing E however (see Twomey (1965) and Phillips (1962) for discussions of this approach.) Several other similar approaches exist for choosing α which constrain some functional $\Omega(\tilde{f}_\alpha)$ to take on a required value (Gordonova and Morozov (1973)), but the problem of the arbitrary choice of the constraint value which should be placed on $\Omega(\tilde{f}_\alpha)$ persists.

Other methods have been proposed such as the Discrepancy Principle described by Morozov (1968) (see Groetsch (1984) pp 43). This requires that given

$$\sum_{i=1}^{N} |y(t_i) - g(t_i)|^2 \leq \delta^2 \tag{4.1}$$

α should be selected such that

$$\sum_{i=1}^{N} |(S\tilde{f}_\alpha)(t_i) - y(t_i)|^2 = \delta^2 \tag{4.2}$$

or, in other words, so that the 'noise' on the solution should be of roughly the same order as that on the data. The Discrepancy Principle requires however, quite strong a-priori knowledge about the original noise process $\varepsilon(t)$ and such knowledge may often not be available.

Fortunately, techniques which estimate α - or estimate both α and an optimal order of regularisation without such a-priori knowledge - exist (see Davies (1982) and Titterington (1985)) including methods such as the Maximum-Likelihood approach. One such technique is Weighted Cross-Validation.

5. WEIGHTED CROSS-VALIDATION

The principles of Weighted Cross-Validation (WCV) (Wahba (1977), Golub, Heath, Wahba (1979)) are well known in the statistics literature and details of the technique may be found in the references. It has not been widely applied in the field of signal processing however, and it is worthwhile to briefly review WCV with particular reference to the bandlimited signal extrapolation problem.

Ideally, we would like to choose α so as to minimise the cost function $T(\alpha)$ where

$$T(\alpha) = \frac{1}{N} \sum_{i=1}^{N} |(Sf_\alpha^N)(t_i) - g(t_i)|^2 \tag{5.1}$$

Essentially, α is chosen so that the Euclidean distance between the vector $\underline{(S\tilde{f}_\alpha)}$ and the true data vector $\underline{g}$ is minimised given that $\tilde{f}_\alpha$ is only permitted to move along the regularised solution trajectory subject to α.

Obviously it is not possible to evaluate $T(\alpha)$ directly since $f(t_i) : t_i \in [t_1 \ldots t_N]$ are not available. Instead, WCV seeks to generate an alternative cost function whose minimiser estimates the value of α which minimises $T(\alpha)$ but which is directly calculable from the available data $y(t_i) : t_i \in [t_1 \ldots t_N]$. This is achieved as follows:

Let $\tilde{f}_{\alpha,k}$ be the solution to

$$\tilde{f}_{\alpha,k} = \min_{h} \sum_{\substack{i=1 \\ i \neq k}}^{N} |(Sh)(t_i) - y(t_i)|^2 + \|h\|_F^2 \qquad (5.2)$$

that is, let $\tilde{f}_{\alpha,k}$ be the regularised minimum-norm extrapolation or reconstruction resulting from a data set in which $y(t_k)$ has been omitted.

The process of calculating $\tilde{f}_{\alpha,k}$ is repeated in turn for $k = 1,2,\ldots N$ until it is possible to provide an estimate for each data point $y(t_k)$ formed from the remaining N-1 data points. This permits the underlying principle of WCV to be tested - that is, the best value of α will be that which allows each data point to be predicted most closely by the other data points. This value of α is found by constructing the function $V(\alpha)$ as a weighted mean-square prediction error of the form

$$V(\alpha) = \frac{1}{N} \sum_{k=1}^{N} |(S\tilde{f}_{\alpha,k})(t_k) - y(t_k)|^2 W_k(\alpha) \qquad (5.3)$$

where $W_k(\alpha)$ are weighting functions.

Justification for the use of $V(\alpha)$ to estimate the minimiser of $T(\alpha)$ is provided by the fact that if α_T is the minimiser of $E[T(\alpha)]$ and α_V is the corresponding minimiser of $E[V(\alpha)]$, then under certain smoothness conditions (see Wahba (1977) for details of these conditions and Sanz and Huang (1984) for a discussion of their significance)

$$\alpha_V = \alpha_T\,(1 + o(1)) \qquad \left(\lim_{N\to\infty} o(1) = 0\right) \qquad (5.4)$$

where $E[\,.\,]$ represents the expectation operator over the data. α_V is thus seen to be a biased estimate of α_T.

The technique of WCV is therefore to plot the function $V(\alpha)$ and to select the minimum value of $V(\alpha)$ by inspection (although this could obviously be achieved by some one-dimensional optimisation method such as polynomial approximation.) Thereafter, the final reconstruction is computed using the value of α corresponding to that minimum.

One drawback to this algorithm lies in its computational complexity. Fortunately, there exists a simple scheme for calculating the function $V(\alpha)$ which avoids the necessity of solving (5.2) N times and then evaluating (5.3) for each value of α. This scheme - which is based upon an eigensystem expansion of the operator SS* and is detailed in Wahba (1977)- converts (5.3) intc a rational polynomial expression in α which is readily computable.

6. COMPUTATIONAL RESULTS

In order to evaluate the effectiveness of WCV in selecting an optimal regularised solution, reconstruction simulations were carried out for a variety of test signals under varying conditions of signal-noise ratio.

The frequency-domain representation of a test signal typical of those which were used appears in Fig. 1. This 'object' - consisting of two side lobes and one central lobe- was bandlimited to 0.06Hz and was sampled at the rate of 1Hz to provide the available data as 17 samples at $t_i = i = [-8,-7... 6,7,8]$ s.

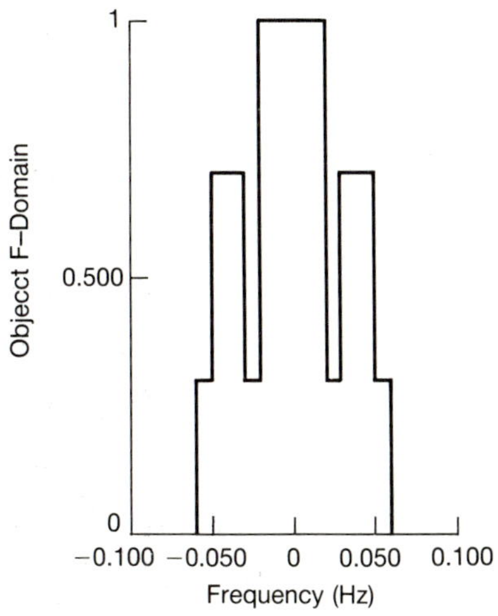

Fig. 1 Original Signal - F-domain

If no additive noise is present, and only finite word length effects are considered to be affecting the data, a very small value of α may be used and the resulting reconstruction will be close to the ideal minimum-norm least-squares reconstruction. The solution obtained in these circumstances appears in Fig. 2, where it may be observed that the central and side lobes of the original signal have been resolved and that the solution signal energy is approximately correct.

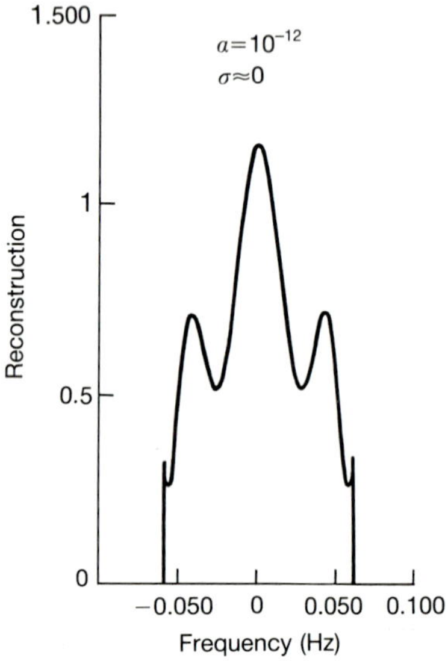

Fig. 2 Noiseless MNLS Reconstruction

The effects of noise, and the resultant problem of choosing α are illustrated in Figs. 3 and 4. Here, noise of standard deviation $\sigma = 1\times10^{-2}$ was added to the available data. The resulting regularised reconstruction for $\alpha = 10^{-1}$ appears in Fig. 3 demonstrating the absence of spurious signal artifacts and lack of resolution which characterises solutions obtained using values of α which are too large. Fig. 4, meanwhile, illustrates the converse case for an excessively small value of $\alpha = 10^{-4}$ used to perform a reconstruction using the same data as that used in Fig. 3. The introduction of spurious peaks into the F-domain reconstruction is apparent.

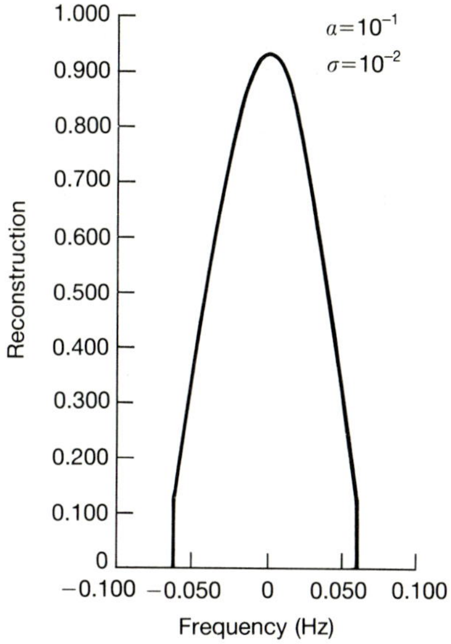

Fig. 3 Over-regularised Reconstruction

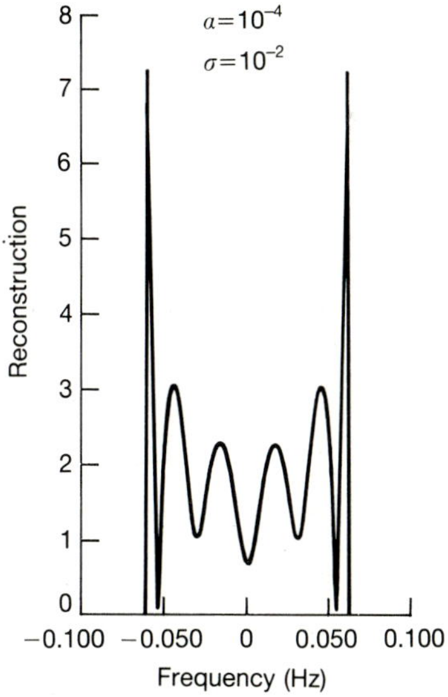

Fig. 4 Under-regularised Reconstruction

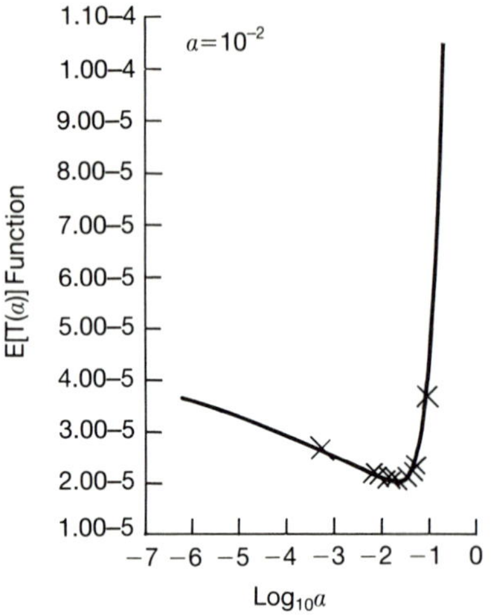

Fig. 5 WCV Estimates - Low SNR

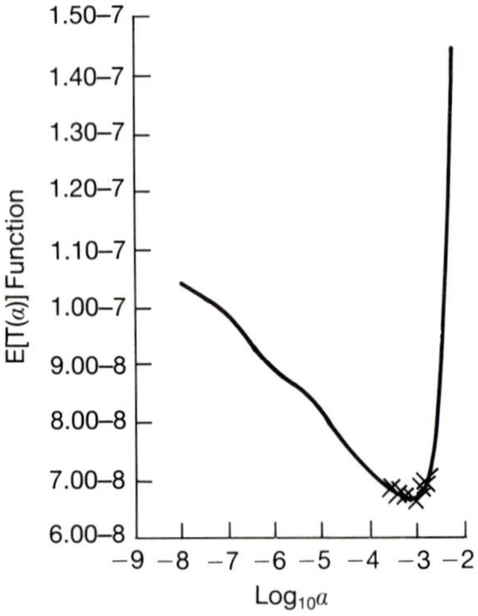

Fig. 6 WCV Estimates - High SNR

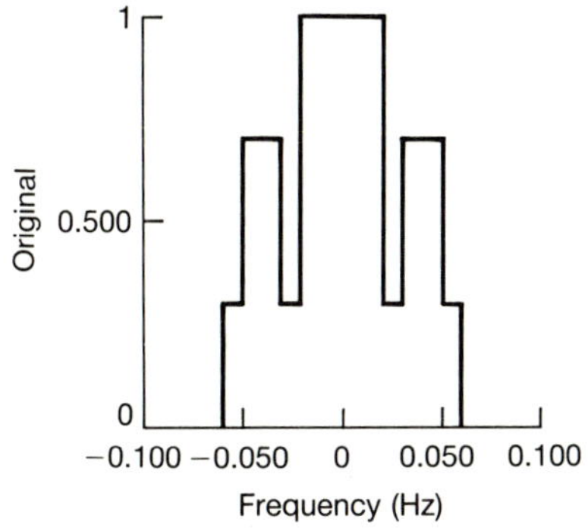

Fig. 7 Original Signal - F-domain

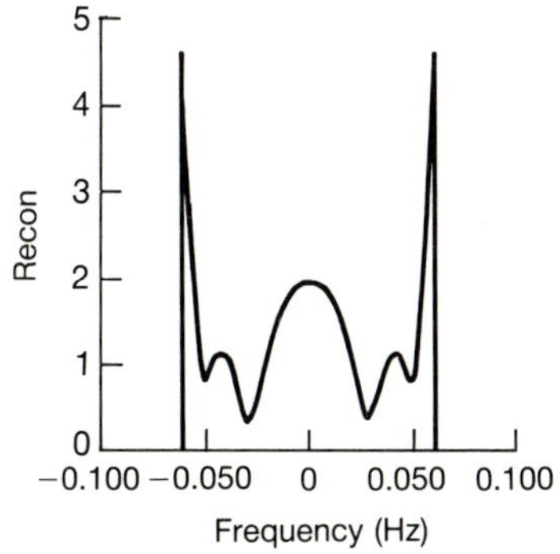

Fig. 8 Reconstruction - α < WCV Estimate

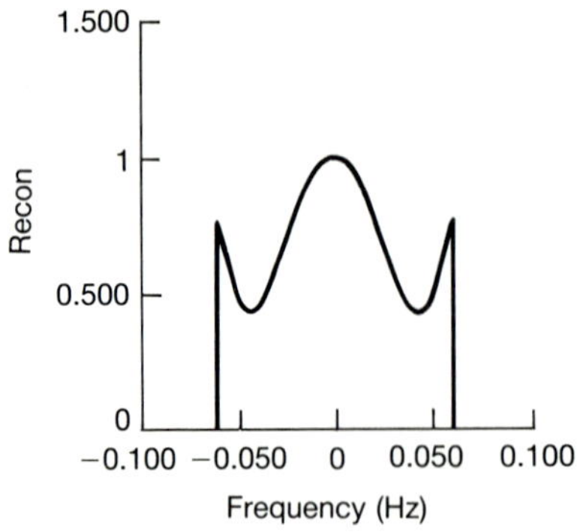

Fig. 9 Reconstruction - α = WCV Estimate

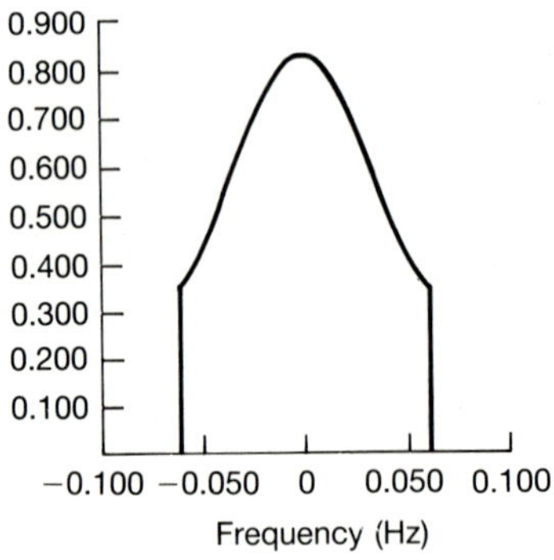

Fig. 10 Reconstruction - α > WCV Estimate

WCV was then used to estimate the optimal α value. Fig. 5 represents the $E[T(\alpha)]$ function for $\sigma = 1x10^{-2}$ using the same 17 data points discussed above. 10 different noisy data sets $\underline{y}$ were generated for this value of noise standard deviation using different noise generator seed values. Each cross on the $E[T(\alpha)]$ curve represents a WCV estimate of the optimal value of α. As may be observed, the WCV estimates predict the minimum location of the $E[T(\alpha)]$ curve well, although the variance of the estimator was not investigated. Fig. 6 illustrates the same experiment for $\sigma = 5x10^{-4}$ and it may be seen that WCV, as expected, predicts a smaller optimal value of α due to the increase in signal-noise ratio.

Figs. 7-10 illustrate the resulting application of WCV to the bandlimited signal extrapolation problem. The same original signal is used as in the experiments described above, however the noise standard deviation ($\sigma=10^{-2}$) was assumed unknown and WCV was used to estimate the optimal reconstructed signal. The predicted value of α provided by WCV was $\alpha = 10^{-2}$ and the reconstruction obtained using this value appears in Fig. 9, whilst the reconstructions obtained using $\alpha = 10^{-4}$ and $\alpha = 10^{-1}$ are represented in Figs. 8 and 10 respectively. It may be observed that the reconstructed signal which exhibits the best compromise between excessive signal energy and low resolution is that generated by WCV.

7. CONCLUSIONS

The technique of Weighted Cross-Validation has been shown to provide a useful means of estimating the optimal solution available from a given regularisation scheme without the imposition of arbitrary solution constraints or the requirement of strong a-priori knowledge, and experimental results have been presented which illustrate the performance of the technique. It is apparent that the procedure could be easily extended to include other signal restoration procedures such as constrained maximum-entropy reconstruction etc.

8. ACKNOWLEDGEMENTS

This work was supported by an M.O.D. Procurement Executive grant, and the authors would like to thank Dr. J.B. Abbiss and Dr. P.G. Earwicker for helpful discussions.

9. REFERENCES

Abbiss, J.B., De Mol, C. and Dhadwal, H.S., (1983) "Regularised Iterative and Non-iterative Procedures for Object Restoration from Experimental Data", *Optica Acta* **30**, 1, 107-124.

Davies, A.R., (1982) "On the Maximum Likelihood Regularisation of Fredholm Convolution Equations of the First Kind". In "Treatment of Integral Equations by Numerical Methods" (C.T.H. Baker and G.F. Miller, Eds.), pp. 95-105, Academic Press, London.

Golub, G.H., Heath, M. and Wahba, G., (1979) "Generalised Cross-validation as a Method for Choosing a Good Ridge Parameter". *Technometrics* **21**, 2, 215-223.

Gordonova, V.I., and Morozov, V.A., (1973) "Numerical Parameter Selection Algorithms in the Regularisation Method". *USSR Comp. Math. Math. Phys.* **13**, 3, 539-545.

Groetsch, C.W., (1984) "The Theory of Tikhonov Regularisation for Fredholm Equations of the First Kind". Pitman, London.

Morozov, V.A., (1968) "The Error Principle in the Solution of Operational Equations by the Regularisation Method", *USSR Comp. Math. Math. Phys.* **8**, 2, 63-87.

Phillips, D.L., (1962) "A Technique for the Numerical Solution of Certain Integral Equations of the First Kind". *J. Asso. Comput. Mach.* **9**, 84-97.

Sanz, J.L.C. and Huang, T.S., (1984) "A Unified Approach to Noniterative Linear Signal Restoration," *IEEE Trans. Acoust. Speech Sig. Proc.* **32**, 2, 403-409.

Tikhonov, A.N., and Arsenin, V.Y., (1977) "Solutions of Ill-Posed Problems". V.H. Winston and Sons, Washington D.C.

Titterington, D.M., (1985) "Common Structure of Smoothing Techniques in Statistics". *Int. Statist. Rev.* **53**, 141-170.

Twomey, S., (1965) "The Application of Numerical Filtering to the Solution of Integral Equations Encountered in Indirect Sensing Measurements", *J. Frank. Inst.* **279**, 2, 95-109.

Wahba, G., (1977) "Practical Approximate Solutions to Linear Operator Equations When the Data are Noisy". *SIAM J. Numer. Anal.* **14**, 4, 651-667.

SUPER RESOLUTION OF SEISMIC DATA USING PRIOR KNOWLEDGE

J.M. Blackledge and M.A. Fiddy
(Physics Department, King's College London)

and

W.A. Ward
(Earth Sciences Department, University of Leeds)

ABSTRACT

An algorithm for extrapolating the spectrum of any signal is presented and discussed in the context of seismic imaging and illustrated using seismic data supplied by the National Coal Board.

1. INTRODUCTION

It is well known that most seismic reflection data are unable to yield the resolution necessary for the quantitative evaluation and interpretation of seismic sections. This is largely due to the strongly coloured nature (i.e., characteristic shape) of the spectrum of the source pulse that is used to interrogate the ground. Thus, there is an increasing demand to design signal processing algorithms that can compensate for the low frequency content of a seismic trace so that high resolution sections can be obtained.

The dominant spectral components of a typical seismic record lie approximately between 30 Hz and 100 Hz. Signals with this frequency content can go some way to developing a geological model for the ground. However, it is now becoming increasingly necessary to have dominant frequencies up to 500 Hz in order to interpret geological structures to within 2 - 3 metres. This is particularly important for the surveys conducted by the National Coal Board in the UK which require detailed information about the geological nature of coal bearing strata to plan both short and long term mining operations. We are therefore looking for an increase in resolution by a factor of 3 - 4. Increasing the resolution by factors of this size is known as super resolution and is obtained by extrapolating the available spectrum of a signal. The degree of resolution enhancement or super resolution is simply given by the extent of spectral extrapolation.

In this paper we examine a method for synthesizing those frequency components that are unavailable in a typical seismic trace (in particular a NCB trace) due to the limits imposed by the 'physics' of a seismic experiment. The main physical factors that are responsible for limiting the frequency content of a signal are:

(i) Attenuation of the high frequency components.
(ii) The low frequency content of the source pulse which is required for adequate earth penetration because of (i).
(iii) The finite band width or frequency response associated with the geophones.
(iv) Loss and corruption of high frequency components by pre-processing such as CDP (Common Depth Point) stacking.

In general, the frequency content of the source pulse increases as the charge size decreases. However, one is still limited to using relatively low frequency high energy sources in order to establish the necessary penetration of seismic energy into the ground.

All these factors combine to yield the fundamental problem of how best to re-establish the spectrum of the impulse response function for the ground at a particular location when a noise corrupted, band limited and convolved form of this function is given. In this paper we present a solution to this problem which is based on the use of prior knowledge to extrapolate a given spectrum from an arbitrary frequency band. The solution is entirely data consistent and for this reason, some degree of spectral "whitening" is undertaken thus avoiding the extrapolation of a coloured spectrum. Stability is obtained by regularization and spectral components are generated where the signal noise ratio (SNR) is originally too poor for the data to be recovered. The technique can be compared to conventional deconvolution methods which change the spectral shape of the data within some band determined by the SNR.

In the time domain, spectral extrapolation not only sharpens the main events but introduces new events into the seismic signal which in many cases are laterally continuous from one trace to another. It is this lateral continuity that provides evidence for the reliability of the extrapolation procedure since each trace is processed individually.

2. THEORETICAL BACKGROUND

The implicit model for the signal is a convolution with additive noise so that the seismic trace data s(t) can be written as

$$s(t) = p(t)*f(t) + n(t) \tag{2.1}$$

where p(t) is the source function, n(t) is the noise function which to a good approximation is random in nature, and f(t) is the impulse response function for the ground. For a continuum model, the impulse response function is given by (Berteussen and Ursin 1983)

$$f(t) = \frac{1}{2} d_t \ln Z(t) \tag{2.2}$$

where Z(t) is the impedance function. Equation (2.1) assumes that data are collected in the far field and is based on the Born approximation (i.e. primary reflections only) and equation (2.2) assumes that the continuum is layered and non-absorbing. Equation (2.2) demonstrates that the strength of the signal s(t) depends on the gradient in the impedance of the ground. In practice, equation (2.1) assumes that the data have been CDP stacked (i.e. the signal s(t) is the average of many different offset measurements) which eliminates most of the second and higher order multiples that occur.

The basic problem is two fold:

(i) Given the data s(t), estimate a high SNR and therefore band-limited form of f(t), $f_1(t)$.

(ii) Given confidence in the estimate $f_1(t)$ with a certain noise-free frequency limit $\pm\Omega/2$ (i.e. given $f_1(\omega)$; $-\Omega/2 \leq \omega \leq \Omega/2$ where ω is the angular frequency), generate a new estimate $f_2(t)$ which is characterized by a frequency band width which is in excess of $\pm\Omega/2$ by a desired amount.

To solve (i) we apply a stochastic deconvolution operator to s(t) in the presence of noise and then filter. This method involves the use of a spectral Automatic Gain Control (AGC) and processes each trace by averaging the shape of the spectrum and applying its inverse in a multiplicative sense to whiten the spectrum. This process can be repeated to obtain optimum whitening of the spectrum which is then low-pass filtered to some value $|\Omega|/2$ giving the first estimate for the impulse response function $f_1(t)$. The solution to (ii) is based on a method of spectral extrapolation which incorporates prior information in the form of a weighting function. From the first estimate we can generate the function $F_1(\omega_n)$; $\omega_n \lessgtr |\Omega|/2$ which is related to the impulse response function by the Fourier transform

$$F_1(\omega_n) = \int_{-T/2}^{T/2} dt\, f(t)\exp(-i\omega_n t/T) \qquad (2.3)$$

Here, $F_1(\omega_n)$ are a set of numbers $1 \lessgtr n \lessgtr N$ spaced $1/T$ apart which are the discrete Fourier coefficients describing the first estimate $f_1(t)$; $t \lessgtr |T|/2$. We now wish to construct a second estimate $f_2(t)$ for the impulse response function $f(t)$ $-\infty \lessgtr t \lessgtr \infty$ from the numbers $F_1(\omega_n)$. A solution for an estimate of this form is equivalent to extrapolating $F_1(\omega_n)$ beyond the band limit $|\Omega|/2$; we shall express $f_2(t)$ as a finite sum of basis functions in the form

$$f_2(t) = \sum_{n=0}^{N} a_n \psi_n(t) \qquad (2.4)$$

If some weighting function $w(t)$ is available which can be constructed from prior information on the structure of $f(t)$, we can generalize (2.4) to include a weighting function by writing

$$f_2(t) = w(t) \sum_{n=0}^{N} a_n \psi_n(t) \qquad (2.5)$$

If we let $\psi_n(t) = \exp(i\omega_n t)$ then the weighted mean square error (Byrne and Fitzgerald 1983)

$$e = \int_{-T/2}^{T/2} w^{-1}(t)\,|f(t) - f_2(t)|^2 dt \qquad (2.6)$$

is minimized by finding the coefficients a_n of the matrix equation

$$\sum_{n=0}^{N} a_n W(\omega_m - \omega_n) = F_1(\omega_m) \qquad (2.7)$$

where $F_1(\omega_m)$ are the spectral data from the first estimate and $W(\omega_n)$ is the discrete Fourier transform of the weighting function. This is the linear weighting method of extrapolation where the weighting function is necessarily positive and real $w(t)>0\forall t$. The inverse weighting of the cost function increases the sensitivity of the second estimate $f_2(t)$ to small deviations between $f(t)$ and $f_2(t)$ in regions where $w(t)$ is small provided $w(t)$ is chosen to be small where $|f(t) - f_2(t)|^2$ is small. If $w(t)$ is chosen to approach zero when $|f(t) - f_2(t)|^2$ does not, then this criterion becomes unstable. The method can be regularized by adding a constant to the main diagonal of the matrix $W(\omega_m-\omega_n)$. This affects how much the second estimate is driven by the weighting function. The addition of a large value to the diagonal will cause $f_2(t)$ to degenerate towards a discrete Fourier transform of the initial spectral components (i.e. to degenerate towards $f_1(t)$) but regularization will stabilize the reconstruction in the presence of noise on the spectral data values $F_1(\omega_m)$. The function $f_2(t)$ is a linear weighted estimate of $f(t)$ which is completely data consistent and optimum in the sense that the inverse weighted mean square error is minimized.

3. WEIGHTING FUNCTIONS

The choice of weighting function is obviously critical in achieving a useful stable estimate $f_2(t)$. We are interested in generating an estimate which is characterized by a sequence of sharp and well defined events. From the algebraic form of the linear estimate (equation 2.5) and since the polynomial part is essentially smooth in behaviour, any sharp or spikey characteristics, if they are known to occur a priori are best encoded in the weighting function.

In many seismic experiments, prior information is available from bore-holes, both from petrophysical electric logs and from a geologist's detailed description of lithological changes seen in the cores. Information from boreholes allows a very detailed impedance model to be constructed for the ground at the localities of the borehole. Thus, from equation (2.2), a highly resolved impulse response function can be developed which in principle can be used as a weighting function for each seismic trace. However, this approach suffers from inaccuracies associated with the conversion of data as a function of depth to data as a function of the two way travel time t. Also, using borehole data to develop weighting functions is a specialized method which must be used in a way that is not over-committed to the ground model and which allows lateral changes to take place as reflection data farther from a borehole site are processed.

To generalize the method of extrapolation and make it acceptable to any form of seismic data and applicable when accurate ground models cannot be developed, we have considered a self weighted method of extrapolation. In this case, the frequency content of the first band limited estimate $f_1(t)$ is first increased by implementing any other suitable extrapolation technique that does not involve prior information. The result is then used as a weight to extrapolate $f_1(t)$ further using the linear weighting method. This argument leads one to consider using Burg's Maximum Entropy Method as a suitable weight generator since in its numerical practice, it can nearly always be expected to emphasize and sharpen the position of the main features on the data. The only danger in using this MEM is the introduction of spurious features in the presence of noise but this can be controlled by regularizing the data and adjusting the number of spectral components that the process is given to work with. This entropy estimate $f_{MEM}(t)$ is given by

$$f_{MEM}(t) = c_o \left| \sum_{n=0}^{N} c_n \exp(i\omega_n t/T) \right|^{-2} \tag{3.1}$$

where the coefficients c_n are found by solving the equation

$$\sum_{n=0}^{N} c_n F(\omega_m - \omega_n) = \begin{cases} 1; & m=0 \\ 0; & m=1,2,\ldots,N \end{cases} \tag{3.2}$$

This estimate is consistent with the data and is based on maximizing the entropy $\int \ln f_{MEM}(t)dt$. The matrix $F(\omega_m-\omega_n)$ is generated from the spectral coefficients of the input data and the MEM estimate can then be used as a weighting function. However, Burg's MEM requires that the input data be positive and yields a positive estimate $f_{MEM}(t)>0\,\forall t$. This constraint is consistent with the one imposed on the weight but is not well suited to seismic trace data which are composed of both positive and negative elements. In principle, there are a number of ways of over-coming this problem. For example, one could first integrate each trace by inverting equation (2.2) to generate an impedance profile which is a positive function $Z(t)>0\,\forall t$ and then differentiate after super resolving so that the result can be displayed in conventional wiggle trace form. The method actually suffers from the fact that Burg's MEM does not make an effective difference in the case of functions whose spectra are characterised by large DC levels (i.e. high values of the coefficients at $\omega = 0$). Thus, a weighting function produced by this method could turn out to be almost identical to the initial data, and therefore minimize the extent of extrapolation. Alternatively, one could choose to work with data on the amplitude envelope. This would involve a method of adequately and simultaneously super resolving the phase of the data so that the impulse response function (which is what we want) could be reconstructed. One could also attempt a solution assuming stationary phase. However, in principle it is not the polarity of the data that is constrained, only the polarity of the weight. We can therefore legitimately use a positive weighting function to super resolve the trace data (i.e. the first estimate $f_1(t)$). We therefore consider the square of the positive elements in the data as a source for Burg's MEM which generates a weight $w(t) = f_{MEM}(t)$ satisfying the condition $w(t)>0\forall t$.

In terms of geophysical prior knowledge, this method of weight generation assumes that a relatively small number of large reflection coefficients (i.e. elements of the impulse response function f(t)) dominate the reflection coefficient series. In practice, taking the square modulus of the data yields a positive function with dominant but not excessively large DC levels which not only satisfies Burg's MEM but also makes it work effectively. By adopting this approach, it is possible to super resolve seismic data using linearly weighted extrapolation but without the need for additional prior information. This approach can therefore be described as a method of spectral extrapolation using self generated weighting functions.

4. THE ALGORITHM

The basic structure of the algorithm is shown in figure 1 which gives a flow diagram of the processing steps involved. The data are first deconvolved using an AGC which is applied successively until the spectrum of s(t) has been completely whitened. The result is then low pass filtered at a frequency which defines the extent of the main components of the original spectrum and where it is safe to assume that the data at frequencies above this value are predominantly noise. This band width can be altered according to the spectral character of the data s(t) and assumes that the noise is predominantly high frequency. In practice, the AGC over amplifies the ultralow frequency components which causes low frequency ringing to occur. Since these components are well below the low frequency response level of geophones (i.e. 25 - 30 Hz), we suppress them using a high pass filter. This gives the first estimate

$f_1(t)$. The square modulus of $f_1(t)$ is then fed into Burg's MEM to generate a weight. Extrapolation of $F_1(\omega)$ using this weight is then undertaken. In practice, the extrapolated spectrum is fed into another AGC to ensure that the final spectrum is white. The result is then filtered and shaped to produce the second estimate $f_2(t)$. By processing each trace in this way, a super resolved seismic section can be obtained.

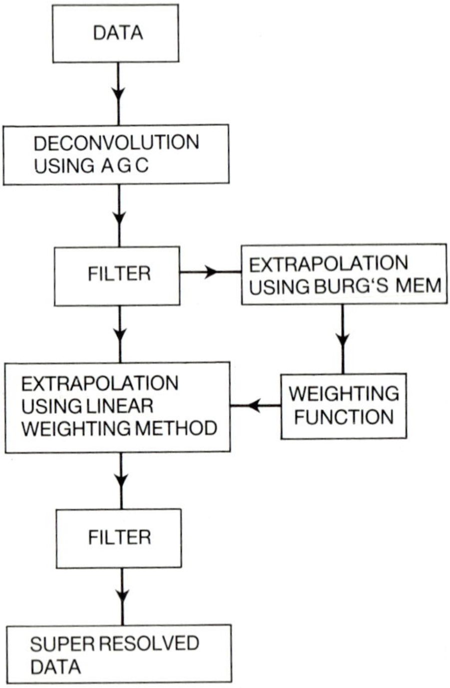

Fig. 1 Flow diagram of processing steps of the extrapolation algorithm.

The algorithm is demonstrated using the data shown in figure 2 which has been supplied by the National Coal Board. These data have usable and predominantly noise free spectral components up to 200 Hz and are therefore filtered at this value after deconvolving. Burg's MEM is then supplied with all the spectral components of the square modulus of the deconvolved function up to 200 Hz, and after extrapolation of this function, its spectrum is filtered back to 300 Hz. The output is shown in figure 3 which is the result of extrapolating each trace by 50%. A Toeplitz matrix invertor

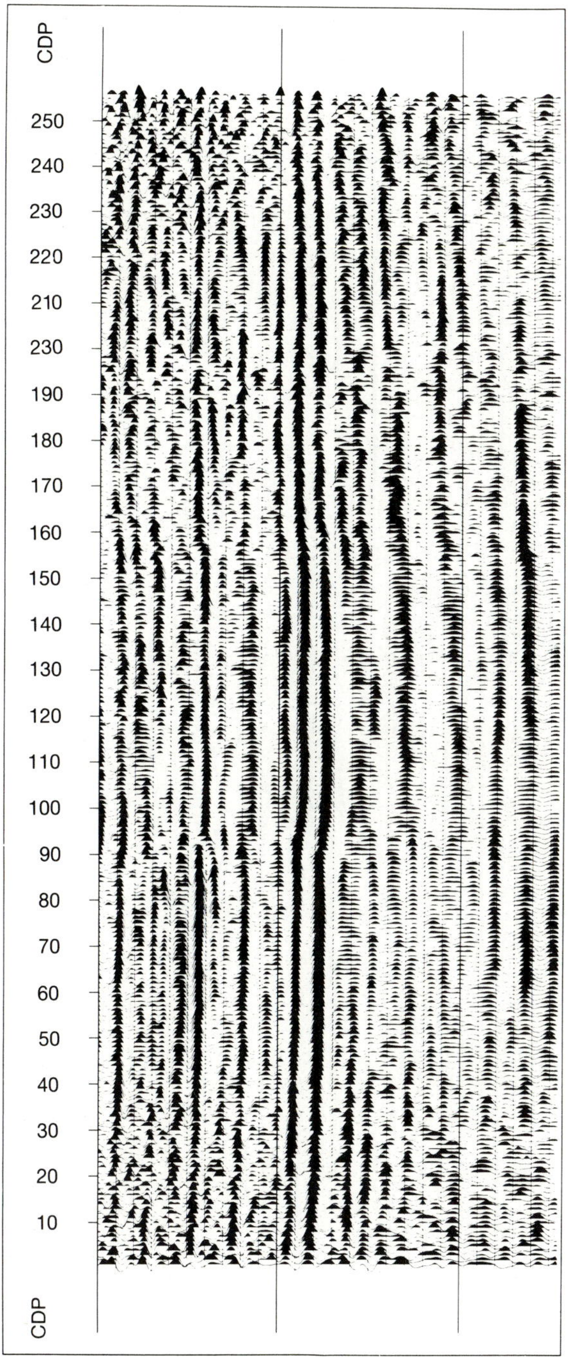
CDP
250
240
230
220
210
230
190
180
170
160
150
140
130
120
110
100
90
80
70
60
50
40
30
20
10
CDP

Fig. 2

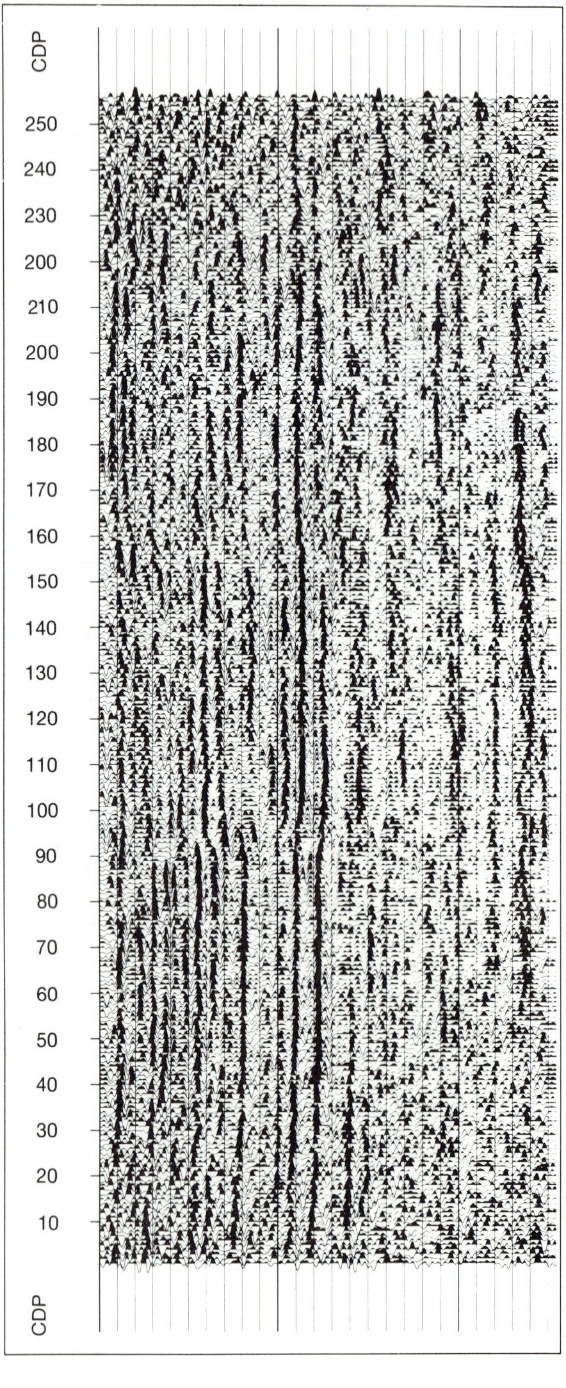
CDP
250
240
230
200
210
200
190
180
170
160
150
140
130
120
110
100
90
80
70
60
50
40
30
20
10
CDP

Fig. 3

is used to solve equations (2.7) and (3.2) and the input data to Burg's MEM and the weight are regularized to produce an optimum result. The dramatic increase in the vertical resolution is illustrated by the addition of new events, many of which are laterally continuous. This result is data consistent to within 10% due to regularization. The increase in resolution between the first estimate $f_1(t)$ and the second estimate $f_2(t)$ can be defined by the degree of extrapolation (i.e. from 200 Hz to 300 Hz) which in this case is 0.5. However, the original data in figure 2 have dominant frequency components only up to 100 Hz. Thus, in comparison with these data, the resolution can be said to have been increased by a factor ~ 3.

5. DISCUSSION

The main dilemma that is inevitably associated with any extrapolation algorithm of this form lies in answering the question: How many of the new spectral components that are generated should we believe? In principle, since any estimate is theoretically data consistent, we could continue to extrapolate further by re-sampling the result and iterating. However, this is only useful if lateral continuity from trace to trace is maintained; otherwise non-physical and trace invariant artifacts are produced. Although used in a specific way, the algorithm is completely general for extrapolating band limited signals in either the time or frequency domains.

ACKNOWLEDGEMENTS

We wish to thank the National Coal Board for their financial support and for making data readily available. The views expressed herein are those of the authors and not necessarily those of the NCB.

REFERENCES

Berteussen, K.A. and Ursin, B., (1983) "Approximate computation of the acoustic impedance from seismic data", *Geophysics*, **48**(10): 1351-1358.

Byrne, C.L., and Fitzgerald, R.M., (1982) "Reconstruction from partial information with application to tomography", *SIAM. Appl. Maths.* **42**: 933-940.

Burg, J.P., (1975) "Maximum entropy spectral analysis", Ph.D. dissertation, Stanford University, Palo Alta, Calif.

USE OF THE Z-TRANSFORM IN BORE MEASUREMENTS OF BRASS INSTRUMENTS

A.M. Duffield and M.C. Jones
(University of Surrey, Guildford)

ABSTRACT

For plane waves propagating in tubular structure, changes in cross-sectional area, or bore, cause reflections. Measurements of the reflected (or transmitted) wave resulting from an input pulse thus contain information on the variation of bore with distance along the tube. If the wave also undergoes loss, due to various mechanisms, retrieval of the bore and damping from the reflected signal is not straightforward.

It is shown that, using the z-transform, algorithms may be developed both for synthesis of the reflected (or transmitted) impulse response from bore/damping data and also for recovery of these from the measured reflected impulse response.

These algorithms depend on a recurrence relation for $H_k(z)$ the ratio of the z-transforms of backward and forward travelling waves at position k Δ along the tube. This leads to a rational function representation of $H_k(z)$ in the form $P_k(z^{-2})/Q_k(z^{-2})$ where P_k and Q_k are polynomials of degree n-k, with nΔ the length of the tube.

A method for finding P_0 and Q_0 from the measured response is described, and an iterative method for thereafter obtaining P_k and Q_k is developed. Results of these algorithms are shown for test data. The synthesis algorithm is also tested against real data.

1. INTRODUCTION

For propagation of an acoustic wave in a tubular structure with rigid walls the fundamental mode is a plane wave (Morse and Ingard (1968)). Changes in cross-section result in reflections, so that a plane-wave incident (from the left) will produce both forward and backward (left and right) travelling waves. If the variations in cross-section are approximated by steps at distances $k\Delta$ ($k = 0,..n$) along the tube, the reflection coefficient, r_k, is related to the change in cross-sectional area (Skudrzyk (1971)) as given by equation (1.1).

$$r_k = \frac{c_k - c_{k+1}}{c_k + c_{k+1}} \tag{1.1}$$

where c_k is the cross-sectional area of the k_{th} element. Thus, knowledge of $\{r_k\}_{k=0}^{n}$ enables the reconstruction of the tubular cross-section or bore.

Several methods (Sondhi and Resnick (1983), Lefèvre et al. (1983)) have been given for recovering $\{r_k\}_{k=0}^{n}$ from analysis of the reflected signal, $g_0(t)$, at the input end of the tube, and in § 2 a method based on the z-transform is derived. The method assumes the input signal, $f_0(t)$, is a finite pulse, but the deconvolution technique used to obtain the impulse response, $h_0(t)$, does not assume this to be finite (FIR), unlike direct or Fourier transform methods (Duffield and Jones (1985)).

The effects of damping are also taken into account in an empirical fashion which assumes that all the damping mechanisms combine to produce a damping factor d_{k+1} over the interval $(k\Delta, (k + 1)\Delta)$.

In § 3 the results of § 2 are applied to model data and the effects of noise examined. Finally, in § 4, synthesised output is compared with real experimental data for brass wind instruments.

2. ITERATIVE EQUATIONS FOR THE Z-TRANSFORM OF THE RESPONSE FUNCTION

The forward and backward waves incident at $x = k\,\Delta$ are $f_k^F(t)$ and $f_k^B(t)$ respectively (see Figure 1). If the reflection coefficient is r_k, the backward and forward outgoing waves, $g_k^B(t)$ and $g_k^F(t)$ are given by (Stansfield and Bogner (1973), Descout et al. (1976))

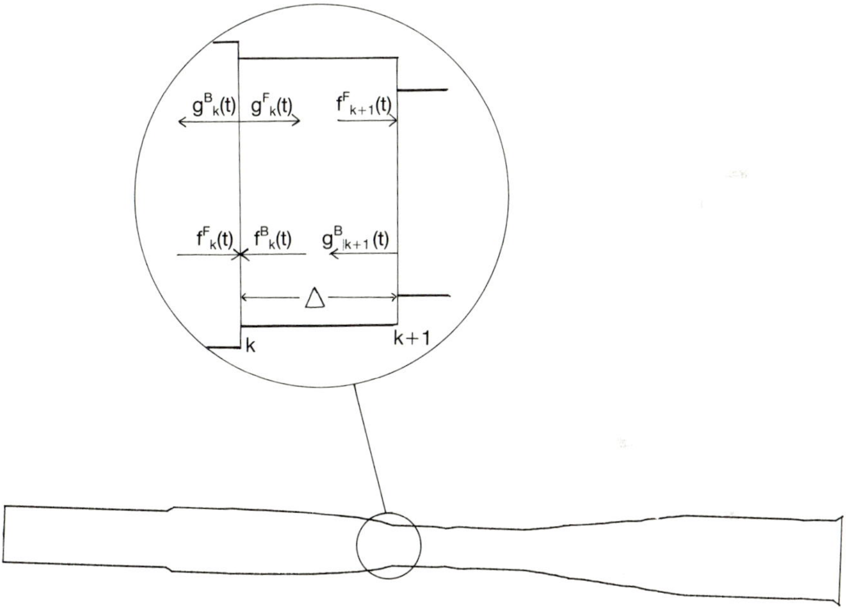

Fig. 1

$$\begin{bmatrix} g_k^B\ (t) \\ g_k^F\ (t) \end{bmatrix} = \begin{bmatrix} r_k & 1-r_k \\ 1+r_k & -r_k \end{bmatrix} \begin{bmatrix} f_k^F\ (t) \\ f_k^B\ (t) \end{bmatrix} \qquad (2.1)$$

Assuming that in propagating through the interval $(k\Delta,\ (k+1)\Delta)$ a wave is damped by a factor d_{k+1}, there are the additional relations

$$f^F_{k+1}(t) = d_{k+1} g^F_k \left(t - \frac{\Delta}{v}\right)$$
$$f^B_k(t) = d_{k+1} g^B_{k+1} \left(t - \frac{\Delta}{v}\right) \quad (k < n) \tag{2.2}$$

where v is the velocity.

If the signals are sampled at intervals of Δ/v then, denoting $q(m\,\Delta/v)$ by $q[m]$,

$$f^{F,B}_k[m] = g^{F,B}_k[m] = 0 \quad (m < k) \tag{2.3}$$

if $f^B_n(t) = 0$, and $f^F_o(t) = 0$ for $t < 0$. Thus the z-transforms are given by

$$F^{F,B}_k(z) = \sum_{m=k}^{\infty} z^{-m} f^{F,B}_k[m], \quad G^{F,B}_k(z) = \sum_{m=k}^{\infty} z^{-m} g^{F,B}_k[m] \tag{2.4}$$

The reflective response of the part of the system in the region $x \geq k\Delta$ is given by

$$H^R_k(z) = G^B_k(z)/F^F_k(z) \tag{2.5}$$

From (2.1) - (2.4)

$$\begin{bmatrix} G^B_k(z) \\ zd^{-1}_{k+1}\, F^F_{k+1}(z) \end{bmatrix} = \begin{bmatrix} r_k & 1-r_k \\ 1+r_k & -r_k \end{bmatrix} \begin{bmatrix} F^F_k(z) \\ d_{k+1}\, z^{-1}\, G^B_{k+1}(z) \end{bmatrix} \tag{2.6}$$

so that

$$(H_k(z)-r_k)(zd^{-1}_{k+1} + r_k d_{k+1} z^{-1} H_{k+1}(z)) = (1+r_k)(1-r_k)H_{k+1}(z) \tag{2.7}$$

which can be written

$$H^R_k(z) = \frac{r_k + z^{-2} d^2_{k+1} H_{k+1}(z)}{1+r_k z^{-2} d^2_{k+1} H_{k+1}(z)} \tag{2.8}$$

This recurrence formula may be compared with those arising from lattice filters (for example, in Rao and Kailath (1984)). Since $H_n(z) = r_n$, it may be seen from (2.8) that $H_k(z)$ is a rational function $P_k(w)/Q_k(w)$ of $w = z^{-2}$, and that the polynomials $P_k(w)$, $Q_k(w)$ satisfy

$$\begin{aligned} P_k(w) &= r_k Q_{k+1}(w) + w d_{k+1}^2 P_{k+1}(w) \\ Q_k(w) &= Q_{k+1}(w) + r_k\, w d_{k+1}^2 P_{k+1}(w) \end{aligned} \qquad (2.9)$$

or, alternatively,

$$\begin{aligned} P_{k+1}(w) &= \frac{P_k(w) - r_k Q_k(w)}{w d_{k+1}^2 (1 - r_k^2)} \\ Q_{k+1}(w) &= \frac{Q_k(w) - r_k P_k(w)}{1 - r_k^2} \end{aligned} \qquad (2.10)$$

with $P_n(w) = r_n$ and $Q_n(w) = 1$. (The constant term in $Q_k(w)$ is always taken to be 1.)

From (2.9) it may be seen that $P_k(w)$, $Q_k(w)$ are polynomials of degree n-k. By writing

$$\begin{aligned} P_k(w) &= p_o^{(k)} + p_1^{(k)} w + \ldots + p_{n-k}^{(k)} w^{n-k} \\ Q_k(w) &= q_o^{(k)} + q_1^{(k)} w + \ldots + q_{n-k}^{(k)} w^{n-k} \end{aligned} \qquad (2.11)$$

the iterative equations

$$\left.\begin{aligned} p_o^{(k)} &= r_k \quad & p_j^{(k)} &= r_k q_j^{(k+1)} + d_{k+1}^2 p_{j-1}^{(k+1)} \\ q_o^{(k)} &= 1 \quad & q_j^{(k)} &= q_j^{(k+1)} + r_k d_{k+1}^2 p_{j-1}^{(k+1)} \end{aligned}\right] \; j=1,\ldots,n-k-1 \qquad (2.12)$$

may be obtained from (2.9). For known $\{r_k\}_{k=0}^{n}$, $\{d_{k+1}^2\}_{k=0}^{n-1}$ (2.12) enable $P_0(w)$ and $Q_0(w)$ to be constructed. The reflective response of the whole tube will then be given by

$$H_o^R(z) = P_o(w)/Q_o(w) = \sum_{j=o}^{\infty} h_j w^j \tag{2.13}$$

when the $\{h_j\}$ are found by solving

$$\begin{aligned}
&h_o && = p_o^{(o)} \\
&h_1 + h_o q_1^{(o)} && = p_1^{(o)} \\
&\vdots && \vdots \\
&h_n + h_{n-1} q_1^{(o)} + \ldots + h_o q_n^{(o)} && = p_n^{(o)} \\
&h_{n+1} + h_n q_1^{(o)} + \ \ldots + h_n q_n^{(o)} && = 0 \\
&\vdots && \\
&h_{2n} + h_{2n-1} q_1^{(o)} + \quad + h_n q_n^{(o)} && = 0
\end{aligned} \tag{2.14}$$

An example of this procedure is given in § 3.

The inverse problem, that of calculating $\{r_k\}_{k=o}^{n}$, $\{d_{k+1}^2\}_{k=0}^{n-1}$ given time-sampled values of $g_o^B(t)$ and $f_o^F(t)$, requires somewhat more effort. First it may be seen from (2.13) that a time sampling interval $2\Delta/v$ will be sufficient to reconstruct $\{r_k\}$ and $\{d_{k+1}\}$. To use (2.14) to find $\{p_j^{(o)}\}_{j=o}^{n}$ and $\{q_j^{(o)}\}_{j=o}^{n}$ requires $\{h_j\}_{j=o}^{2n}$ to be found from $\{g_j\}$ and $\{f_j\}$ (where $g_j = g_o^B(z_j\Delta/v)$ etc.). Since $\{h_j\}$ is infinite, it is not a good approximation to assume $h_j = 0(j > 2n)$ and so values g_j for j considerably greater than 2n may be required to be measured. However, this is not necessary since, from (2.5) and (2.13),

$$
\begin{array}{ll}
g_o & =f_o p_o^{(o)} \\
g_1+g_o^{(o)} & =f_1 p_o^{(o)}+f_o p_1^{(o)} \\
\ldots+\ldots & \ldots\ \ldots\ \ldots \\
g_m+g_{m-1}q_1^{(o)}+\ldots+g_o q_m^{(o)} & =f_m p_o^{(o)}\ \ldots\ \ldots+f_o p_m^{(o)} \\
\ldots\ \ldots\ \ldots\ \ldots\ \ldots & =\ldots\ \ldots\ \ldots\ \ldots \\
g_n+g_{n-1}q_1^{(o)}+\ldots+g_{n-m}q_m^{(o)}+\ldots+g_o q_n^{(o)} & =f_m p_{n-m}^{(o)}+\ldots\ \ldots+f_o p_n^{(o)} \\
\ldots\ \ldots\ \ldots\ \ldots\ \ldots\ \ldots & \ldots\ \ldots\ \ldots\ \ldots \\
\ldots\ \ldots\ \ldots\ \ldots\ \ldots & \ldots\ \ldots \\
g_{m+n}+\ldots\ \ldots+g_n q_m^{(o)}+\ldots+g_m q_n^{(o)} & =f_m p_n^{(o)} \\
g_{n+m+1}+\ldots+g_{n+1}q_m^{(o)}+\ldots+g_{m+1}q_n^{(o)} & =0 \\
\ldots\ \ldots\ \ldots\ \ldots\ \ldots\ \ldots & \ldots \\
g_{m+n+k}+\ldots\ \ldots+g_{n+k}q_m^{(o)}+g_{m+k}q_n^{(o)} & =0 \\
\ldots\ \ldots\ \ldots\ \ldots\ \ldots\ \ldots & =\ldots \\
g_{m+2n}+\ldots\ \ldots+g_{2n}q_m^{(o)}+\ldots+g_{n+k}q_n^{(o)} & =0
\end{array}
\tag{2.15}
$$

The last of n of these equations may be solved for $\{q_j^{(o)}\}_{j=1}^{n}$, and an iterative deconvolution procedure, as described in Duffield and Jones (1985) used to find $\{p_j^{(o)}\}_{j=o}^{n}$. In solving for $\{q_j^{(o)}\}_{j=1}^{n}$ a fast $(O(n^2))$ bordering method may be used. This is described in the appendix.

3. MODEL RESULTS

The model bore profile used is shown in Figure 1. The associated reflection coefficients $\{r_k\}$ along the structure have been calculated using equation (1.1). In turn, $\{r_k\}$ has been used to synthesise $\{h_k\}$, in the absence of damping ($d_k = 1$). The results are shown in Figure 2. Inverting the problem for this synthesised $\{h_k\}$ reproduces the original bore profile with high precision.

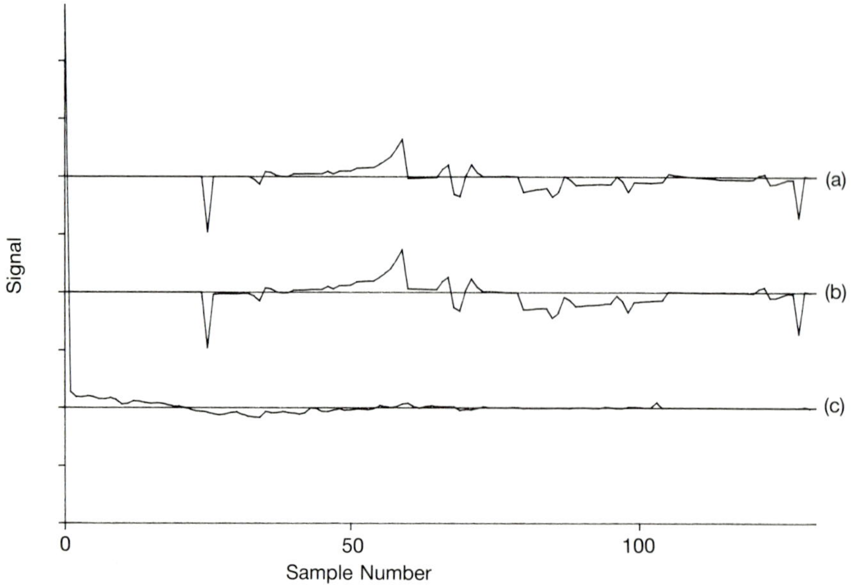

Fig. 2

Addition of even small amounts of noise to h leads to significant numerical inaccuracies in p and q which in turn lead to progressively larger errors in $\{r_k\}$ and $\{d^2_{k+1}\}$, and the reconstructed radius. Figures 3 and 4 show how the calculated $\{p_k\}$ and $\{q_k\}$ are changed when 1% Gaussian noise is added to the same synthesised h. In spite of the error most of the significant features of the originals are retained.

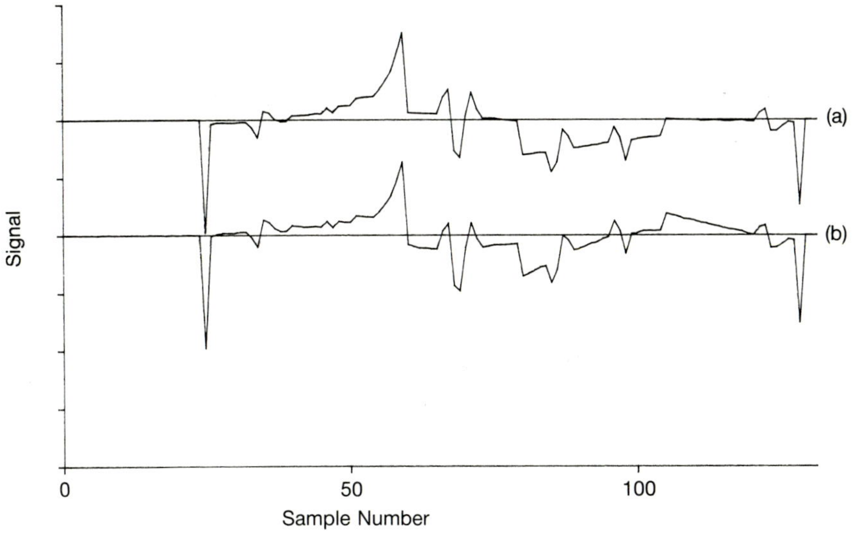
(a)
(b)
Signal
0
50
100
Sample Number

Fig. 3

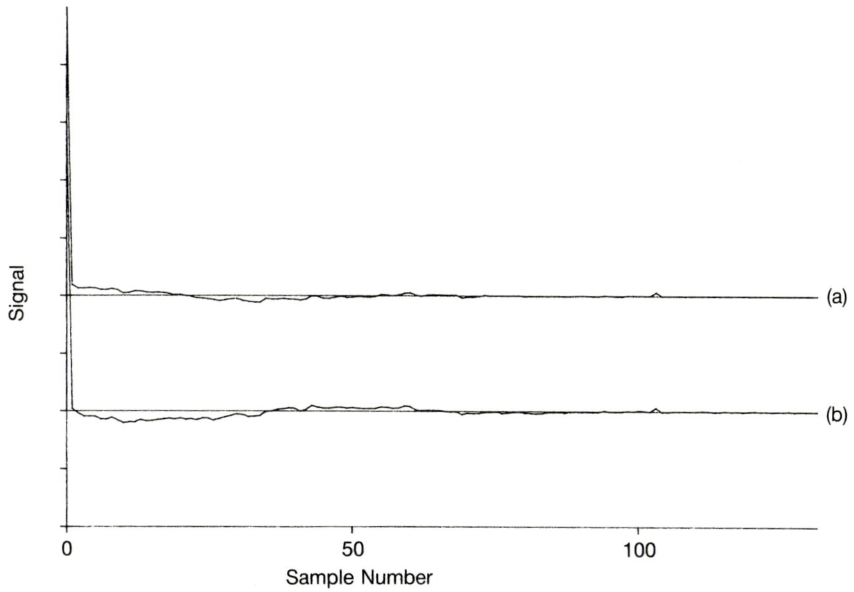
(a)
(b)
Signal
0
50
100
Sample Number

Fig. 4

Other results indicate that slightly modifying just one element of {h} is sufficient to produce noticeable errors in {p} and {q}. It seems therefore that without modification the analysis procedure will produce poor results when applied to real data in the presence of noise.

4. RESULTS OF MEASUREMENTS ON BRASS INSTRUMENTS

The synthesis procedure has been shown to generate a very reasonable impulse response from the bore profile of a trumpet. When this generated impulse response is convolved with the known input impulse, the resulting response bears considerable similarity to the experimentally measured response, as shown in Figure 5. Further work on retrieving the reflectivities from the impulse response in a way which is less sensitive to noise might therefore be expected to give reasonable results. As yet we have not been able to do this, but continuing effort is being exerted in this direction.

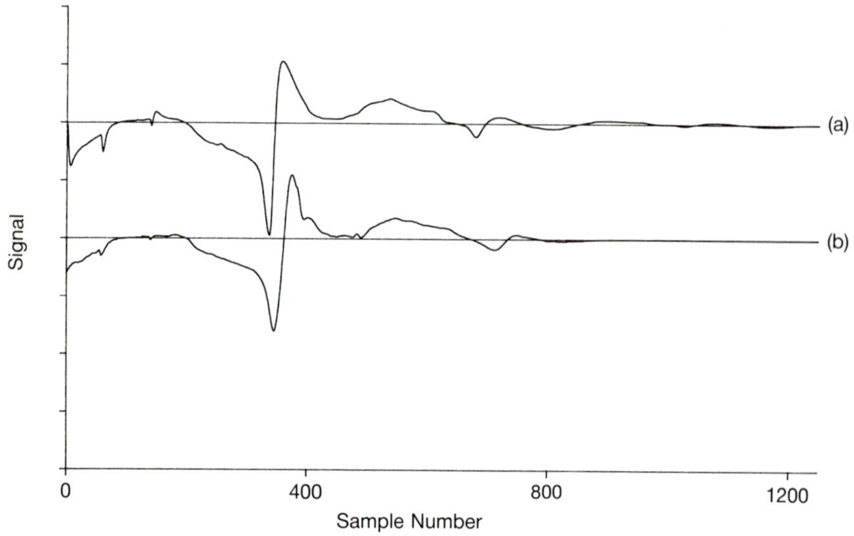

Fig. 5

REFERENCES

Descout, R., Tousignant, B. and Lecours, M. (1976), "Vocal tract area measurements: Two time-domain methods", ICASSP, Philadelphia, 75-78, IEE, New York.

Duffield, A.M. and Jones, M.C. (1985), "Use of the Gerchberg Algorithm in obtaining the response function of linear systems" (Poster presented at Mathematics in Signal Processing, Bath).

Lefèvre, J. -P., Tousignant, B. and Lecours, M. (1983), "Study of French vocallic configurations from acoustical measurements", *Acustica* **52**, 227-231 (in French).

Morse, P.M. and Ingard, K.U. (1968), "Theoretical Acoustics", Chapter 9, McGraw Hill.

Rao, S.K. and Kailath, T. (1984), "Orthogonal Digital Filters for VLSI Implementation", *IEEE Trans. Circuits Syst.*, CAS-31, 933-945.

Skudrzyk, E. (1971), "The Foundations of Acoustics", Chapter 17, Springer-Verlag, N.Y.

Sondhi, M.M. and Resnick, J.R. (1983), "The inverse problem for the vocal tract: Numerical Methods, Acoustical Experiments, and Speech Synthesis", *JASA* **73**, 985-1002.

Stansfield, E.V. and Bogner R.E. (1973), "Determination of vocal tract area function from transfer impedence", *Proc. IRE* **120**, 153-158.

APPENDIX

To solve the equations

$$\begin{bmatrix} a_{n+1} \\ a_n \\ \cdot \\ \cdot \\ \cdot \\ \cdot \\ a_{2n} \end{bmatrix} + \begin{bmatrix} a_n & a_{n-1} & \cdots & a_1 \\ a_{n+1} & a_n & \cdots & a_2 \\ \cdot & & & \\ \cdot & & & \\ \cdot & & & \\ \cdot & & & \\ a_{2n-1} & a_{2m-2} & \cdots & a_n \end{bmatrix} \begin{bmatrix} q_1 \\ q_2 \\ \cdot \\ \cdot \\ \cdot \\ \cdot \\ q_n \end{bmatrix} = \begin{bmatrix} 0 \\ 0 \\ \cdot \\ \cdot \\ \cdot \\ \cdot \\ 0 \end{bmatrix} \qquad (A.1)$$

the following notation is introduced

$$\begin{bmatrix} a_{n+1} \\ \cdot \\ \cdot \\ \cdot \\ \cdot \\ \cdot \\ a_{n+k} \end{bmatrix} = \mathbf{a}_k \quad \begin{bmatrix} a_n & a_{n-1} & \cdots & a_{n-k+1} \\ a_{n+1} & a_n & \cdots & a_{n-k+2} \\ \cdot & \cdot & & \\ \cdot & \cdot & & \\ \cdot & \cdot & & \\ \cdot & \cdot & & \\ a_{n+k-1} & a_{n+k-2} & \cdots & a_n \end{bmatrix} = \mathbf{A}_k$$

and

$$\begin{bmatrix} q_1^{(k)} \\ \\ \\ \\ q_k^{(k)} \end{bmatrix} = \mathbf{q}_k^{(k)} \quad \text{for the solution of}$$

$$\mathbf{a}_k + \mathbf{A}_k \mathbf{a}_k^{(k)} = \mathbf{0}$$

Now

$$\mathbf{a}_{k+1} = \begin{bmatrix} \mathbf{a}_k \\ \\ a_{n+k+1} \end{bmatrix} \quad \text{and} \quad \mathbf{A}_{k+1} = \begin{bmatrix} \mathbf{A}_k & \mathbf{u}_k \\ \\ \mathbf{v}_k^T & a_n \end{bmatrix},$$

where

$$\mathbf{u}_k = \begin{bmatrix} a_{n-k} \\ a_{n-k+1} \\ \cdot \\ \cdot \\ \cdot \\ \cdot \\ a_{n-1} \end{bmatrix} \quad \text{and} \quad \mathbf{v}_k = \begin{bmatrix} a_{n+k} \\ a_{n+k+1} \\ \cdot \\ \cdot \\ \cdot \\ \cdot \\ a_{n+1} \end{bmatrix},$$

so that, partitioning

$$\mathbf{q}_{k+1}^{(k+1)} = \begin{bmatrix} \mathbf{q}_k^{(k+1)} \\ \\ q_{k+1}^{(k+1)} \end{bmatrix},$$

$$\begin{bmatrix} \mathbf{a}_k \\ \\ a_{n+k+1} \end{bmatrix} + \begin{bmatrix} \mathbf{A}_k & \mathbf{u}_k \\ \\ \mathbf{v}_k^T & a_n \end{bmatrix} \begin{bmatrix} \mathbf{q}_k^{(k+1)} \\ \\ q_{k+1}^{(k+1)} \end{bmatrix} = \begin{bmatrix} \mathbf{0} \\ \\ 0 \end{bmatrix} \qquad (A.2)$$

Expanding (A.2)

$$\mathbf{a}_k + \mathbf{A}_k\mathbf{q}_k^{(k+1)} + \mathbf{u}_k q_k^{(k+1)} = \mathbf{0}$$

$$a_{n+k+1} + \mathbf{v}_k^T\, \mathbf{q}_k^{(k+1)} + a_n q_{k+1}^{(k+1)} = 0 \qquad (A.3)$$

From the first of these

$$\mathbf{A}_k^{-1}\,\mathbf{a}_k + \mathbf{q}_k^{(k+1)} + \mathbf{A}_k^{-1}\,\mathbf{u}_k q_{k+1}^{(k+1)} = \mathbf{0} \tag{A.4}$$

and substituting for $q_k^{(k+1)}$ from (A.4) in the second

$$a_{n+k+1} - \mathbf{v}_k^T\,(\mathbf{A}_k^{-1}\,\mathbf{a}_k + \mathbf{A}_k^{-1}\,\mathbf{u}_k\,q_{k+1}^{(k+1)}) + a_n q_{k+1}^{(k+1)} = \mathbf{0} \tag{A.5}$$

giving

$$q_{k+1}^{(k+1)} = (a_{n+k+1} - \mathbf{v}_k^T\,\mathbf{A}_k^{-1}\,\mathbf{a}_k)/(\mathbf{v}_k^T\mathbf{A}_k^{-1}\,\mathbf{u}_k - \mathbf{a}_n)$$

$$= (a_{n+k+1} + \mathbf{v}_k^T\,\mathbf{q}_k^{(k)})/(\mathbf{v}_k^T\,\mathbf{x}_k^{(k)} - a_n) \tag{A.6}$$

where

$$\mathbf{A}_k\,\mathbf{x}_k^{(k)} = \mathbf{u}_k,$$

$$\text{with}\quad \mathbf{x}_k^{(k)} = \begin{bmatrix} x_{n-k}^{(k)} \\ \\ \vdots \\ \\ x_{n-1}^{(k)} \end{bmatrix} \tag{A.7}$$

(A.7) in (A.4) gives

$$\mathbf{q}_k^{(k+1)} = \mathbf{q}_k^{(k)} - \mathbf{x}_k^{(k)}\,q_{k+1}^{(k+1)} \tag{A.8}$$

From (A.7)

$$\mathbf{A}_{k+1}\,\mathbf{x}_{k+1}^{(k+1)} = \mathbf{u}_{k+1} \tag{A.9}$$

so that, partitioning

$$\begin{bmatrix} a_n & \mathbf{b}_k^T \\ & \\ \mathbf{a}_k & \mathbf{A}_k \end{bmatrix} \begin{bmatrix} x_{n-k-1}^{(k+1)} \\ \\ \mathbf{u}_k^{(k+1)} \end{bmatrix} = \begin{bmatrix} a_{n-k-1} \\ \\ \mathbf{u}_k \end{bmatrix} \tag{A.10}$$

where

$$b_k = \begin{bmatrix} a_{n-1} \\ \cdot \\ \cdot \\ \cdot \\ a_{n-k} \end{bmatrix}$$

Expanding (A.10)

$$a_n x_{n-k-1}^{(k+1)} + \mathbf{b}_k^T \mathbf{x}_k^{(k+1)} = a_{n-k-1}$$

$$\mathbf{a}_k x_{n-k-1}^{(k+1)} + \mathbf{A}_k \mathbf{x}_k^{(k+1)} = u_k \tag{A.11}$$

The second of these gives

$$\mathbf{x}_k^{(k+1)} = \mathbf{x}_k^{(k)} + \mathbf{q}_k^{(k)} x_{n-k-1}^{(k+1)} \tag{A.12}$$

and substituting for $x_k^{(k+1)}$ from (A.12) in the first gives

$$x_{n-k-1}^{(k+1)} = (a_{n-k-1} - \mathbf{b}_k^T \mathbf{x}_k^{(k)})/(a_n + \mathbf{b}_k^T q_k^{(k)}) \tag{A.13}$$

Putting

$$\alpha_k = \mathbf{v}_k^T q_\kappa^{(k)}, \quad \beta_k = \mathbf{v}_k^T \mathbf{x}_k^{(k)}, \quad \gamma_k = \mathbf{b}_k^T \mathbf{q}_k^{(k)}, \quad \delta_k = \mathbf{b}_k^T \mathbf{x}_\kappa^{(k)}$$

gives

$$\beta_{k+1} = \beta_k + (a_{n+k-1} + \alpha_k) \mathbf{x}_{n-k-1}^{(k+1)}$$

and

$$\gamma_{k+1} = \gamma_k + (a_{n-k-1} - \delta_k)q_{k+1}^{(k+1)}$$

so that only α_k and δ_k have to be calculated fully at each step.

OPTIMAL FEASIBLE SOLUTIONS FOR SIGNAL RESTORATION AND RECONSTRUCTION

R.M. Leahy
(Signal and Image Processing Institute, University of Southern California, USA)

and

C.E. Goutis
(Department of Electrical Engineering, Patras University, Patras, Greece)

ABSTRACT

In this paper a general technique is described for solving a noisy, sampled version of the Fredholm integral equation of the first kind where prior knowledge of the unknown function such as upper and lower bounds and spatial or temporal extent and also of the statistics of the noise such as variance, mean, and covariance are used as additional constraints. A unique solution is obtained under certain conditions by employing the Fenchel duality theorem which reduces the problem to an equivalent dual problem of reduced dimensionality. The method is demonstrated in an application to tomographic image reconstruction and compared with a minimum mean square error estimation technique.

1. INTRODUCTION

This paper introduces a general technique for finding optimal solutions to constrained signal restoration or reconstruction problems given a set of data corrupted by additive noise. The new approach employs the concept of duality to optimize a convex or concave cost function subject to a set of constraints. This yields a unique function consistent with all our prior knowledge of the problem. The technique is demonstrated in an application to tomographic image reconstruction and compared with a minimum variance estimator based on the prior assumption that both the image and noise are drawn from known multivariate Gaussian processes.

A large class of problems in signal processing require the approximate solution of the Fredholm integral equation of the first kind:

$$g(t) = \int_D f(r)w(r,t)dr \qquad (1)$$

where r and t are multidimensional coordinate vectors for the functions g and f, and D denotes the region of support of f. Since the data is usually collected in a sampled form and corrupted by noise a more accurate formulation of the problem is to estimate f(r) given a finite set of noisy samples of g(t) defined by

$$g_k^o = \int_D f(r).w_k(r)dr + n_k \qquad k = 1,2,\ldots,N \qquad (2)$$

where n_k is an additive noise process and the kernels $w_k(r)=w(r,t_k)$ Problems of this kind are ill posed [1] since the solution space is infinite dimensional while the data set is finite. Thus it is necessary to restrict the solution space in someway so that the problem reduces to the solution of a problem of dimension at most N.

The most common approach in image restoration/reconstruction is to assume the function is piecewise constant and approximate f(r), using a discrete function representation, as a vector f of length P. The vector f could also denote the parameters of an alternative linear model representing the weightings of a set of P basis functions. Under this assumption the integral equations (2) may be replaced with a linear system. Writing the g_k^o as a vector $g^o = (g_1^o, g_2^o, \ldots, g_N^o)$ gives

$$g^o = Wf + n \qquad (3)$$

where W denotes the P X N system matrix. Choice of the method for solving equation (3) should be determined by the eigen-structure of W, the signal to noise ratio and the relative dimensions of f and g^o. One technique for solving equation (3) is to ignore the presence of noise, find the pseudoinverse W^+ of W and form the solution $f = W^T g^O$. The resulting estimate will often be poor due to noise effects particularly if W has a high condition number. The performance can sometimes be improved using regularization techniques which effectively reduce the dimensionality of the solution below N to allow reduction in the noise sensitivity [1]. Alternatively iterative techniques may be used for forming the approximate pseudoinverse of W; terminating the iterations at the point at which the error and noise energies match [15] will have a similar effect to the regularization techniques. These procedures are characterized by an implicit trade-off between resolution (higher dimensionality) and robustness to noise (lower dimensionality).

An alternative class of estimators for f may be derived by assuming that f and n are drawn from two independent stochastic processes. An estimate $\hat{f}$ of f may then be made using estimators based on maximum likelihood [5], minimum variance [3] or Bayesian [8] formulations; the choice of estimator depends on the degree of a priori knowledge of the statistics of f and n. If it is assumed that f is drawn from a multivariate Gaussian process with known stationary or non-stationary mean and covariance then the minimum variance and maximum likelihood estimators are equivalent. The major problem with these techniques is that assumptions concerning the distribution of f are required, in many cases no such prior assumptions are available. The consequences of this problem are discussed in section 5.

A popular approach which avoids the need for a statistical model for f(r) is to search for a vector f subject to a set of constraints based on a prior knowledge of the properties of the solution and the statistics of the additive noise process. Problems of this type are commonly solved using the method of projection onto convex sets (POCS) [16]. In this method each constraint is used to define a convex set in the solution space. The intersection of all the constraint sets, termed the 'feasible set' [16], forms the set of possible solutions. An element of this set is found by taking an initial estimate of the function and projecting successive estimates in the solution space onto each of the convex sets in turn. At convergence this method yields some $f \in \bigcap_i C_i$ where the C_i denote the constraint sets. This technique allows the matching of an arbitrary number of constraints, provided the feasible set is not empty. However, the method is computationally expensive, particularly when used in image processing [13], and does not yield a unique solution since all members of the feasible set are considered equally valid. As with the statistical techniques this method also requires the adoption of some finite dimensional approximation to f(r).

An alternative to POCS which does yield a unique or 'optimal feasible' solution is to introduce a cost function representing some desired quality of the function. The problem can then be posed as a constrained optimization:

$$\min h(f) \text{ such that } f \in \bigcap_i C_i \qquad (4)$$

where h(f) denotes the cost function. Many techniques exist for solving problems of the form of (4) [11]; these may be classified for our purposes into two main types; primal and dual methods. The primal methods perform an iterative search for the solution directly in the solution space whereas the dual methods perform a search in a dual space, the solution of the problem is then found via a model relating the two spaces. Since the primal technique perform a search in the solution space a prior finite dimensional model must be adopted for f(r). Conversely the dual methods may be formulated in an infinite dimensional setting, a finite dimensional model then arises naturally from applying the duality theorems. It is these latter techniques that are of interest in this paper.

2. THE OPTIMAL FEASIBLE SOLUTION

In this section the reconstruction/restoration problem is formulated in a vector space setting, the types of constraints which may arise are described and several specific examples are given.

A reasonable choice of solution space is $C[\mathbf{D}]$, the space of continuous functions defined on the compact set $\mathbf{D}$, with norm $\|f\|=\max f|f(r)|$ for all $r \in \mathbf{D}$. The integrals defined in equation (1) can be rewritten as a transformation from the solution space to an N-dimensional (N-D) constraint space. Expressing the data and noise as the vectors $g^o = (g_1^o, g_2^o, \ldots, g_N^o)$ and $n = (n_1, n_2, \ldots, n_N)$ in R^N allows us to write

$$g^o = G(f)+n \qquad (5)$$

where G denotes a linear transformation from $C[\mathbf{D}]$ to R^N. For most problems the prior information available may be separated into two groups of constraints: (i) those governing particular properties of the solution such as upper and lower bounds; and (ii) those based on the statistical properties of the noise. Each of the constraints may then be used to define a subset of the solution space which contains all those functions obeying that constraint. Let R_i denote constraint sets based on the solution properties and S_i denote those based on the noise statistics. The feasible set is then given by the intersection of all the R_i's and S_i's.

To select a single element of the feasible set we introduce a cost function representing some desired quality of the solution and optimize this over the feasible set. In this paper this optimization is achieved by applying the Fenchel duality theorem [11]. This theorem equates the minimization of the difference of a convex and concave functional defined on two intersecting convex sets to an equivalent dual problem. Due to space limitations the theorem is not quoted fully here however we use the form of the theorem as proven in [11]. This technique was first applied in signal processing by McClellan and Lang [12] for solving the multidimensional maximum entropy spectral estimation problem, here we give a general procedure for solving constrained reconstruction and restoration problems.

There are two nice features of this theorem. Firstly, the primal problem may be formulated in an infinite dimensional space; a finite dimensional model, which relates the solutions of the primal and dual problems, then arises naturally from the application of the theorem. Secondly, since the theorem involves optimization over the intersection of two convex sets it is possible to separate the constraints into the two groups defined in the preceding section, this greatly simplifies the problem.

Consider the case where we wish to minimize the convex functional $h(f)$ over the feasible set $R \cap S$ where R is the convex set containing all functions consistent with our prior knowledge of the form of the solution and S is the convex set containing all functions such that the error $(g^{o}-G(f))$ matches constraints based on the noise statistics. Applying Fenchel's theorem and setting the concave functional $k(f) = 0$ (zero is both a convex and concave functional) gives

$$\inf_{f \in R \cap S} [h(f)] = \max_{f^{*} \in R^{*} \cap S^{*}} [k^{*}(f^{*})-h^{*}(f^{*})] \qquad (6)$$

where the conjugate functionals k^{*} and h^{*} and conjugate sets R^{*} and S^{*} are defined below. The dual problem is therefore given by the right hand side of equation (6); the functional form of the dual problem is given by the solution of two subproblems which yield $h^{*}(f^{*})$ and $k^{*}(f^{*})$ respectively.

The functions f^{*} belong to the dual of the solution space $X = C[\mathbf{D}]$. The dual of $C[\mathbf{D}]$ is the space of the functions of normalized bounded variation defined on $\mathbf{D}$ and denoted $X^{*} = NBV[\mathbf{D}]$, [22]. The conjugate sets, R^{*} and S^{*}, are subsets of X^{*} and k^{*} and f^{*} are functionals on R^{*} and S^{*}.

Consider first the set S^{*}. For our case with $k(f)=0$ this is defined as

$$S^* = \{f^* \in X^* : \inf_{f \in S} [\langle f, f^* \rangle > -\infty]\} \tag{7}$$

Since S is defined entirely in terms of functions of the linear operator G(f) it follows, from an argument in [24], that the functional in equation (7) must be expressible in the form

$$\langle f, f^* \rangle = \int_D f(r).\lambda^T \underline{w}(r)dr \tag{8}$$

where λ is a vector in R^n. This can be verified by considering the function $f^O = f+d$ where $d \in N(g)$, the null space of G; $\langle f^O, f^* \rangle$ may be made infinite by multiplying d by an arbitrarily large constant unless $\langle d, f^* \rangle = 0$ which implies that the functional must be expressible in the above form.

The corresponding conjugate functional, h^*, is defined as

$$k^*(f^*) = \inf_{f \in S} \langle f, f^* \rangle = \inf_{f \in S} \int_D f(r).\lambda^T \underline{w}(r)dr \tag{9}$$

Using the definition of the set Γ as the image of S in R^n under the transformation G(f) and noting that $\int_D f(r)\underline{w}(r)dr = g$ this becomes

$$k^*(f^*) = \inf_{g \in \Gamma} \lambda^T g \tag{10}$$

This is a constrained optimization problem in R^N and may be tackled using the primal-dual Lagrange multiplier method described in [22, p.299].

A similar procedure may be followed to produce R^* and h^*. The solution of the dual problem lies in the intersection $R^* \cap S^*$ which is given by

$$R^* \cap S^* = \{f^* \in S^* : [\sup_{f \in R} \int_D f(r).\lambda^T \underline{w}(r) - h(f) < \infty]\} \tag{11}$$

In some cases it will be found that $R^* \cap S^* = S^*$ thus all values of λ are allowed and the problem is therefore unconstrained. However this depends on the constraints and cost function in the primal problem and hence the condition in equation (11) requires investigation for each particular problem to determine the allowed range of values of λ. The conjugate functional, h^*, on R^* is given by

$$h^*(f^*) = \sup_{f \in R} \{\int_D f(r).\lambda^T \underline{w}(r)dr - h(f)\} \qquad (12)$$

This second subproblem may be solved analytically given the constraints which define the convex set R to yield a model $\hat{f}(r)$ for $f(r)$. This model relates the finite dimensional parameter λ of the dual problem to the desired image $f(r)$ of the primal problem.

The solution $h^*(f^*)$ to the second subproblem, equation (12), when combined with $k^*(f^*)$, equation (10), in the expression for the dual functional, equation (6), will yield the dual equivalent of the primal problem as a function of the unknown parameters λ'. This result is very important since the solution of the infinite dimensional problem is given via an N-D vector, furthermore the model is independent of the particular form of the noise constraints.

The above procedure is summarized in the general technique as follows:

[1] Select convex set R and S to include all prior knowledge of the particular problem.

[2] Solve subproblem (10) to obtain $k^*(f^*)$, if this cannot be achieved analytically use an estimate of λ and perform numerical optimization.

[3] Solve subproblem (12) to obtain $h^*(f^*)$ and the model for $f(r)$ in terms of λ and $\underline{w}(r)$.

[4] Combine h^* and k^* to form the dual functional $\phi(\lambda)=k^*(f^*)-h^*(f^*)$, i.e. the right hand side of equation (6), and perform numerical optimization to find the optimal vector $\lambda \in R^N$. Note: if $k^*(f^*)$ cannot be found analytically use successive estimates λ^n of λ to update the solution of equation (10), i.e. minimize $\lambda^{n^T} g$ over $g \in \Gamma$, which is then used in turn to update λ^n

[5] Substitute the optimal λ in the model found in step [3] to yield the solution to the primal problem.

To obtain a solution for a problem with a concave cost function, such as entropy, simple modifications in the general technique are required as described in [10].

Before applying this technique to specific problems we consider the conditions under which the method yields a unique solution to the primal problem. From the statement of the theorem in the appendix the following conditions must be fulfilled if the primal and dual problems are to be equivalent and their solutions unique: (i) the infimum of h(f)-g(f) over $R \cap S$ must be finite; (ii) the epigraph [h,R] or hypograph [k,S] must have non-empty interiors; and (iii) the intersection $R \cap S$ must have points in the relative interior of R and S. The first of these conditions is fulfilled by a suitable choice of cost function and constraint sets, if the infimum is not bounded then the problem is clearly not well formulated. The existence of interior points in the epigraph or hypograph must be investigated for each particular problem however this condition requires only that the functionals are continuous on their convex sets [11, p.194]. The third condition requires not only that $R \cap S$ is non-empty, i.e. that there is a feasible solution, but also there must be a solution in the relative interior of R and S.

3. THE MINIMUM VARIANCE ESTIMATOR

In this section we include a brief description of the minimum mean square error (MMSE) estimator with which the above technique is compared. The MMSE estimate of an image f is defined as that which minimizes the expected error between the true and estimated image, i.e. $E\| f - \hat{f}\|^2$.

We shall assume that both the image and the noise are drawn from multivariate Gaussian processes with the probability density functions:

$$p(f) = K_f \exp\left(-\frac{1}{2}(f - f^o)^T \mathbf{R}_{ff}^{-1}(f - f^o)\right) \tag{13}$$

and

$$p(n) = K_n \exp\left(-\frac{1}{2}(n - n^o)^T \mathbf{R}_{nn}^{-1}(n - n^o)\right) \tag{14}$$

where we know a priori the following values:

$$\begin{aligned} E(f) &= f^o \\ E(n) &= n^o = 0 \\ E((f - f^o)(f - f^o)^T) &= \mathbf{R}_{ff} \\ E(nn^T) &= \mathbf{R}_{nn} \end{aligned} \tag{15}$$

Applying the minimum mean square error criteria to the above we can derive the resulting solution [11]

$$\hat{f} = f^{O} + \mathbf{R}_{ff}\mathbf{W}^{T}[\mathbf{W}\mathbf{R}_{ff}\mathbf{W}^{T} + \mathbf{R}_{nn}]^{-1}g \qquad (16)$$

In the following section this method is applied to the problem of image reconstruction from projections in the manner proposed by Bounocore [3].

4. APPLICATIONS TO COMPUTER ASSISTED TOMOGRAPHY

In this section we will describe an application of the above techniques to image reconstruction from projections. In this problem the integral equations (2) define the process of integrating the function f(r) over the regions through which each ray passes. Thus the kernels $w_k(r)$ are strip like functions equal to unity over the area through which the kth projection beam passes and zero elsewhere. In the fan beam case the functions include a weighting by the inverse of the distance from the source to compensate for the diverging nature of the beam. Consider the problem of finding the minimum energy given a set of projection samples where a prior estimate of the additive noise covariance is available and prior bounds on the solution are known. Using this information it is possible to apply the optimal technique described in section 2. If in addition we assume the image is a Gaussian random field for which we have a prior estimate of the mean and covariance then the MMSE estimator can also be applied.

4.1 The optimal feasible solution

In CT the data vector g denotes a collection of M projections each consisting of P samples such that N = P x M. The notation $\underline{g}_i$ will be used to denote the subvector of g corresponding to a single projection oriented at the ith projection angle; thus $g = (\underline{g}_1, \underline{g}_2, \ldots \underline{g}_M)$. The notation $\underline{w}_i(r)$ and $\underline{\lambda}_i$ is similarly defined.

We will make the assumption that each projection subvector is statistically independent and the noise on each projection is drawn from a multivariate Gaussian process with P x P covariance matrix C. We can define an appropriate constraint set by exploiting the fact that $n^{T}C^{-1}n$ is χ_2^2 distributed [2]; thus we can define some 95% confidence region such that if the error $(\underline{g}_i - \underline{g}_i^{O})$ has the same covariance as the noise then there is a 95% probability that the quadratic $(\underline{g}_i-\underline{g}_i^O)^T C^{-1}(\underline{g}_i-\underline{g}_i^O)$ lies within this confidence region. The corresponding constraint set S is defined as

$$S = \{f \in X: (\underline{g}_i-\underline{g}_i^O)^T C^{-1}(\underline{g}_i-\underline{g}_i^O) \leq \delta \text{ for } i = 1,2,\ldots,M\} \tag{17}$$

where δ is chosen from χ^2 tables since $n^T C\text{-}1\, n$ is χ_2^2 distributed [2]. Using upper and lower bounds α and β respectively the set R is defined as

$$R = \{f \in X: \alpha \leq f(r) \leq \beta, \text{ for all } r \in \mathbf{D}\} \tag{18}$$

The general method of section 2 can now be applied to the primal constrained energy minimization problem.

Substitution of the above constraint sets in the subproblems equation (17) and (20) results in the constrained model for the image

$$\hat{f}(r) = \max[\min[\lambda^T \underline{w}(r), \beta], \alpha] \tag{19}$$

and the corresponding dual functional:

$$\max_{\lambda \in R^N} \phi(\lambda) = \lambda^T g^O - \int_D \frac{\hat{f}(r)^2}{2}\, dr - \sum_{i=1}^{M} (\underline{\lambda}_i{}^T C \underline{\lambda}_i)^{\frac{1}{2}} \tag{20}$$

In equation (19) the image is given by a weighted sum of the $w_k(r)$'s truncated within the limits α and β. This is a truncated form of the well known 'backprojection' model in which each element λ_k of λ is backprojected or smeared along the path of the kth projection beam as defined by $w_k(r)$; in the fanbeam case this backprojection also involves a weighting by the inverse of the distance from the source.

Consider equation (11), since f(r) is bounded it follows that it is of finite energy thus the condition in equation (11) is satisfied for all $\lambda \in R^N$ and hence the dual problem is an unconstrained optimization over λ.

The dual functional $\phi(\lambda)$ is optimized when its gradient is zero with respect to λ. Thus the solution may be found by an iterative gradient search method such as described in [11]. Difficulties are encountered in finding the derivative of the term $\int_D \hat{f}(r)^2/2$ in equation (20) because the model $\hat{f}$ is a non-linear function constrained by the bounds α and β. However a local gradient may be defined as

$$\nabla\left[\int_D \hat{f}(r)^2/2\right] = \int_D \hat{f}(r)\underline{w}(r)dr = G(\hat{f}) = g \qquad (21)$$

since in the neighbourhood of a particular value of λ the corresponding image $\hat{f}(r)$ is independent of λ for those regions of **D** for which $\lambda^T \underline{w}(r)>\beta$ or $\lambda^T \underline{w}(r)<\alpha$. Thus the gradient of the dual functional is defined as

$$\nabla_i\phi(\lambda) = \underline{g}_i^0 - \underline{g}_i + \left[\frac{\delta}{\underline{\lambda}_i^T C\underline{\lambda}_i}\right]^{\frac{1}{2}} . C\underline{\lambda}_i \qquad (22)$$

where $\nabla_i\phi$ denotes the gradient with respect to the ith subvector $\underline{\lambda}_i$. To perform the iterative search it is necessary to estimate the projection subvector $\underline{g}_i^n$ corresponding to the nth estimate of the parameter subvector, $\underline{\lambda}_i^n$. This is a computationally expensive task for the constrained image since the regions lying outside the bounds α,β must be determined and truncated. However in the absence of these bounds the vectors λ and g are related by a nonlinear system and thus the image need not be estimated to update the $\underline{g}_i^n$.

We shall consider the case where the optimal λ yields an image lying within the specified bounds. Thus setting $\nabla\phi=\theta$, the zero vector, yields a set of nonlinear equations which may be expressed in the form

$$g = B'\,\lambda \qquad (23)$$

where B' is an MP x MP matrix consisting of M^2 blocks B'_{ij} equal to

$$B'_{ij} = \begin{cases} \int_D \underline{w}_i(r)\underline{w}_j^T(r)dr & i\neq j \\ \int_D \underline{w}_i(r)\underline{w}_i^T(r)dr + \left[\dfrac{\delta}{\underline{\lambda}_i^T C\underline{\lambda}_i}\right]^{\frac{1}{2}} . C & i=j \end{cases} \qquad (24)$$

This nonlinear matrix has a structure which allows very fast inversion. It is in fact the sum of a block circulant or permuted block circulant matrix for fanbeam or parallel systems respectively [6] and a term, added to the diagonal blocks only, equal to the noise covariance weighted by a function of the unknown vector λ. For the parallel geometry the permuted block circulant section of the matrix $\mathbf{B}'$ consists of blocks $\mathbf{B}_{ij}$ corresponding to the areas of intersection of the ith and j^{th} projection beams. The permuted block circulance arises from the fact that the area of intersection of two projection beams is dependent only on the angle of separation of the ith and j^{th} projection beams, thus the elements of $\mathbf{B}_{ij}$ are dependent only on $|i-j|$. Another property arising from the imaging geometry is that the non-zero elements of the matrix form approximately elliptical regions in the submatrix. These regions are constant inside, zero outside and have a relatively short transition region of varying values on the border of the ellipse. This structure lends itself to both fast computation and efficient storage. By subtracting successive rows from each other and storing the non-zero differences only the storage requirement is drastically reduced as is the time required to calculate the projections corresponding to some estimate of the parameter λ[4,6].

4.2 The minimum variance estimator

The MMSE estimator in section 3 is employed here for tomographic reconstruction in the manner described by Bounocore [3]. In this paper the 'natural pixel' model is adopted as an alternative to the conventional square pixel model employed in the ART algorithms [7]. This model decomposes the image into the regions formed by the intersection of the projection beams, i.e. the intersection of the $w_k(r)$. The parameters of the model form a vector λ which on backprojection yield this natural pixel image: $\hat{f} = \lambda^T \underline{w}(r)$. Thus the natural pixel model is identical to the minimum energy model, equation (19), derived in the previous subsection (except the latter contains the bounds $[\alpha,\beta]$). To apply the MMSE estimator the following assumptions were made: the noise and image are zero mean multivariate random processes, the noise autocovariance is denoted by $\mathbf{C}$ and the image autocovariance is given by $\mathbf{R}_{ff}(r,s)=\delta(r-s)$, i.e. the image is uncorrelated. Applying these assumptions to equation (16) where the unknown vector f is replaced with the natural pixel model parameter λ it can be shown [3] that equation (16) reduces to the solution of the linear problem:

$$g^{O} = \mathbf{B}^{*}\lambda \qquad (25)$$

where the matrix $\mathbf{B}^{*}$ is an MP x MP block matrix with blocks

$$\mathbf{B}^{*}_{ij} = \begin{cases} \int_D \underline{w}_i(r)\underline{w}_j^T(r)dr & i\neq j \\ \int_D \underline{w}_i(r)\underline{w}_i^T(r)dr + \mathbf{C} & i=j \end{cases} \qquad (26)$$

Comparing equations (25) and (23) we see that the two problems differ only in the weighting attached to the noise covariance terms added to the diagonal blocks. We suggest here that the solution to equation (25) is an approximation of the non-linear system in equation (23); this will be discussed below following the presentation of some computer simulations.

5. FAST ALGORITHMS AND COMPUTED RESULTS

5.1 The algorithms

Two algorithms are defined below for solving equations (23) and (25). Although the system (23) is nonlinear we have verified experimentally that the solution may be found using a block-type Gauss-Seidel iteration provided the noise is broadband. In this technique each subvector in equation (23) is updated in turn according to:

$$\underline{\lambda}_i^{n+1} = \underline{\lambda}_i^{n} + \gamma^{n}(\mathbf{B'}^{n}_{ii})^{-1}(\underline{g}_i^{O}-\underline{g}_i^{n} - \left[\frac{\delta}{\underline{\lambda}_i^{T}\mathbf{C}\underline{\lambda}_i}\right]^{\frac{1}{2}}.\mathbf{C}\underline{\lambda}_i^{n}) \quad i=1,2,\ldots.M \qquad (27)$$

where $\mathbf{B'}^{n}_{ii}$ is the ith diagonal block of $\mathbf{B'}$ evaluated using equation (24) with $\lambda=\lambda^{n}$; γ^{n} denotes a steplength in the range $0<\gamma<2$.

Similarly equation (25) may be solved using the block updating procedure

$$\underline{\lambda}_i^{n+1} = \underline{\lambda}_i^{n} + \gamma^{n}(\mathbf{B}^{*n}_{ii})^{-1}(\underline{g}_i^{O} - \underline{g}_i^{n} - \mathbf{C}\underline{\lambda}_i^{n}) \qquad i=1,2,\ldots.M \qquad (28)$$

5.2 *Simulation Results*

The image used in the simulations was the Shepp and Logan head phantom [14] generated by overlaying 10 ellipses. The simulated projection data was corrupted in separate tests with both white and coloured additive Gaussian noise, results for white noise only are presented here. The signal to noise ratio (SNR) was calculated as

$$\mathrm{SNR} = 10.\mathrm{Log}_{10} \left[\frac{\sum_{i=1}^{N} (g_i^o)^2}{\sum_{i=1}^{N} (n_i)^2} \right] \tag{29}$$

The expected value of the constraint δ was found by averaging the calculated values $\underline{n}_i^T C^{-1} \underline{n}_i$ where $\underline{n}_i$ is the noise added to the ith projection. The robustness of the algorithm to a poor choice of δ remains to be tested.

All tests were performed for M=16 projections and P=49 points per projection since this was found to give reasonable reconstructions of the original image, Fig. 1a, for noiseless data. Results are given as both photographic plates and tables of the relative Q-values of the reconstruction. The Q-value has been widely used as a measure of reconstructed image quality and is defined as

$$Q = \frac{\sum_{j=1}^{K^2} (f_j^n - f_j^o)^2}{\sum_{j=1}^{K^2} (f_j^o - \bar{f})^2} \tag{30}$$

where K^2 is the number of pixels in the reconstruction, K=49 below; f^o and f^n are the original and reconstructed image vectors respectively and $\bar{f}$ is the mean value of the original image.

Fig. 1a shows the original phantom image, Figs. 1b-1d are the reconstructed images for a SNR of 23dB. Fig 1b shows the reconstruction for the convolution backprojection (CBP) algorithm in [14], Fig 1c is the MMSE estimate and Fig 1d is the optimal feasible solution (OFS). Table 1 gives the Q-values for several different SNRs. From these results we can conclude that the OFS is consistently better than the MMSE estimate of the head phantom, which in turn is superior to the CBP reconstruction. We attribute this result to the fact that the OFS makes no prior assumptions about the image distribution other than that we wish to find an image, consistent with the data, of minimum energy, the MMSE estimate however makes the assumption that the image is totally uncorrelated. This later assumption is clearly unjustified since this implies a total lack of structure, and hence information, in the image. It is possible that the performance may be improved by introducing a non-zero, non-stationary mean f^o for the image using some prior estimate for f; it is shown in [9] that with a suitable choice of the prior the resulting variations of the true image about this mean are uncorrelated, however we contend that; we are looking for <u>unknown</u> structures and thus any prior estimate <u>must</u> be missing these structures and hence the error cannot be uncorrelated. Performance may also be improved by allowing for a correlation between each pixel and its neighbours, however the experience of applying Bayesian estimators in CAT using various ad hoc covariance approximations described in [8] indicates that this approach may also be unsatisfactory.

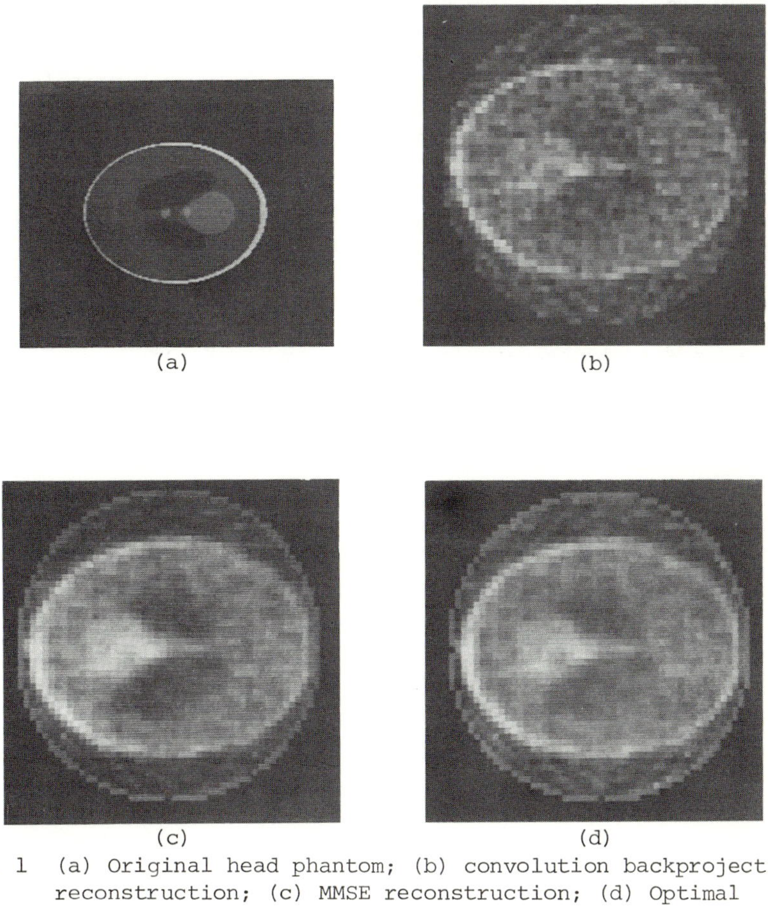

Fig. 1 (a) Original head phantom; (b) convolution backprojection reconstruction; (c) MMSE reconstruction; (d) Optimal feasible solution reconstruction.

SNR (dB)	Convolution Backprojection	MMSE estimator	Optimal Feasible Solution
23.0	1.42	1.26	1.18
20.0	1.50	1.30	1.27
14.0	1.63	1.45	1.41
5.0	1.78	1.66	1.62

Table 1 Q-values, equ. (30), for reconstructed images from projection data corrupted by additive white noise with various SNRs.

6. CONCLUSIONS

A method was described for solving the constrained restoration/reconstruction problem using a general convex cost function. By formulating the problem in a continuous, infinite dimensional setting and applying the Fenchel duality theorem the problem reduces to the optimization of a dual functional over a finite dimensional vector. The primal and dual problems are related by a model which arises naturally from applying Fenchel's theorem; the form of the model is determined by the choice of cost function and the prior constraints on the solution, the parameters of the model are the elements of the vector which solve the dual problem. The major advantages of this technique are that no prior model need be adopted for the unknown image and any combination of convex constraint sets can, in theory, be included (in practice however the resulting dual problem may become computationally intractible for large numbers of constraint sets). The method was applied to the problem of reconstruction of images from their projections using the minimum energy cost criterion and compared with the MMSE estimator in [3]. Results verify that, for the example chosen, the former method performs noticeably better than the later; this is attributed to the more realistic formulation of the problem. The similarity between the two systems of equations for the two methods was noted, equs. (23) and (25), and it was suggested that the MMSE estimator, equ. (25), gives an approximate linearized solution of the constrained minimum energy solution, equ. (23).

7. REFERENCES

[1] Tikhonov, A. and Arsenin, V. (1977) Solutions of ill-posed problems, Winston, Washington DC.

[2] Brillinger, D.R., (1975) Time Series: Data Analysis and Theory, Holt, Reinhart and Winston.

[3] Buonocore, M.H., Brady, W.R. and Macovski, A., (1981) "A Natural Pixel Decomposition for Two-dimensional Image Reconstruction", *IEEE Trans. Biomed. Eng.*, Vol. BME-28, pp. 69-78.

[4] Durrani, T.S. and Goutis, C.E., (1980) "Optimization Techniques for Digital Image Reconstructions from their Projections", *Proc. IEE,* Vol. 127, Pt. E., No. 5, pp. 161-169.

[5] Frieden, B.R., (1972) "Restoring with Maximum Likelihood and Maximum Entropy", *J. Optical Soc. Amer.*, Vol. 62, No. 4, pp. 511-517.

[6] Goutis, C.E., Leahy, R.M. and Drossos, S.N., (1985) "A Unified Approach to Tomographic and Spectral Analysis", submitted to Proc. IEEE.

[7] Herman, G.T., Lent, A. and Rowland, S.W., (1973) "ART: Mathematics and Application", *J. Theor. Biol.* Vol. 42, pp. 1-32.

[8] Herman, G.T. and Lent, A., (1976b) "A Computer Implementation of a Bayesian Analysis of Image Reconstruction", *Information and Control,* Vol. 31, No. 4, pp. 364-384.

[9] Hunt, B.R. and Cannon, T.M., (1976) "Nonstationary Assumptions for Gaussian Model of Images", *IEEE Trans. SMC,* Vol. SMC-6, pp. 876-882.

[10] Leahy, R.M., (1984) "Optimal Techniques for tomographic and spectral analysis", *Ph.D. Thesis,* University of Newcastle upon Tyne, England.

[11] Luenberger, D.G., (1969) Optimization by Vector Space Methods, New York, Wiley, 1st Edition.

[12] McClellan, J.H. and Lang, S.W., (1983) "Duality for Multi-dimensional MEM Spectral Analysis", *IEE Proc.*, Pt. F, Vol. 130, No. 3, pp. 230-235.

[13] Sezan, M.I. and Stark, H., (1982) "Image Restoration by the Method of Convex Projections", *IEEE Trans. Med. Imaging,* Vol. MI-1, pp. 95-101.

[14] Shepp, L.A. and Logan, B.F., (1974) "The Fourier Reconstruction of a Head Section", *IEEE Trans. on Nuclear Science,* Vol. NS-21, pp. 21-43.

[15] Trussel, H.J., (1983) "Convergence Criteria for Iterative Restoration Methods", *IEEE Trans. Acoustics, Speech and Signal Processing,* Vol. ASSP-31, pp. 129-136.

[16] Trussel, H.J. and Civanlar, M.R., (1984) "The Feasible Solution in Signal Restoration", *IEEE Trans. Acoustics, Speech and Signal Processing,* Vol. ASSP-32, No. 2, pp. 201-213.

[17] Youla, D.C. and Webb, H., (1982) "Image Restoration by the Method of Convex Projections: Part I - Theory", *IEEE Trans. Med. Imag.,* Vol. ME-1, No. 2, pp. 81-94.

IV IMAGE RECONSTRUCTION

APPLICATION OF CONVEX PROJECTIONS TO IMAGE RECOVERY IN DIRECT FOURIER TOMOGRAPHY FROM INCOMPLETE VIEWS

M. Ibrahim Sezan and H. Stark*
(Rensselaer Polytechnic Institute, Electrical, Computer, and Systems Engineering Department, Troy, New York)

ABSTRACT

We consider the problems of reconstructing CAT imagery by the direct Fourier method when not all view data is available. To restore the missing view data we use the method of convex projections. Convex projections is a recursive image restoration technique that finds a solution consistent with the measured data and a priori known constraints in both the space and Fourier domain. Because direct Fourier reconstruction is a frequency-domain technique it is ideally matched to convex projection restoration when, for one reason or another, we are forced to generate an image from a less than complete set of view data.

We design and apply an algorithm which interpolates/ extrapolates the missing Fourier domain information by convex projections and reconstructs an image by the direct Fourier method. A simulated human thorax cross-section is restored and reconstructed. The restorations using convex projections are compared with the Gerchberg-Papoulis extrapolation method and shown to be superior.

1. INTRODUCTION

In this paper we show how the method of projections onto convex sets (POCS) can be applied to a significant problem in industrial and medical imaging, namely, how to reconstruct good quality imagery from tomographic projection data when one has an incomplete set of views. We shall assume (and later justify) that the direct Fourier method (DFM) is the algorithm

---

*This work is supported by the National Science Foundation under Grant ECS-81-11364.

of choice for reconstruction. The method of POCS† was discussed in two recent papers in the literature [1], [2] in connection with image restoration when only partial frequency or space domain data was available to reconstruct an arbitrary image. Lent and Tuy [3] appear to be the first to restore an image using the POCS algorithm. Tuy [4] used POCS to restore missing projection data in connection with reconstruction via filtered convolution back-projection (FCBP). Although they demonstrated impressive results, their algorithm required the time-consuming artifice of generating projection data at each iteration that was consistent with *a priori* constraints. Since constraints are generally applied to the image or its spectrum (but not its projections) we were motivated to study the restoration from incomplete projections problem from a different viewpoint.

It is important for the reader to distinguish between the processes of reconstruction and restoration. As commonly used, by reconstruction is meant the activity of generating an image (analytically or otherwise) from an essentially complete set of data which is not in image format to begin with e.g. the way a TV image is reconstructed from an electrical signal, or the way an image is reconstructed from a hologram.

In contrast, by restoration we mean the activity of improving the quality of an image, or the data from which the image is reconstructed by combining *a priori* known properties of the image with measured data. Restoration generally implies an incomplete data set e.g. incomplete view data in tomography, a low-pass blurred image in photography, or loss of phase as in X-ray diffraction and interferometric astronomy.

Of course the distinction between reconstruction and restoration is the authors' own and is done here so that the reader will know what the authors have in mind when they use these terms. Both reconstruction and restoration can either be ill-posed or well-posed; both can sometimes be described using the same underlying mathematics. And finally there are numerous instances where the distinction between restoration and reconstruction is blurred e.g. if we are given only the Fourier magnitude of a signal x(t) and we enjoy the *a priori* knowledge that x(t) is minimum-phase, is the retrieval of phase a reconstruction or restoration?

---

† Projection onto convex sets refers to mathematical projections while projections in CAT refer to a set of parallel line integrals (the Radon transform) that forms the set of data used in parallel-beam computed tomography. It should be clear from the context which projection is meant.

It might be well to consider why in computer aided tomography (CAT) - a process which is highly controllable - one should be forced to work with incomplete view data†. In X-ray diffraction, optical astronomy and indeed many other scientific imaging type measurements one cannot obtain all the data (directly at least) needed to obtain the desired image. For example in X-ray diffraction, the phase that furnishes depth information is lost while in optical astronomy both atmospheric turbulence and finite-aperture optics limit the resolving power of the instrument. In CAT there are several reasons why an incomplete view data situation might prevail. In medical CAT there may be insufficient time for 360-degree data acquisition of time-varying anatomical situations. In industrial applications e.g. inspection, incomplete view data can result when obstructions prevent 360 degree viewing; or when imaging a moving part or a time-varying event.

Since image reconstruction from limited-view data is not merely of academic interest one next has to deal with the choice of restoration/reconstruction. Why DFM/POCS? With regard to the method of restoration POCS offers two noteworthy advantages: (i) it enables any number of *a priori* convex-type constraints to be incorporated in the algorithm; and (ii) it guarantees convergence: weak in general, strong in practice. Having chosen POCS as the restoration algorithm, the logical reconstruction method is the direct Fourier method (DFM). Not only is DFM fast as a reconstruction algorithm but because it involves transformations between space and Fourier domains, it enables space-domain and Fourier-domain constraints to be applied in their respective spaces directly. Finally because exact interpolation formulas exist -- which can be realized in practice to a high-degree of accuracy -- DFM-reconstructed images be made of the same quality as by the method of filtered convolution back projection (FCBP).

The organization of this paper is as follows: in sections 2 and 3 we briefly review the POCS and DFM algorithms. In Section 4 we describe the PRDF algorithm which combines POCS and DFM into a single reconstruction/restoration module. In Sections 5 and 6 we describe the experimental data and the results of applying PRDF. And in Section 7 we summarize the paper.

2. REVIEW OF THE METHOD OF PROJECTIONS ONTO CONVEX SETS (POCS)

The method of POCS has been discussed in some detail in References [1] and [2]. Strictly for the reader's

† Parallel-ray projection geometry is assumed. The view data at angle ϕ is the (sampled) set of projections along lines at that angle passing through the receiver detector array.

convenience we review here only the central idea of POCS and give the projection operators (without derivation) of use in CAT.

The image to be restored, $f(x,y)$ is assumed to be an element of the Hilbert space $\mathcal{H}$ of square-integrable functions.

The basic idea of POCS is as follows: Every known property of the unknown $f\varepsilon\mathcal{H}$ will restrict f to lie in a closed convex set C_i in $\mathcal{H}$. Thus, for m known properties there are m closed convex sets C_i, $i=1,2,\ldots,m$ and $f\varepsilon C_0 \triangleq \bigcap_{i=1}^{m} C_i$. Then the problem is to find a point of C_0 given the sets C_i and projection operators P_i projecting onto C_i, $i=1,2,\ldots,m$. The convergence properties of the sequence $\{f_k\}$ generated by the recursion relation

$$f_{k+1} = P_m P_{m-1} \cdots P_1 f_k \quad ; \quad k=0,1,\ldots \tag{4}$$

or more generally by

$$f_{k+1} = T_m T_{m-1} \cdots T_1 f_k \quad ; \quad k=0,1,\ldots \tag{5}$$

with $T_i \triangleq I+\lambda_i(P_i-I)$, $0 < \lambda_i < 2$ are based on fundamental theorems given by Opial [5] and Gubin et al. [6] . The λ_i's $i=1,\ldots,m$ are relaxation parameters and can be used to accelerate the rate of convergence of the algorithm. However, the λ's that are effective in the absence of noise will often be ineffective when noise is present [7] . Thus a single set of λ's is not effective at all signal-to-noise ratios. The following sets and their associated projection operators are among the ones used in Refs. [2] and [7] and they will be used here.

1. C_1: The set of all functions in $\mathcal{H}$ that vanish a.e.† outside a prescribed region $S\subset\Omega$. Given an arbitrary f in $\mathcal{H}$ its projection onto C_1 is realized by

$$P_1 f = \begin{cases} f, & (x,y)\,\varepsilon S \\ 0, & (x,y) \notin S \end{cases} \tag{6}$$

† a.e. is short for "almost everywhere."

2. C_2: The set of all function in H whose Fourier transforms assume a prescribed value G over a closed region L in the u-v Fourier plane. The projection of an arbitrary f in H is realized by

$$P_2 f \leftrightarrow \begin{cases} G(u,v), & (u,v)\in L \\ F(u,v), & (u,v)\notin L \end{cases} \tag{7}$$

where $F(u,v)=\mathcal{F}[f(x,y)]$ and $\mathcal{F}$ is the Fourier transform operator. In the case of incomplete view data G(u,v) is known in a data cone with subtended angle of less than 180 degrees.

3. C_3: The set of all real-valued nonnegative functions in H that satisfy the energy constraint

$$\iint_\Omega |f(x,y)|^2 \, dxdy \leq E \overset{\Delta}{=} \rho^2. \tag{8}$$

To describe the projection onto C_3, we write $f \overset{\Delta}{=} f_1 + jf_2$ where $f_1=\mathrm{Re}(f)$ and $f_2=\mathrm{Im}(f)$ and let f_1^+ be the rectified portion of f_1, i.e., $f_1^+=f_1$ when $f_1>0$ and $f_1^+=0$ when $f_1\leq 0$. Then the projection of an arbitrary $f\in H$ onto C_3 is realized by

$$P_3 f = \begin{cases} 0, & f_1<0 \\ f_1^+, & E_1^+\leq E \\ \sqrt{\dfrac{E}{E_1^+}}\, f_1^+, & E_1^+>E \end{cases} \tag{9}$$

where E_1^+ is the energy in f_1^+; i.e.,

$$E_1^+ \overset{\Delta}{=} \iint_\Omega (f_1^+)^2 \, dxdy. \tag{10}$$

4. C_4: The set of all real valued f's in H whose amplitudes must lie in a prescribed closed interval $[a,b]$; $a\geq 0$, $b>0$, $a<b$. The projection onto C_4 is realized by the following rule

$$P_4 f = \begin{cases} a \,, & f(x,y) < a \\ f(x,y) \,, & a \leq f(x,y) \leq b \\ b \,, & f(x,y) > b \end{cases} \tag{11}$$

Operators P_1, P_2, P_3 and P_4 were used in Equations (4) and (5) in restoring an image from its noiseless partial frequency spectrum [2]. They were also used together with a noise smoothing operator to restore images from a noisy partial frequency spectrum [7]. In all cases the method of POCS outperformed the Gerchberg-Papoulis (G-P) algorithm [8]. We point out in passing that the G-P algorithm also involves projections onto convex sets (actually linear subspaces) but generally ignores most a priori known constraints except space-limitedness and band-limitedness.

3. REVIEW OF THE DIRECT FOURIER METHOD (DFM) OF RECONSTRUCTION FROM PROJECTIONS

The DFM is based on the central slice projection theorem [9]. Let $\mu(x,y)$ represent a function whose image is desired (e.g. in CAT, $\mu(x,y)$ is the X-ray absorptivity). Let $\mu_\phi(\hat{x},\hat{y})$ represent the same distribution in a coordinate system $\hat{x}$-$\hat{y}$ rotated from x-y by an angle ϕ. The projection data at view angle ϕ is defined as

$$p_\phi(\hat{x}) \triangleq \int_L \mu_\phi(\hat{x},\hat{y})\, d\hat{y} \tag{12}$$

where L is the beam path. The central slice projection theorem states that

$$P_\phi(\rho) = M(\rho,\phi) \quad , \quad \rho \geq 0 \quad 0 < \phi \leq 2\pi \tag{13a}$$

where

$$P_\phi(\rho) = \int_{-\infty}^{\infty} p_\phi(\hat{x})\, e^{-i2\pi\rho\hat{x}} d\hat{x} \tag{13b}$$

and $M(\rho,\phi)$ denotes the 2-D Fourier transform of the cross-section distribution $\mu(x,y)$ in polar coordinates. In practice $p_\phi(\hat{x})$ is the data actually obtained at the location of detectors $\{\hat{x}_j\}$. In place of Eq. (13b) the discrete Fourier transform is used to obtain $P_\phi(\rho)$ at the discrete set of spatial frequencies $\{\rho_n\}$. Since the view angles are also a discrete set $M(\rho,\phi)$ is known at points on a polar grid $\{\rho_n, \phi_k\}$, $n=1,2,\ldots,N$; $k=1,2,\ldots,2K+2$ where N is the number of detectors, K is the highest harmonic in the angularly band-limited image and $\phi_k = \frac{2\pi}{2K+2} k$.

At this point, direct Fourier methods require interpolating $M(\rho_n, \phi_k)$ to $M_c(u_\ell, v_m)$ where $M_c(.)$ is the Fourier transform of $\mu(x,y)$ on a Cartesian raster. The interpolation from a polar to Cartesian raster is required in order to use the FFT routine. However inexact interpolation is a major source of error in reconstruction by direct Fourier methods even when the interpolation errors are not large (Lewitt [10] discusses this point in some detail). Attempts at implementing DFM using nearest neighbour and linear interpolation were done by Mersereau and Oppenheim [11]. Interpolation of temporal signals (that contain spatial frequency information) for NMR reconstruction is discussed by Cho et al. in Ref. [12].

Recently an exact interpolation method for certain broad classes of signals was discussed [9, 13, 14] which, in direct comparison and under practical constraints, produced images equal or superior in quality to convolution back-projection methods. The method is based on the polar sampling theorem.

$$M(\rho,\phi) = \sum_{n=-\infty}^{\infty} \sum_{k=0}^{2K+1} M\left(\frac{n}{2A}, \frac{\pi k}{k+1}\right) \text{sinc}\left[2A\left(\rho - \frac{n}{2A}\right)\right] \sigma\left(\phi - \frac{\pi k}{K+1}\right) \tag{14a}$$

where 2A is the diameter that bounds the object in space, K is the highest angular frequency in the ϕ-periodic function $M(\rho,\phi)$ and $\sigma(\phi)$ is an azimuthal interpolating function given by

$$\sigma(\phi) \triangleq \frac{\sin[(K+1)\phi]}{2K+2 \sin[(1/2)\phi]} \tag{14b}$$

The proof of Eq. (14a) is given in [14]. Its use is as follows: to compute $M_c(u_\ell, v_m)$, Eq. (14a) is used with

$$\rho_{\ell m} \triangleq \sqrt{u_\ell^2 + v_m^2} \tag{15}$$

$$\phi_{\ell m} \triangleq \cos^{-1}\left(\frac{u_\ell}{\rho_{\ell m}}\right) . \tag{16}$$

Thus the set $\{u_\ell, v_m\}$ can be computed directly from the set $\{\rho_n, \phi_k\}$. In practice a truncated interpolation involving a finite number of cardinal functions is used. Thus Eq. (14a) is replaced by

$$\hat{M}(\rho,\phi) = \sum_{n=n_\rho-L_\rho}^{n_\rho+L_\rho} \sum_{k=k_\phi-L_\phi}^{k_\phi+L_\phi} M(\frac{n}{2A}, \frac{\pi k}{K+1}) \cdot \text{sinc}(2A(\rho - \frac{n}{2A}))\sigma(\phi - \frac{\pi k}{K+1}), \tag{17}$$

where $[\,2A\rho\,] \stackrel{\Delta}{=} n_\rho$ and $[\,(K+1)\phi/\pi\,] \stackrel{\Delta}{=} k_\phi$†, are the nearest neighbours to the point about which the expansion is done. In practice, image quality is improved when the truncation is tapered. The taper normally used is the Cartesian product of two identical one-dimensional triangular windows, $\omega(n)$, of the form

$$\omega(n) = \max(1-|n|/M,\ 0) \quad : \quad n=0,\pm 1,\pm 2,\ldots \tag{18}$$

$M=\infty$ is the abrupt truncation case. If $L_\rho = L_\phi = 0$ the above practical approximation reduces to nearest neighbour approximation. In this paper we use Eq. (17) i.e., exact interpolation with truncation and tapering (M=5).

After interpolating the 2-D Fourier spectrum from a polar to a Cartesian raster, a 2-D inverse FFT is computed to obtain the desired image. The direct Fourier method is illustrated in Figure 1.

4. PRDF ALGORITHM

If projection data is available over a limited angular range, then the image Fourier transform $M(\rho,\phi)$ can be interpolated over a Cartesian grid within this angular range. The pixels remaining outside this angular range are set to zero as a first approximation. The missing frequency information is then restored by POCS and the desired image is obtained by inverse transforming. The complete algorithm is called PRDF and it is illustrated in Figure 2.

† The operation indicated by the brackets indicates the closest integer to the value within the brackets, i.e. rounding.

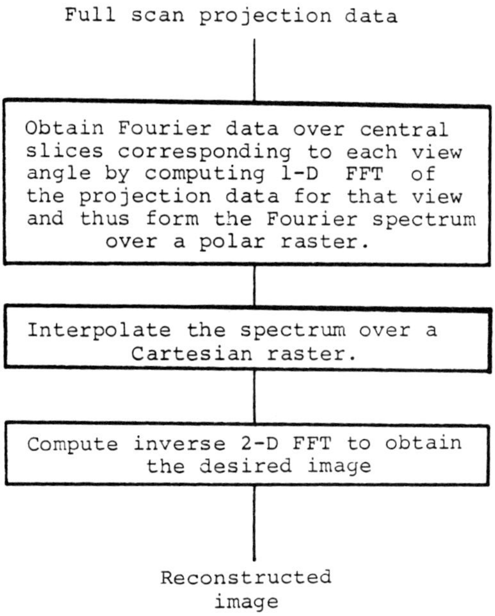

Fig. 1 Direct Fourier inversion method (DFM).

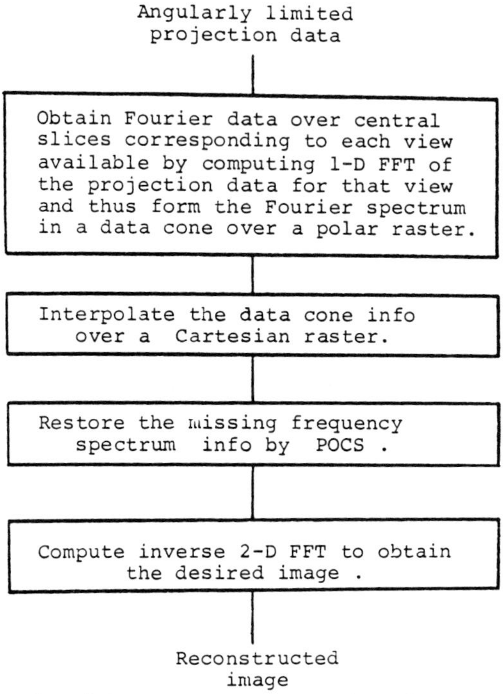

Fig. 2 PRDF algorithm for reconstructing from angularly limited view data

In applying PRDF the four constraints sets and their associated projectors listed in Section 2 were used. It was found experimentally that C_4 had little influence on the outcome and, hence, was generally not used. On the other hand C_3 was the major factor in enabling our results to be superior to the Gerchberg-Papoulis algorithm. The C_3 energy constraint can be accurately estimated from reference images. Moreover it furnishes regularization since it forces all the iterates to satisfy $||f_k|| \leq E$ where E is the energy bound. This is the reason that convex projection algorithms are inherently stable. In contrast, the Gerchberg-Papoulis algorithm is ill-posed and the divergence of its iterates is often observed. A study of the effects of noise on image extrapolation in which the Gerchberg-Papoulis algorithm was compared with convex projections is furnished in [7] .

Convex projection algorithms can use pure (P_i) or relaxed (T_i) projectors. In PDRF, algorithms using pure projectors were given the name UNIRELAX; those using relaxed projectors were called RELAX. The RELAX algorithms involved the relaxation parameters λ_i, i = 1,2,3. It can be shown [15] that the optimum λ_i's must satisfy $\lambda_i \geq 1$. By optimum we refer to those λ_i's that furnish a most rapid convergence to a solution. Our choice of λ_i's were largely determined by experimental trial and error although we could have incorporated optimization subroutines given in [15, 16] that would have furnished optimum λ_i's at each step. In that case the results of the simulations would no doubt have improved (perhaps only slightly) but the running time of the algorithms would have increased.

5. RECONSTRUCTING A THORAX PHANTOM FROM ANGULARLY LIMITED X-RAY PROJECTION DATA

The 128 by 128 pixels thorax phantom† is shown in Figure 3. The simulated parallel beam projection data assumes an X-ray energy of 70 Kev, 128 detectors which are spaced apart by .3 cm and 360 views over 360 degrees. The interpolated transform with $L_\rho=3$, $L_\phi=1$ i.e., 21 interpolation points and resulting reconstruction from complete view data is shown in

† The thorax phantom is widely used in the literature. It was designed by G. Herman and A. Lakshminarayanan.

Figure 4.

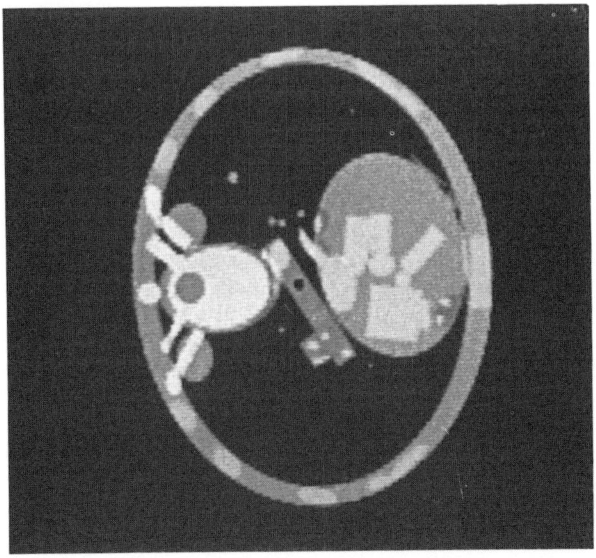

Fig. 3 Thorax Phantom

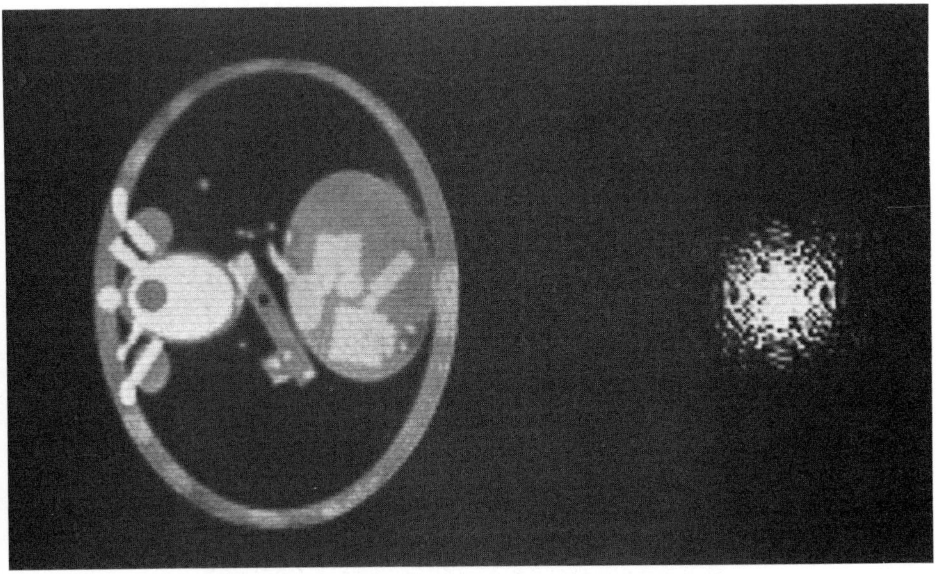

Fig. 4 DFM reconstruction (left) and interpolated spectrum (L_ρ=3, L_ϕ=1) (right).

The 11 percent reconstruction error between the reconstructed image and the original is due to 1) inadequate sampling of the projection data, 2) inadequate sampling in radial frequency, 3) insufficient number of views. Since the object is of finite support the projection data at each view will also have finite support. Hence instead of obtaining $P_\phi(k/2A)$ one obtains a periodic extension

$$\hat{P}_\phi(k/2A) = \sum_{n=-\infty}^{\infty} P_\phi((k+nN)/2A) \qquad k=1,2,\ldots,N \tag{18}$$

in which the n=0 term cannot be isolated without containing contamination from the n≠0 higher orders. However this unavoidable effect can be reduced by choosing N (number of detectors) large enough. In practical situations the finite width of the X-ray beam produces smoothened projection data which is somewhat equivalent to low-pass filtering the projection data. Hence, in practice the effect of aliasing artifacts can be made tolerably small. The other two effects are discussed at length in [9].

Reconstruction by PRDF from angularly limited projection data is attempted for the following three cases; projection data is limited to the view range of i) [-80°,80°], ii) [-67°,67°], iii) [-45°,45°]. The formula of Eq. (17) with $L_\rho=3$, $L_\phi=1$ is used to interpolate $\hat{M}(\rho,\phi)$ from polar to Cartesian points. The Fourier transform for full-view data and the above three cases is shown after interpolation in Figure 5.

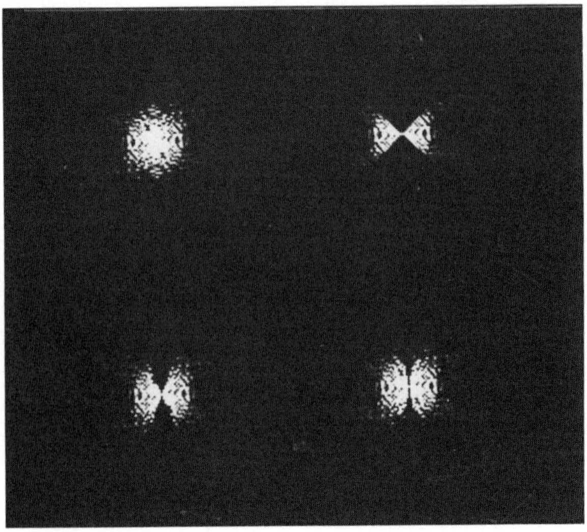

Fig. 5 Interpolated frequency spectra: Clockwise from upper left: full view projection data, view limited to [-45°,45°] view limited to [-80°,80°] view limited to [-67°,67°].

6. RESULTS

The basis for comparing results was the percent error e_k at the k'th iteration defined by

$$e_k \overset{\Delta}{=} 100 \cdot \frac{||f-f_k||}{||f||} \tag{19}$$

where $f \overset{\Delta}{=} f(x,y)$ is the full scan (360 degrees) reconstructed image. While e_k is a standard measure for evaluating the efficacy of an image restoration algorithm, its relationship to subjective or clinical evaluation has not been established.

A summary of the a priori known facts and assumptions made about the image are given in Table I. Table II furnishes a summary of the experimental results. The first (leftmost) column of Table II gives the angular field of view over which data is available; in the second column is the name of the algorithm; in the third and fourth columns are the recursion relations and the associated values of the relaxation paramaters respectively; the next column is the error between the full scan image and the image immediately reconstructed from the limited view data (the so-called zero-order or "naive" solution); and the last column is the error at the 30th iteration for the three cases of angularly limited projection (view) data.

Table I

SUMMARY OF A PRIORI ASSUMED CONSTRAINTS

a priori constraints	Actual
(1) Image support confined to rectangular region of length 124 pixels and width 103 pixels.	(1) Image support is elliptical (Figure 3).
(2) Gray levels f satisfy $0 \leq f \leq 0.4$	(2) Gray levels f satisfy $0 \leq f \leq 0.38$
(3) Energy over 128x128 square pixel field cannot exceed ρ^2=284.000	(3) Energy over 128x128 square pixel field is 282.74

Table II

Summary of significant results

Available View Range	Algorithm			Percent Error	
	Name	Description	Relaxation Parameters	e_0 (zero-order) solution	e_{30}
[-45,45]	GP	$P_2P_1f_k=f_{k+1}$	$\lambda_1=\lambda_2=1.0$		47.511
	UNIRELAXL	$P_4P_2P_3P_1f_k=f_{k+1}$	$\lambda_1=\lambda_2=1.0$ $\lambda_3=\lambda_4=1.0$	52.419	42.057
[-67,67]	GP	$P_2P_1f_k=f_{k+1}$	$\lambda_1=\lambda_2=1.0$		22.203
	UNIRELAX	$P_2P_3P_1f_k=f_{k+1}$	$\lambda_1=\lambda_2=\lambda_3=1.0$	45.000	17.837
	RELAX	$T_2T_3T_1f_k=f_{k+1}$	$\lambda_1=\lambda_3=1.9995$ $\lambda_2=1.0$		16.184
[-80,80]	GP	$P_2P_1f_k=f_{k+1}$	$\lambda_1=\lambda_2=1.0$		15.485
	UNIRELAX	$P_2P_3P_1f_k=f_{k+1}$	$\lambda_1=\lambda_2=\lambda_3=1.0$	41.361	12.100
	RELAX	$T_2T_3T_1f_k=f_{k+1}$	$\lambda_1=\lambda_3=1.9995$ $\lambda_2=1.0$		9.352

a) Results for When Projection Data is Limited to [-45°,45°]

In this case the reconstructed image at the end of 30 iterations is of low visual quality for all algorithms. Still the DFM reconstruction using UNIRELAXL (i.e. level constraints applied) restoration outperforms the DFM reconstruction using GP restoration. The error performance of the reconstructions are compared in Figure 6.

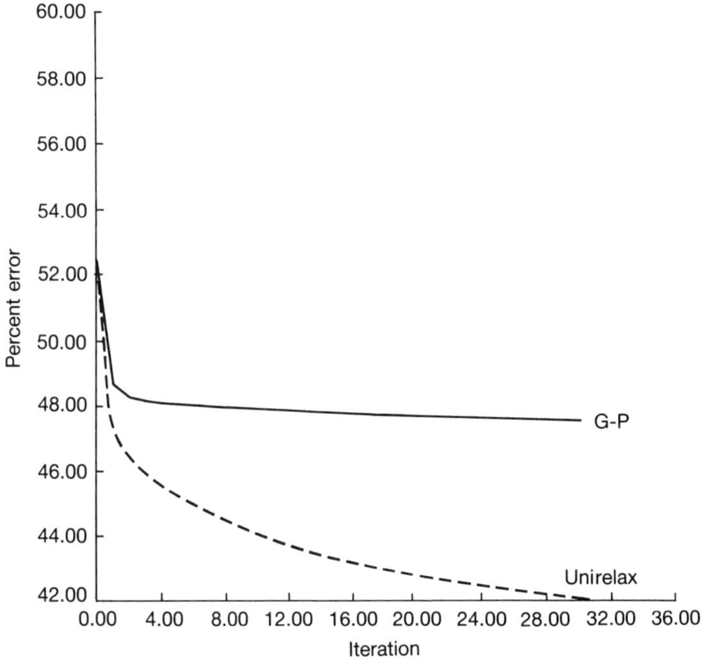

Fig. 6 Performance comparison of GP, and UNIRELAXL in reconstructing from projection data available over [-45°,45°].

b) Results for When Projection Data is Limited to [-67°,67°]

In this case the three different restoration algorithms GP, UNIRELAX, and RELAX are separately combined with DFM reconstruction. The naive solution, which is merely inverse Fourier transforming the polar to Cartesian interpolated spectrum having zeroes for the entries corresponding to unavailable angular region (Figure 5) results in an error of 45 per cent. The error performance of the GP, UNIRELAX, and RELAX reconstructions are compared in Figure 7. With the RELAX restoration an error less than 17 per cent is reached at the end of 30 iterations and the resulting reconstruction has fairly good visual quality. As can be seen from Figure 7, PDRF reconstruction incorporating UNIRELAX and RELAX restoration outperforms the reconstruction based on GP restoration.

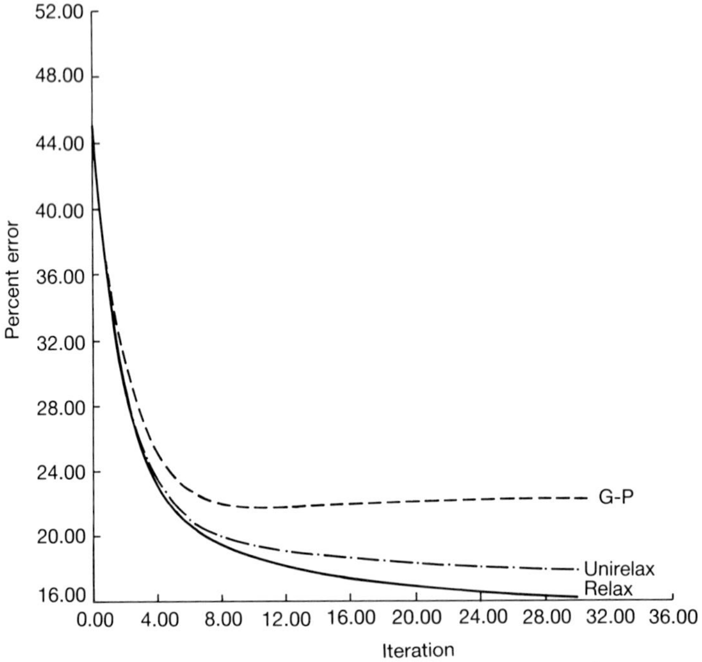

Fig. 7 Performance comparison of GP, UNIRELAX and RELAX in reconstructing from projection data available over $[-67^\circ, 67^\circ]$.

c) Results for When Projection Data is Limited to $[-80^\circ, 80^\circ]$

The naive reconstruction (e_o = 41 percent) and the reconstructed images at the end of 30 iterations using RELAX, UNIRELAX and GP restorations are compared at the bottom of Table II. In Figure 8 we compared the error performance of the reconstructions using RELAX, UNIRELAX and GP algorithms; the errors are respectively 9.3, 12.1 and 15.5 percent at the end of 30 iterations. As before, reconstructions using a priori restoration is superior to the naive reconstruction. We note from Fig. 8 that the error history of the reconstruction is not strictly monotonic. Both the GP and UNIRELAX show an increase in the error beyond a minimum as k increases. This "ill-posed" behaviour is very quickly arrested in the UNIRELAX algorithm because of the energy constraint. The GP algorithm having no build-in constraint on the energy of its iterates is more likely to diverge. The reconstruction by the RELAX restoration outperforms the reconstruction performed by the classical GP restoration.

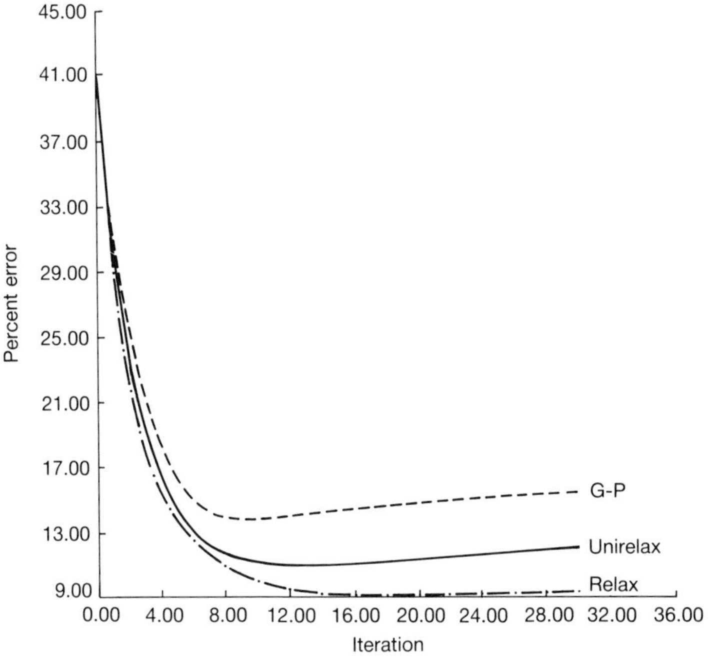

Fig. 8 Performance comparison of GP, UNIRELAX and RELAX in reconstructing from projection data available over [-80°,80°].

The GP and RELAX reconstructions are compared in Figure 9. The superior performance of RELAX is especially evident in the reconstruction of the elliptical structure (note the obliterations in the GP case) and the reduced plume-like clouding from the tips of bright objects.

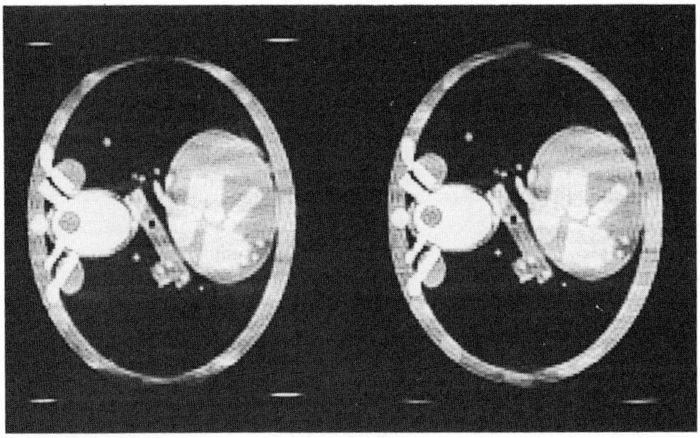

Fig. 9 GP (left) and RELAX (right) reconstructions at 30th iteration.

7. CONCLUSIONS AND SUMMARY

In this paper we presented a reconstruction algorithm (PRDF) that is synthesized from a recent restoration algorithm called projections onto convex sets (POCS) and from a direct Fourier inversion method (DFM) based on an exact polar-to-Cartesian interpolation formula. The PRDF algorithm is capable of performing reconstruction from angularly-limited projection data. If the projection data is a collected via fan beam geometry, rebinning combined with interpolation [17] is applied prior to reconstruction to perform the conversion to equivalent parallel beam projection data. A new sampling theorem, described in [17], enables exact interpolation from a broad category of non-uniformly spaced samples. This theorem enables non-uniformly spaced rebinned data to be interpolated into uniformly spaced data. A remaining complication, however, is that missing view divergent data does not translate exactly into missing view parallel data.

As noted before DFM is a computationally faster algorithm than filtered convolution back projection (FCBP). It is shown in reference [9] that DFM requires $O[N^2 \log N]$ operations compared with $O(N^3)$ operations for FCBP to reconstruct a similar quality (NxN) image. However, one can use POCS with the FCBP algorithm as well to reconstruct from angularly limited projection data. Tuy [4] used POCS with the FCBP algorithm to solve the limited angular view problem. At the end of each iteration, the projection data over the missing angular range is generated for that iterate. This calculation can be time consuming and susceptible to numerical errors that may be introduced during the calculation.

We foresee numerous applications to the method of convex projections in other areas such as NMR tomographic imaging, synthetic aperture radar, diffraction tomography and ultrasound. It has already been successfully used in recovering a signal from its Fourier phase [15] or Fourier magnitude [16] and signal recovery subject to moment constraints [18].

This paper did not explicitly address the issue of noise: how does noise affect the quality of the restoration? Such a study is furnished in [7]. A very important constraint set when noise is present is C_5, the set of all functions {f} such that the distance of f from a known reference, say f_R, is less than or equal to some radius ρ. Not only can one easily relate ρ to the noise variance but this constraint furnishes a build-in regularization. Constraints imposed by noise as well as other constraints of use in convex projections are treated in a nice paper by Trussell and Civanlar [19].

ACKNOWLEDGEMENT

Many thanks are due to Dr. Heang Tuy who supplied the thorax phantom data. We would also like to acknowledge our valuable technical discussions with him. This work is supported by National Science Foundation under Grant ECS-81-11264.

REFERENCES

[1] Youla, D.C. and Webb, H., (1982) "Image restoration by the method of convex projections: Part 1 - Theory", *IEEE Trans. on Medical Imaging,* Vol. MI-1, pp. 81-94, No. 2.

[2] Sezan, M.I. and Stark, H., (1982) "Image restoration by the method of convex projections: Part 2 - Applications and Numerical Results", *IEEE Trans. on Medical Imaging,* Vol. MI-1, pp. 95-101, No. 2.

[3] Lent, A. and Tuy, H., (1981) "An iterative method for the extrapolation of band limited functions", *J. Math. Anal. Appl.,* Vol. 83, pp. 554-565.

[4] Tuy, H.K., (1983) "An Algorithm for incomplete range of views reconstruction", Technical Digest of Topical Meeting on Signal Recovery and Synthesis with Incomplete Information and Partial Constraints, pp. FA1-1, FA1-4, Nevada.

[5] Opial, Z., (1967) "Weak convergence of the sequence of successive approximations for nonexpansive mappings", *Bull. Am. Math. Soc.* **73**, pp. 591-597.

[6] Gubin, L.G., Polyak, B.T. and Raik, E.V., (1967) "The method of projections for finding the common point of convex sets", *U.S.S.R. Comput. Math. and Math. Phys.,* Vol. 7, No. 6, pp. 1-24.

[7] Sezan, M.I. and Stark, H., (1983) "Image restoration by convex projections in the presence of noise", *Applied Optics,* Vol. 22, pp. 2781-2789.

[8] Papoulis, A., (1975) "A new algorithm in spectral analysis and band limited extrapolation", *IEEE Trans. on Circuits and Systems,* Vol. CAS-22, pp. 735-742.

[9] Stark, H., Woods, J.W., Paul, I. and Hingorani, R., (1981) "An investigation of computerized tomography by direct Fourier inversion and optimum interpolation", *IEEE Trans. Biomed. Eng.*, Vol. BME-28, pp. 496-505.

[10] Lewitt, R.M., (1983) "Reconstruction Algorithms: Transform Methods", *IEEE Proc.*, Vol. 71, No. 3.

[11] Mersereau and Oppenheim, A., (1974) "Digital Reconstruction of Multidimensional Signals From Their Projections", *IEEE Proc.*, Vol. 62, No. 10, pp. 1319-1338.

[12] Cho et al., (1984) "Direct Fourier Reconstruction Techniques in NMR Tomography", to be published in Selected Topics in Image Science, Springer-Verlag Lecture Notes in Medical Informatics, Vol. 23.

[13] Stark, H., (1979) "Sampling theorems in polar coordinates", *J. Opt. Soc. Amer.*, Vol. 69, pp. 1519-1525.

[14] Stark, H. and Wengrovitz, M., (1983) "Comments and Corrections on the Use of Polar Sampling Theorems in CT", *IEEE Trans. on Acoust., Speech and Signal Proc.*, Vol. ASSP-31, No. 5, pp. 1329-1331.

[15] Levi, A. and Stark, H., (1983) "Signal Restoration from Phase by Projections Onto Convex Sets", *J. Opt. Soc. Am.*, Vol. 73, No. 6, pp. 810-822.

[16] Levi, A. and Stark, H., (1984) "Image Restoration by the Method of Generalized Projections with Application to Restoration from Magnitude", *J. Opt. Soc. Am.*, Vol. 1, No. 9, pp. 932-943.

[17] Clark, J.J., Palmer, M.R. and Lawrence, P.D., (1985) "A Transformation Method for the Reconstruction of Functions from Nonuniformly Spaced Samples," *IEEE Trans. Acoust., Speech, and Signal Proc.*, Vol. ASSP-33, No. 4, pp. 1151-1165.

[18] Sezan, M.I. and Stark, H., "Incorporation of A Priori Moment Information Into Signal Recovery and Synthesis Problems", to appear in *J. Math. Anal. Application*.

[19] Trussell, H.J. and Civanlar, M.R., (1984) "The Feasible Solution in Signal Restoration," *IEEE Trans. Acoust., Speech, and Signal Proc.* Vol. ASSP-32, No. 2, pp. 201-212.

MAXIMUM ENTROPY DATA ANALYSIS

J. Skilling
(Department of Applied Mathematics and Theoretical Physics, University of Cambridge)

ABSTRACT

Maximum entropy is the preferred method of assigning distributions of proportions subject to incomplete constraints. Several arguments lead to this conclusion, from the combinatoric results of Jaynes (1979) through ideas about distributions with minimum information (Kullback 1959) to the axiomatic abstractions of Shore and Johnson (1980). In this tutorial paper, we develop the latter approach, and show why maximum entropy is the compelling choice among methods of assigning distributions constrained by their marginals. We then illustrate a few of the complex inter-relationships which exist between apparently diverse problems, to justify the assumption that maximum entropy should be used for any form of constraint.

The importance of maximum entropy in practical data analysis arises because many physical systems are represented by distributions of proportions. Thus an optical image is a distributed pattern of radiation intensity, as is a radio image or an X-ray image in astronomy. Medical X-ray scanning uses absorption measurements to generate distributed patterns of absorptivity. Spectral measurement yield information on the number density of oscillators distributed along a spectrum. In these and other fields, maximum entropy has been successfully used to obtain clear, high-quality reconstructions.

1. THE KANGAROO PROBLEM

It can be instructive to analyse some data. Suppose we are given the following information, first given in similar form by Gull and Skilling (1984a).

2/5 of kangaroos are male, and 3/5 are female:

1/4 of kangaroos are left-handed, and 3/4 are right-handed.

On the basis of this information alone, estimate the proportion of left-handed male kangaroos. A single answer is required.

This problem is equivalent to completing the 2x2 contingency table of the relative proportions of kangaroos in each category (Table 1). We avoid Diophantine complications by considering the desired answer, x, to be a continuous variable.

	Left	Right	
Male	x	$\frac{2}{5} - x$	$\frac{2}{5}$
Female	$\frac{1}{4} - x$	$\frac{7}{20} + x$	$\frac{3}{5}$
	$\frac{1}{4}$	$\frac{3}{4}$	

Table 1

Positively restricts x to the range $0 \leq x \leq 1/4$, but the data are incomplete and strict deductive logic can take us no further.

Unable to deduce x, can we infer an optimal "best" value? Each of the extreme values of x implies correlations which it would be difficult to defend on the basis of the information given. Thus x=0 implies that all male kangaroos are right-handed, which is surely unwarranted by the data. Indeed it would be difficult to defend any particular degree of correlation between sex and handedness, and we are led to prefer the uncorrelated result $x = 2/5 \times 1/4 = 1/10$. This is the most non-committal choice, and we claim that this is the "best" answer we can give to the assigned problem.

The argument generalises to larger, mxn, contingency tables p_{ij}. Suppose we are given the following information:

$$\sum_j p_{ij} = u_i \quad (i=1,2,\ldots,m) \text{ and } \sum_i p_{ij} = v_j \quad (j=1,2,\ldots,n).$$

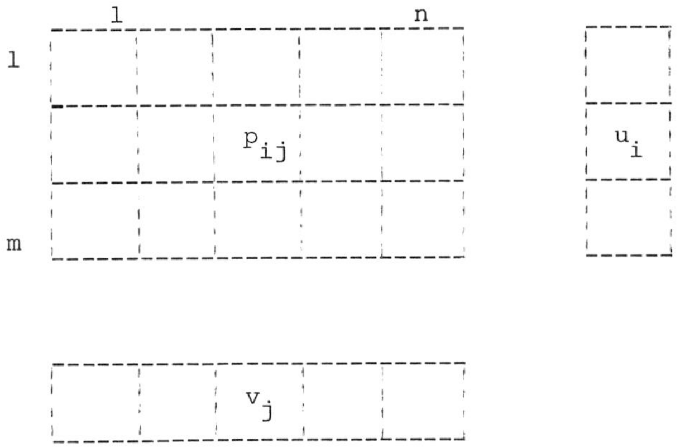

If required to complete the contingency table, our optimal (least committal) answer must be the direct product

$$p_{ij} = u_i v_j$$

in which the vertical data u tell us nothing about the horizontal structure, and vice versa. Similarly in more than two dimensions, marginal data

$$\sum_{jk} p_{ijk} = u_i \, , \; \sum_{ik} p_{ijk} = v_j \, , \; \sum_{ij} p_{ijk} = w_k$$

would lead us to the selection

$$p_{ijk} = u_i v_j w_k$$

and so on.

It is a remarkable fact that these answers can always be obtained by maximising the entropy functional

$$S(p) = - \Sigma \, p \log p$$

subject to the given contraints. Furthermore, the entropy (or a monotonic function of it) is the only functional with this property (Shore and Johnson 1980, Livesey and Skilling 1985). Accordingly, we are led to consider the maximum entropy method (MEM) in which p is chosen by maximising its entropy.

Our analysis can now be connected with standard regularization theory (e.g. Tikhonov and Arsenin 1977). When we are required to select a single distribution p from the many which are possible, we need to set up relationships which express our relative preferences. The consistency requirement (that if p is preferred to q, and q is preferred to r, then p must be preferred to r) allows us to express our preferences as inequality relationships between values of some functional R, so that "p is preferred to q" becomes "R(p)>R(q)". Selecting the best distribution p then amounts to maximising R subject to whatever constraints are available: the choice of p is said to be regularized by R.

Many different functionals have been proposed to express preferences for different types of distribution. However, we have seen that the entropy S has a special status in that it is the only functional which does not induce correlations for which there is no evidence either in the data or in one's prior knowledge. We will return briefly to the question of prior knowledge later.

2. OTHER DATA

From the discussion above, it might seem that MEM was only applicable to marginal data. We argue that MEM is much more widely applicable, because of close relationships between apparently different forms of data.

Fourier Data

Suppose we have marginal data u and v on a mxn contingency table, where m and n are coprime. Its mn cells can be ordered onto a one-dimensional sequence by indexing diagonally with wraparound.

p_0	p_6	p_{12}	p_3	p_9	u_0
p_{10}	p_1	p_7	p_{13}	p_4	u_1
p_5	p_{11}	p_2	p_8	p_{14}	u_2

v_0	v_1	v_2	v_3	v_4

The data u and v are now equivalent to incomplete Fourier data on the sequence. Defining Fourier coefficients by

$$F_k = \sum_{q=0}^{mn-1} p_q e^{2\pi ikq/mn}$$

we have

$$F_{jm} = \sum_{q=0}^{n-1} v_q e^{2\pi ijq/n} \qquad (j=0,1,\ldots,n-1)$$

$$F_{jn} = \sum_{q=0}^{m-1} u_q e^{2\pi ijq/m} \qquad (j=0,1,\ldots,m-1)$$

so that the marginal data could equivalently have been given as the particular Fourier coefficients F_0, F_m, F_{2m}, ...; F_n, F_{2n}, ... of the one-dimensional sequence. Fourier data in more than one dimension can be obtained by re-ordering appropriately larger contingency tables. Thus 4-dimensional tables can be re-ordered to give 2-dimensional Fourier data, and so on. Of course, the original 2-dimensional marginal data are also equivalent to the particular Fourier coefficients $F_{j,0}$ and $F_{0,k}$ of the 2-dimensional distribution, which is a different selection of components.

Because MEM should be used for marginal data, it should also be used for the equivalent one-dimensional problem. Given some incomplete Fourier data, it would be peculiar to use MEM only if the sequence length had coprime factors and the data followed the particular sampling we discussed. The natural action is to use the same regularization method (MEM) for any Fourier data.

Convolution data

Convolution of a distribution p with a spatially invariant point-spread-function b is equivalent to multiplying the Fourier transform F of the distribution by the transform B of the point-spread-function. Thus convolution data are equivalent to multiples $B_k F_k$ of the transform components. The incomplete Fourier data discussed above are a special case of this with some of the B_k being zero. Because MEM should be used for the latter problem, it is natural to use MEM for all convolution data.

Tomographic data

In two dimensions, tomographic data may be idealized as line integrals

$$\iint p(x,y)\ \delta(\ x \sin\theta + y \cos\theta - h)\ dx\ dy$$

For each orientation θ, data regularly sampled in offset h may be transformed in h to Fourier coefficients $F_{k \sin\theta,\ k \cos\theta}$.

Because we have argued that MEM should be used for this type of data, it is then natural to use MEM for all tomographic data, whether or not the sampling is regular.

Generalization

The techniques of re-ordering, re-scaling and transforming can be used to show that other data besides our chosen examples may be made equivalent to an appropriate form of marginal data. Thus we now assume, as an axiom, that the same regularization functional should be used for all types of data. Entropy happens to be the uniquely compelling choice for marginal data, so we should also use it for reconstructing distributions of proportions from any type of data. Our examples have all been linear, but any convex nonlinear constraint can be replaced by linear inequality constraints on its supporting tangent hyperplanes, so that it becomes natural to use MEM for nonlinear data also. (Nonconvex constraints may yield multiple local maxima of entropy, but that complication is inevitable.)

3. PRIOR KNOWLEDGE

In our analysis of the kangaroo problem, we implicitly assumed that all four possible classifications of kangaroos were a priori equivalent. Having no further information, we could make no other consistent choice. However, in practice we might have relevant prior information. Suppose we are told that there are six types of kangaroo. Type A is male and left-handed, and so on up to tye F.

	Left	Right
Male	A B	C
Female	D	E F

Types A and B, and E and F, differ in other ways, so that each type is a priori equivalent.

Given some observational data (possibly incomplete), estimate the proportions of kangaroos of each type. From our previous discussion, it is clear that we should maximise the entropy

$$S = -p_A \log p_A - p_B \log p_B - p_C \log p_C - p_D \log p_D - p_E \log p_E - p_F \log p_F$$

of the proportional distribution $(p_A, p_B, p_C, p_D, p_E, p_F)$ of the types of kangaroo, subject to whatever data we are given. If we observe merely the marginals of sex and handedness, then S will (very reasonably) be maximized with $p_A = p_B$ and $p_E = p_F$.

Writing the proportions in the 2x2 table as $p_1 = p_A + p_B$, $p_2 = p_B$, $p_3 = p_C$, $p_4 = p_E + p_F$, we reach

$$S = -p_1 \log (p_1/2) - p_2 \log p_2 - p_3 \log p_3 - p_4 \log (p_4/2)$$

The maximum entropy selection of the free parameter x, given the original data, is now x=0.1624 instead of x=0.1. As it should be, this is biassed away from 1/10 towards a positive correlation, because we are seeking a minimally committal distribution over types whose characteristics suggest such correlation.

More generally, if cell i of our distribution is divided into m_i sub-cells , and each sub-cell is a priori equivalent the entropy becomes

$$S(p;m) = - \Sigma\, p_i \log (p_i/m_i)$$

This, and its continuous limit (Jaynes 1968),

$$S(p;m) = - \int p(x) \log(p(x)/m(x))\, dx$$

is the proper coordinate-invariant form we should use.

Many types of prior knowledge do not correspond to straightforward, fixed prior models m. For example, a preference for local correlations between cells, being translation invariant, can not be coded as a fixed pre-assigned model. In such cases it will be necessary to use adaptive models, in which m is adjusted in the light of a preliminary reconstruction p.

4. ACCURACY OF THE SELECTED RECONSTRUCTION

A difficulty of principle arises when we attempt to discuss the accuracy of a regularised distribution. Suppose the distribution has M cells, but only r data constraints are available ($r < M$). There will, in general, remain a manifold of distributions, of dimension $M-r$, all of which are consistent with the data. Certainly we can select an optimal distribution by maximizing a functional, and we can quantify the cost of selecting any other acceptable distribution in terms of the decrease in value of the functional. However, this may or may not be helpful.

On the positive side, entropy does have a well-known interpretation in terms of bits of information, so that a decrease S corresponds to an information cost S/log2 bits. There is also the theorem that if distribution p maximizes S over linear equality constraints and distribution q also satisfies the constraints, then

$$S(q) = S(p) - x$$

where

$$x = \Sigma\, q \log(q/p)$$

is the cross-entropy of q relative to p. Thus the cost of q relative to p can be found easily, without explicit reference to the data or to the underlying model m. Convex constraints impose

$$S(q) \leqslant S(p) - x$$

so that x becomes a lower bound on the cost of selecting q instead of p.

On the negative side, the entropy cost $S(p)-S(q)$ is, necessarily, a global functional of the new distribution q. Especially when M is large (and continuous distributions have $M \to \infty$), individual cells can fluctuate substantially without significantly affecting the data or imposing a significant regularization cost. Thus it becomes difficult to display, or to put local bounds on, the fluctuations which a given extra cost allows. It is only by deliberately discarding the full size of the manifold and working in some smaller subspace that local error bounds can sometimes be obtained.

This difficulty of principle can be traced back to the initial formulation of our problem, when we were required to select a single distribution. For large problems with many degrees of freedom, it is impossible in practice to sample adequately the space of distributions which fit the constraints, and thus to explore the fluctuations. It is for this reason that it has become common practice to seek a single optimal result. The whole regularization approach is set up precisely for those problems in which the fluctuations cannot be displayed.

5. EXPERIMENTAL NOISE

Formal derivations of maximum entropy as a preferred method of assigning distributions p require hard constraints, whereby any trial distribution may be classified as either "feasible" (obeying the constraints) or not. The latter are rejected, and entropy is then used to select the preferred feasible distribution. Although experimental data are inevitably noisy, we still wish to construct a feasible set, comprising only those distributions which fit the data tolerably.

Whatever the type of experiment, the data and their noise characteristics may be subsumed into the likelihood function

$$L = \mathrm{prob}(\text{ data } | \text{ p})$$

and it is this conditional probability which we use as the basis for constructing the feasible set. If the data D_k $(k=1,2,\ldots,r)$ are subject to normal errors $\pm\sigma_k$, then L becomes related to chisquared

$$L \propto \exp(-\chi^2/2)$$

$$\chi^2 = \sum_k (F_k(p) - D_k)^2/\sigma_k^2$$

where $F_k(p)$ is the "mock" datum which would have been observed in the absence of noise if p were the correct distribution. Fortunately, we expect χ^2 to peak strongly near r, the number of independent data, so we may reject any distribution having $\chi^2 \gtrsim r$. Conversely, we accept any p with $\chi^2 \lesssim r$.

The maximum entropy method for practical data analysis then consists of maximizing S(p) over $\chi^2(p) \lesssim r$. Occasionally it is useful to restrict the feasible set further by more sophisticated testing of the normalized residuals $(F_k - D_k)/\sigma_k$, but such cases are rare.

Note that this approach bypasses any problems concerned with the value of the Lagrange multiplier λ in the equivalent variational problem

$$\delta \text{ (regularization function} - \lambda \cdot \text{constraint function)} = 0.$$

It is not even necessary to assume that the constraint function is differentiable.

CONCLUSION

The maximum entropy method rests on a firm basis as the preferred method of reconstructing distributions of proportions, regardless of the type of data that one is given. It is also widely successful (e.g. Gull and Skilling 1984b). All other methods induce extra correlated structure beyond that required by the data. If (but only if) suitable prior knowledge is available, some such correlations can be expected. If so, they can (and should) be incorporated into the maximum entropy formalism through a model of the expected distribution. This gives maximum entropy great flexibility as well as power, which the construction of data-adaptive models may be expected to enhance.

REFERENCES

Gull, S.F. and Skilling, J., (1984a) "The maximum entropy method" in "Indirect imaging" (Roberts, J.A., ed.) Cambridge University Press, Cambridge.

Gull, S.F. and Skilling, J., (1984b) "Maximum entropy method in image processing" IEE Proc. **131** (**F**) 646-659.

Jaynes, E.T., (1968) "Prior probabilities" IEEE Trans. SSC-4 227-241.

Jaynes, E.T., (1979) "Concentrations of distributions at entropy maxima", reprinted in "Papers on probability, statistics and statistical physics" (Rosencrantz, R., ed.) Reidel, Dordrecht, 1983.

Kullback, S., (1959) "Information theory and statistics" Dover, New York.

Livesey, A.K. and Skilling, J., (1985) "Maximum entropy theory" *Acta Cryst.* **A41**, 113-122.

Shore, J.E. and Johnson, R.W., (1980) "Axiomatic derivation of maximum entropy and the principle of maximum cross-entropy" IEEE Trans. **IT-26** 26-37 (correction IEEE Trans **IT-29** 942-943 1983).

Tikhonov, A.N. and Arsenin, V.Y., (1977) "Solutions of ill-posed problems" Winston, Washington D.C.

UNIFIED FREQUENCY-DOMAIN IMAGE RECONSTRUCTION

M. Kaveh
(Department of Electrical Engineering, University of Minnesota)

and

M. Soumekh
*(Department of Electrical Engineering,
Worcester Polytechnic Institute)*

ABSTRACT

This paper presents a new view of the problem of image reconstruction in the spatial Fourier domain based on the generalized projections of a two-dimensional image. The algorithm that is presented is derived from the exact continuous relations between the Fourier transforms of a spatially limited image and those of its generalized projections. A Fourier domain reconstruction formula is derived that is a natural consequence of these relations giving values of the two-dimensional transform on a regular grid useful for an inverse FFT. This approach is contrasted with the more ad hoc methods of Fourier domain interpolation and with the spatial domain approaches such as the backprojection and backpropagation methods.

1. INTRODUCTION

This paper addresses the problem of efficiently reconstructing a two-dimensional image when data are presented on regular, arbitrarily-shaped contours in the frequency plane. The contours are assumed to result from known or estimable transformations of Cartesian coordinates. This type of data is often a consequence of an image reconstruction formulation which is best expressed in the Fourier domain. Image reconstruction is widely used in remote-sensing and non-invasive applications involving multiple measurements in a lower dimension (generalised projections) with varying experimental parameters. The most common examples are found in the general area of computer-assisted tomography in which the parameter that is varied is usually the angle of view. These parameters, along with the physical model of the generation of the generalised projections, determine the frequency domain contours and the transformation of frequency variables from the cartesian coordinates to points on the contours.

Much has been reported in the past decade on a special case of this problem, namely reconstruction based on straight-path projections [1,2]. The Fourier domain approach, with notable exceptions [3], usually has consisted of an ad hoc polynomial-based method of interpolating the values on the contours to points on a uniform grid suitable for fast Fourier transformation. Such schemes have often been considered inaccurate leading to the popularity of the well-defined, spatially interpolated scheme of convolution-backprojection.

More recently, extensions of these same techniques have been developed for the case of diffraction tomography as presented in, for example [7]. We believe that the premise on which these Fourier-based interpolation schemes have been based is not correct, giving rise to the perceived superiority of their spatial domain counterparts. The correct formulation leads to results comparable to the spatial interpolation schemes with substantially higher computational efficiency in many cases.

In this paper we first introduce the problem in its most general form in the continuous two-dimensional space. We then express the reconstruction formula in terms of the Jacobian of the transformations that represent the frequency-plane profile contours. Introduction of the appropriate Fourier-domain sampling at this stage yields a natural interpolation formula to the cartesian coordinates. This algorithm, denoted the Unified Frequency-Domain Reconstruction Algorithm (UFR), includes in its interpolation kernel a priori information about the support of the desired reconstructed image.

2. PROBLEM FORMULATION

Consider a two-dimensional function, $f(x,y)$ of cartesian spatial coordinates (x,y). Let (μ,λ) be the corresponding spatial Fourier coordinates and denote the Fourier transform of $f(x,y)$ by $\tilde{f}(\mu,\lambda)$. We assume that the measured (computed) information about $f(x,y)$ is given in the Fourier domain in terms of an observed function $O(\alpha,\beta)$, where (α,β) are related to the cartesian coordinates via a known transformation matrix $\underline{T}$. That is:

$$\tilde{f}(\mu',\lambda') = O(\alpha,\beta) \qquad (1),$$

with

$$\begin{bmatrix}\mu' \\ \lambda'\end{bmatrix} = \underline{T}(\alpha,\beta) = \begin{bmatrix}T_1(\alpha,\beta) \\ T_2(\alpha,\beta)\end{bmatrix} \qquad (2),$$

T_1 and T_2 are assumed to be differentiable. We call the loci defined by $\underline{T}$ the Reconstruction CONtours (RCONs). Examples of (α,β) are:

1) The polar coordinates found in conjunction with the Fourier-slice property of straight-path tomography, with

$$\underline{T}(\alpha,\beta) = \begin{bmatrix} \text{Cos}\alpha & \text{Sin}\alpha \\ \text{Sin}\alpha & \text{Cos}\alpha \end{bmatrix} \begin{bmatrix} 0 \\ \beta \end{bmatrix} \quad \text{for} \quad \begin{matrix} 0 \leq \alpha < 2\pi \\ 0 \leq \beta < \infty \end{matrix} \qquad (3),$$

2) The semicircles with wavenumber radius k_o associated with diffraction tomography [5,8]. In this case, $\underline{T}$ is given by

$$\underline{T}(\alpha,\beta) = \begin{bmatrix} \text{Cos}\alpha & -\text{Sin}\alpha \\ \text{Sin}\alpha & \text{Cos}\alpha \end{bmatrix} \begin{bmatrix} -k_o + \sqrt{k_o^2 - \beta^2} \\ \beta \end{bmatrix} \quad \text{for} \quad \begin{matrix} 0 \leq \alpha < 2\pi \\ 0 \leq |\beta| \leq k_o \end{matrix} \qquad (4)$$

Normally, $O(\alpha,\beta)$ is only available at a finite number of sample values (α_n,β_m). The objective of the reconstruction algorithm is to obtain an estimate of $f(x,y)$ based on the set of available $O(\alpha_n,\beta_m)$. It is at this point that questions of interpolation have traditionally come up. Those who have been interested in Fourier domain reconstruction have usually interpolated $O(\alpha_n,\beta_m)$ onto a uniform grid, using one of the traditional polynomial interpolation functions, to obtain an estimate of $\tilde{f}(\mu_i,\lambda_j)$. This function is then inverse transformed via a two-dimensional FFT to result in an estimate of $f(x,y)$. This approach ignores the specific and intimate relation between (μ',λ') and (α,β). In the following, we present a technique that is based on the above mentioned relation and operates on the exact integral relation between $f(x,y)$, $O(\alpha,\beta)$ and $\tilde{f}(\mu',\lambda')$. We note in passing that the Fourier-slice and its generalizations are based on Fourier integrals, leading one to continue using the integral relations in formulating the interpolations as well.

Consider the inverse Fourier transform equation of

$$f(x,y) = \frac{1}{4\pi^2} \int_{-\infty}^{\infty} \int_{-\infty}^{\infty} \tilde{f}(\mu',\lambda').\exp[\,j(\mu'x + \lambda'y)]\,d\mu'd\lambda' \qquad (5)$$

For simplicity, let us assume that the mapping of the transform function is one-to-one. Therefore, with the help of equation (1), we can make variable transformations in equation (5) to obtain:

$$f(x,y) = \frac{1}{4\pi^2} \int_{-\infty}^{\infty} \int_{-\infty}^{\infty} O(\alpha,\beta) . \exp[\, j\{T_1(\alpha,\beta) x + T_2(\alpha,\beta) y\}]$$

$$J(\alpha,\beta)\, d\alpha d\beta \qquad (6)$$

where $J(\alpha,\beta)$ is the Jacobian of transformation. Hence, the Fourier transform of the left hand side of (6) evaluated at (μ_i,λ_j) is:

$$\tilde{f}(\mu_i,\lambda_j) = \frac{1}{4\pi^2} \iint_D \int_{-\infty}^{\infty} \int_{-\infty}^{\infty} J(\alpha,\beta) ; O(\alpha,\beta) .$$

$$\exp[\, j\{T_1(\alpha,\beta)\}x + T_2(\alpha,\beta)y\}] \; d\alpha d\beta \; \exp[-j(\mu_i x + \lambda_j y)]\, dxdy \qquad (7)$$

where D is the finite support of the object function, $f(x,y)$, in the spatial domain. Equation (7) can be written as

$$\tilde{f}(\mu_i,\lambda_j) = \frac{1}{4\pi^2} \int_{-\infty}^{\infty} \int_{-\infty}^{\infty} J(\alpha,\beta) , O(\alpha,\beta)$$

$$\iint_D \exp[-j\{\mu_i - T_1(\alpha,\beta)\}x - j\{\lambda_j - T_2(\alpha,\beta)y\}]\, dxdy \; d\alpha d\beta \qquad (8).$$

Let $I(x,y)$ be the indicator function for the support of the object function (i.e., $I(x,y) = 1$ when $(x,y)\varepsilon D$, zero elsewhere), and $\tilde{I}(\mu,\lambda)$ be its Fourier transform. We can rewrite equation (8) as

$$\tilde{f}(\mu_i,\lambda_k) = \frac{1}{4\pi^2} \int_{-\infty}^{\infty} \int_{-\infty}^{\infty} \tilde{I}[\, \mu_i - T_1(\alpha,\beta), \lambda_j - T_2(\alpha,\beta)]\, O(\alpha,\beta) J(\alpha,\beta)\, d\alpha d\beta . \qquad (9)$$

Equation (9) is an exact consequence of the basic relation expressed in (6). At this point we express the fact that $O(\alpha,\beta)$ is only available over a finite number of points (α_n,β_m), through an ideal sampling representation

$$O(\alpha,\beta) = \sum_{n=1}^{K} \sum_{m=1}^{N} O(\alpha_n,\beta_m)\, \delta(\alpha-\alpha_n, \beta-\beta_m) \qquad (10)$$

where $\delta(x,y)$ is the two-dimensional Kronecker delta function Substituting (10) into (9), the exact (within the sampling errors offered by (α_n,β_m)) reconstruction formula in the Fourier domain is obtained. We have

$$\tilde{f}(\mu_i,\lambda_k) = \sum_{n=1}^{K} \sum_{m=1}^{N} J(\alpha_n,\beta_m) O(\alpha_n,\beta_m)$$

$$\tilde{I}[\mu_i - T_1(\alpha_n,\beta_m),\ \lambda_k - T_2(\alpha_n,\beta_m)\ ]. \qquad (11)$$

The form of (11) is reminiscent of Nyquist's interpolation formula for a sampled signal. The interpolation kernel in the present case has two distinct components. The scale-factor $J(\alpha_n,\beta_m)$ appropriately normalizes the density of the irregularly.spaced available data and is crucial to the success of this algorithm. $\tilde{I}(\ ,\ )$ is the continuous frequency-space interpolation function containing information about the support of the function to be reconstructed. An example of $I(\underline{r})$ is a disk of radius R. In this case,

$$\tilde{I}(\mu,\lambda) = \frac{J_1(R\sqrt{\mu^2+\lambda^2})}{R\sqrt{\mu^2+\lambda^2}}$$

where J_1 is a first order Bessel function. When R is on the order of half the image dimensions, the main lobe of $\tilde{I}$ has a width on the order of the grid separation in the frequency domain. Thus the main contribution of the kernel $\tilde{I}$ in (11), for a given data point at (α_n,β_m), is to the grid points that are closest to (α_n,β_m). Consequently for computational efficiency, we approximate $\tilde{f}(\mu_i,\nu_k)$ in terms of a small number of its neighbouring data points. This assignment is obviously data driven. Thus

$$\tilde{f}(\mu_i,\lambda_k) \simeq \sum_n \sum_m J(\alpha_n,\beta_m) O(\alpha_n,\beta_m) \tilde{I}[\mu_i - T_1(\alpha_n,\beta_m), \lambda_k - T_2(\alpha_n,\beta_m)], \qquad (12)$$

(n,m) s.t. (μ_i,ν_k) is one of L nearest neighbours of $T(\alpha_n,\beta_m)$

The computational requirements for (12) is on the order of LN^2 operations followed by a two-dimensional FFT where L is typically 4 or 9. The computation time for the nine neighbour case is slightly higher than the bilinear interpolation algorithm [1] and substantially lower than the backpropagation algorithm [6]. The error of this algorithm is usually substantially less than that for the bilinear method and comparable to that of the convolution-backprojection or backpropagation methods [8,9].

3. CONCLUSIONS

We formulated the problem of image reconstruction in its most general form under the condition that data are available in the Fourier space on arbitrary but known contours. Two types of contours were given as the most common examples (i) radial contours obtained in straight-path tomography and (ii) semi-circular contours encountered in diffraction tomography. An exact interpolation formula was derived for obtaining samples of the object Fourier transform on a regular cartesian grid. This was followed by suggestion for approximating the interpolation scheme to achieve computational efficiency. Reference was made to examples that have shown accuracy of reconstruction with this interpolation method comparable to those obtained by the spatial interpolation schemes of convolution-backprojection and backpropagation.

4. ACKNOWLEDGEMENT

This work was supported by the U.S. National Science Foundation under Grant ECS-7926008.

5. REFERENCES

[1] Mersereau, R.M. and Oppenheim, A.V., (1974), "Digital Reconstruction of Multi-dimensional Signals from Their Projections", Proc. IEEE, Vol. 62, 1319-1338.

[2] Kak, A.C., (1979), "Computerized Tomography with X-Ray, Emission, and Ultrasound Sources", Proc. IEEE, Vol. 67, 1245-1272.

[3] Ramachandran, G.N. and Lakshminarayanan, A.V., (1971), "Three Dimensional Reconstruction from Radiographs and Electron Micrographs: Application of Convolutions Instead of Fourier Transforms", Proc. National Acad. Sci., Vol. 68, 2236-2240.

[4] Stark, H., Woods, J.W., Paul, I. and Hingorani, R., "Fourier Reconstruction in Computer Tomography", IEEE Trans. Acoust., Speech and Signal Process., Vol. 29 237-245.

[5] Kaveh, M., Mueller, R.K. and Iverson, R.D., (1979), "Ultrasonic Tomography Based on a Perturbation Solution of the Wave Equation", Comput. Graphics and Image Processing, Vol. 9, 105-116.

[6] Devaney, A.J., (1982), "A Filtered Backpropagation Algorithm for Diffraction Tomography", Ultrasonic Imaging, Vol. 4, 336-350.

[7] Pan, S.X. and Kak, A.C., (1983), "A Computational Study of Reconstruction Algorithms for Diffractive Tomography: Interpolation vs. Filtered-Backpropagation", IEEE Trans. Acoust. Speech and Signal Process, Vol. 31, 1262-1275.

[8] Kaveh, M. and Soumekh, M., (1986), "Computer-Assisted Diffraction Tomography", Chapter in Image Recovery, Theory and Applications, Stark, H., Ed. (to be published).

[9] Kaveh, M., Soumekh, M., and Greenleaf, J.G., (1984), "Signal Processing for Diffraction Tomography", IEEE Trans. on Sonics and Ultrasonics, Vol. 31, 230-238.

RADIAL GEOMETRY FOR TOMOGRAPHIC IMAGE RECONSTRUCTION AND DISPLAY

M.C. Jones
(Department of Physics, University of Surrey, Guildford)

ABSTRACT

It is shown how an image display in the form of a rotating raster scan can be used to reconstruct images from tomographic data.

1. INTRODUCTION

In tomographic imaging, the density, $m(\mathbf{r})$, satisfies the projection equations

$$\int m(\mathbf{r})h_p(\mathbf{r})\, d\,\mathbf{r} = z_p, \tag{1.1}$$

where p runs over all beam positions, $h_p(\mathbf{r})$ is a geometrical factor describing the shape of the beam, and the integral is over the support of m.

To find $m(\mathbf{r})$ from (1.1), $m(\mathbf{r})$ can be expanded as a series using some set of functions, $\{k_j\}$, complete over the support of m:

$$m(\mathbf{r}) = \sum_{q=1}^{N} m_q k_q(\mathbf{r}) \tag{1.2}$$

In (1.2) $m(\mathbf{r})$ has been approximated by a finite sum as only a finite number of $\{m_q\}$ will be determinable from (1.1), since $\{z_p\}$ is finite. But if this approximation is made, the completeness condition may also be relaxed, and then a solution in the space spanned by any set

$$\{k_q\}_{q=1}^{N}$$

may be sought. This requires a solution to

$$\sum_q \left\{ \int h_p(\mathbf{r}) k_q(\mathbf{r})\, d\mathbf{r} \right\} m_q = z_p \quad \text{or, in matrix form, } \mathbf{D}\,\mathbf{m} = \mathbf{z} \tag{1.3}$$

be found, where $\mathbf{D}$ has p,q^{th} element

$$d_{pq} = \int h_p(\mathbf{r}) k_q(\mathbf{r})\, d\mathbf{r}$$

In the standard pictorial representation of $m(\mathbf{r})$, over a rectangular array of pixels, the set $\{u_{i,j}\}$, with

$$\begin{aligned} u_{ij}(\mathbf{r}) &= 1 \qquad r \text{ in } ij^{th} \text{ pixel} \\ &= 0 \qquad \text{otherwise,} \end{aligned} \tag{1.4}$$

is used. This gives rise to the matrix

$$\mathbf{D} = \left[d_{p;i,j} \right] = \left[\int h_p(\mathbf{r})\, u_{i,j}(\mathbf{r})\, d\mathbf{r} \right] \tag{1.5}$$

in the linear equations (1.3). For a standard scanning sequence with a parallel beam,

$$\begin{aligned} h_p(\mathbf{r}) = v_{t,\alpha}(\mathbf{r}) &= 1 \qquad \mathbf{r} \text{ in the } t,\ \alpha^{th} \text{ strip} \\ &= 0 \qquad \text{otherwise.} \end{aligned} \tag{1.6}$$

Here $1 \leqslant t \leqslant N$ and $0 \leqslant \alpha \leqslant 2K - 1$ (see figure 1). $\mathbf{D}$ is then a sparse matrix of an irregular nature.

It is possible to choose $\{k_q\}$ so that $\mathbf{D}$ has its sparseness organized, making the equations (1.3) more easily soluble. One such set is $\{w_{s,\beta}\}$ where

$$\begin{aligned} w_{s,\beta}(\mathbf{r}) &= 1 \qquad \mathbf{r} \text{ in } s,\ \beta^{th} \text{ cell} \\ &= 0 \qquad \text{otherwise} \end{aligned} \tag{1.7}$$

(see figure 2). In this case

$$d_{t,\alpha;s,\beta} = \int v_{t,\alpha}(\mathbf{r})\, w_{s,\beta}(\mathbf{r})\, d\,\mathbf{r} \quad \text{and} \quad d_{t,\alpha;+n;s,\beta+n} = d_{t,\alpha;s,\beta} \tag{1.8}$$

so that **D** is block circulant. A Fourier transformation then makes **D** block diagonal (see Davis (1979)), significantly simplifying the solution of (1.3).

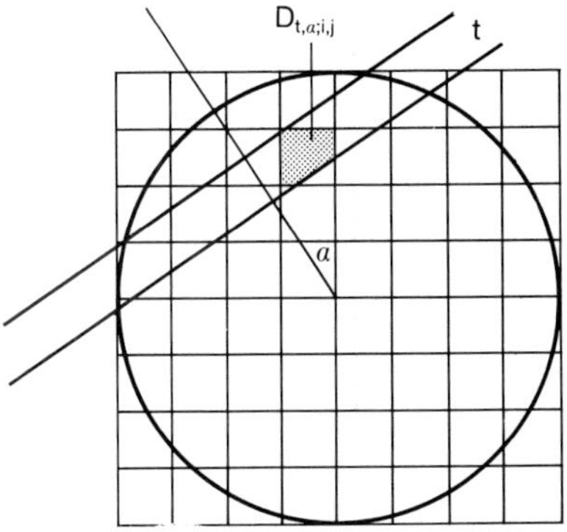

Fig. 1 The standard rectangular display using functions {u} given by equations (1.4).

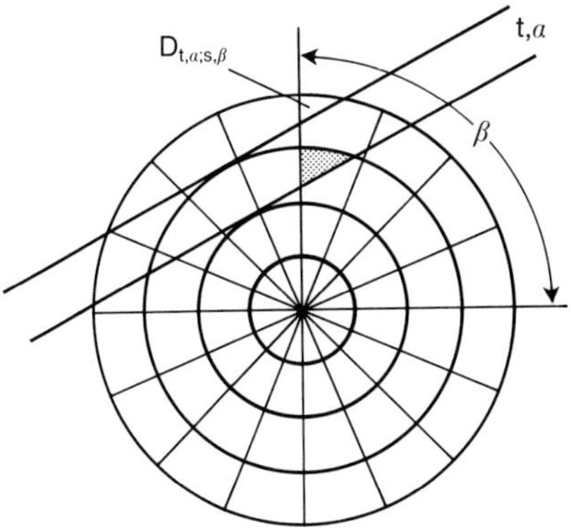

Fig. 2 A radial geometry for reconstruction using the functions {w} of equations (1.7). There is no obvious way of display in this geometry other than on a much higher resolution rectangular display.

However, displaying

$$\Sigma\, m_{s,\beta}\, w_{s,\beta}\,(\mathbf{r})$$

on any conventional display requires a transformation, back to the functions $\{u_{i,j}\}$, for example (Stark (1982)). It is well known that various artefacts can appear in tomographic reconstruction, and some of these no doubt result from errors introduced by these geometrical factors. This has also been discussed by Buonocore, Brody and Macovski (1981) and Wood and More (1981). Gindi and Gmitro (1984) have suggested Randon-transforming pictures to facilitate pattern recognition, and have designed an optical system to achieve this. It is therefore of some interest to consider $\{v_{t,\alpha}\}$ as a set of functions which are of value in, not only tomography, but image-processing generally, and which have the additional advantage of being directly displayable.

2. THE ROTATING RASTER SCAN

If the functions $\{v_{t,\alpha}\}$ given by (1.6) are chosen for $\{k_q\}$, then the sum

$$\sum_{t\alpha} m_{t,\alpha} v_{t,\alpha}\,(\mathbf{r})$$

can be written

$$\sum_{\alpha}\sum_{t} m_{t,\alpha}\, v_{t,\alpha}\,(\mathbf{r}) \qquad (2.1)$$

The inner sum can be implemented as a raster scan at an angle α with each line of uniform brightness, $m_{t\alpha}$ for the tth line. The outer sum is achieved by superimposing rotated raster scans at different angles.

There are a number of ways of achieving such a rotating raster scan. It can be done optically, as in Nishi et al. (1984), using holographic lenses, or by driving an oscilloscope from a microcomputer as described by Vincent (1984). In those cases superposition is achieved by the human visual system. The results presented in §3 are simulated on a medium resolution standard rectangular display.

An important point to be noticed in (2.1) is that some of the $\{m_{t,\alpha}\}$ may be negative, so that in display the zero of brightness is set to the middle of the grey scale.

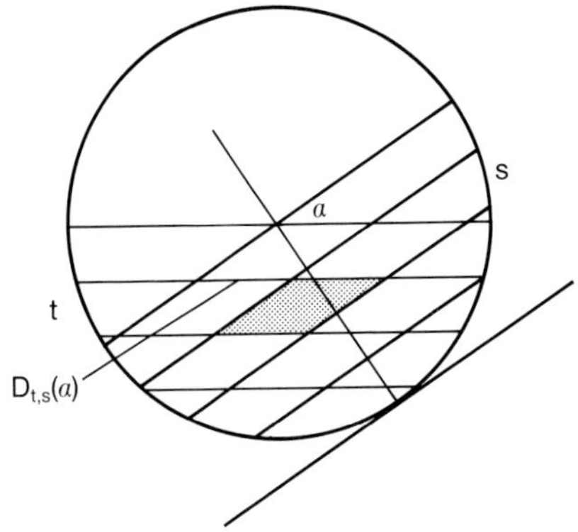

Fig. 3 A radial geometry that is suitable for both reconstruction and display, using projection functions of equations (1.6).

3. SOLUTION OF THE EQUATIONS

The $\{m_{t\alpha}\}$ in (2.1) are determined from (1.3) with

$$d_{t,\alpha;s,\beta} = \int_C v_{t,\alpha}(\mathbf{r})\, v_{s,\beta}(\mathbf{r})\, d\,\mathbf{r}$$

where C, the support of $m(\mathbf{r})$, is a disc.

$\mathbf{D}$ is partitioned

$$\left[\mathbf{D}_{t,s}\right]$$

and $\mathbf{m}$, $\mathbf{z}$ partitioned $[\mathbf{m}_s]$, $[\mathbf{z}_t]$ respectively, where $\mathbf{D}_{t,s}$ is symmetric and, by virtue of (1.8), circulant. A Fourier transform gives

$$\sum_s \Omega^{-1}\mathbf{D}_{t,s}\,\Omega\,\Omega^{-1}\,\mathbf{m}_s = \Omega^{-1}\,\mathbf{z}_t \qquad (3.1)$$

(where Ω is the Fourier transform matrix with $\alpha\beta^{th}$ element $e^{-2\pi i\alpha\beta/2K}$ which reduces to

$$\sum_s \Lambda_{ts}(\alpha)\; M_s(\alpha) = Z_t(\alpha) \quad \alpha = 0,\ldots,2K-1, \qquad (3.2)$$

with

$$\Lambda_{t,s}(\alpha) = \sum_{\beta} d_{to,s\beta}\, w_{2K}^{\alpha\beta}, \tag{3.3}$$

where

$$w_{2K} = e^{-2\pi i/2K}.$$

Equations (3.2) can now be solved by standard methods and **m** obtained by a further Fourier transform. However it should be noted, though often overlooked, that equations (1.3) with $\{h_p\} = \{v_{to}\}$ are singular for

$$\sum_t d_{t,\alpha;q} + d_{t,\alpha+K;q} = \int_C k_q(\mathbf{r})\, d\,\mathbf{r} \tag{3.4}$$

and

$$\sum_t z_{t,\alpha} + z_{t,\alpha+K} = \int_C m(\mathbf{r})\, d\,\mathbf{r} \tag{3.5}$$

are independent of α. This is due to an over-completeness in the set $\{v_{t,\alpha}\}$ and results in some arbitrariness in the brightness $\{m_{t,\alpha}\}$ to be used in the rotating raster display. Here this arbitrariness has been removed by choosing the minimum norm solution.

4. RESULTS AND DISCUSSION

In figure 4 an image taken from Sezan and Stark (1982) is displayed on a 512 x 512 rectangular display. In figure 5 its reconstruction with N = 16 and K = 8 is simulated using the rotating raster scan. The rotating raster scan reconstruction with N = 32 K = 16 is shown in figure 6.

These show that quite good edge resolution is achieved, there is a marked granularity. This could be removed by giving up the minimum norm solution which results when m is reconstructed in the subpsace spanned by $\{v_{t,\alpha}\}$. A sensible constraint to impose is non-negativity of m, and further investigation on this basis will be pursued.

A comparison with the standard rectangular reconstruction, for the model shown in figure 4, is made in Figure 7. This shows considerably more artefacts than in Figure 6.

Fig. 4 The image model used

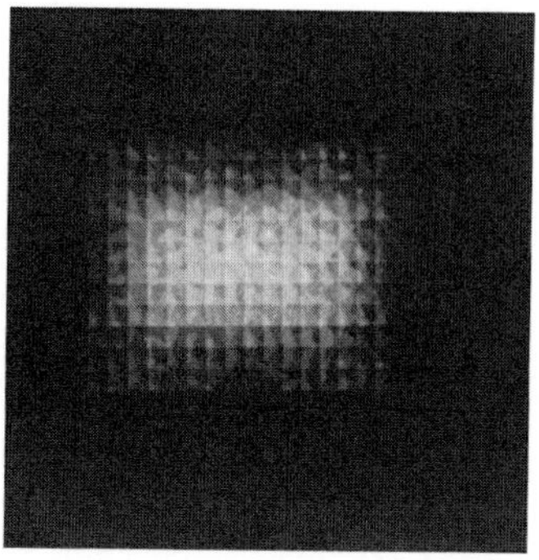

Fig. 5 Rotating raster scan reconstruction using N = 16 K = 8

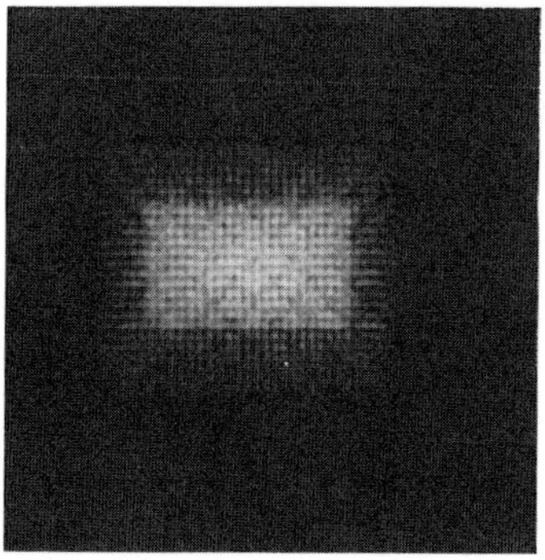

Fig. 6 Rotating raster scan reconstruction using N = 32 K = 16

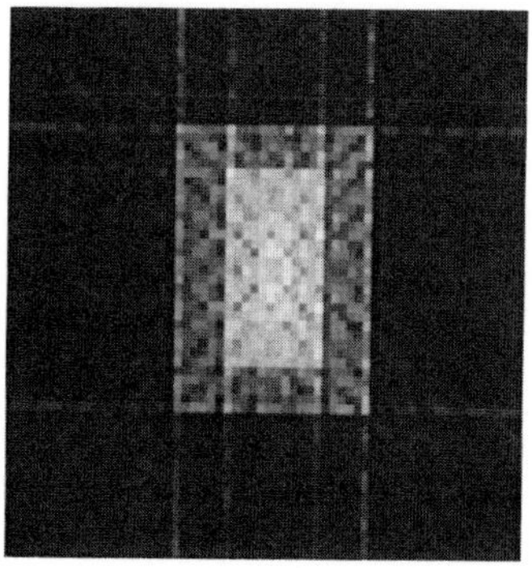

Fig. 7 A standard rectangular filtered back-projection reconstruction using the same projection data as Figure 6.

REFERENCES

Buonocore, M.H., Brody, W.R. and Macovski, A., (1981) "A natural pixel decomposition for two-dimensional image reconstruction" *IEEE Trans. Biomed. Eng. BME-28*(2), 69-78.

Davis, P.J., (1979) "Circulant Matrices", Wiley.

Gindi, G.R. and Gmitro, A.F., (1984) "Optical feature extraction via the radon transform", *Opt. Eng.* **23** (5), 499-506.

Nishi, K., Kurahashi, K. and Kubo, T., (1984) "Multidirectional holographic scanner for point-of-scale bar-code symbol reader", *Opt. Eng.* **16**, 784-787.

Stark, H., (1982) "Polar sampling theorems of use in optics", Applications of Mathematics in Modern Optics, William H. Carter (Ed) *Proc. SPIE* **358**, 24-30.

Sezan, M.I. and Stark, H., (1982) "Image restoration by the method of convex projections: Part 2 - Applications and numerical results" *IEEE Trans. Med. Imaging MI-1,* 95-101.

Vincent, J.M., "A technique for tomographic reconstruction and display", Internal report, University of Surrey, Physics Department.

Wood, S.L., and Morf, M., (1981) "A fast implementation of a minimum variance estimator for computerized tomography image reconstruction", *IEEE Trans. Biomed. Eng. BME-28(2),* 56-68.

COMPARISON BETWEEN A FACTORIZATION METHOD AND A PARTITIONING METHOD TO DERIVE INVARIANT KALMAN FILTERS FOR FAST IMAGE RESTORATION

X.C. Du, D. Saint-Félix and G. Demoment
(Laboratoire des Signaux et Systèmes (CNRS/ESE), France)

ABSTRACT

We are interested in the fast computation of the steady-state or asymptotic gain of a Kalman filter used in a recursive and suboptimal estimation procedure. The dimension of the state being very large, we must find a faster technique than the direct solution of a Riccati equation to compute the asymptotic gain. Two possible methods are compared here: one is the factorization technique of Chandrasekhar-type, and the other the so-called partitioning method. The factorization method runs better than the partitioning one for a scalar observation and a high dimension state, which is a situation often encountered in 2-D image restoration. In order to be able to efficiently use the Chandrasekhar equations with arbitrary initial conditions, we have developed the discrete dual-Chandrasekhar equations for dealing with any initial conditions.

1. INTRODUCTION

The standard linear state-space model for a discrete-time process is:

$$\underline{x}_{i+1} = F\underline{x}_i + G\underline{u}_i \tag{1.1}$$

$$\underline{y}_i = H\underline{x}_i + \underline{b}_i, \quad i=1,2,\ldots \tag{1.2}$$

where

$$E \begin{vmatrix} \underline{u}_i \\ \underline{b}_i \\ \underline{x}_1 \end{vmatrix} (\underline{u}^*_j, \underline{b}^*_j, \underline{x}^*_1) = \begin{vmatrix} Q & O & O \\ O & R & O \\ O & O & P_x \end{vmatrix} \delta_{i,j} \tag{1.3}$$

Matrices F, G, H, Q, R and P_x are assumed to be known and constant, and the vectors $\underline{x}_i$, $\underline{y}_i$ and $\underline{u}_i$ are of dimension n, m and p respectively. The Kalman gain matrix necessary for an optimal recursive estimation of the state $\underline{x}$ is given by

$$K_{gi} = FP_iH^*(HP_iH^* + R)^{-1} = FP_iH^*(R_i^e)^{-1} = K_i(R_i^e)^{-1} \quad (1.4)$$

and this gain is calculated by solving the following Riccati equation:

$$P_{i+1} = FP_iF^* + GQG^* - K_i(R_i^e)^{-1}K_i^* \quad (1.5)$$

where P_{i+1} is the prediction error covariance matrix:

$$P_{i+1} = E\{(\underline{x}_{i+1} - \hat{\underline{x}}_{i+1|i})(\underline{x}_{i+1} - \hat{\underline{x}}_{i+1|i})^*\}$$

This Riccati equation arises naturally in a variety of situations and many studies have been done over the past twenty years to establish stable solutions, and, if possible, to do it rapidly. In many applications we are interested in the steady-state or asymptotic solution of the Riccati equations, in order to design a sub-optimal Kalman filter with a constant gain. The advantage lies evidently in reducing the volume of computations. One can distinguish two families of methods in the literature to compute the Kalman gain: the iterative methods and the non-iterative ones. The non-iterative methods consist of solving an algebraic equation by computing the eigenvalues of an Hamiltonian system. Use of these methods is restricted as soon as the state dimension becomes very large. So we are interested here in iterative methods because our image model is of large dimension ($n \geq 1000$).

Certainly, the computation of the asymptotic gain can be done by directly iterating (1.5) with:

$$P = \lim_{i \to \infty} P_i \quad (1.6)$$

the asymptotic gain being then given by:

$$K = FPH^*(HPH^* + R)^{-1} \quad (1.7)$$

But the direct computation of (1.6) via (1.5) requires a large amount of arithmetical operations. It is necessary for us to use fast algorithms which are capable of exploiting

the constancy properties of matrices F, G, H, Q, R and P_x.
The aim of this article is to compare two possible methods for computing the asymptotic Kalman gain when the state dimension is very large such as in image restoration (see Saint-Félix et al. (1985)).

2. CHANDRASEKHAR TYPE ALGORITHMS

To avoid computation of the entire P_i at each recursion via the Riccati equation, Chandrasekhar equations compute the variations of vector-gain and other related quantities. When the system (1.1,1.2) is invariant, the variation of the error covariance matrix P_i can be easily factorized as:

$$\delta P_i = P_i - P_{i-1} = Y_{i-1} M_{i-1} Y_{i-1}^* \tag{2.1}$$

Defining similarly $\delta K_i = K_i - K_{i-1}$, the Kalman gain can be calculated with the following Chandrasekhar equations (Kailath et al. (1973, 1974)):

$$Y_i = [F - K_i (R_i^e)^{-1} H] Y_{i-1} \tag{2.2}$$

$$M_{i+1} = M_i + M_i Y_i^* H^* (R_i^e)^{-1} H Y_i M_i \tag{2.3}$$

$$R_{i+1}^e = R_i^e + H Y_i M_i Y_i^* H^* \tag{2.4}$$

$$K_{i+1} = K_i + F Y_i M_i Y_i^* H^* \tag{2.5}$$

Although P_i does not appear in these equations, it may be as well calculated if needed at each recursion by:

$$P_{i+1} = P_i + Y_i M_i Y_i^* \tag{2.6}$$

The initial values of (2.4) and (2.5) are given by:

$$K_1 = F P_x H^* \ , \quad R_1^e = R + H P_x H^* \tag{2.7}$$

but for (2.3) and (2.4) some preliminary computations are needed:

$$\delta P_2 = F P_x F^* + G Q G^* - K_1 (R_1^e)^{-1} K_1^* - P_x = Y_1 M_1 Y_1^* \tag{2.8}$$

The matrix M_1 is a signature matrix of dimension (a,a) where a is the rank of δP_i which may often be much lower than the state dimension n. In many applications such as on-line deconvolution, fast image restoration (Demoment et al.(1985)), etc., the observation is scalar. In these conditions, a will be equal to unity, and the total number of arithmetic operations at each recursion for a Chandrasekhar algorithm is given in the following table.

Terms	operations	
	adds	mults.
HY_i	n-1	n
FY_i	n(n-1)	n^2
eq.(2.2)	n	n+2
eq.(2.3)	1	3
eq.(2.4)	1	1
eq.(2.5)	n	n
total	n^2+2n+1	n^2+3n+6

Table I: Number of arithmetic operations for Chandrasekhar algorithm.

According to table I, the number of elementary operations at each iteration is:

$$N_c = (n^2 + 2n + 1) \text{ adds} + (n^2 + 3n + 6) \text{ mults.}$$

Let T_c represent the total number of iterations required to obtain the asymptotic gain defined by the condition $||\delta P_{Tc}|| \leq \xi ||P_{Tc}||$ where ξ is a priori given. The total number of arithmetic operations required by this algorithm is then:

$$N_{T_c} = T_c \, N_c = T_c[\,(n^2 + 2n + 1) \text{ adds} + (n^2 + 3n + 8) \text{ mults.}] \tag{2.9}$$

3. PARTITIONING ALGORITHMS

This method is also based on the invariance property of the system (1.2,1.2). We compute $P(2^k)$ from $P(2^{k-1})$ instead of sequentially computing P(O), P(1), P(2)... This idea comes out simply from the fact that the solution of the algebraic Riccati equation can be obtained with the help of the solution of a 2n-dimensional linear equation (Hamiltonian) of type $A_k = \Phi A_{k-1} = \Phi^k A_o$. This equation gives immediately a doubling solution A^{2k}, because the computation of the transition matrix $\Phi^{2k} = \Phi^k \Phi^k$ is done by doubling at each time instant the length of the current iteration step k (Anderson (1978)). This algorithm was also developed with a different approach by Ljung et al. (1976) using scattering theory. We use here the general expressions of Lainiotis (1976), which are claimed to present advantages when changing initial conditions. The equations are given by:

$$P(2^{k+1}\Delta) = P_o(2^k\Delta) + \psi_o(2^k\Delta)[\,I+P(2^k\Delta)O_o(2^k\Delta)\,]^{-1}P(2^k\Delta)\psi_o^*(2^k\Delta) \tag{3.1}$$

where the "nominal" parts $P_o(.)$, $\psi_o(.)$ and $O_o(.)$ are computed by:

$$P_o(2^{k+1}\Delta) = P_o(2^k\Delta) + \psi_o(2^k\Delta)[\,I+P_o(2^k\Delta)O_o(2^k\Delta)\,]^{-1}P_o(2^k\Delta)\psi_o^*(2^k\Delta) \tag{3.2}$$

$$\psi_o(2^{k+1}\Delta) = \psi_o(2^k\Delta)[\,I+P_o(2^k\Delta)O_o(2^k\Delta)]^{-1}\psi_o^*(2^k\Delta) \tag{3.3}$$

$$O_o(2^{k+1}\Delta) = O_o(2^k\Delta) + \psi_o^*(2^k\Delta)[\,I+P_o(2^k\Delta)O_o(2^k\Delta)\,]^{-1}P_o(2^k\Delta)\psi_o(2^k\Delta) \tag{3.4}$$

for k = 0,1,2,... and where Δ is the basic interval.

Initialization of (3.1) to (3.4) is trivial : $P(1)=P_x$, $\psi_o(1)=I$, $O_o(1)=O$ and $P_o(1)=O$. But initialization of $P(\Delta)$, $\psi_o(\Delta)$, $O_o(\Delta)$ and $P_o(\Delta)$ needs solving directly:

$$\psi_o(i) = [I-K_o(i)H]F\psi_o(i-1), \quad \psi_o(1) = I \tag{3.5}$$

$$O_o(i) = O_o(i-1) + \psi_o^*(i-1)F^*H^*P_{\tilde{z}_o}^{-1}(i|i-1)HF\psi_o(i-1), \; O_o(0) = 0 \tag{3.6}$$

$$P_o(i|i-1) = FP_o(i-1)F^* + Q \tag{3.7}$$

$$K_o(i) = P_o(i|i-1)H^*P_{\tilde{z}_o}^{-1}(i|i-1) \tag{3.8}$$

$$P_o(i) = [I-K_o(i)H]P_o(i|i-1) \tag{3.9}$$

$$P_{\tilde{z}_o}(i|i-1) = HP_o(i|i-1)H^* + R, \quad i=1,2..\Delta \tag{3.10}$$

To simplify the initialization step, let $\Delta=1$. The solution $P(2^k)$ is computed with nominal parts $P_o(.)$, $O_o(.)$ and $\psi_o(.)$, and the convergence is quadratic (see Anderson (1978)). Taking here the same initial covariance matrix $P(0)=P_x$ as the Chandrasekhar algorithm, the solution $P(2^k)$ exactly coincides with the solution P_i obtained by the Chandrasekhar algorithm at $i=2^k$. After $T_p = \text{Log}_2(T_c)$ iterations, the solution $P(2^{Tp})$ verifies also the convergence condition precedently defined. The amount of arithmetic operations at each iteration is given in table II.

Remark: We have neglected in this table the initialization step. Matrices $P(.)$, $P_o(.)$ and $O_o(.)$ are symmetrical, but matrix $\psi_o(.)$ is not. The multiplication of two symmetrical matrices requires $(n^3+n^2)/2$ operations. Note also that the inversion of two matrices in (3.1) to (3.4) needs at each time instant $O(n^2)$ operations, for these matrices are also symmetrical. The total number of elementary operations after T_c iterations is:

$$N_{T_p} = T_p[(8n^3 + 2n^2 + O(n^2))\text{mults.} + (6n^3 + 4n^2 + O(n^2))\text{adds}] \tag{3.11}$$

Terms	operations	
	adds	mults.
$P(.)O_o(.)$	$(n^3+n^2)/2$	$(n^3+n^2)/2$
$[I+P(.)O_o(.)]^{-1}$	$O(n^2)$	$O(n^2)$
eq. (3.1)	n^2	$(n^3+n^2)/2+2n^3$
$P_o(.)O_o(.)$	$(n^3+n^2)/2$	$(n^3+n^2)/2$
$[I+P_o(.)O_o(.)]^{-1}$	$O(n^2)$	$O(n^2)$
$\psi_o(.)[I+P(.)O_o(.)]^{-1}$	n^3	n^3
eq. (3.3)	n^3	n^3
$[I+P(.)O_o(.)]^{-1}P_o(.)$	$(n^3+n^2)/2$	$(n^3+n^2)/2$
eq. (3.2)	n^2+2n^3	$2n^3$
eq. (3.4)	$(n^3+n^2)/2$	
total	$6n^3+4n^2+O(n^2)$	$8n^3+2n^2+O(n^2)$

Table II: Number of arithmetic operations for partitioning algorithm.

4. COMPARISON

Comparison between the two algorithms is difficult to do in the general case, for the global number of operations required by each algorithm depends in a non-linear fashion on the state dimension n. But if we take as criterion the total amount of computations, a relative comparison is possible. Neglecting additions, we can define a performance index A as the ratio of the total amount of operations required by the Chandrasekhar algorithm to that required by the partitioning algorithm:

$$A = A(T_c, n) = \frac{T_c(n^2+3n+6)}{\mathrm{Log}_2(T_c)\ (8n^3+2n^2+O(n^2))} \tag{4.1}$$

which can be simplified as (n being large):

$$A = \frac{T_c}{8\mathrm{Log}_2(T_c)\ n} = \frac{f(T_c)}{n} \qquad (4.2)$$

where $f(T_c) = T_c/(8\mathrm{Log}_2(T_c))$, is a monotone increasing function of T_c. For the purpose of comparison, the total number of iterations T_c must be taken as an entire power of 2, i.e. $T_c = 2^k$. Fig. 1 shows the curve A = 1 in the plane: dimension n/ iteration number k ($T_c = 2^k$). It separates the plane into two parts. In part 1, the Chandrasekhar algorithm is faster than the partitioning one, for A is lower than 1.

In practical applications, the iterations number has the same order of magnitude as the state dimension (Saint-Félix et al. (1985)), i.e., $T_c = qn$ ($n>1000$), with $2<q<10$, depending on the prior convergence criterion. In this case, $A(T_c,n)$ is always lower than 1 (approximately $A \leq 0.1$). The dotted line in Fig. 1 represents the performance index A for q = 10. This line is far below the separating line A=1. So we can conclude that the Chandrasekhar algorithm is always faster than the partitioning algorithm when the observation is scalar. This allows us to correct the proposition of Lainiotis that partitioning algorithms are always faster than Chandrasekhar algorithms, even when they have their greatest computational advantages (i.e. rank a is small). The partitioning algorithm may be advantageous only if the state dimension is not too large, for example n=10.

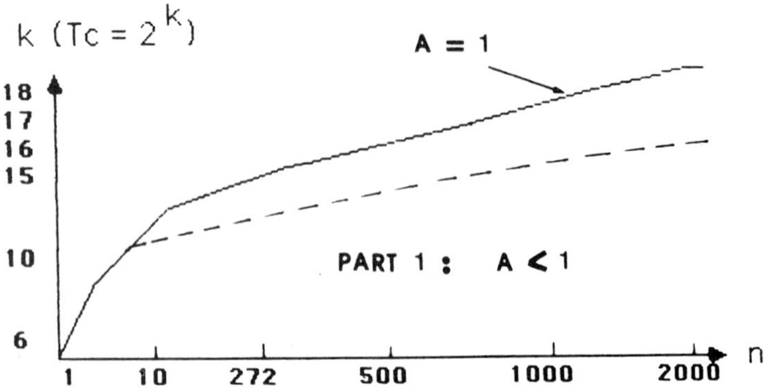

Fig. 1 Relative performance index as a function of state dimension and iterations number.

5. DISCRETE DUAL-CHANDRASEKHAR EQUATIONS

The partitioning methods are robust with respect to initial conditions, since they can apply regardless of the nature of the initial conditions. Conversely, the possible computational advantages of Chandrasekhar algorithms depend essentially on the low-rank property which corresponds to particular initial conditions leading to a low rank for the increment δP_2. In order to remedy this drawback, we can, however, use the so called dual-Chandrasekhar equations. They were developed in the continuous-time case by Ljung and Kailath (1977), and in the discrete-time cases detailed by Du et al. (1985) from Friedlander et al. (1976).

The basic idea consists of decomposing the actual initial covariance matrix P_{xr} into two parts P_x and δP_x:

$$P_{xr} = P_x + \delta P_x \tag{5.1}$$

where P_x is chosen with a low-rank property, and thus suitable for applying the Chandrasekhar algorithm. Let K_{og} represent the generalized asymptotic Kalman gain corresponding to P_x and K_{gi} the generalized Kalman gain corresponding to P_{xr}, and define:

$$\delta K_{gi} = K_{gi} - K_{ogi} \tag{5.2}$$

We obtain the following discrete dual-Chandrasekhar equations:

$$\delta K_{gi} = \Phi(i+1,1)\,\delta P_x M_o^*(i,1)\,(R_{oi}^e)^{-1} \tag{5.3}$$

$$M_o(i+1,1) = \{M_o(i,1)+M_o(i,1)Y_1[M_1^{-1}-N_o(i,1)Y_1]^{-1}N_o(i,1)\}\tilde{F},$$

$$M_o(1,1) = H\ ,\ \dim(M_o(.,.)) = (1,n) \tag{5.4}$$

$$N_o(i+1,1) = N_o(i,1) - Y_1^* M_o^*(i,1)\,(R_{oi}^e)^{-1} M_o(i,1),$$

$$N_o(1,1) = O,\quad \dim(N_o(.)) = (1,n) \tag{5.5}$$

$$\tilde{F} = F(I+P_x H^* H)^{-1} \tag{5.6}$$

Proof: (See appendix)

From equation (5.3) one obtains after T_c iterations:

$$\delta K_{gTc} = \Phi(T_c+1,1)\,\delta P_x M_o^*(T_c,1)\,(R^e_{oT_c})^{-1} \qquad (5.7)$$

where index T_c represents the total iteration number previously defined. $M_o^*(T_c,1)$ in (5.7) can be computed by repeated use of (5.4) and (5.5). The number of arithmetic operations at each iteration is given in table III.

Terms	operations	
	adds	mults.
$N_o(.)Y_1$	n-1	n
$M_o(.)Y_1$	n-1	n
eq. (5.4)	n^2+1	n^2+n+3
eq. (5.5)	n	n+1
total	n^2+3n-1	n^2+4n+4

Table III: Number of arithmetic operations for dual-Chandrasekhar algorithm.

Computation of $\Phi(T_c+1,1)$, which is the transition matrix of $(F-K_{gi}H)$, is rather complicated. We have:

$$\Phi(T_c+1,1) = \Phi_o(T_c+1,1)[\,I-(P_{xr}-P_x)W_o(T_c+1,1)\,]^{-1} \qquad (5.8)$$

$W_o(.)$ may be computed by:

$$W_o(i+1,1) = W_o(i,1) - M_o^*(i,1)(R^e_{oi})^{-1}M_o(i,1) \qquad (5.9)$$

Repeating (5.9) from i=0 to $i=T_c$, we finally obtain $W_o(T_c+1,1)$. $n(n+1)/2$ operations are required to compute this equation at each iteration. Defining:

$$Y = [\ I - \delta P_x W_o(T_c+1,1)\ ]^{-1} \delta P_x M_o^*(T_c,1)(R^e_{oT_c})^{-1} \tag{5.10}$$

we have:

$$\delta K_{gT_c} = \Phi_o(T_c+1,1)Y \tag{5.11}$$

which can be recursively computed by:

$$\delta K_{gi} = (\ F - K_{ogi}H\ )\delta K_{gi-1},\quad \delta K_{go} = Y \tag{5.12}$$

where Y is an n-dimensional vector. Computation of (5.10) needs in general $O(n^3)$ operations. The amount of arithmetic operations for computing (5.12) at each iteration is equal to n^2. Then the total number of arithmetic operations needed by the discrete dual-Chandrasekhar equations is given by:

$$N_{cd} = T_c[\ (n^2+4n+4) + 0.5(n+1)n + n^2] + O(n^3) = T_c(2.5n^2+4.5n+4)+O(n^3) \tag{5.13}$$

Remark: Equation (5.12) requires proportional to n^3 operations, but it is used only once. In the general case where an arbitrary initial condition P_{xr} is taken, we have to add the number of arithmetic operations required by the dual-Chandrasekhar equations to the number required by the Chandrasekhar equations. The performance index A is then modified as:

$$A_m(T_c,n) = \frac{T_c(n^2+3n+8) + T_c(2.5n^2+4.5n+4) + O(n^3)}{\mathrm{Log}_2(T_c)\ \ (8n^3+2n^2+O(n^2))}$$

$$= 3.5A(T_c,n) + \frac{O(n^3)}{\mathrm{Log}_2(T_c)\ 8n^3} \tag{5.14}$$

In practical situations where $A \leq 0.1$, we have $A_m \leq 0.35$ according to (5.14), and the conclusions given in the preceding section remain valid.

6. CONCLUSION

When the state dimension is large, which is a situation encountered in 2-D image restoration (see Saint-Fèlix et al.

(1985)), and when the system is invariant, and when the observation is scalar, the Chandrasekhar and dual-Chandrasekhar algorithms perform better than the partitioning one to compute the asymptotic Kalman gain with arbitrary initial conditions.

7. APPENDIX

Discrete Dual-Chandrasekhar Equations

By definition $K_{gi}=K_i(R_i^e)^{-1}$, $K_{ogi}=K_{oi}(R_{oi}^e)^{-1}$, where K_{gi} and K_{ogi} correspond to the gains computed with the initial conditions P_{rx} and P_x respectively. Defining $\delta K_{gi} = K_{gi} - K_{ogi}$, we have:

$$\delta K_{gi} = (F-K_{gi}H)\delta P_i H^* (R_{oi}^e)^{-1} \tag{7.1}$$

where $\delta P_{i+1} = P_{i+1} - P_{oi+1}$. Note that:

$$\delta R_i^e = R_i^e - R_{oi}^e = H\delta P_i H^* \text{ , and } P_{oi} = P_i - \delta P_i$$

We can show that:

$$\delta P_{i+1} = \Phi(i+1,1)\delta P_x \Phi_o^*(i+1,1) \tag{7.2}$$

where $\Phi(.,.)$ and $\Phi_o(.,.)$ are the transition matrices of $(F-K_{gi}H)$ and of $(F-K_{ogi}H)$ respectively. So the equation (5.3):

$$\delta K_{gi} = (F-K_{gi}H)\Phi(i,1)\delta P_x M_o^*(i,1)(R_{oi}^e)^{-1} \tag{7.3}$$

with here: $M_o(i,1) = H\Phi_o(i,1)$. We have the following backward and forward equations:

$$\Phi_o(i+1,1) = \Phi_o(i,1)[I-\tilde{Q}W_o(i,1)]^{-1}\tilde{F} \tag{7.4}$$

where: $\tilde{Q} = Y_1 M_1 Y_1^*$, $\tilde{F} = F(I+P_x HH^*)^{-1}$

$$W_o(i+1,1) = W_o(i,1) - \Phi_o^*(i,1)H^*H(R+P_{oi}H^*H)^{-1}\Phi_o(i,1) \tag{7.5}$$

Multiply (7.4) by H from the left and define:

$$N_o(i,l) = Y_1^* W_o(i,l) \tag{7.6}$$

We have:

$$M_o(i+1,l) = \{M_o(i,l) + N_o(i,l)Y_1[M_1^{-1} - N_o(i,l)Y_1]^{-1} N_o(i,l)\}\tilde{F} \tag{7.7}$$

$$N_o(i+1,l) = N_o(i,l) - Y_1^* M_o^*(i,l)(R_{oi}^e)^{-1} M_o(i,l) \tag{7.8}$$

$$W_o(i+1,l) = W_o(i,l) - M_o^*(i,l)(R_{oi}^e)^{-1} M_o(i,l) \tag{7.9}$$

and according to the equation (7) given in Friedlander et al. (1976), the transition matrix for initial conditions is given by:

$$\Phi(i+1,l) = \Phi_o(i+1,l)[I-(P_{xr}-P_x)W_o(i+1,l)]^{-1} \tag{7.10}$$

8. REFERENCES

Anderson, B.D.O., (1978) "Second-order Convergent Algorithms for the Steady-State Riccati Equation", *Int. J. Contr.*, Vol. 78, n° 2, 296-306.

Demoment, G., Reynaud, R. and Herment, A., "Fast Minimum Variance Deconvolution". *IEEE Trans.*, Vol. ASSP-33, 1324-1326.

Du, X.C., Saint-Félix, D. and Demoment, G., (1985) "Dual Chandrasekhar equations: The discrete case". L2S Internal Report. Submitted for publication.

Friedlander, B., Kailath, T. and Ljung, L., (1976) "Scattering Theory and Linear Least Squares Estimation Part II. Discrete-Time Problems". *J. Franklin Inst.*, 71-82.

Hewer, G.A., (1971) "An Iterative Technique for the Computation of the Steady-State Gains for the Discrete Optimal Regulator". *IEEE. Trans.*, Vol.AC-16, 382-384.

Kailath, T., (1973) "Some New Algorithms for Recursive Estimation in Constant Linear Systems". *IEEE. Trans.*, Vol. IT-19, 750-760.

Kleinmarn, D.L., (1968) "On an Iterative Technique for Riccati Equation Computation". *IEEE. Trans.*, Vol.AC-13, 114-115.

Lainiotis, D.G., (1976) "Partitioned Riccati Solutions and Integration-Free Doubling Algorithm". *IEEE. Trans.*, Vol. AC-21, 677-688.

Lainiotis, D.G., (1978) "Discrete Riccati Equation: Generalized Partitioned Algorithms". *Int. J. Sci.*, Vol. 15, 169-185.

Laub, J.A., (1979) "A Schur Method for Solving Algebraic Riccati Equations". *IEEE. Trans.*, Vol.AC-24, 913-921.

Ljung, L., Kailath, T., and Friedlander, B., (1976) "Scattering Theory and Linear Least Squares Estimation Part I: Continuous-Time Problems". Proc. *IEEE.*, Vol. 64, 131-139.

Ljung, L. and Kailath, T., (1977) "Efficient Change of Initial Conditions, Dual Chandrasekhar Equations, and Some Applications". *IEEE. Trans.*, Vol.AC-22, 443-447.

Morf, M., Sidhu, G.S. and Kailath, T., (1974) "Some New Algorithms for Recursive Estimation in Constant, Linear, Discrete-Time Systems". *IEEE. Trans.*, Vol.AC-19, 315-323.

Saint-Félix, D., Djafari, A.M. and Demoment, G., (1983) "Quelques problèmes liés à l'amélioration de la restauration d'images". In Proc. 9th Conf. GRETSI, Nice, France, 349-354.

Saint-Félix, D., Du, X.C. and Demoment, G., (1985) "Image Restoration Using a Non-Causal State-Space Model and a Fast 2-D Kalman Filter". In this proceedings.

Sandell, N.R., (1974) "On Newton's Method for Riccati Equation Solution". *IEEE. Trans.*, Vol.AC-19, 254-255.

Sidhu, G.S. and Bierman, G.J., (1977) "Integration-Free Interval Doubling for Riccati Equation Solutions". *IEEE Trans.*, Vol.AC-22, 831-834.

Vaughan, D., (1970) "A Nonrecursive Algebraic Solution for the Discrete Time Equations". *IEEE. Trans.*, Vol.AC-15, 597-599.

Willems, J.C., (1971) "Least Squares Stationary Optimal Control and the Algebraic Riccati Equation". *IEEE. Trans.*, Vol.AC-16, 621-634.

IMAGE RESTORATION USING A NON-CAUSAL STATE SPACE MODEL AND A FAST 2-D KALMAN FILTER

D. Saint-Félix[†], X.C. Du and G. Demoment
(Laboratoire des Signaux et Systèmes (CNRS/ESE), France)

ABSTRACT

The ill-posed image restoration problem is treated in a Bayesian framework to deal with both noise and prior information required to stabilize the solution. Direct discretization of the convolution equation provides a state-space model whose huge dimensions make a standard Kalman filter untractable in spite of its recursive nature. That is why the model dimension is usually reduced by introducing dynamics into the state equation, which requires an artificial causality assumption. To avoid this difficulty, we propose a state-space model where the state is taken constant and equal to the entire object to be restored, and where dynamics appear in the observation equation only. The shift properties of the convolution summation allow us to derive a fast Kalman algorithm through factorization techniques. A sub-optimal asymptotic filter with a reduced update is then derived. Finally, experimental results are presented.

1. INTRODUCTION

In this paper, we address the general problem of restoring an image degraded by noise and smeared by a linear, shift-invariant point spread function (PSF). Stabilizing the solution of this ill-posed problem requires the use of "a priori" information on this solution. A Bayesian framework is chosen to handle both this a priori information and the noise on the observation (Hunt (1977), Geman (1984)). The object to be restored and the noise are therefore considered as two independent random processes, and the prior information consists of the knowledge of their probability density functions, which are assumed gaussian here. Estimation of the posterior distributions turns out to be a well-posed problem (see Akaike

† Present address: Thomson-C.G.R., DTS, 283, rue de la Minière, 78530 BUC, France.

(1980)). But for practical reasons, a point estimator of the object is more convenient; the linear minimum variance (LMV) estimator is chosen here since, with our gaussian assumptions, it is optimal and identical to the maximum a posteriori or minimax estimators. The complete solution of the restoration problem then consists of two sub-problems (Kaufman et al. (1983)): (i) estimating the object, which is a smoothing problem and, (ii) estimating the parameters defining the a priori distributions, which is an identification problem. We tackle herein only the important sub-problem (i) of image deconvolution where the a priori parameters are assumed to be known. Some indications are given at the end of the paper on a possible solution to the identification problem (ii).

Deconvolution is then a standard estimation problem whose high dimensions make recursive estimation schemes (e.g. Kalman filters) attractive. But a straightforward derivation from the initial convolution model leads to a degenerate Kalman filter which is untractable because of the high dimension of the state. That is why the usual approaches (see Willsky (1979)) consist of introducing dynamics into the problem to reduce its dimension. But this requires a semi-causality assumption on the state process, which is rather artificial for a 2-D field. The aim of this paper is to derive a fast 2-D Kalman filter that does not require a causality assumption on the 2-D object. A suboptimal filter, using a truncated asymptotic gain vector naturally ensues from it. This reduced update filter gives a satisfactory solution if compared with our previous deterministic regularization methods described in Saint-Félix et al. (1984-a), (1984-b) which require several iterations, i.e. image scannings, to yield a good restoration.

2. DECONVOLUTION AS AN ESTIMATION PROBLEM

Let $\underline{x}$, $\underline{y}$ and $\underline{b}$ be vectors built by concatenating samples of the complex-valued 2-D fields corresponding respectively to the unknown object, the observed image and the measurement noise. The image degradation may be described by the linear system of equations:

$$\underline{y} = H\underline{x} + \underline{b} \tag{2.1}$$

where

$$\underline{y}^t = [\,y(1,1),\ldots,y(1,L'),\ldots,y(L,L')\,]$$

$$\underline{x}^t = [\,x(1,1),\ldots,x(1,M'),\ldots,x(M,M')\,] \tag{2.2}$$

$$\underline{b}^t = [\,b(1,1),\ldots,b(1,L'),\ldots,b(L,L')\,]$$

The structure of the observation matrix H depends on the image scanning used and on the prior information on the image edges (see Saint-Félix et al. (1984-a)). We consider herein that only a portion of a wider object is observed. H is then a block-circulant matrix (LL',MM') whose first line is given by:

$$H : (\tilde{H}_N \;.\;.\; \tilde{H}_1 \;\; O \;.\;.\; O) \tag{2.3}$$

where each block $\tilde{H}_j$ is also a circulant matrix (L',M') formed with PSF samples, and whose first line is:

$$\tilde{H}_j : (h(j,N') \;.\;.\; h(j,1) \;\; O \;.\;.\; O) \tag{2.4}$$

The dimensions of the image, object and PSF are then related as follows:

$$M = L + N - 1 \quad \text{and} \quad M' = L' + N' - 1 \tag{2.5}$$

As mentioned in the introduction, $\{x_{i,j}\}$ and $\{b_{i,j}\}$ are assumed to be realizations of two independent gaussian random processes whose first and second order moments are known a priori:

$$E\{\underline{x}\} = \underline{x}_o \qquad E\{(\underline{x}-\underline{x}_o)(\underline{x}-\underline{x}_o)^*\} = P_x \tag{2.6}$$

$$E\{\underline{b}\} = \underline{O} \qquad E\{\underline{b}.\underline{b}^*\} = R_{bb} \qquad E\{\underline{b}.\underline{x}^*\} = O \tag{2.7}$$

where $\underline{x}^*$ stands for the adjoint of $\underline{x}$.

With our assumptions, the LMV estimator is straightforwardly given by:

$$\hat{\underline{x}} = \underline{x}_o + P_x H^*[HP_x H^* + R_{bb}]^{-1}[\underline{y} - H\underline{x}_o] \tag{2.8}$$

$$\hat{\underline{x}} = [H^* R_{bb}^{-1} H + P_x^{-1}]^{-1} [H^* R_{bb}^{-1} . \underline{y} + P_x^{-1} \underline{x}_o] \tag{2.9}$$

The dimensions of $\underline{x}$ and $\underline{y}$ (about 16,000 for an image 128x128) make the direct inversion of the observation covariance matrix in (2.8) or of the information matrix in (2.9) untractable. Recursive estimation techniques are a well known alternative way. For this purpose, the equation (2.1) may be written:

$$y_i = H_i \underline{x} + b_i \qquad \text{for } i = 1,2,\ldots,MM' \tag{2.10}$$

where y_i is the ith coordinate of $\underline{y}$ and H_i the ith row of H. The Kalman filter providing the estimate $\hat{\underline{x}}_i$ knowing $y_1,\ldots,y_{i-1}$ for a constant state $\underline{x}$ is easily derived. No matrix inversion is required since the observation and therefore the innovation covariance are scalar quantities. But the prediction error covariance matrix, whose dimensions are (MM',MM'), must be updated at each recursion. The usual way to deal with this difficulty consists of putting into question the state dimension only and, consequently, reducing it by introducing dynamics into the state equation. But all the dynamic state-space models used for a recursive processing are based on a semi-causality assumption for the 2-D field, and often on a separability assumption for the covariance.

3. CHOICE OF A STATE-SPACE MODEL

A first family of state-dynamic models is due to Nahi (1972) and Powell et al. (1974). The fields corresponding to the image (assumed to be degraded by noise only) and the noise are transformed into scalar-valued stochastic processes by ordering the field components according to the scanning order. But the problem of stochastic realization which follows has no solution since P_x does not verify a necessary condition of separability, and the optimality is lost with all the approximations that have been proposed (see Willsky (1979)). Moreover, the model obtained is causal: x_{mn} depends on those x_{ij} only situated in a quadrant or in an extended half-plane. The causality effects can be reduced by smoothing the estimates provided by several Kalman filters scanning in opposite directions.

Besides these purely one-dimensional approaches, two-dimensional models have been developed since Habibi (1972). The most famous is Woods' model (1977),(1981) where the 2-D random field is described by:

$$x_{mn} = \sum_{i=1}^{K} \sum_{j=-K}^{K} a_{ij} x_{m-i,n-j} + \sum_{j=1}^{K} b_{oj} x_{m,n-j} + w_{mn} \tag{3.1}$$

It is readily seen that x_{mn} depends only on pixels located in a non-symmetric half-plane. When the object is only degraded by noise : $y_{mn} = x_{mn} + b_{mn}$, the scanning operation transforms the 2-D problem into an equivalent 1-D problem by defining a global state of dimension K(M'+1) which extends the state defined in (3.1) to the whole width of the image. A Kalman strip filter and a reduced update filter are then derived. Willsky (1979) has suggested an extension for the case $H \neq I$ by introducing a delay in the observation. Note that these models are non constant to take into account initialization at the beginning of a new row of the image and thus do not easily lend themselves to the derivation of fast algorithms.

The causality assumption, artificial for 2-D fields, is the main cause of difficulty when extending 1-D recursive techniques to 2-D problems. Non-recursive techniques exist that allow the use of non causal dynamic models (see Jain (1978) or Chellappa (1982)), but many approximations must be made in order to reduce the computational burden.

Our approach is based on two fundamental ideas. First, the strategy chosen to scan the image depends on the observer only. Thus it ought to condition the observation equation only and not the "state" equation through artificial causality. Secondly, the total amount of computation in a Kalman filter not only depends on the state vector dimension, but also on the nature of the equations which are used. Taking advantage of the convolution shift properties should lead to a substantial simplification. Therefore we use the following state-space model:

$$\underline{x}_i = \underline{x}_{i-1} = \underline{x} \qquad (3.2)$$

$$y_i = H_i \underline{x} + b_i \qquad (3.3)$$

$$H_i = H_{i-1} D_i \qquad \text{and} \qquad H_i D_i^t = H_{i-1} \qquad (3.4)$$

where the "state" $\underline{x}$ is constant and equal to the entire object to be restored. (3.4) is introduced to model the non-constancy of H_i and the observation transition matrix D_i depends on the image scanning mode. Note that D_i is generally not invertible (see next paragraph), so that dynamics cannot be moved into the state equation (3.2) by a parameter change.

4. FAST 2-D DECONVOLUTION ALGORITHM

4.1 Derivation of the algorithm

The difficulty encountered when computing the prediction error covariance matrix P_i can be solved by replacing the Riccati equations of the standard Kalman filter by Chandrasekhar equations. The basic idea is to compute the increments of the quantities involved in the filter rather than the quantities themselves and to factorize the increment of the prediction error covariance matrix. Since our model (3.2-4) is not invariant but shift invariant, the factorization technique requires a definition of the quantities' increments that incorporates the shifting dynamics through D_i:

$$\Delta K_i = K_i - D_i^t K_{i-1} \tag{4.1.1}$$

$$\Delta R_i^e = R_i^e - R_{i-1}^e \tag{4.1.2}$$

$$\Delta P_i = P_i - D_i^t P_{i-1} D_i \tag{4.1.3}$$

These definitions differ from the original ones given by Kailath (1980) for a constant state space-model. Assuming a stationary noise and a constant observation transition matrix ($D_i = D$) for the purpose of factorization, we obtain the fast algorithm:

$$\hat{\underline{x}}_{k+1} = \hat{\underline{x}}_k + K_k (R_k^e)^{-1} (y_k - H_k \hat{\underline{x}}_k) \tag{4.1.4}$$

$$A_k = H_k Y_{k-1} \tag{4.1.5}$$

$$B_k = A_k M_{k-1} \tag{4.1.6}$$

$$R_k^e = R_{k-1}^e + A_k B_k^* \tag{4.1.7}$$

$$K_k = D^t K_{k-1} + Y_{k-1} B_k^* \tag{4.1.8}$$

$$Y_k = Y_{k-1} - K_k (R_k^e)^{-1} A_k \tag{4.1.9}$$

$$M_k = M_{k-1} + B_k^* (R_{k-1}^e)^{-1} B_k \tag{4.1.10}$$

(See Demoment et al. (1985) for details in the real-variable case). Extension to the complex-variable case is done without difficulty. Initialization is given by:

$$K_1 = P_x H_1^* \tag{4.1.11}$$

$$R_1^e = H_1 P_x H_1^* + s^2 \qquad \text{with } E\{b_i b_j^*\} = s^2 . \delta_{ij} \tag{4.1.12}$$

$$\Delta P_2 = P_x - K_1 (R_1^e)^{-1} K_1^* - D^t P_x D \tag{4.1.13}$$

where ΔP_2 must be factorized as:

$$\Delta P_2 = Y_1 M_1 Y_1^* \qquad \text{with } M_1 = M_1^* \tag{4.1.14}$$

The computational advantage of this method with respect to a standard Kalman filter is made clear by the following observation. If "a" is the rank of ΔP_i, the dimensions of matrices M_i and Y_i are respectively (a,a) and (MM',a). When the initial covariance matrix P_x is , or may be chosen, such that a << MM', there is then a significant saving in using the Chandrasekhar equations instead of the standard Riccati equations.

4.2 Interpretation of the observation transition matrix constancy

Before studying the factorization of ΔP_2, let us investigate the consequences of the D_i constancy assumption. The observation matrix H_i is defined by:

$$H_i = [\ 0\ |\ldots|\ 0\ |\ \tilde{H}_N^k |\ \tilde{H}_{N-1}^k |\ldots| \tilde{H}_1^k |\ 0\ |\ldots|\ 0\ ] \tag{4.2.1}$$

where each block $\tilde{H}_j^k$ is a (1,M') dimension matrix formed by the kth row of $\tilde{H}_j$. In a raster scan, when y_i and y_{i+1} belong to the same image row, H_{i+1} is obtained from H_i by a single shift since the blocks $\tilde{H}_j$ are circulant. D_i=D is then a shift matrix (D = (d_{mn}) with: $d_{mn} = \delta_{m+1,n}$) which verifies (3.4). But when y_i and y_{i+1} are respectively the end and beginning of two successive rows, we have $H_{i+1} = H_i . D^{N'}$. The constancy condition of the transition matrix is no longer verified. This condition can be reached again if the image is assumed to be coiled around a cylinder with its rows concatenated on a helix. H then becomes circulant instead of block-circulant. This is the usual assumption made when using fast Fourier transform algorithms for image restoration (see Willsky (1979), Andrews and Hunt (1977)). Uncontrolled correlation between "left" and "right" edges of the object through the observation equation can be avoided by extending the corresponding edges of the measured image before coiling it. The number of added columns must be at least greater than the PSF width or than the object correlation distance. Of course, the state estimate is not updated with (4.1.4) for recursions corresponding to the missing observations in the extended part.

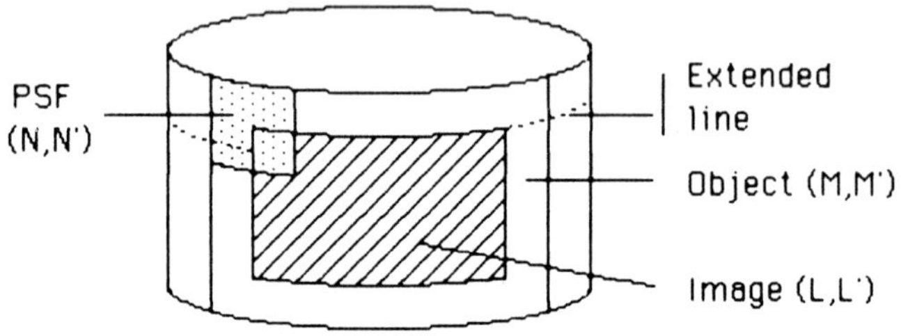

Fig. 1 Image coiled on a helix

4.3 Initial factorization

Initial factorization of ΔP_2 depends on the matrix $Q = P_x - D^t P_x D$. In the general case, P_x is a block matrix defined from the blocks $R_{ij}=E\{\underline{x}_i.\underline{x}_j^*\}$ where $\underline{x}_i$ is the ith row of the object. Suppose that the process $\{x_{m,n}\}$ is "vertically" stationary ($R_{ij} = R_{i-j}$). P_x is then block-Toeplitz :

$$P_x = \begin{bmatrix} R_o & R_{-1} & \cdot & \cdot & R_{-d} & O & \cdot & \cdot & \cdot & O \\ \cdot & & & & & \cdot & & \cdot & & \cdot \\ \cdot & & & & & & & \cdot & & O \\ R_d & & & & & & & & & R_{-d} \\ O & \cdot & & & & & & & & \cdot \\ \cdot & \cdot & & & & & & & & \cdot \\ O & \cdot & \cdot & \cdot & O & R_d & \cdot & & \cdot & R_o \end{bmatrix} \tag{4.3.1}$$

where d is the correlation distance between rows ($R_{ij} = O$ for $|i-j|>d$) and:

$$R_i = (r^i_{mn}) = (E\{x_{j,m}.x^*_{j-i,n}\}) = R^*_{-i} \tag{4.3.2}$$

Suppose next that d' is the correlation distance between columns ($r^i_{jk} = O$ if $|j-k|>d'$ for all i) and that the process

$\{x_{m,n}\}$ is "horizontally" stationary. The elements of R_i must take into account the row extensions and their coiling on a helix. Assuming that the extended part of the object belongs to the same 2-D field, i.e. has the same statistical properties, R_i may be written as:

$$R_i = \begin{bmatrix} r_o^i & \cdot & \cdot & r_{-d'}^i & 0 & \cdot & \cdot & 0 & r_{d'}^{i-1} & \cdot & \cdot & r_1^{i-1} \\ \cdot & & & & \cdot & & & & & & \cdot & \cdot \\ \cdot & & & & & & \cdot & & & & & r_{d'}^{i-1} \\ r_{d'}^i & & & & & & & & & & & 0 \\ 0 & \cdot & & & & & & & \cdot & & & \cdot \\ \cdot & & & & & & & & & \cdot & & 0 \\ 0 & & \cdot & & & & & & & & & r_{-d'}^i \\ r_{-d'}^{i+1} & & & & \cdot & & & & & & & \cdot \\ \cdot & \cdot & & & & & \cdot & & & & & \cdot \\ r_{-1}^{i+1} & \cdot & \cdot & r_{-d'}^{i+1} & 0 & \cdot & \cdot & 0 & r_{d'}^i & \cdot & \cdot & r_o^i \end{bmatrix} \tag{4.3.3}$$

Note that R_{-d-1} (resp. R_{d+1}) is an inferior (resp. superior) triangular matrix. Therefore P_x is a Toeplitz matrix whose non-zero elements are located in 2d+1 bands parallel to the main diagonal.

Let (p_{ij}) and (q_{ij}) be the elements of P_x and D^tP_xD respectively. We get:

$$q_{ij} = p_{i-1,j-1} \quad \text{for } i=2,\ldots,MM' \text{ and } j=2,\ldots,MM'$$

$$q_{1j} = q_{i1} = 0 \quad \text{for } i=1,\ldots,MM' \text{ and } j=1,\ldots,MM' \tag{4.3.4}$$

Since $p_{i+1,j+1} = p_{i,j}$, all the elements of Q are zero except those located on the first row and column. Thus the rank of Q is at most 2. Since K_1 is a vector (scalar observation), the rank of ΔP_2 verifies:

$$\text{rank}(\Delta P_2) = \text{rank}(P_x - D^tP_xD - K_1(R_1^e)^{-1}K_1^*) \leq 3 \tag{4.3.5}$$

Defining the vectors $\underline{f}$ and $\underline{g}$ of dimension MM' as:

$$\underline{f}^* = [r,0,\ldots,0] \quad \text{with } r^2 = r_o^o \tag{4.3.6}$$

$$\underline{g}^* = r^{-1}[0, r^o_{-1}, \ldots, r^o_{-d}, 0, \ldots, r^{-1}_{d}, \ldots, r^{-d}_{-d}, 0, \ldots, 0] \quad (4.3.7)$$

then ΔP_2 may be easily factorized as :

$$\Delta P_2 = [\ \tilde{\underline{K}}_1 \mid \underline{f} \mid \underline{g}\ ] \begin{bmatrix} -1 & 0 & 0 \\ 0 & 1 & 1 \\ 0 & 1 & 0 \end{bmatrix} \begin{bmatrix} \tilde{\underline{K}}_1^* \\ \underline{f}^* \\ \underline{g}^* \end{bmatrix} = Y_1 M_1 Y_1^* \quad (4.3.8)$$

with : $\tilde{\underline{K}}_1 = (R^e_1)^{-1/2} K_1$.

Therefore, when the object is assumed to be stationary, the rank a of ΔP_i is 3, but taking into account the sparseness of M_1, and the relation: $H_k \underline{f} = 0$ for k>1, reduces the effective computations required by (4.1.4-4.1.10).

If the process $\{x_{m,n}\}$ is further assumed to be non-correlated, P_x is diagonal and $\underline{g} = \underline{0}$. Thus the rank of M_1 equals 2 and:

$$M_k = \begin{bmatrix} m_K & 0 \\ 0 & 1 \end{bmatrix} \qquad Y_k = [\,\tilde{\underline{Y}}_k \mid \underline{f}\,] \quad (4.3.9)$$

and the effective "rank" for computation in this case is 1.

Introducing an a priori correlation in P_x according to (4.3.1) naturally extends the operating part of the gain beyond the x_{ij} involved in the convolution summation for a given observation. This natural extension of the gain support toward the "future" is one of the main features of our non-causal state modelization if compared with other approaches such as Woods (1977),(1981). However, in this last case the computational burden is approximately increased by a factor of 2.

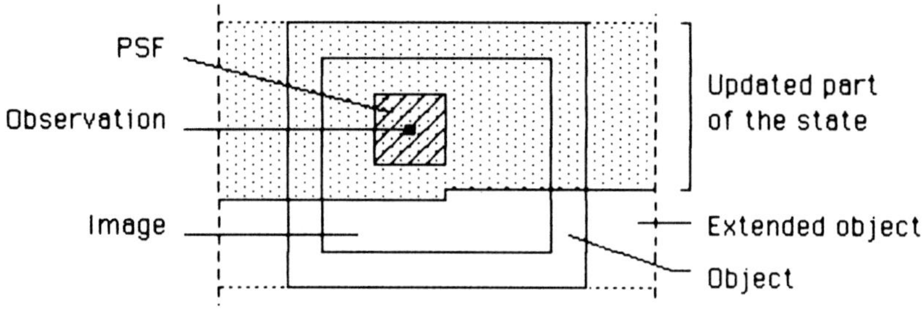

Fig. 2 Updated part of the state

5. FAST ASYMPTOTIC REDUCED-UPDATE FILTER

The previous filter requires computing a gain vector at each recursion and to update an increasing number of object pixels. Two different approaches make computational savings possible. When $\{x_{m,n}\}$ and $\{b_{m,n}\}$ are assumed to be stationary, the increment of the generalized Kalman gain vector ΔK_i converges rapidly to the nul vector. As the gain vector is data-independent, its asymptotic value can be computed in a first step by repeating (4.1.5 - 4.1.10) until its increment is below a given threshold. We have shown (Du et al. (1985)) that, from a computational point of view, this method is more efficient here than the partitioning methods. The computational requirement can be reduced by replacing (4.1.8) with:

$$K_{g,k} = D^t K_{g,k-1} \qquad (5.1)$$

where $K_{g,1}$ is the asymptotic gain vector previously computed and stored.

If the PSF is regular, the gain components are negligible when situated far from the current observation y_i. Hence, we can truncate the Kalman gain to its efficient part and constrain the state updating (4.1.4) to a sliding windown whose dimensions, (K,K'), are approximately those of the PSF.

Thus when both conditions are fulfilled, the suboptimal filter using a truncated asymptotic gain vector requires only (NN'+KK') multiplications per pixel and it turns out that this fast asymptotic filter does not induce any significant degradation into the restoration when compared to the optimal one.

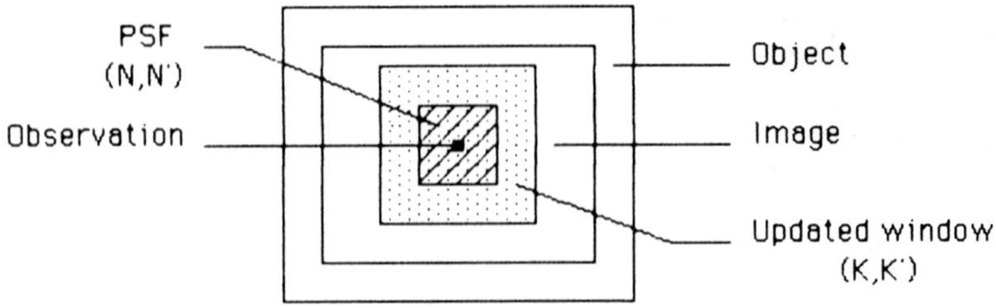

Fig. 3 Asymptotic reduced-update filter

6. RESULTS

Both optimal and suboptimal methods were used to restore simulated and actual images. Fig. 4 shows a simulated example. The original "H" shaped object (31x31) (Fig. 4-a) is convolved with a sinc function PSF (15x15) to give the blurred image (Fig. 4-b). This image is then restored using our optimal (Fig. 4-c) and suboptimal methods (Fig. 4-d). It turns out that a very good restoration can be achieved in this noise-free case. In Fig. 4-e, a white noise is added to the blurred image with a 10dB signal to noise ratio (SNR). Restorations with both filters using an a priori diagonal P_x and a non-diagonal P_x are shown in Figs. 4-f-g and 4-h-i respectively. In these examples, the a priori SNR has been overestimated voluntarily in order to emphasize the influence of the a priori correlation on the restoration quality.

Fig. 5-a represents an actual ultrasonic image (80x100) of 3 nylon wires, measured with a 5dB SNR. The transducers used in this imaging technique have such dynamic properties that their PSF have high dimensions (in this example: (51x75)). Results of restoration with both filters (Figs. 5-b-c) show that the resolution can be improved in both directions with a ratio of approximately 10.

Note also that, in these two examples, the dimensions of the update window used in the suboptimal filter and those of the PSF are the same and that the solution quality is not significantly degraded when compared to that of the optimal one.

(a)
Original object

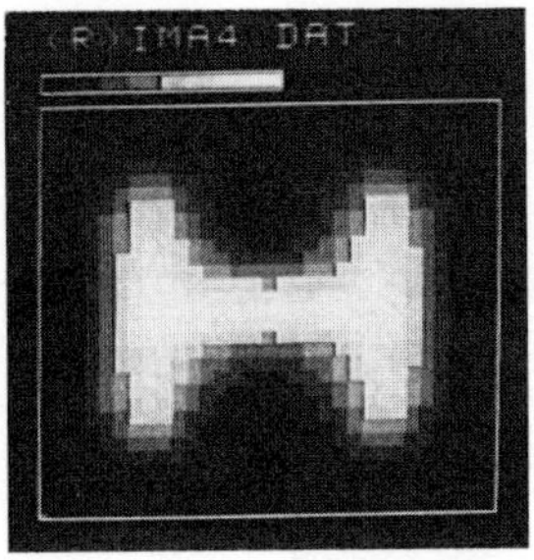

(b)
Blurred image

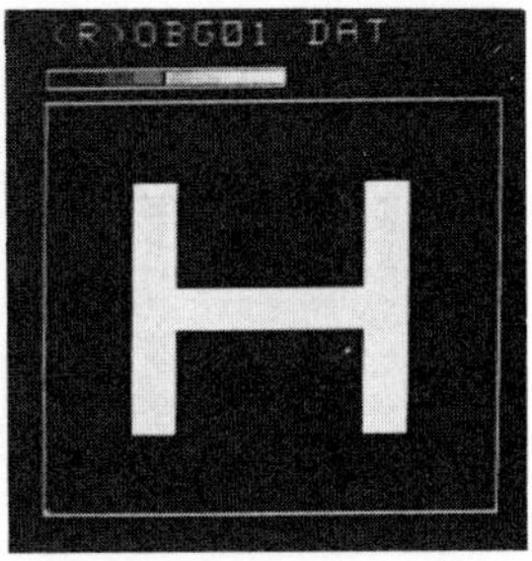

(c)
Optimal filter

(d)
Suboptimal filter

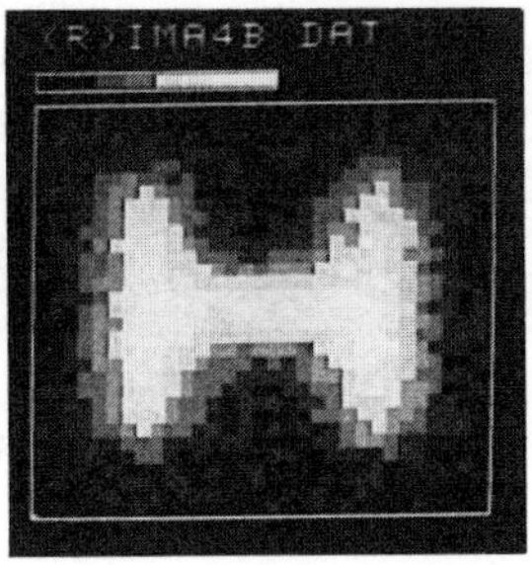

(e)
Noisy blurred image

(f)
Optimal filter
$P_x = I$

(g)
Suboptimal filter
$P_x = I$

(h)
Optimal filter
$P_x \neq I$

(i)
Suboptimal filter
$P_x \neq I$

Fig. 4 Simulated image restoration

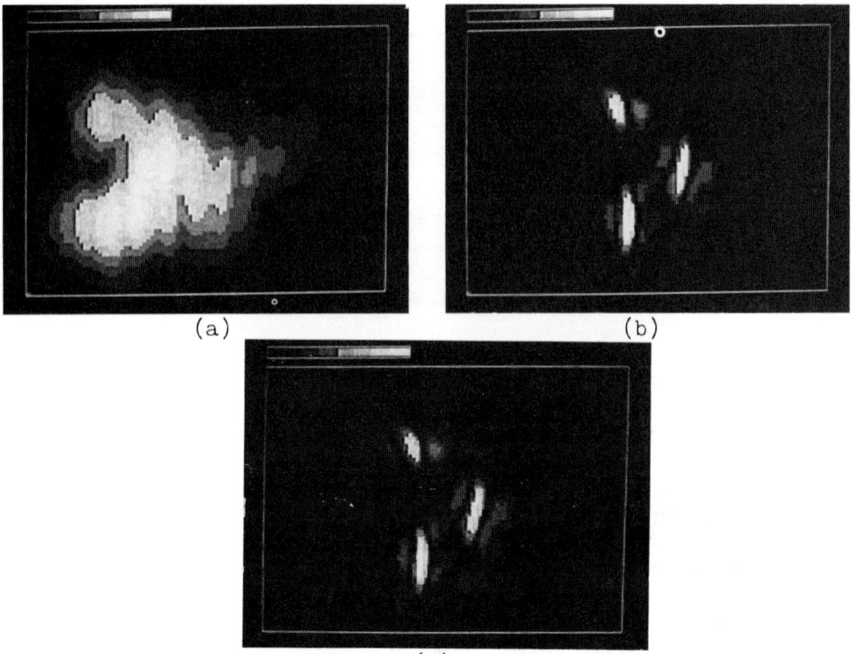

(a) Original image (80x100) of 3 point reflectors (rectified and filtered). (b-c) Same images obtained after optimal and suboptimal restoration.

Fig. 5 Ultrasonic B-scan echographic image restoration

7. CONCLUSIONS

The main features of our fast 2-D Kalman filters are the absence of a causal description of the object to be restored and an easy implementation due to factorization techniques. Use of a non-causal state-space model allows the Kalman gain to extend naturally into the "future" of the observed pixel. The filters can also take into account an a priori correlation in the object, yielding a further extension of the gain. Results show that use of the fast reduced update filter does not significantly alter restoration quality.

Of course the assumptions we used to derive these filters may be discussed. As far as the stationarity is concerned, it should be noticed that the method can process non-stationary objects. Taking into account a non-stationarity of the only mean through $\underline{x}_o$ does not increase the computational burden.

However a non-stationarity of the centered covariance makes one lose the computational advantage of the factorization technique. In this latter case, an alternative solution would consist of processing sub-blocks of the object with local second-order stationarity assumptions.

As far as the prior information is concerned, $\underline{x}_o$, P_x and R_{bb} are hyper-parameters that ought to be estimated from the image in a "total" deconvolution procedure. This could be done using a maximum likelihood approach applied to a parameter vector containing all their elements. Practically, physical considerations and approximations enable one to reduce the effective dimension of this parameter vector and make this estimation problem tractable. For example, if the object and the noise are assumed to be uncorrelated, the SNR is the only parameter left. Its log-likelihood can be easily computed from the innovation and its covariance which are two by-products of our Kalman filters. Nevertheless a difficult maximization problem remains, whose solution can be approximated by computing this log-likelihood on a finite set of parameter values and determining the maximum on this set.

8. REFERENCES

Aboutalib, A.O., Murphy, M.S., Silverman, L.M., (1977) "Digital restoration of images degraded by general motion blurs," *IEEE Trans.,* AC-22, pp. 294-302.

Akaike, H., (1980) "Likelihood and the Bayes procedure", in Bayesian statistics, Bernardo J.M. et al, Eds, University Press, Valencia, Spain, pp. 141-166.

Andrews, H.C., Hunt, B.R., (1977) "Digital image restoration" Prentice-Hall, Englewood Cliffs, NJ.

Chellappa, R., Kashyap, R.L., (1982) "Digital image restoration using spatial interaction models," *IEEE Trans.,* ASSP-30, pp. 461-472.

Demoment, G., Reynaud, R., Herment, A., (1985) "Fast minimum variance deconvolution," *IEEE Trans.,* ASSP-33, pp. 1324-1326.

Du, X.C., Saint-Félix, D., Demoment, G., (1985) "Comparison between factorization methods and partition methods in the derivation of invariant Kalman filters for fast image restoration". These proceedings.

Geman, S., Geman, D., (1984) "Stochastic relaxation, Gibbs distributions and the Bayesian restoration of images," *IEEE Trans.,* PAMI-6, pp. 721-741.

Habibi, A., (1972) "Two-dimensional Bayesian estimate of images," *Proc. IEEE,* Vol. 60, pp. 878-883.

Hunt, B.R., (1977) "Bayesian methods in nonlinear digital image restoration," *IEEE Trans.*, C-26, pp. 219-229.

Jain, A.K., Jain, J.R., (1978) "Partial differential equations and finite difference methods in image processing - Part 2: Image restoration," *IEEE Trans.*, AC-23, pp. 817-833.

Kailath, T., (1980) "Some alternatives in recursive estimation," *Int. J. Contr.*, Vol. 32, pp. 311-328.

Kaufman, H., Woods, J.W., Dravida, S., Tekalp, A.M., (1983) "Estimation and indentification of two-dimensional images," *IEEE Trans.*, AC-28, pp. 745-756.

Nahi, N.E., (1972) "Role of recursive estimation in statistical image enhancement," *Proc. IEEE*, Vol. 60, pp. 872-877.

Powell, S.R., Silverman, L.M., (1974) "Modelling of two-dimensional covariance functions with application to image restoration," *IEEE Trans.*, AC-19, pp. 8-13.

Saint-Félix, D., Djafari, A.M., Demoment, G., (1984-a) "Restauration d'images dans C par une méthode rapide de régularisation," in Proc. Premier Colloque Image, pp. 349-354, Biarritz, France.

Saint-Félix, D., Djafari, A.M., Demoment, G., (1984-b) "Iterative generalized inverse image restoration," in Proc. ICASSP 84, pp. 37.9.1-37.9.4, San-Diego, Cal.

Willsky, A.S., (1979) "Digital signal processing and control and estimation theory," Chapt. 4, MIT Press, Cambridge, Mass.

Woods, J.W., Radewan, C.H., (1977) "Kalman filtering in two dimensions", *IEEE Trans.*, IT-23, pp. 473-482, and Correction to, (1979), *IEEE Trans.*, IT-25, pp. 628-629.

Woods, J.W., Ingle, V.K., (1981) "Kalman filtering in two dimensions : further results," *IEEE Trans.*, ASSP-29, pp. 188-197.

THE SPATIALLY COHERENT PROCESSING OF RADAR RETURNS*

A. Cantoni
(Department of Electrical and Computer Engineering, University of Newcastle, New South Wales, Australia)

and

G. Corsini, R. Reggiannini and E. Dalle Mese
(Istituto di Elettronica e Telecomunicazioni, Università degli Studi di Pisa, Italy)

ABSTRACT

The paper illustrates in a simple but rather general manner how the reflectivity function of a target can be recovered by the spatially coherent processing of returns collected at different relative positions of radar and target. Unlike many other derivations available in the literature, the approach taken does not have to rely upon the presence of Doppler frequencies in the received echoes to derive the imaging equations.

The most significant performance measures, such as attainable range and cross-range resolutions, are estimated for rectilinear target motion trajectories and for different signal formats.

Finally, the effect of deviations from the straight iso-range assumption made in the theoretical analysis is investigated in terms of the distortions induced in the reconstructed image. Some numerical results are provided to check the approximate analytical evaluations.

1. INTRODUCTION

The reconstruction of a target image from its radar returns has been considered previously using a description in terms of the distribution of Doppler frequencies over the target surface, [1], or for specific examples of motion [2]. The approach based on Doppler frequencies is misleading since it obscures the basic mechanism which enables the target reflectivity function to be recovered. Not only is instantaneous motion not necessary to enable an image to be obtained, but it can invalidate the signal representation used to develop the imaging equations. Furthermore, for even simple rectilinear

*Work supported by Selenia S.p.A., Rome, Italy and M.P.I., Italy.

motion the Doppler frequency of a point scatterer is not constant and hence its interpretation is conceptually difficult.

This paper illustrates in a conceptually simple but very general manner how the reflectivity function of a target can be recovered by the spatially coherent processing of returns obtained for different relative positions of radar and target, [3]. The approach taken makes clear the basic underlying mechanism which enables the reflectivity function to be recovered for arbitrary target trajectories. Furthermore, the effect of instantaneous motion is easily identified. The range and cross-range resolution obtainable is estimated for rectilinear target motion trajectories and for various transmitted signal formats. The case of rectilinear motion of the target is singled out for special consideration since it is of practical interest and also because it is a situation amenable to mathematical analysis.

One of the crucial assumptions made in the development of the received signal representation which permit the reflectivity function to be recovered via a two dimensional inverse Fourier transform operation is the so-called straight iso-range approximation. The effect of this approximation on the imaging system two dimensional impulse response is illustrated. Numerical results which confirm the approximate analytical results are provided.

2. SIGNAL REPRESENTATION

For a stationary distributed target, the received signal is represented in terms of the equivalent reflectivity function according to the following equation

$$s_R(t) = \iint_{\text{object}} f(x,y)\, s_T(t-2R(x,y)/c)\, dx\, dy \tag{2.1}$$

where f(x,y) is a complex function defined over the target relative to a set of co-ordinates fixed to the target, $s_T(.)$ is the transmitted signal and R(x,y) is the distance from the radar of the point on the target of coordinates x,y. This approach has received considerable attention in the literature as a method of representing the received signal for a distributed rigid target, [4,5,6]. The set of co-ordinates used throughout this paper to define the reflectivity function is illustrated in Fig. 1. Let f(x,y) be the reflectivity function associated with the target, defined relative to a cartesian co-ordinate system fixed to the target and such that the x co-ordinate makes an angle $\theta(t)$ to the positive axis. Let the (u,v) co-ordinate system be defined as shown in Fig. 1, such that u is aligned with the line joining the radar and a reference point on the target at which the origin of the (x,y) co-ordinate system is located. In this system of co-ordinates, (2.1) becomes

$$s_R(t) = \iint_{\text{object}} f(u\cos d - v\sin d, u\sin d + v\cos d)\, s_T(t-2R(u,v)/c)\,du\,dv \tag{2.2}$$

where $d \triangleq \psi - \theta$.

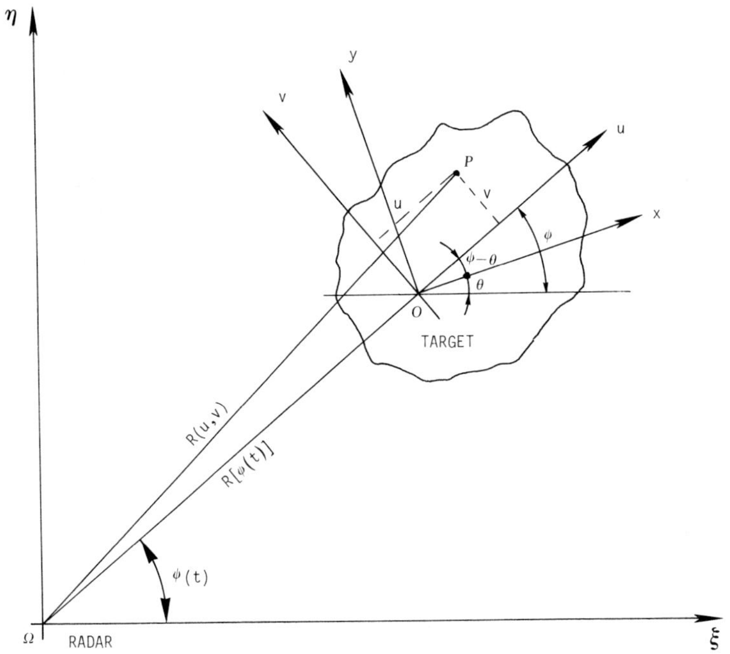

Fig. 1

Equation (2.2) provides a good approximation of $s_R(t)$ under the general conditions on R and d, which are time dependent functions since the object is normally in motion. Under the hypothesis that $\max_{\text{object}} |v| << R(t)+u$, the range function $R(u,v) \triangleq \sqrt{(R(t)+u)^2+v^2}$ (see Fig. 2.1) can be approximated by $R(u,v) \tilde{=} R(t)+u$. This will be referred to as the straight iso-range approximation to the range function, and will prove a crucial step in the subsequent discussions.

3. IMAGING EQUATIONS FOR PULSED RADARS

3.1 *Single and Multifrequency Pulsed Signals*

A single frequency pulse transmitted signal of duration T at time instant t_n is defined by $s_T(t) \stackrel{\Delta}{=} a(t-t_n)\exp[j(\omega_o(t-t_n)+\phi_n)]$, where ω_o is the carrier frequency and a(t) is the pulse gating function defined by a(t)=a for $0 \le t \le T$ and a(t)=0 elsewhere.

Substituting the previous expression in (2.1) and frequency down-shifting yields:

$$s'_R(t) = \iint_{\text{object}} a(t-t_n-2R(u,v)/c)\,u(t-t_n)\,f(.,.)\,\exp(j2\omega_o R(u,v)/c)\;dudv \qquad (3.1)$$

where u(t) is a gate function defined as 1 in $(0,T_\ell)$ and zero elsewhere, T_ℓ being the length of the locally generated reference pulse.

Assuming that the target lies in the range interval (R_{min}, R_{max}), it can be easily shown that for

$$2R_{max}/c < (t-t_n) < (2R_{min}/c)+T; \quad (2R_{min}\ c) + T < T_\ell \qquad (3.2)$$

the signal $s'_R(t)$ can be expressed as

$$s'_R(t) = a \iint_{\text{object}}^{R} f(.,.)\exp(-j(2\omega_o R(u,v)/c))\;dudv \qquad (3.3)$$

A convenient physical interpretation of condition (3.2) is that it indicates that the target must be totally illuminated in order for (3.3) to hold. In fact (3.2) imposes a lower limit on T, which is given by $T \ge 4L/c$, L being the space extent of the target.

Using the straight iso-range approximation and dropping the multiplicative phase term (range compensation) enables one to express (3.3) as

$$s''_R(X,Y) \stackrel{\Delta}{=} s''_R(t) \stackrel{\Delta}{=} s'_R(t)\;\exp(j\,2\omega_o R(t)/c)$$

$$= \iint_{\text{object}} f(x,y)\exp(-j(xX+yY))\;dxdy\ , \qquad (3.4)$$

where

$$X(t)\triangleq(2\omega_o/\cos d(t); \qquad Y(t)\triangleq(2\omega_o/c)\sin d(t) \qquad (3.5)$$

Equation (3.4) can be interpreted as the two dimensional Fourier transform (F.T.) of f(x,y). Note that, however, this transform is available only for a locus of points on a circle of radius $2\omega_o/c$ in the X,Y plane defined by eqs. (3.5).

If the target is in rectilinear motion and it is assumed that $\theta(t)=0$, eqs. (3.5) become

$$X(t)=(2\omega_o/c)\cos\psi(t); \qquad Y(t)=(2\omega_o/c)\sin\psi(t) \qquad (3.6)$$

and describe an arc of radius $(2\ \omega_o)/c$ and width 2Δ centred on the angle ψ_o as the target moves from $\psi_o-\Delta$ to $\psi_o+\Delta$. Consideration of a generic $\theta(t)$ does not alter the quality of this result since, as can be shown, it may only involve a rotation of the cited locus of points around the origin of the (X,Y) plane, and possibly a different value for Δ. Hence $\theta(t)$ can be assumed zero in all that follows. It is seen from eqs. (3.6) that when $\psi_o=90^\circ$ one has the maximum resolution in the x direction (azimuth), while practically no resolution is obtained in the y direction (distance).

In fact in this case a single point scatterer located at point (x_o,y_o) of the (x,y) system produces a response which is given approximately by

$$w(x,y)=\exp[\,j2\pi(y-y_o)/c]\,\mathrm{sinc}[\,4\pi(x-x_o)\sin\Delta/\lambda\,], \qquad (3.7)$$

where $\mathrm{sinc}(.)\triangleq\sin(.)/(.)$.

In order to obtain resolution also along the y axis a number of different transmitted frequencies are required. One method is to include in the pulsed signal all the discrete frequencies at one time, that is,

$$s_T(t)=a(t-t_n)\sum_{i=0}^{2M}\exp(j(\omega_i(t-t_n)+\phi_{ni})) \qquad (3.8)$$

Another method is to sequentially transmit the discrete frequency signals. The performance limits of these methods have been derived by means of the same approach employed in the previous case. These limits, along with the constraints that must be posed on the various parameters involved (frequency separation, observation interval length etc.) cannot be considered in this report, but are extensively discussed in [3].

3.2 Chirp Signal

Another method to achieve resolution along the y axis (range) is to transmit a chirp signal $s_T = a(t)\ \exp(j\ \omega_o t + \gamma t^2)$, where γ is a constant value that determines the maximum frequency deviation.

The received signal $s_R(t)$, after being mixed with a reference chirp, can be expressed as (recall eq. (2.2)):

$$s_R(t) = \iint_{\text{object}} a(t-2R(u,v)/c)\,\bar{a}(t-2\bar{R}(\psi)/c)\,f(.,.)\exp\{-j[R(u,v)-\bar{R}(\psi)] \cdot [(2\omega_o/c)+4(\gamma/c)(t-2\bar{R}(\psi)/c)+4(\gamma/c^2)(\bar{R}(\psi)-R(u,v))]\}du\,dv \quad (3.9)$$

where $\bar{a}(t)$ is a gate function, with time duration not necessarily equal to T and $2\bar{R}(\psi)/c$ is a delay time that must be chosen in a suitable way.

In [3] it is shown that if certain conditions on pulse and observation lengths hold and if the straight iso-range approximation is made, then eq. (3.9) can be written as follows:

$$s'_R(t) = \exp\{-j[R(t)-\bar{R}(\psi)][(2\omega_o/c)+(4\gamma/c)(t-2\bar{R}(\psi)/c)]\} \cdot \iint_{\text{object}} f(.,.)\exp\{-ju[(2\omega_o/c)+(4\gamma/c)(t-2\bar{R}(\psi)/c)]\}du\ dv \quad (3.10)$$

The above equation can be expressed in a form which resembles the imaging equations already derived by introducing the following variables

$$X \triangleq q(t)\cos(d(t)); \qquad Y \triangleq q(t)\sin(d(t)) \quad (3.11)$$

$$q(t) \triangleq 2\omega_o/c + (4\gamma/c)(t-2\bar{R}(\psi)/c) \quad (3.12)$$

Using (3.11)-(3.12) in equation (3.10) yields

$$s'_R(X,Y) = F(X,Y)\ \exp\{-jq(t)[R(t)-\bar{R}(\psi)]\}, \quad (3.13)$$

where F(X,Y) is the two-dimensional F.T. of f(x,y).

If the trajectory of the object is known, the phase term in (3.13) due to R(t) can be eliminated. This and other conditions derived in [3] allow one to express eq. (3.13) as:

$s''_R(X,Y)=F(X,Y)W(X,Y)$, which indicates that the signal $s''_R(X,Y)$ can be interpreted as the two dimensional F.T. of the reflectivity function gated by a window function $W(X,Y)$. This window is an annular sector bounded by two circular arcs with radii Y_ℓ and Y_u and two radial lines with angles to the X axis d_1 and d_2, where

$$Y_\ell \overset{\Delta}{=} 2\omega_o/c+4\pi B(1-k_1)/c, \quad k_1 < 1; \qquad Y_u \overset{\Delta}{=} 2\ \omega_o/c+4\pi B/c \qquad (3.14)$$

B being the maximum frequency deviation of the chirp signal and d_1,d_2 the minimum and maximum values of the angle $d(t)$ during the observation time.

For the case of rectilinear motion of the target in the region around $\psi=90^\circ$ and $\theta(t)\equiv 0^\circ$ the window function generated is shown in Fig. 2; also, Fig. 3 shows the typical disposition of the points at which the Fourier samples are known, for a chirp signal. The points located on a generic radius correspond to different samples taken within the same pulse length, whereas different radii correspond to different pulses.

The above derivation is similar to that presented in [4] for the spotlight SAR scenario.

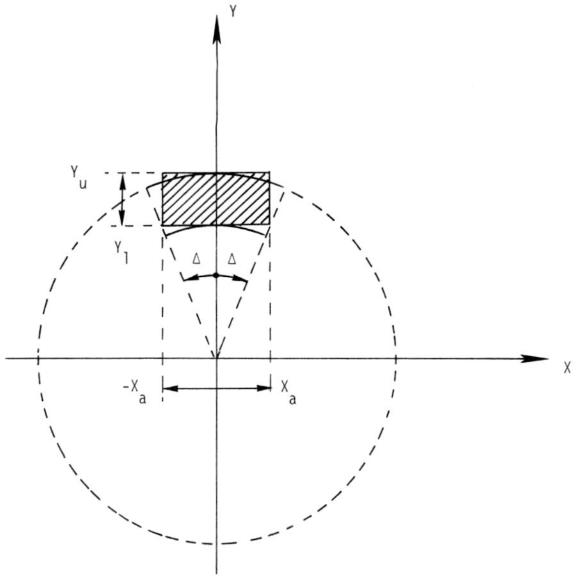

Fig. 2

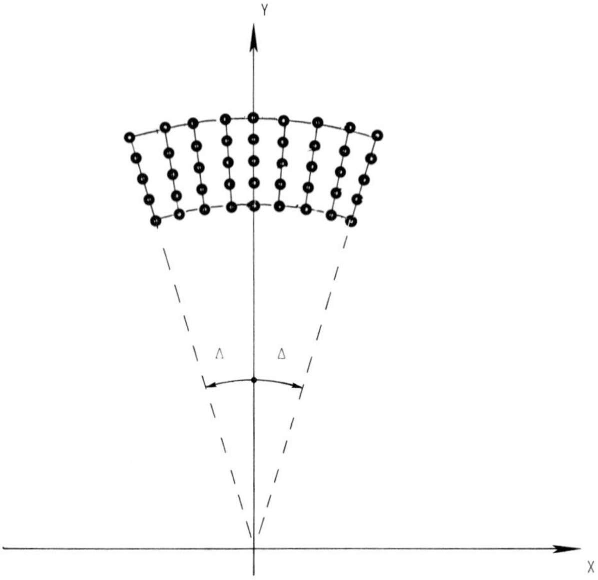

Fig. 3

3.3 Resolution

As is seen in the previous sections, resolution along range and cross-range directions can be achieved if the region of the (X,Y) plane in which the samples of the two dimensional F.T. of f(x,y) are available exhibits non-zero dimension respectively along the X and Y axes. Accordingly, both monochromatic and chirp signals can achieve cross-range resolution, since cross-range resolution essentially depends on the average transmitted frequency and the observation angle 2Δ, and can be shown to have a very weak relation with the signal bandwidth. If one defines the resolution r_x as the distance between the main peak and the first null of the sinc function in (3.7), it follows that:

$$r_x \simeq \lambda/4\sin\Delta \tag{3.15}$$

Use of a chirp signal permits to obtain resolution also along the range direction, since in this case bandwidth is the parameter that matters. An approximate expression for the resolution achievable r_y is derived in [3]; as expected, r_y turns out to be inversely proportional to the exploited bandwidth k_1B(see (3.14)):

$$r_y \tilde{=} c/2k_1B \tag{3.16}$$

Using (3.15) and (3.16) one obtains, for example, for f_o=3 GHz and $\Delta=3^\circ$, an azimuthal resolution of 0.47 m independent of the range, and a range resolution of 3m for k_1B=100 MHz.

4. EFFECTS OF THE STRAIGHT ISO-RANGE APPROXIMATION

In this section the effects of the straight iso-range approximation are investigated in terms of the distortions introduced into the image of an ideal, isolated scatterer.

Most of the mathematics is missing in this presentation; the interested reader is referred to the technical report [8] for more details. The analysis will be carried out under the assumptions that the signal is a chirp, range compensation is exact and the target motion is rectilinear and centred on $\psi=90^\circ$; without loss of generality, $\theta(t)$ will be also supposed to be zero. It is worth noting that the assumption of a wideband signal is not restrictive since the monochromatic signal can be thought of as the limiting case of a chirp (as the frequency rate approaches zero).

As can be easily derived from eq. (3.10), the received signal without use of the straight iso-range approximation can be expressed as:

$$s_R(t) = \iint_{\text{object}} f(x,y)\exp\{-j[xX(t)+yY(t)]\}\exp[-jq(t)\varepsilon(x.y)]dx\,dy \tag{4.1}$$

where X(t), Y(t) and q(t) are defined by eqs. (3.11)-(3.12). The product $q(t)\varepsilon(x,y)$ can be expressed in terms of X and Y recalling [3] that the error in the straight iso-range approximation is given approximately by:

$$\varepsilon(u,v) \tilde{=} v^2/[R(t)+R(u,v)+u] \tag{4.2}$$

It is now assumed that during the observation interval the denominator of (4.2) can be regarded as a constant and can be approximated by some distance value 2R, relative to some intermediate return (such as assumption could be objectionable for some types of motion), that is:

$$\varepsilon(u,v) \tilde{=} v^2/(2R) \tag{4.3}$$

Using (4.3) and passing from the (u,v) to the (x,y) coordinate system enables one to express the received signal in the following form:

$$s_R(t) = \iint_{\text{object}} f(x,y)\exp\{-j[xX(t)+yY(t)]\}\exp\{-j[yX(t)-xY(t)]^2/[2Rq(t)]\}dx \tag{4.4}$$

Recalling that $q(t) = \sqrt{X^2(t)+Y^2(t)}$, one is led to:

$$s_R(X,Y) = \iint_{\text{object}} f(x,y)\exp[-j(xX+yY)]\exp\{-j(yX-xY)^2/[2R\sqrt{X^2+Y^2}]\}dx\,dy \quad X, Y \in \text{(set of available observations)} \tag{4.5}$$

where the dependence of X and Y on time has been dropped and it has been pointed out that the domain in which $s_R(X,Y)$ is known depends on the available observations.

The distortive effect due to the second exponential in the integrand of eq. (4.5) can be simply evaluated if the following assumptions are made:

a) the target is a single ideal scatterer of coordinates (x_o,y_o), that is: $f(x,y)=h\delta(x-x_o,y-y_o)$, where h is normally a complex constant.

b) the X-Y plane domain in which the samples of F(X,Y) are known is sufficiently small to permit F(X,Y) to be approximated by its Taylor series expansion truncated at the quadratic terms;

c) the above domain is approximately rectangular.

Assumption a) is too crude a model to represent practical targets, but has theoretical significance, however, since it permits to derive what could be called the imaging system impulse response, i.e., the response to a single point scatterer.

Assumptions b) and c) are usually acceptable since the chirp signal has narrow relative band and the observation sector is restricted, for practical reasons, to a few degrees.

The above assumptions lead to the following expression of the estimate f(x,y) of the target reflectivity function:

$$\hat{f}(x,y)=G\ \text{sinc}\,[(\Omega/c)(y-y_d)].L(x)\ , \tag{4.6}$$

where:

$\Omega \triangleq 2\pi B$ is the angular frequency deviation of the chirp

$$G \triangleq (h/4\pi^2)(\Omega/c)\exp[-j(y-y_d)Y_c],\ Y_c \triangleq (Y_u+Y_\ell)/2, \tag{4.7}$$

$$y_d \stackrel{\Delta}{=} y_o + x_o^2/(2R) \tag{4.8}$$

$$L(x) \stackrel{\Delta}{=} \int_{-X_a}^{X_a} \exp\{j[\,(x-x_d) - \frac{2y_o^2-x_o^2}{4RY_c}\,X]X\}dX \tag{4.9}$$

$$x_d \stackrel{\Delta}{=} x_o - x_o y_o/R \tag{4.10}$$

$X_a \stackrel{\Delta}{=} Y_\ell \sin\Delta$ is the integration limit along the x axis (see fig. 2).

As is seen from (4.6), as far as the various approximations and assumptions hold, the reflectivity function is proportional to the product of two functions
x alone, the other on y alone.

The part depending on y exhibits a spread due to the bandwidth finiteness of the functions used in the inverse transformation, and is maximum at the ordinate y_d defined by (4.8). Thus the scatterer position appears shifted along the positive direction of the y axis by the amount $x_o^2/2R$. The width of the main lobe of the sinc function is not affected by the scatterer position and hence the range resolution remains to a good approximation that given by (3.16). From (4.8) and (3.16) it could be possible to derive some constraints on the maximum allowable size of the target. Unfortunately, it is not possible to go deeper into this point due to the space limitations; anyway, a detailed analysis can be found in [8].

The portion of f(x,y) depending on x is more difficult to deal with, since it requires the evaluation of the integral specified in (4.9). It is seen that one of the effects of the straight iso-range approximation is to introduce a quadratic term in the exponent of the integrand function. If such a term were missing or negligible, L(x) would simply reduce to the form: sinc $[\,(x-x_d)X_a)\,]$, which peaks at the abscissa x_d defined by (4.10). Thus the scatterer position appears shifted also along the x-axis direction by the amount: $-x_o y_o/R$.

Constraints posed on the amount of this shift entail additional bounds to be put on the target dimensions; details can be found in [8].

As an example, fig. 4 shows the system response to a single scatterer of coordinates (50m, 30m); the image peak is shifted along the x and y axes by the amounts predicted by (4.8) and

(4.10). True scatterer position and the response peak are also indicated as intersections of segments of orthogonal straight lines. The peak emerging in the lower part of the figure is an aliased replica of the main peak due to the low sampling rate used in the simulation, and should be ignored.

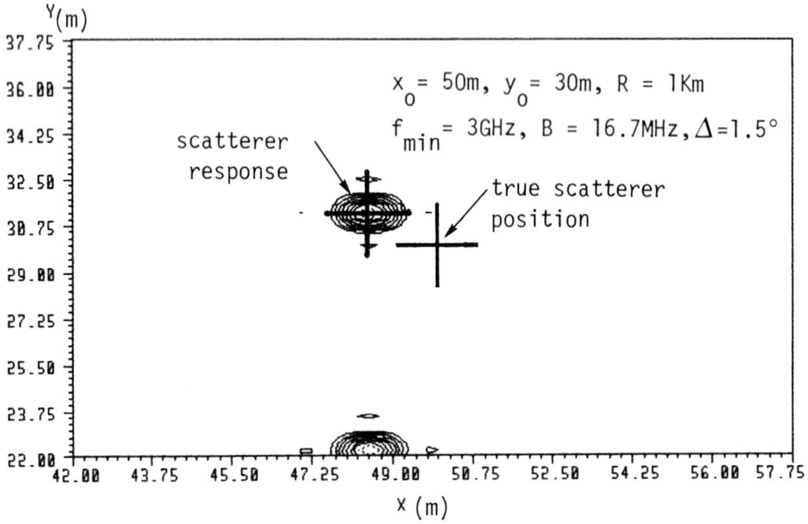

Fig. 4

Fig. 5 illustrates in a qualitative fashion the type of distortion a cartesian grid (thin line) is subject to. The reconstructed image (thick line) appears to be compressed in the upper part of the figure and expanded in the lower part; moreover, straight lines originally parallel to the x axis are transformed into upward concavity curves.

It is worth noting that the expression (4.10) of the shift along the x-axis has been derived in the assumption that the quadratic term at the exponent in the integrand of (4.9) has negligible effect in the evaluation of the integral; this requires, of course, that further conditions be associated with the results of the previous simplified analysis.

To be exact it would be necessary to evaluate the integral specified by (4.9); this is done in [7] where some numerical results are also provided. An alternative approximate approach to the evaluation of the integral (4.9) can also be found in [7]; it is shown that L(x) can be put in the form:

$$L(x)=2X_a\{\text{sinc}\,[\,(x-x_d)X_a\,]+E(x,x_d)\}\quad , \qquad (4.11)$$

where $E(x,x_d)$ represents an error term affecting the sinc function.

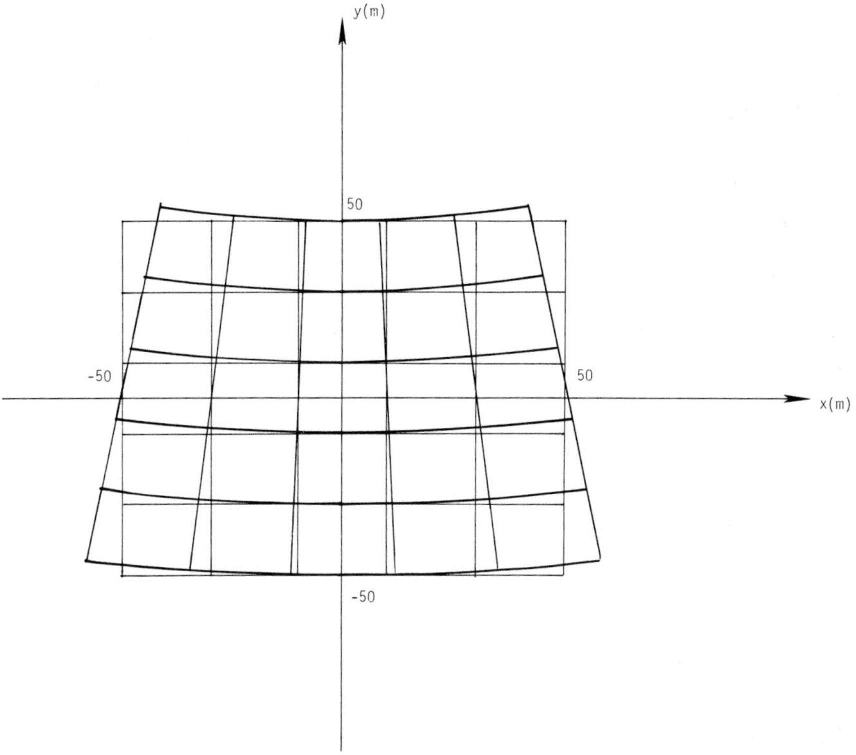

Fig. 5

It can be shown that the additive term $E(x,x_d)$ causes no further shifts in the system response, but only a spread of its main lobe, provided that the modulus of $E(x,x_d)$ is not large with respect to the sinc term. The presence of the term $E(x,x_d)$ in (4.11) implies, however, that there will be an error in the system gain, that is, in the response at the point (x_d,y_d), and also leakage from the nominal resolution cell into the adjacent ones. It can be shown that if the signal bandwidth and the target range are fixed, then any bound put on the modulus of

$E(x,x_d)$ requires that target size also be bounded; conditions can be identified under which the distortions which affect the system response to a single scatterer can simply be thought to consist of a shift (whose components are governed by eqs. (4.8) and (4.10)) plus a spread of the response peak whose extent is directly related to the two dimensional bandwidth available.

5. CONCLUSIONS

In this paper the equations necessary to illustrate how an image can be obtained by the coherent processing of returns for different relative orientations of target and radar are developed for a number of different transmitted signal formats. The various assumptions and limitations necessary to obtain an image are discussed and for each case the resolution performance attainable is indicated.

Further, the effects of the straight iso-range approximation are investigated in terms of the distortions which arise in the reconstructed image as signal bandwidth, observation angle and target distance and dimensions are varied. The case of a single scatterer has been selected due to its simple mathematical tractability.

It is pointed out that if the target dimensions are properly bounded, then the distortions can be regarded as mere shifts and sinc-function-like spreads of the image points; expressions for these distortions are provided.

REFERENCES

[1] Chen, C.C. and Andrews, H.C., (1980) "Target Motion Induced Radar Imaging", *Trans. IEEE AES, vol.. 16,* 1, pp. 2-14.

[2] Mensa, D.L., (1981) "High Resolution Radar Imaging", Artech House Inc.

[3] Cantoni, A. and Corsini, G., (1984) "Spatially Coherent Processing of Radar Returns", Technical Report 84SEL/3, Instituto di Elettronica e Telecomunicazioni, University of Pisa.

[4] Munson, D.C., O'Brien, J.D. and Jenkins, W.K., (1983) "A Tomographic Formulation of Spotlight-mode Synthetic Aperture Radar", *Proc. IEEE, vol. 21,* 8, pp. 917-925.

[5] Walker, J.L., (1980) "Range-Doppler Imaging of Rotating Objects", *Trans. of IEEE AES, vol. AES-16,* pp. 23-51.

[6] Mitchell, R.L., (1974) "Models of Extended Targets and their Coherent Radar Image", *Proc. IEEE,* vol. 62, pp. 254-258.

[7] Cantoni, A., Corsini, G. and Reggiannini, R., (1984) "The Spatially Coherent Processing of Radar Returns - Motion Compensation and Tracking Errors", Technical Memorandum 2, Institute di Elettronica e Telecomunicazioni, University of Pisa.

[8] Cantoni, A. and Reggiannini, R., (1985) "Analisi della approssimazione" straight isorange" per segnalazioni a larga banda", Rapporto Tecnico 85SEL/6, Istituto di Elettronica e Telecomunicazioni, University of Pisa.

FUZZY CLASSIFICATION IN SIGNAL PROCESSING

J.T. Kent and K.V. Mardia
(Department of Statistics, University of Leeds)

ABSTRACT

Recently there has been considerable statistical work on the spatial classification of signal data when the pixels are pure. In this paper, we extend this work to include mixed pixel data. The approach we use leads to estimates of a fuzzy classification rather than of a hard classification. A logistic autonormal process is proposed to model the membership process. Using this model, a Bayesian estimate of the fuzzy classification is given. The results of some simulations are presented to study the properties of our approach. For illustrative purposes the discussion here is limited to real-valued signals on a one-dimensional lattice. However, in practice, the extensions to higher dimensional signals on a higher dimensional lattice, which can be made in a natural way, will be more useful.

1. INTRODUCTION

Let $x_1, x_2, \ldots, x_n$ be observed real-valued signals in one dimension corresponding to n 'pixels' at the regularly spaced sites $1, \ldots, n$. Suppose there are k classes, $C_1, \ldots, C_k$, to which each site can belong. To simplify the discussion we shall take $k = 2$, but the mathematical presentation can be extended. We can describe the membership at each site using an indicator variable,

$$y_i = 0 \text{ if the } i^{th} \text{ pixel belongs to } C_1,$$
$$= 1 \text{ if the } i^{th} \text{ pixel belongs to } C_2.$$

The observed process $\{x_i\}$ will depend in some way on the membership process $\{y_i\}$, and the objective in spatial discriminant analysis is to estimate the $\{y_i\}$ process from the

$\{x_i\}$ process. Examples are given in Switzer (1980, 1983), Mardia (1984) and Besag (1986).

The purpose of this paper is to consider a modification to this problem. We shall replace the pure or hard membership process described above by a "fuzzy" membership process satisfying

$$0 \leq y_i \leq 1 \quad , \quad i=1,\ldots,n.$$

Thus each pixel may now be "mixed" and so proportionally belong to both classes. Models for the observed signal and for the membership process will be considered in the next sections.

Note that for interpretative purposes a fuzzy classification can be reduced to a hard classification by the transformation

$$0 \leq y_i < 1/2 \quad \Rightarrow \quad i^{th} \text{ pixel belongs to } C_1$$
$$1/2 < y_i \leq 1 \quad \Rightarrow \quad i^{th} \text{ pixel belongs to } C_2 \ ,$$

with ambiguity when $y = 1/2$.

2. A STATISTICAL MODEL

Let us assume that μ_j is the true mean for pure signals from the jth class, $j=1,2$,. If y_i is known and represents the proportion of pixel i belonging to class C_1, the mean signal for pixel i is

$$\mu_1 y_i + \mu_2(1-y_i).$$

This suggests assuming that the observed signals are distributed as

$$x_i \sim N(\mu_1 y_i + \mu_2(1-y_i),\ \sigma^2), \tag{2.1}$$

given $(y_1,\ldots y_n)$, where σ^2 represents the magnitude of the noise. Here $N(\mu,\sigma^2)$ denotes the normal distribution with mean μ and variance σ^2. Also, given $\underset{\sim}{y} = (y_1,\ldots,y_n)'$, we will suppose that the noise is uncorrelated between pixels, so that given $(y_1,\ldots,y_n)$,

$$x_1,\ldots,x_n \text{ are independently distributed.} \tag{2.2}$$

Further, we also need to model the membership process $\{y_i\}$. The simplest type of process to work with is a Gaussian process. However, the y_i are restricted by $0 \le y_i \le 1$, so that the proportions themselves cannot have a normal distribution. Instead we use the logit transformation to define

$$v_i = \text{logit}(y_i) = \log\{y_i/(1-y_i)\}, \quad i = 1,\ldots,n. \qquad (2.3)$$

Note that $v_i = -\infty$ for $y_i = 0$ and $v_i = \infty$ for $y_i = 1$, so that $v_i = -\infty$ implies that the ith pixel belongs to C_1 and $v_i = \infty$ implies that it belongs to C_2. To get a proper stochastic model we assume that these extreme values are unattainable, so $0 < y_i < 1$ for all i.

As a model for the $\{v_i\}$ process we shall suppose that

$$\{v_i\} \text{ is a stationary Gaussian Markov random field (MRF).} \qquad (2.4)$$

This process allows the presence of a non-trivial auto-correlation structure to reflect the fact that nearby sites have similar structure. Details of this model will be discussed in the next section. We write throughout

$$\underset{\sim}{x} = (x_1,x_2,\ldots,x_n)', \quad \underset{\sim}{y} = (y_1,y_2,\ldots,y_n)', \text{ and } \underset{\sim}{v} = (v_1,v_2,\ldots,v_n)'.$$

The assumptions above allow us to derive the posterior distribution of $\underset{\sim}{y}$ regarding the signal $\underset{\sim}{x}$ as data. Hence, the posterior mode of the density of $\underset{\sim}{v}$ given $\underset{\sim}{x}$ can be obtained. We defer details until §4.

The modelling approach of this section is an extension to the mixed pixel framework of the pure pixel models of Besag (1983, 1986) and Geman and Geman (1984).

3. DETAILS OF THE MODEL

A Gaussian Markov random field is specified by giving the conditional mean and variance of the process at each site given its value at all other sites. In the simplest model we assume

$$v_i|\text{rest} \sim N(\alpha + \beta\{(v_{i-1}-\alpha) + v_{i+1}-\alpha)\},\tau^2) \qquad (2.5)$$

where $|\beta| < 1/2$. At boundaries we take $v_o = v_{n+1} = \alpha$.

The conditional mean depends here only on the two nearest neighbours. For details of such models see Bartlett (1975, pp. 23-24) and Besag (1974).

Markov random field models have been used before to model images (Woods, 1972, and Chellappa and Kashyap, 1982). However, in our case it is not the true image y_i, $i=1,\dots,n$ which follows a Gaussian process, but instead the nonlinear transformation $v_i = \text{logit}(y_i)$, $i=1,\dots,n$. Hopefully this approach will provide a more realistic modelling strategy for mixed classification problems.

One question which arises is whether such processes generate realistic patterns for certain values of α, β and τ^2. Some simulated patterns are given in Fig. 1 where the parameter β is replaced by $\lambda = (1-\sqrt{1-4\beta^2})/2\beta$. These patterns were generated by an algorithm for the continuous state space based on Geman and Geman (1984) and Metropolis et al., (1953). Further, these simulations were conditional in the sense that before assigning grey levels, we used the $v_i - \bar{v} + \alpha$ in place of the v_i. To plot the results in Figure 1, we took $y_i < 0.2$ as black, $0.2 \le y_i \le 0.8$ as grey and $y_i > 0.8$ as white.

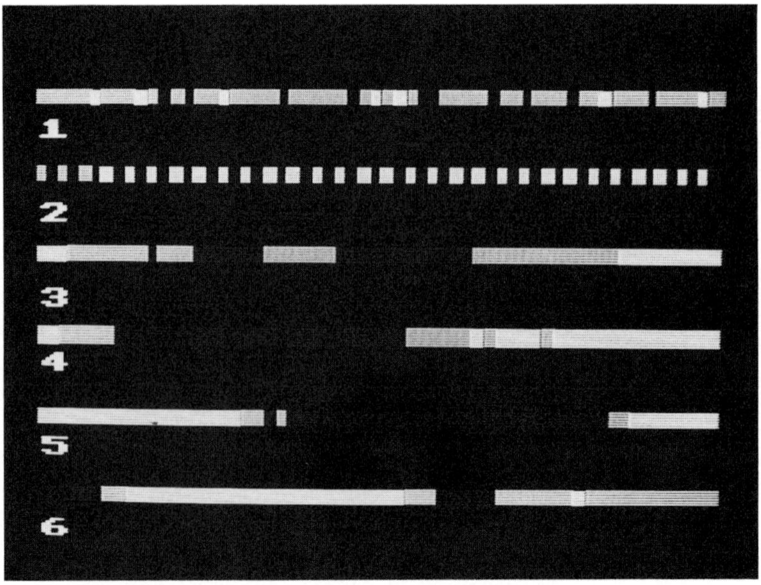

Fig. 1 Patterns from the fuzzy membership process n=60. For patterns 1-5, $\alpha=0$; for pattern 6, $\alpha=3$. In pattern (1) $\lambda=0$, $\tau^2=4$. (2) $\lambda=-0.9999$, $\tau^2=4$. (3) $\lambda=0.99$, $\tau^2=1$. (4) $\lambda=0.999$, $\tau^2=1$. (5) $\lambda=0.999$, $\tau^2=4$. (6) $\lambda=0.999$, $\tau^2=1$.

Figure 1 illustrates that the parameters α, β and $\sigma_v^2 = \tau^2/(1-4\beta^2)^{1/2}$ have the following interpretations

(i) α = measure of the prevalence of class 2 over class 1

(ii) β = measure of the interaction amongst neighbours

(iii) σ_v^2 = measure of hardness.

That is, a small σ_v^2 indicates a large number of mixed pixels (fuzziness) whereas a large σ_v^2 indicates a large number of relatively pure pixels (hardness). For $\beta = 0$ we have a white noise fuzzy pattern.

Figs. 2a, 2b plot the density of y where

$$v = \text{logit}(y) \sim N(\mu_v, \sigma_v^2), \quad \alpha = \mu_v, \qquad (2.6)$$

so that the probability density function of y is given by

$$f(y) = \{(2\pi\sigma_v^2)^{1/2} y(1-y)\}^{-1} \exp -\tfrac{1}{2}\{[\text{logit}(y) - \mu_v]^2/\sigma_v^2\}, \quad 0 < y < 1,$$
$$= 0 \qquad \text{otherwise.}$$

We have $f(0+) = f(1-) = 0$. The density has a stationary point at $y = 1/2$ for $\mu_v = 0$. For $\sigma_v^2 < 2$, the density is unimodal on (0,1) with mode at $y = 1/2$. In contrast, for $\sigma_v^2 > 2$, we get for some μ_v a U-shaped distribution with modes near 0 and 1 and a local minimum at $y = 1/2$. This illustrates why σ_v^2 is a measure of hardness. Note that the unimodal distribution (2.5) will have large mean but small variance for nearly pure pixels; the large mean will push the distribution towards the hard classification, but τ^2 (σ_v^2) will still allow some fuzziness. Of course, this logisitic normal distribution is well known (Johnson, 1949), but its use in classification problems is new.

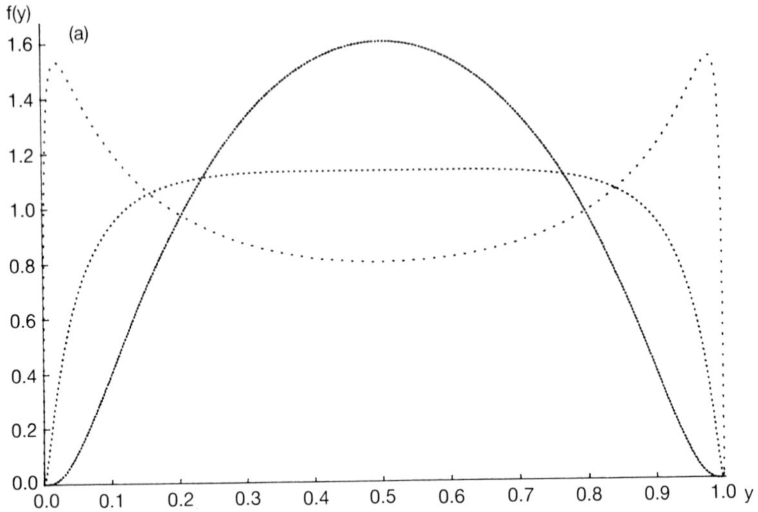

Fig. 2a. The logistic-normal density for $\mu_v = 0$, $\sigma_v^2 = 1$ (———) $\sigma_v^2 = 2$(....), $\sigma_v^2 = 4$ (· · ·).

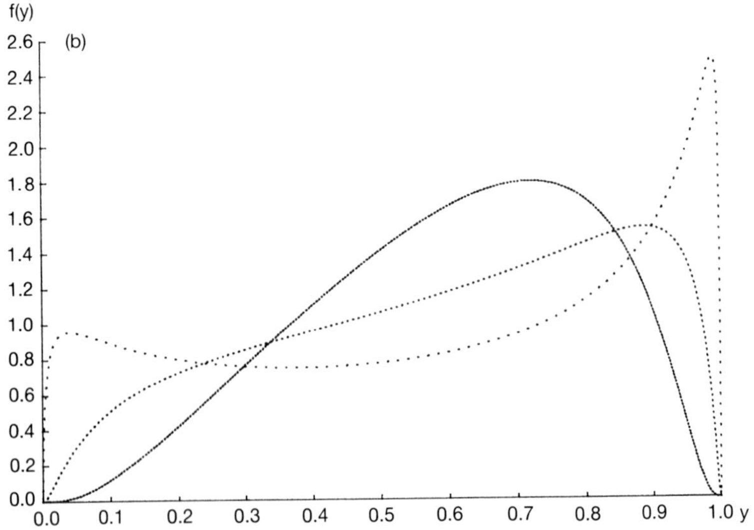

Fig. 2b. The logistic-normal density for $\mu_v = .5$, $\sigma_v^2 = .1$ (———) $\sigma_v^2 = 2$(.....), $\sigma_v^2 = 4$(· · · ·).

We now give a much stronger result concerning behaviour as $\sigma_v^2 \to \infty$, namely that in this limit the fuzzy classification tends to a hard classification, i.e. $y_i = 0$ or 1.

Let $\Phi(.)$ be the distribution function corresponding to a N(0,1) variate. Without any loss of generality, consider (2.6) only. If μ_v is fixed,

$$\lim_{\sigma_v \to \infty} \Pr(a < v < b) = \lim[\Phi(\frac{b-\mu_v}{\sigma_v}) - \Phi(\frac{a-\mu_v}{\sigma_v})] = 0.$$

Thus we observe in the limit $v = \infty$ or $v = -\infty$, each with probability 1/2. More generally, if $\mu_v/\sigma_v \to c$ as $\mu_v \to \infty$ and $\sigma_v \to \infty$ then

$$\lim \Pr(v = \infty) = \Phi(c) = \lim \Pr(y = 1).$$

Thus for larger values of c, we expect more signals from C_2. Again this is confirmed for patterns in Fig. 1.

4. A BAYESIAN SOLUTION

Regarding $\underset{\sim}{v}$ as a Bayesian parameter vector with the prior auto-normal process given by (2.4) (or (2.5) in our simple case) we can write the posterior density of $\underset{\sim}{v}$ given $\underset{\sim}{x}$. We wish to find the mode of this posterior density. The mode satisfies the following equation:

$$\tau^2\{x_i - \mu_1 y_i - \mu_2(1 - y_i)\}(\mu_2 - \mu_1)y_i(1 - y_i)$$
$$+ \sigma^2[v_i - \{\alpha + \beta(v_{i-1} - \alpha) + \beta(v_{i+1} - \alpha)\}] = 0. \tag{4.1}$$

To solve (4.1) for the mode, we proceed as follows. If y_i^* is near y_i and we let $v_i^* = \text{logit}(y_i^*)$, then a first order Taylor series expansion gives

$$y_i \simeq y_i^* + y_i^*(1-y_i^*)(v_i - v_i^*). \tag{4.2}$$

After a little manipulation, we get an iterative formula for calculating the mode from (4.1) and (4.2). Namely, given $v_i^{(r-1)}$, $1 \leq i \leq n$, set

$$v_i^{(r)} = v_i^* + [\tau^2 y_i^{*2}(1-y_i^*)^2(\mu_1-\mu_2)^2 + \sigma^2]^{-1} \times [\tau^2 y_i^*(1-y_i^*)(\mu_1-\mu_2)$$

$$\{x_i - y_i^*\mu_1 - (1-y_i^*)\mu_2\} - \sigma^2(v_i^* - \alpha) + \sigma^2\beta(v_{i-1}^* + v_{i+1}^* - 2\alpha)] , \tag{4.3}$$

where $y_i^* = y_i^{(r-1)}$, $v_i^* = v_i^{(r-1)}$, $i = 1,..,n$, and we interpret $v_i^{(r-1)} = 0$ for $i = 0, n+1$. Suitable initial values are $v_i = \alpha$. For $\alpha = \beta = 0$ and large τ, it can be seen that the above classification becomes, for $\mu_1 < \mu_2$,

if $x_i > (\mu_1 + \mu_2)/2$ assign the ith pixel to C_2,

otherwise assign the ith pixel to C_1. This is an important property.

The above solution extends immediately for more complicated neighbourhood structure in (4.1).

5. NUMERICAL COMPARISONS

We now compare the fuzzy classification method with other methods which have been mainly developed for hard pictures. The aim is to assess whether the fuzzy method will be comparable to these hard classification methods when the fuzzy classification is hardened.

Let us take $n = 60$. Some simple deterministic patterns are generated by alternating c observations from $C_1 : N(\mu_1,\sigma^2)$ with c observations from $C_2 : N(\mu_2,\sigma^2)$ until the 60 sites are filled. Here we refer to c as the cluster size and we take $c = 1,5,6,10$ and 12. Hence the true membership process is hard a priori, and therefore we might expect the hard methods to perform better. The mean values μ_1 and μ_2 ($\mu_1 < \mu_2$) and the variance σ^2 are treated as known here.

In assessing the following methods, the two end points are ignored.

(i) Naive method. Assign the ith site to C_1 if $x_i < (\mu_1+\mu_2)/2$.

Otherwise the site is assigned to C_2.

This method ignores any spatial/neighbourhood information and has error rate

$$E = \Phi\{-\tfrac{1}{2}|\mu_1 - \mu_2|/\sigma\}.$$

(ii) Mean method. Assign the ith site to C_1 if

$$\bar{x}_i = (x_{i-1} + x_i + x_{i+1})/3 < (\mu_1 + \mu_2)/2,$$

otherwise assign to C_2.

(iii) Median method. Use the median in place of the mean in (ii).

(iv) Combined method. (Mardia and Gill, 1986)
If $\sum_{s=-1}^{1} (x_{i+s} - \bar{x}_i)^2/\sigma^2 > \chi^2_{2,\alpha}$ at $\alpha = 0.05$, we use (i), otherwise we use (ii). Here $\chi^2_{2,\alpha}$ denotes the upper α critical value of the χ^2_2 distribution.

(v) Besag's hard method. Use for $y^*_i = 0,1$,

$$\Pr(y^*_i | y^*_{i-1}, y^*_{i+1}) =$$
$$\exp\{y^*_i(-\alpha^* + \beta^*(y^*_{i-1} + y^*_{i+1}))\}/\{1 + \exp[-\alpha^* + \beta^*(y^*_{i-1} + y^*_{i+1})]\}. \qquad (5.1)$$

Here $y^*_i = 0$ or 1 implies the ith site belongs to C_1 or C_2, respectively. Note that if $\alpha^* = \beta^*$, then the pattern is symmetric in the two classes C_1 and C_2.

(vi) The fuzzy method. We take the model of §4, with $\alpha = 0$ a priori. The partitions are hardened by assigning the ith site to C_1 if $y_i > 1/2$, otherwise we assign to C_2. It can be shown that as $\sigma_v \to \infty$ for fixed β the fuzzy membership process converges to a hard membership process. The first-order conditional probabilities can be matched to those in (5.1), giving

$$\beta^* = \operatorname{logit}(\gamma), \qquad (5.2)$$

where $\gamma = \frac{1}{2}[\pi + 2(2\sin^{-1}\lambda + \sin^{-1}\lambda^2)][\pi + 2\sin^{-1}\lambda^2]^{-1}$

and $\lambda = \{1 - (1 - 4\beta^2)^{1/2}\}/2\beta, \qquad |\lambda| < 1.$

Table 1 gives the results of 600 simulations for $\mu_1 = 0$, $\mu_2 = 2$, $\sigma^2 = 1$. We have calculated the exact classification error for the mean method from the formula

$$\frac{1}{n-2}\left[2\left(\frac{n}{c}-1\right)\Phi\left(-\frac{\sqrt{3}|\mu_1-\mu_2|}{6\sigma}\right)+\frac{n}{c}(c-2)\Phi\left(-\frac{\sqrt{3}|\mu_2-\mu_1|}{2\sigma}\right)\right].$$

The accuracy of these methods can be roughly ordered as fuzzy $\geq$ hard $\geq$ mean $\geq$ meadian $\geq$ combined.

However, Besag's hard method and the fuzzy method require optimum selection of parameters. These have to be selected a priori in the right direction, e.g. for the fuzzy method, we have to take $\beta \approx -0.49$ for the repulsion/attraction case and $\beta \approx 0.49$ for the clustering case (and $\tau^2 \approx 1$ in both cases) to get acceptable results. We note that iterations sometimes show small oscillations for the fuzzy and hard methods - in such cases we classify by the iteration which maximises the posterior probability. It is interesting to note that for $\beta^* = -1.5, 1.5, 3$ in Table 1, we get the corresponding values of β from (5.2) as -.498, .43 and .498. These values of β compare well with those in Table 1.

The same conclusions emerge when we increase σ^2 and thus increase the error rate. It is pleasing to note that for hard pictures, the fuzzy method is comparable with the above methods especially designed for hard pictures.

6. DISCUSSION

First, note that all this work can be generalised to multivariate signal data from multi-dimensional spatial co-ordinates, although more technical work is required. Switzer (1980) gives an example of remotely sensed data with different varieties of rocks in two dimensions with four variates. Further details and examples are given in Kent and Mardia (1986). Secondly, in practice the parameters α, β, τ^2, etc. are unknown which raises the question of how best to estimate them. Further, we have made assumptions of normality which may not be valid for some sets of data. However, given previous training data, we may be able to transform to normality. Note, in §5, that the error of classification may not be the best way of assessing various methods if the objective is to identify some sort of pattern in the observed signal.

Cluster Size	Mean	Med.	Comb.	Hard(α^*,β^*)	Fuzzy (β,τ^2), α = 0
1	.72	.75	.52	.068 (-1.5,-1.5)	.001 (-.499,.1)
5	.14	.14	.12	.13 (1.5,1.5)	.11 (.45,1)
6	.12	.13	.11	.13 (1.5,1.5)	.097 (.45, 1)
10	.087	.11	.093	.085 (3,3)	.066 (.49, 1)
12	.073	.086	.081	.084 (3,3)	.049 (.49, 1)

Table 1

Comparison of classification errors of various methods for clusters of different sizes with 600 simulations, n = 60, $\mu_1 = 0$, $\mu_2 = 2$, $\sigma^2 = 1$ and E = 0.16. Overall simulation standard error of each entry = 0.003. Parameters for hard and fuzzy classifications are in their optimal range.

7. ACKNOWLEDGEMENTS

The authors are grateful to Alan Watkins for his help. Professor Mardia would also like to acknowledge the support from the ONR Contract NOOO 14-84-K-0421 to the Department of Statistics, Princeton University; he also benefited from a Research-Workshop in Edinburgh supported by the SERC.

8. REFERENCES

Bartlett, M.S., (1975) The Statistical Analysis of Spatial Pattern, Chapman and Hall, London.

Besag, J., (1974) Spatial Interaction and the Statistical Analysis of Lattice Systems (wtih Discussion), *J. Royal Statist. Soc. B,* **36**, 192-236.

Besag, J., (1983) Spatial Discriminant Analysis. Discussion to Switzer, P., *Bull. of the Inter. Statist. Inst.* **50**, 962-972.

Besag, J., (1986) On the statistical analysis of dirty pictures. (with discussion). *J. Royal Statist. Soc., B,* to appear.

Chellappa, R. and Kashyap, R.L., (1982) Digital Image Restoration Using Spatial Interaction Models, *IEEE Trans. Acoust., Speech, Signal Proc.,* ASSP-30, 461-472.

Geman, S. and Geman, D., (1984) Stochastic Relaxation, Gibbs Distributions and the Bayesian Restoration of Images, *IEEE Trans. PAMI.* **6**, 721-741.

Johnson, N.L., (1949) Systems of Frequency Curves generated by Methods of Translation, *Biometrika* **36**, 149-176.

Kent, J.T. and Mardia, K.V., (1986) Spatial Classification using Fuzzy Membership Models. To appear.

Mardia, K.V., (1984) Spatial Discrimination and Classification aps. *Commun. Statist.* **13**, 2181-2197.

Mardia, K.V. and Gill, C.A., (1986) A Combined Rule for Classification Maps. To appear.

Metropolis, N., Rosenbluth, A.W., Rosenbluth, M.N., Teller, A.H., Teller, E. (1953) Equations of State Calculations by Fast Computing Machines, *J. Chem. Phys.* **21**, 1087-1091.

Switzer, P., (1980) Extension of Linear Discrimination Analysis for Staistical Classification of Remotely Sensed Satellite Imagery, *J. Inter. Assoc. Mathl. Geol.* **12**, 367-376.

Switzer, P., (1983) Some Spatial Statistics for the Interpretation of Satellite Data (with Discussion), *Bull. of the Inter. Statist. Inst.* **50**, (Book 2), 962-972, **50**, (Book 3) 409-426.

Woods, J.W., (1972) Two-Dimensional Discrete Markovian Fields, *IEEE Trans. Inform. Theory*, IT-18, 232-240.

THE PRINCIPLE OF DYADIC SYMMETRY AND THE GENERATION OF SIMPLE, HIGH EFFICIENCY TRANSFORMS FOR IMAGE PROCESSING

R.J. Clarke
(Department of Electronic and Electrical Engineering, Heriot-Watt University, Edinburgh)

and

W.K. Cham
(Science Centre, Chinese University of Hong Kong)

ABSTRACT

Transform operations are frequently applied in the field of image processing for such purposes as pattern recognition, filtering and image data compression. A drawback of transform processing at high data rates is, however, the complexity of the mathematical manipulations involved when real valued transforms (for example, the discrete cosine transform) are used to achieve high compression efficiency. If, alternatively, simple binary valued transforms (the Walsh-Hadamard transform, for example) are employed to allow easy processing then the obtainable degree of data compression is significantly smaller, the reason being the poor match between the rectangular shapes of the WHT basis vectors and those of typically occurring image vectors.

In this paper we initially demonstrate how the principle of dyadic symmetry may be employed to define the WHT in a straightforward way, and then show how the associated theory may be applied to allow the replacement of certain WHT basis vectors by others having a restricted set of element values (+/-1, +/-c) to produce a new orthogonal transform matrix with better coding efficiency and yet, employing the value c=1/2, easy implementation. Finally, it is shown how the principle of dyadic symmetry may be applied to decompose the transform matrix in such a way as to allow the development of a fast algorithm for its computation.

1. INTRODUCTION

During the past 15 years or so, the technique of image data manipulation using transform operators has reached a high degree of sophistication and, in the data compression field, it is now possible to code monochrome and colour images at rates in the region of 1 bit per element with wholly acceptable reconstructed quality. When processing image data at very high rates, however, the employment of real valued transforms still poses something of a problem, and this has meant that a compromise is often necessary, and the high degree of compression efficiency which is possible using the discrete cosine transform, for example, has to be sacrificed to allow easy implementation by the use of a poorer operator such as the Walsh-Hadamard transform. It is of interest, then, to attempt to develop transform basis matrices which contain entries having only a restricted range of values, but whose shapes still match those of commonly occurring image data vectors as well as possible in order to maintain compression efficiency (Cham et al. 1984). In the present paper this is carried out by application of the principle of dyadic symmetry which is initially defined and described and then used to develop a conventional WHT matrix in a straightforward way. Subsequently it is employed to develop a new orthogonal transform matrix derived from the WHT matrix but whose element values are either + /-1 or +/-c. From the point of view of implementation it is obviously useful to employ the value c=1/2 and, under that condition, it is found that the new transform has a compression efficiency which lies between that of the WHT and that of the DCT. Finally the principle of dyadic symmetry is applied to demonstrate how the transform matrix may be decomposed in order to generate a fast algorithm for its implementation.

2. THE PRINCIPLE OF DYADIC SYMMETRY

A vector of 2^m elements, $a_0, a_1, \ldots a_{2^m-1}$ is said to have the i'th dyadic symmetry if

$$a_j = s.\ a_{j(+)i} \qquad (2.1)$$

where s takes on the values +1 or -1 for even and odd symmetries respectively, i lies in the range 1 to 2^m-1, and (+) implies the exclusive OR operation (Cham 1983). Taking the example of an 8-vector (m=3) and s = +1 -

if i = 1 (001)

j	0	1	2	3	4	5	6	7
	000	001	010	011	100	101	110	111
j(+)001	001	000	011	010	101	100	111	110
	1	0	3	2	5	4	7	6

and so $a_0 = a_1$, $a_2 = a_3$, $a_4 = a_5$, $a_6 = a_7$, and the vector can be written

$$h_1 \quad h_1 \quad h_2 \quad h_2 \quad h_3 \quad h_3 \quad h_4 \quad h_4$$

neighbouring elements having the same values in pairs.

Similarly if $i = 5$ (101)

j	0	1	2	3	4	5	6	7
	000	001	010	011	100	101	110	111
$j(+)101$	101	100	111	110	011	000	011	010
	5	4	7	6	1	0	3	2

and so $a_0 = a_5$, $a_1 = a_4$, $a_2 = a_7$, $a_3 = a_6$, and the vector can be written

$$h_1 \quad h_2 \quad h_3 \quad h_4 \quad h_2 \quad h_1 \quad h_4 \quad h_3$$

The complete 8th order result is

$$\begin{array}{c} \underline{i} \\ 1 \\ 2 \\ 3 \\ 4 \\ 5 \\ 6 \\ 7 \end{array} \quad \begin{array}{c} \underline{\text{Vector Elements}} \\ \left[\begin{array}{cccccccc} h_1 & h_1 & h_2 & h_2 & h_3 & h_3 & h_4 & h_4 \\ h_1 & h_2 & h_1 & h_2 & h_3 & h_4 & h_3 & h_4 \\ h_1 & h_2 & h_2 & h_1 & h_3 & h_4 & h_4 & h_3 \\ h_1 & h_2 & h_3 & h_4 & h_1 & h_2 & h_3 & h_4 \\ h_1 & h_2 & h_3 & h_4 & h_2 & h_1 & h_4 & h_3 \\ h_1 & h_2 & h_3 & h_4 & h_3 & h_4 & h_1 & h_2 \\ h_1 & h_2 & h_3 & h_4 & h_4 & h_3 & h_2 & h_1 \end{array}\right] \end{array} \qquad (2.2)$$

3. WALSH MATRIX GENERATION

If a vector has dyadic symmetries $I_1, I_2 \ldots I_r$ it also has symmetry I_k where

$$I_k = I_1 \; (+) \; I_2 \; (+) \; . \; . \; . \; I_r \tag{3.1}$$

and so a vector with symmetries 1 and 2 has, in addition, symmetry 3 but no others -

$$\left.\begin{array}{l} 001 \; (+) \; 010 \; = \; 011 \\ 001 \; (+) \; 011 \; = 010 \\ 010 \; (+) \; 011 \; = 001 \end{array}\right\} \tag{3.2}$$

whereas a vector with symmetries 1,3 and 7, for example, has all the possible symmetries. In the first case the symmetries are mutually dependent whilst in the second they are independent, and so can be used to generate a Walsh matrix by appropriate choice of even or odd symmetry as follows, where the presence of even or odd dyadic symmetry 1, 3 or 7 is indicated by '0' or '1' respectively in the column headed 'dyadic symmetry' -

i	dyadic symmetry 001	011	111	Walsh matrix basis vectors							
-	0	0	0	1	1	1	1	1	1	1	1
1	0	0	1	1	1	1	1	-1	-1	-1	-1
2	0	1	0	1	1	-1	-1	-1	-1	1	1
3	0	1	1	1	1	-1	-1	1	1	-1	-1
4	1	0	0	1	-1	-1	1	1	-1	-1	1
5	1	0	1	1	-1	-1	1	-1	1	1	-1
6	1	1	0	1	-1	1	-1	-1	1	-1	1
7	1	1	1	1	-1	1	-1	1	-1	1	-1

(3.3)

To illustrate the above development consider the cases i = 1 and 6. For i = 1 the three symmetries are 1(even), 3(even) and 7(odd). The first implies that neighbouring elements are of the same magnitude and sign -

$$h_1 \quad h_1 \quad h_2 \quad h_2 \quad h_3 \quad h_3 \quad h_4 \quad h_4$$

the second that (see Eq. 2.2) $a_0 = a_3$, $a_1 = a_2$ etc., giving

$$h_1 \quad h_1 \quad h_1 \quad h_1 \quad h_3 \quad h_3 \quad h_3 \quad h_3$$

and the third that $a_0 = -a_7$, $a_1 = -a_6$ etc., giving

$$h_1 \quad h_1 \quad h_1 \quad h_1 \quad -h_1 \quad -h_1 \quad -h_1 \quad -h_1 \tag{3.4}$$

i.e. Walsh basis vector 1 (apart from an overall scale factor).

For i = 6 we have symmetries and relations -

1(odd)	h_1	$-h_1$	h_2	$-h_2$	h_3	$-h_3$	h_4	$-h_4$
3(odd)	h_1	$-h_1$	h_1	$-h_1$	h_1	$-h_1$	h_1	$-h_1$
7(even)	h_1	$-h_1$	h_1	$-h_1$	$-h_1$	h_1	$-h_1$	h_1

(3.5)

i.e., basis vector 6, when scaled by h_1.

Since all vector pairs have at least one similar symmetry, in one case even, in the other odd, their components are mutually orthogonal.

4. NEW TRANSFORM GENERATION

In order to improve the data compression efficiency of the WHT it is necessary to modify the basis vectors to smooth out, as far as possible, any sharp discontinuities which exist between neighbouring elements. In this way the vector shapes will tend to those of the discrete cosine transform (DCT) which has the best efficiency of all data-independent transforms for reasonably high values of inter-element correlation coefficient (those of the DCT themselves become identical with the vectors of the optimum - Karhunen Loève - transform for a first order Markov source as the correlation coefficient tends to unity - (Clarke 1981)). We therefore replace pairs of WHT basis vectors with other more suitable ones, ensuring that at each stage orthogonality is maintained. For effective separation of DC and AC transform components, it is desirable that any transform operator has a constant DC basis vector, and we therefore leave the first unchanged, i.e.,

$$H_0 = 1 \quad 1 \quad 1 \quad 1 \quad 1 \quad 1 \quad 1 \quad 1$$

the next is

$$H_1 = 1 \quad 1 \quad 1 \quad 1 \quad -1 \quad -1 \quad -1 \quad -1$$

which has its step change replaced to give -

$$T_1 = 1 \quad 1 \quad c \quad c \quad -c \quad -c \quad -1 \quad -1$$

where c is a positive constant less than unity.

This vector has even first and odd seventh symmetries but lacks the third (see Eq. 2.2). The other vector with these properties is -

$$H_3 = 1 \quad 1 \quad -1 \quad -1 \quad 1 \quad 1 \quad -1 \quad -1$$

which must therefore be replaced by -

$$T_3 = c \quad c \quad -1 \quad -1 \quad 1 \quad 1 \quad -c \quad -c$$

The third vector is smoothed out as follows -

$$H_2 = 1 \quad 1 \quad -1 \quad -1 \quad -1 \quad -1 \quad 1 \quad 1$$

$$T_2 = 1 \quad c \quad -c \quad -1 \quad -1 \quad -c \quad c \quad 1$$

This vector has odd third and even seventh symmetries but lacks the first, necessitating alteration of H_6 to -

$$T_6 = c \quad -1 \quad 1 \quad -c \quad -c \quad 1 \quad -1 \quad c$$

H_4, as H_0, is unchanged, and H_5 and H_7 are replaced by -

$$T_5 = 1 \quad -1 \quad -c \quad c \quad -c \quad c \quad 1 \quad -1$$

$$T_7 = c \quad -c \quad 1 \quad -1 \quad 1 \quad -1 \quad c \quad -c$$

giving orthogonal transform matrix -

$$[\,HCT\,] = \begin{bmatrix} 1 & 1 & 1 & 1 & 1 & 1 & 1 & 1 \\ 1 & 1 & c & c & -c & -c & -1 & -1 \\ 1 & c & -c & -1 & -1 & -c & c & 1 \\ c & c & -1 & -1 & 1 & 1 & -c & -c \\ 1 & -1 & -1 & 1 & 1 & -1 & -1 & 1 \\ 1 & -1 & -c & c & -c & c & 1 & -1 \\ c & -1 & 1 & -c & -c & 1 & -1 & c \\ c & -c & 1 & -1 & 1 & -1 & c & -c \end{bmatrix} \quad (4.1)$$

Extension to block sizes other than eight is straightforward.

The transform is designated the HCT (high correlation transform) since it is one of a pair of transforms (the HCT and LCT) having improved performance at high and low values of data correlation, respectively. Using a first order Markov process as a test source, it is easy, for various values of correlation coefficient, to determine the effect of varying the value of c on the energy packing performance of the transform. It turns out that, for a wide range of values of correlation, the performance is nearly optimum for $c = 1/2$, which may be implemented by a simple binary shift. Typically, if we retain 25-50% of the transform coefficients the efficiency is closer to that of the DCT than to that of the WHT. Specifically, if in the case of the 8th order transform, three coefficients are retained, the percentage energy remaining in those coefficients is, for the DCT 94.3, for the HCT 93.7 and for the WHT 92.0.

5. FAST TRANSFORM DECOMPOSITION

a) The WHT

Using the first dyadic symmetry, which is even for H_0-H_3, and odd for H_4-H_7, [C], the coefficient array, is given by -

$$[C] = \begin{bmatrix} & & \\ & \text{WHT} & \\ & \text{MATRIX} & \\ & & \end{bmatrix} \begin{bmatrix} x_0 \\ x_1 \\ x_2 \\ x_3 \\ x_4 \\ x_5 \\ x_6 \\ x_7 \end{bmatrix} \tag{5.1}$$

where $[x_0, x_1 \ . \ . \ . \ x_7]^T$ is the data vector. We define

$$[U] = \begin{bmatrix} u_0 \\ u_1 \\ u_2 \\ u_3 \end{bmatrix} = \begin{bmatrix} x_0 + x_1 \\ x_2 + x_3 \\ x_4 + x_5 \\ x_6 + x_7 \end{bmatrix} \qquad [V] = \begin{bmatrix} v_0 \\ v_1 \\ v_2 \\ v_3 \end{bmatrix} = \begin{bmatrix} x_0 - x_1 \\ x_3 - x_2 \\ x_4 - x_5 \\ x_7 - x_6 \end{bmatrix}$$

then -

$$[C] = \begin{matrix} \begin{bmatrix} 1 & 1 & 1 & 1 \\ 1 & 1 & -1 & -1 \\ 1 & -1 & -1 & 1 \\ 1 & -1 & 1 & -1 \end{bmatrix} \begin{bmatrix} u_0 \\ u_1 \\ u_2 \\ u_3 \end{bmatrix} \\ \begin{bmatrix} 1 & 1 & 1 & 1 \\ 1 & 1 & -1 & -1 \\ 1 & -1 & -1 & 1 \\ 1 & -1 & 1 & -1 \end{bmatrix} \begin{bmatrix} v_0 \\ v_1 \\ v_2 \\ v_3 \end{bmatrix} \end{matrix} = \begin{matrix} \begin{bmatrix} 1 & 1 \\ 1 & -1 \end{bmatrix} \begin{bmatrix} u_0 + u_1 \\ u_2 + u_3 \end{bmatrix} \\ \begin{bmatrix} 1 & 1 \\ 1 & -1 \end{bmatrix} \begin{bmatrix} u_0 - u_1 \\ u_3 - u_2 \end{bmatrix} \\ \begin{bmatrix} 1 & 1 \\ 1 & -1 \end{bmatrix} \begin{bmatrix} v_0 + v_1 \\ v_2 + v_3 \end{bmatrix} \\ \begin{bmatrix} 1 & 1 \\ 1 & -1 \end{bmatrix} \begin{bmatrix} v_0 - v_1 \\ v_3 - v_2 \end{bmatrix} \end{matrix} \tag{5.2}$$

which leads to the algorithm shown in Figure 5.1. Redefinition of [U] and [V] according to the chosen dyadic symmetry allows a decomposition for each of the 2^m-1 symmetries.

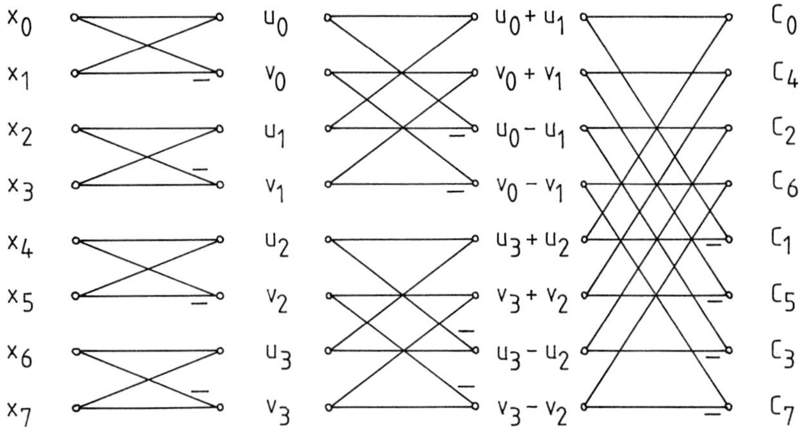

Fig. 5.1 Fast WHT algorithm

b) The HCT

The matrix has seventh even (vectors 0, 2, 4, 6) and odd (vectors 1, 3, 5, 7) dyadic symmetry. Thus we define -

$$[U] = \begin{bmatrix} u_0 \\ u_1 \\ u_2 \\ u_3 \end{bmatrix} = \begin{bmatrix} x_0 + x_7 \\ x_1 + x_6 \\ x_2 + x_5 \\ x_3 + x_4 \end{bmatrix} \qquad [V] = \begin{bmatrix} v_0 \\ v_1 \\ v_2 \\ v_3 \end{bmatrix} = \begin{bmatrix} x_0 - x_7 \\ x_1 - x_6 \\ x_2 - x_5 \\ x_3 - x_4 \end{bmatrix}$$

then

$$[C] = \begin{matrix} \begin{bmatrix} 1 & 1 & 1 & 1 \\ 1 & c & -c & -1 \\ 1 & -1 & -1 & 1 \\ c & -1 & 1 & -c \end{bmatrix} \begin{bmatrix} u_0 \\ u_1 \\ u_2 \\ u_3 \end{bmatrix} \\ \begin{bmatrix} 1 & 1 & c & c \\ c & c & -1 & -1 \\ 1 & -1 & -c & c \\ c & -c & 1 & -1 \end{bmatrix} \begin{bmatrix} v_0 \\ v_1 \\ v_2 \\ v_3 \end{bmatrix} \end{matrix} \qquad \begin{matrix} \begin{bmatrix} 1 & 1 \\ 1 & -1 \end{bmatrix} \begin{bmatrix} u_0 + u_3 \\ u_1 + u_2 \end{bmatrix} \\ \begin{bmatrix} 1 & c \\ c & -1 \end{bmatrix} \begin{bmatrix} u_0 - u_3 \\ u_1 - u_2 \end{bmatrix} \\ \begin{bmatrix} 1 & 1 \\ 1 & -1 \end{bmatrix} \begin{bmatrix} v_0 + cv_3 \\ v_1 + cv_2 \end{bmatrix} \\ \begin{bmatrix} 1 & 1 \\ 1 & -1 \end{bmatrix} \begin{bmatrix} cv_0 - v_3 \\ cv_1 - v_2 \end{bmatrix} \end{matrix} \qquad (5.3)$$

giving the algorithm of figure 5.2.

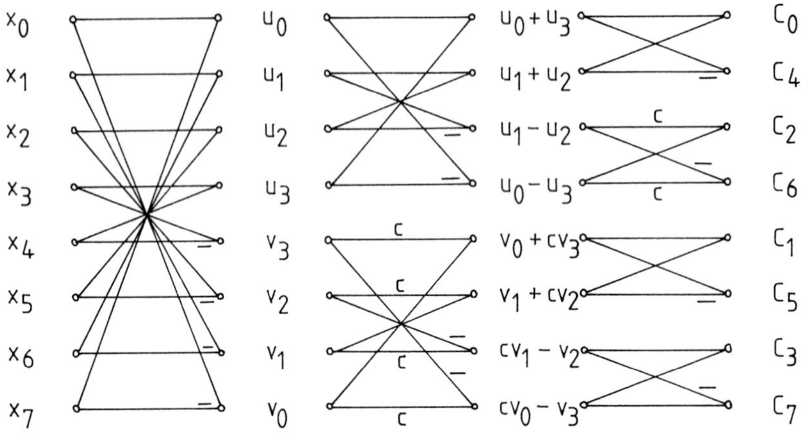

Fig. 5.2 Fast HCT algorithm

6. CONCLUSIONS

It has been shown that, using the principle of dyadic symmetry, it is possible to generate orthogonal transform matrices which are simple to implement and yet have processing efficiencies which are significantly better than those of the WHT of various block sizes. In addition, it has also been demonstrated that the principle allows straightforward generation of conventional WHT matrices and of fast implementations for both transforms.

7. ACKNOWLEDGEMENT

The authors are pleased to acknowledge the financial support provided by British Telecom Research Laboratories, Martlesham, Suffolk, for the work reported here.

8. REFERENCES

Cham, W.K., (1983) "Transform coding of pictorial data", Unpublished doctoral thesis, Loughborough University of Technology.

Cham, W.K., Allott, D. and Clarke, R.J., (1984) "Block classification coding with combined transforms" Proceedings of the International Conference on Acoustics, Speech and Signal Processing, San Diego, Ca., pp. 29-10-1 to 29-10-4.

Clarke, R.J., (1981) "On the relation between the Karhunen-Loève and cosine transforms". *IEE Proceedings,* Part F, Vol. 128, pp. 359-6o.

GOAL DRIVEN PARAMETER EVALUATION FOR THE DETECTION OF OBJECTS IN S.A.R. DATA

D.B. Sharman and T.S. Durrani
(Department of Electronic and Electrical Engineering, University of Strathclyde, Glasgow)

ABSTRACT

This paper tackles the problem of selecting appropriate values for the parameters (features) of an algorithm when a precise mathematical model of the problem is not available, but imprecise knowledge of the factors affecting performance, robustness etc, are known. The choice of these values should attempt to produce a result that is optimal in terms of all the available a-priori knowledge.

The approach employs a Bayesian rule set to produce symbolic descriptors of the image through statistical evidence collected from the image. These descriptors form the pre-conditions to a second level of rules that produce parameter values. This has the effect of making the algorithm context sensitive, ie. the parameters are 'tuned' as the spatial characteristics of the image change.

The technique has been used to control the application of an algorithm designed to detect anomalous regions (objects) in image data. Some initial results are presented. These show the effect of using this approach on the parameters of the algorithm processing S.A.R. images.

1. INTRODUCTION

It has been suggested that the front end processing employed by a vision system could be improved by using the symbolic knowledge contained within the heuristic approach required to interpret the processed image (Davis et al. (1983)); (Milios, Nawab (1985)); (Sharman, Durrani (1985)).

One contributory factor for this assertion is that it is often true that more is known about an imaging environment/ process than can be incorporated into the strict confines of a numerical algorithm based on a model of the process. This is due mainly to the image model being designed in the metric domain whilst there exists knowledge about the area of interest whose relationship to the metric domain is not available. As an example, although a trained radar operator is able to distinguish between true targets and false alarms, it may be that 'state of the art' image models can not emulate this process to produce a 'cleaner' image. This has led the authors to develop a method that integrates imprecise domain specific knowledge with the knowledge that is embodied within the mathematical formalism of an image formation model.

The objective of this approach is to control and plan the use of an algorithm, that is based on some theoretical low level model, through a system that can form symbolic descriptions of the image to provide a context in which to process. The context of the area to be processed is defined by high level descriptors such as 'close to an edge', 'homogeneous region', 'interesting region', etc. These descriptors will afffect the parameter values used in the algorithm so that improvements in performance, robustness, and a reduction in the amount of post-processing interpretation can be achieved. The technique presented uses a form of goal directed reasoning to guide an analytic procedure by using all the available a-priori information.

Often this prior knowledge is available as a simple, 'common sense' set of rules that can be incorporated directly into an analytically based algorithm. This is in contrast to the conventional use of prior knowledge in low level models in which the knowledge is represented typically as prior-probabilities or as analytic constants. In this paper both high and low level prior knowledge is used. Bayesian prior-probabilities are used in constructing local image descriptions whilst high level rules act on these descriptions of the image to direct the analytic processes.

The strength in such an approach results from the feedback that is introduced between high level interpretation and low level image processing as opposed to the conventional bottom up philosophy. A common pattern of development of image processing programs is a strong reliance on a mathematical model at the outset, followed by testing and informal modification. These modifications can often be an important source of the programs performance. It is exactly this process of 'tuning' the model that is being formalised here by using domain specific knowledge to interact with the program directly. The knowledge rules not only attempt to overcome weaknesses in the model, they can also be designed to bias the operating characteristics towards results one would expect from available real world knowledge of the environment. This paper outlines a preliminary attempt at designing such a processing framework.

Section II deals with the data abstractions (factual knowledge) and the rule structures (procedural knowledge). It also outlines the overall system architecture and discusses the various elements used.

Sections III and IV deal with the specific problem of improving detection of anomalous regions (objects) in aerial images and present the results of using such an approach to process the image.

2. PROPOSED SYSTEM

The objective of this work is to control and/or guide the use of a low level, data driven, spatially local algorithm given high level contextual information extracted from the image.

A Bayesian inferencing mechanism is proposed to make inexact inferences from statistical evidence collected from the image. This evidence takes the form of numerical features resulting from the application of simple signal processing routines. The inferences produce symbolic descriptors of the image which in turn are used to provide the processing context. To allow the system to interpret this context, it is necessary to construct a set of a-priori 'performance rules'. These rules will embody knowledge about the interaction of the algorithm and the data; where the assumptions in the algorithm break down; and what action to take, given the context, to improve the performance of the process. The approach thus iterates through a number of basic steps:

1) Choose area to process.
2) Extract numeric features from the image.
3) Infer symbolic descriptors of the image to provide the context.
4) Use 'performance rules' to interpret the context and thus obtain parameter values for the algorithm and/or a processing strategy.
5) Use the algorithm to process the image.

A more detailed description of the features of the proposed approach now follows.

2.1 Data Abstractions

The proposed system uses two levels of data abstraction (Milios, Nawab (1985)). They are:

(1) Signal Features
(2) Image Descriptors

Signal features result from a numerical analysis of the image and are a very common way of characterising signal properties. Typically they have a symbolic part that identifies the property of the signal and one or more numeric parts that assign values to that property eg. mean, variance. Signal features form the building blocks of many image formation models and are also the basic data structures out of which the more abstract image descriptors are constructed.

The image descriptors are a more compact way of characterising the image than signal features but greater ambiguity is introduced as the descriptors do not have an analytic definition. It is thus necessary to produce the descriptors by applying an a-priori rule set to the signal features. The image descriptors are purely symbolic and carry no numeric information eg. 'close to edge', 'homogeneous region'. They provide the context in which to process and this forms the basis for obtaining the parameter values of the data driven algorithm. Parameters are an identical data type to the signal features and result by applying a-priori 'performance rules' to the image descriptors. Fig. 1 illustrates the feedback introduced into the process and the different levels of data abstraction used.

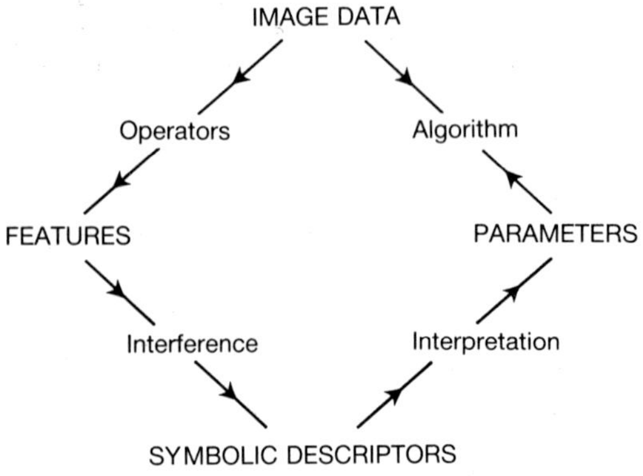

Fig. 1 Data Abstractions

2.2 Knowledge Rules

There are two types of rules in the proposed system, one to refer descriptors from features and one to infer parameter values from descriptors. These rules embody generic knowledge about the nature of the signals, the interpretations being sought and the properties of the image formation model used to process the image. Each rule has a set of pre-conditions called the 'evidence' and one post-condition called the 'goal'.

The rules used to infer descriptors from features are designed to deal with inexact and difficult inferences where the relationships between evidence and goal is rather diffuse, (Forsyth (1984)). A description of the structure of these rules and their inferencing mechanism is now given.

Goal	Prior Prob.	No. of Precons	Ref	Prob. True	Prob. False
'Descriptor	P(H)	n	1	P(E :H)	P(E :notH)
			.		
			.		
			n	P(E :H)	P(E :notH)

Fig. 2

Fig. 2 illustrates the data format required to represent each rule. The first item is the goal, which in this case is the image descriptor. The second item, P(H), is the prior probability of that descriptor being applicable before any evidence has been collected. The third item is the number of pieces of evidence (features) which are relevant to that descriptor - either as supportive or contradictory evidence. After this there are a series of three element fields corresponding to each piece of evidence. The first item in the triple is a reference identifying the piece of evidence to be considered. The second item, P(E:H), is the probability that the evidence will be observed given that the goal is true. The third item of the triple, P(E:notH), is the probability that the evidence will be observed given that the goal is not true.

Using any piece of evidence in the triple, it is possible to calculate the posterior probability of the descriptor by Bayes Theorem as:

$$P(H:E) = \frac{P(E:H)P(H)}{P(E:H)P(H)+P(E:notH)(1-P(H))}$$

When using signal features as evidence to support some goal, the evidence is not simply 'true' or 'false' but can take on a range of values. In this situation a pre-determined minimum and maximum value of the feature corresponds to 'true' and 'false' (or vice versa) and for intermediate values we interpolate linearly between P(H:E) and P(H:notE). This corresponds to uncertainty in the users response to a question asked by a conventional rule based system.

The prior probabilities may themselves be less than accurate but it is thought that this will not adversely affect the performance of the system. From a practical point of view a rough estimate using P(E:H) and P(E:notH) is probably good enough. This is because only the relative magnitudes, and not the exact values, of the posterior probabilities are required.

The inferencing engine, whose job it is to decide what piece of evidence to collect next chooses the one that could cause the maximum amount of fluctuation in the posterior probabilities in all of the goals. This introduces an element of reasoning into the system as it is always the most useful piece of evidence that is calculated next.

A goal is satisfied when its minimum posterior probability using all the remaining evidence is greater than some fraction of its maximum posterior probability using all of the evidence.

The second rule set, the 'performance rules', have descriptors as their pre-conditions and parameter values as their goals. These rules are simply 'IF pre-conditions THEN postcondition' structures as the relationship between descriptors and parameters is much clearer and bears little uncertainty.

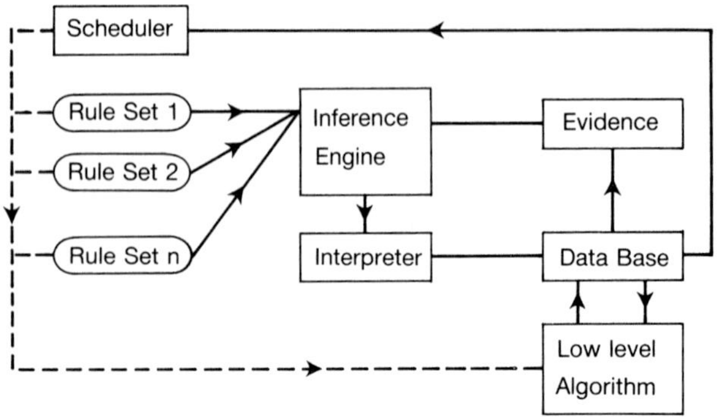

Fig. 3 System Architecture

2.3 *System Architecture*

Fig 3 illustrates in block diagram form the overall system architecture as implemented in software. A brief description of each part of the system is now given.

The database holds all available data about the image. This will include the original image, the current area being processed, parameter values as they are made available and the results of processing the image using the low level model.

The scheduler accesses the database to monitor the 'state' of the system. Depending on the contents of the database the scheduler can invoke any one of a number of rule sets, or alternatively if all the parameter values are available, the low level process itself. In the example given in this paper, the scheduler runs the rule sets in a purely sequential fashion. Each rule set only produces one image descriptor from the evidence used.

The evidence (signal features) is accessed through a collection of signal processing operators that in turn access the original image in the database. These operators are used, as required, by the inferencing engine when a rule set is invoked by the scheduler.

The interpreter contains the performance rules used to infer parameter values from descriptors and places these in the database ready for use with the image model.

3. OBJECT DETECTION

The following section describes a low level process that is to be controlled using the strategy outlined above.

The detecting of small anomalous regions in images has aroused much interest in such areas as optical aerial reconnaissance, radar analysis, terrain classification, and medical diagnosis through imagery. An algorithm for detecting small objects or targets in natural terrain from aerial images has been developed, (Quatieri (1983)). A brief description of the algorithm and its underlying assumptions is now given.

3.1 *Theoretical Model*

The image is modelled as a non-stationary, two dimensional Gaussian random process with unknown but estimable statistics. The objective is to locate small regions that may exist within the image whose statistics are not known prior. Using a significance test, this is achieved by identifying areas of the image that are statistically dissimilar to the background. Thus the assumption is made that any test region with a set of statistics similar to the background forms part of the background. If this association has low probability then the hypothesis is rejected and the region is assigned to be an object. Consider a set X of N pixels;

$$\underline{X} = [x_1, x_2, \ldots, x_n]$$

If these pixels are assumed to have a Gaussian distribution i.e. form part of the background, then the joint probability density function can be expressed as;

$$P(\underline{X}) = \frac{1}{(2\pi)^{N/2} \, |K|^{\frac{1}{2}}} \exp[-\tfrac{1}{2}(\underline{x}-\underline{m})^T K^{-1} (\underline{x}-\underline{m})]$$

where $m=E[\underline{X}]$ and $K=E[(\underline{x}-\underline{m})(\underline{x}-\underline{m})^T]$.

Thus any anomalous regions will satisfy $P(\underline{X}) < \lambda$, where λ is dependent on background statistics so that a constant false alarm rate can be maintained. To provide a more efficient and computationally less expensive implementation of the procedure, the problem can be reformulated. By exploiting the symmetric, positive definite properties of the covariance matrix, then a decomposition of this matrix forms the basis for an approximate implementation of the test expressed in terms of the normalised error residuals of a two dimensional linear predictor. This can be written as

$$\sum_m \sum_n \frac{e^2_{mn}}{\sigma^2} > C$$

where e is the error residual from a fixed order predictor in predicting x with its mean subtracted and σ^2 is the prediction error variance. C is a constant threshold independent of the background statistics. In order to estimate the parameters of the model in any region of the image, a window is used and stationarity is assumed over its entirety. The criterion employed is that of minimising

$$E[n_o, m_o] = \sum_m \sum_n e^2(n,m)$$

over the whole window, where $E[n_o, m_o]$ is the sum of the square of the error residuals over a window centred at $x(n_o, m_o)$ within the image. This least squares estimation procedure is the covariance method of linear prediction. To implement the model a window on the image is used to estimate the predictor coefficients and the error variance. The pixel at the centre of the window is then tested to see if it belongs to the background.

3.2 Practical Considerations

From the above it is clear that the model requires a stationary signal to operate effectively, and as almost all real images do not exhibit this property, we are forced to assume quasi-stationarity over some small region of the image. The problem is then to decide what size of window will provide the best possible estimates of the predictor coefficients and the background error variance. If a large window size is chosen then the quasi-stationary assumption may fail. Alternatively, too small a window may produce erroneous parameter estimates if the image has a lot of noise.

In aerial images there are often well defined regions of land types present, each exhibiting distinct statistical properties, and it is often at the boundary between these regions that is of greatest interest. Should the model be applied in these parts of the image then the window would contain two different regions and no reasonable estimate of the model parameters will result. In this situation the presence of any objects would be suppressed by an artificially high error variance.

Large amounts of noise in the image will produce high numbers of false alarms leading to difficulty in interpreting the processed image. Attempts to remove the noise may also remove some of the statistical anomalies that the system is trying to detect. Again contextual information could filter out some of these false alarms at the numerical level of processing by altering the threshold in regions where one would not 'expect' objects to occur or alternatively decrease the threshold in areas of greatest interest. This may introduce an unrealistic bias into the process.

We note that knowledge of the image characteristics in the area being processed could enable better estimates of the model parameters and thus implement a more robust version of the algorithm.

4. RESULTS

The following shows the results of an experiment conducted to monitor the effect of using context to modify the processing characteristics of the object detection algorithm.

Fig. 4 shows the original, contrast enhanced SAR image. A number of distinct segments can be seen. These correspond to different types of land use such as wooded, cultivated, urban etc. The logarithm of the original image was used to produce a second image whose noise was now additive. If the number of speckles in the aperture is large enough then the noise distribution in this second image is approximately Gaussian thus fitting the linear predictive model (April, Arsenault 1976). This image was used in all subsequent processing.

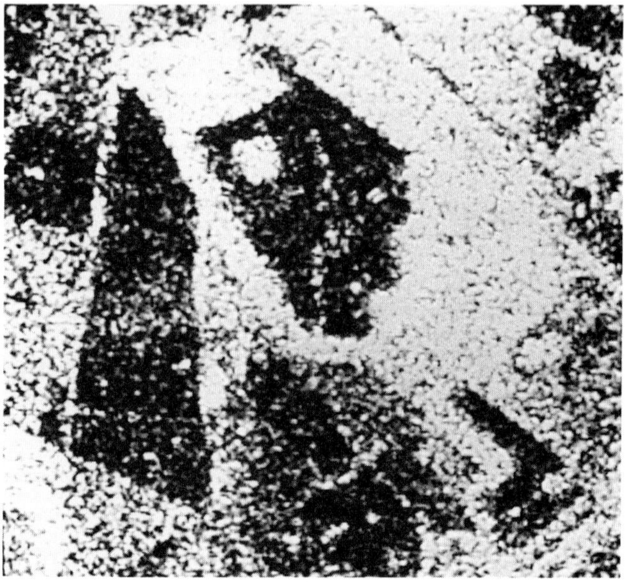

Fig. 4 Original SAR image (Courtesy of RAE Farnborough)

The first model parameter modified by the image context is the size of the window used to estimate the predictor coefficient coefficients and the background error variance. This depends only on the validity of the quasi-stationary assumption over the extent of the window. Fig. 5 shows the rule used to produce either one of three possible descriptors of 'stationarity' of the image. The probabilities used here were developed from training samples with similar types of ground cover. This is then interpreted directly by the performance rules to give one of three window sizes. Fig. 6 shows the result of applying this rule over all parts of the image. As one would expect a large window is used in homogeneous portions of the image whereas a small window is used close to sharp boundaries where the image is very non-stationary.

GOAL	PRIOR PROB	NO. OF PRECONS	REF	PROB TRUE	PROB FALSE
Stationary	0.7	2	1	0.9	0.1
			2	0.9	0.1
Medium-Stationary	0.2	2	1	0.4	0.6
			2	0.4	0.6
Not-Stationary	0.1	2	1	0.1	0.9
			2	0.1	0.9

Preconditions:
1 -> Is the mean stable in this region?
2 -> Is the variance stable in this region?

Fig. 5 Rules for "Stationary" Label

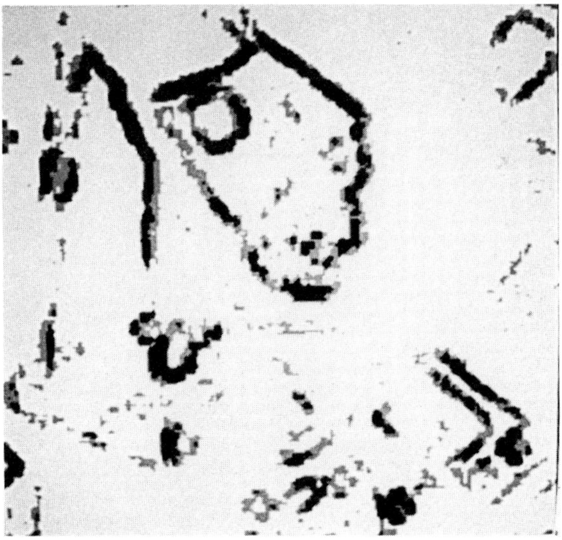

Fig. 6 Result of labelling using 'stationary' rules.

light -> Stationary (Large size of estimation window) Medium -> Slightly Stationary (MEDIUM size of estimation window) Dark -> Not Stationary (SMALL size of estimation window)

Where the window actually includes a boundary between homogeneous segments of the image it would be desirable to move the window to a region with a set of statistics similar to the region containing the test point but containing no edges or boundaries. To do this it is required to know where the edges occur in the image. Fig. 7 shows the result of using an 9x9 pyramid edge filter on the original image. It is worth noting that this may not be an appropriate operator to use as the primary interest is in boundaries between textures and not simply edges. Fig. 8 shows the rule used to produce a descriptor of how the statistics of the image vary at right angles to the edge closest to the test point. The descriptor is interpreted by the performance rules such that the window is migrated away from the edge only if the direction in which it travels is homogeneous. Fig. 9 shows the areas of the image in which the window is moved when applying the rule. Note that windows between two parallel lines have not been moved as there is no better place to move them to.

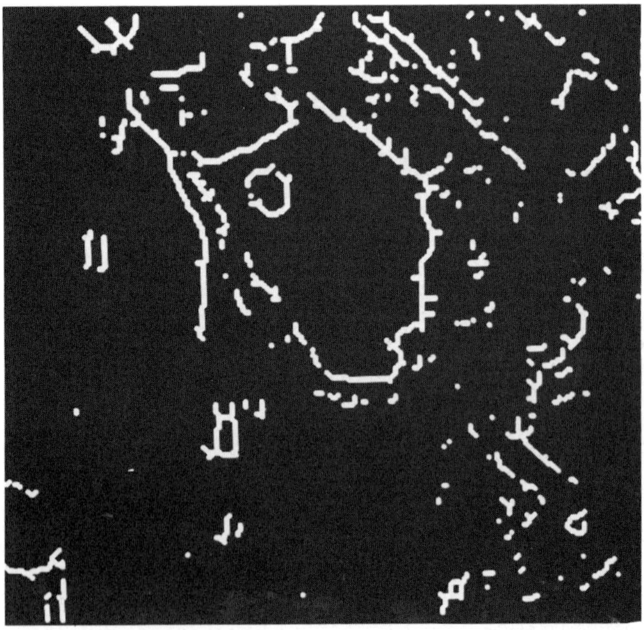

Fig. 7 Edge map from a 9x9 pyramid edge filter

GOAL	PRIOR PROB	NO. OF PRECONS	REF	PROB TRUE	PROB FALSE
Distinct Edge	0.5	3	3	0.1	0.9
			4	0.3	0.9
			5	0.3	0.9
Slight Edge	0.3	3	3	0.9	0.1
			4	0.5	0.5
			5	0.5	0.5
Non Distinct Edge	0.2	3	3	1.0	0.0
			4	0.95	0.1
			5	0.95	0.1

Preconditions:
3 -> Close to edge in edge map?
4 -> Is mean stable at right angles to the edge?
5 ->Is variance stable at right angles to the edge?

Fig. 8 Rules for "Edge" Label

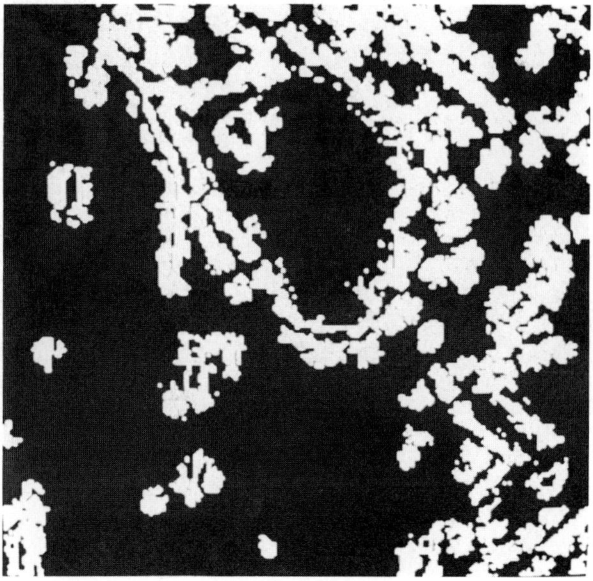

Fig. 9 Areas where the estimation window is moved due to edge labelling.

In Fig. 10 the light regions show the objects identified using the original algorithm on a portion of the image. In Fig. 11 we see the objects identified using the rule system. The main difference between the two is the increased sensitivity in border regions of the image. This was the main consideration when constructing the performance rules as the a-priori knowledge available identified that the model breaks down in these regions. An object is defined as a statistical anomaly in the low level model. In the real world this will correspond to a small group of pixels that is not part of the background. The significance of these anomalies is obviously dependent on the specific application. In Fig. 10 and Fig. 11 they may correspond to farm buildings, small areas of land use, etc.

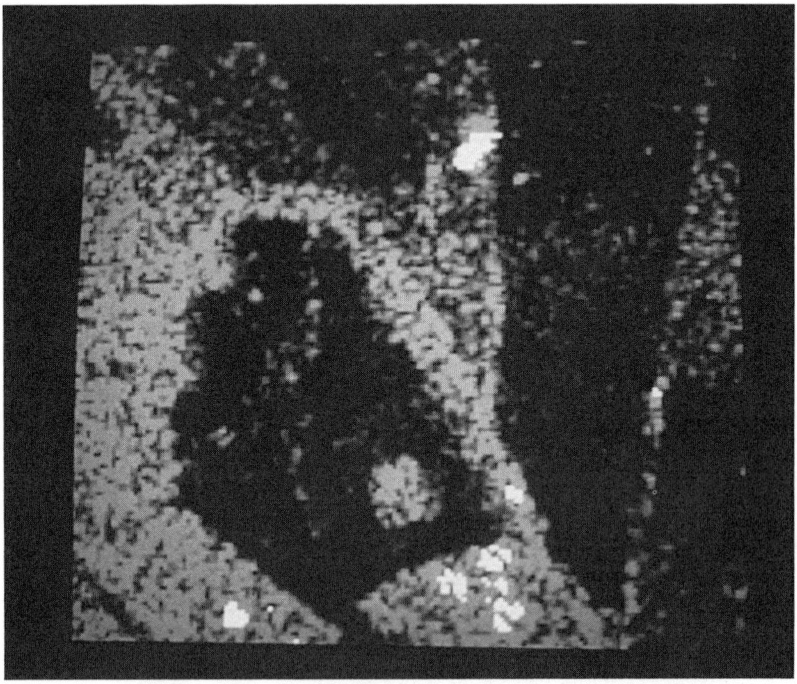

Fig. 10 Light areas are objects identified using the 'blind' algorithm

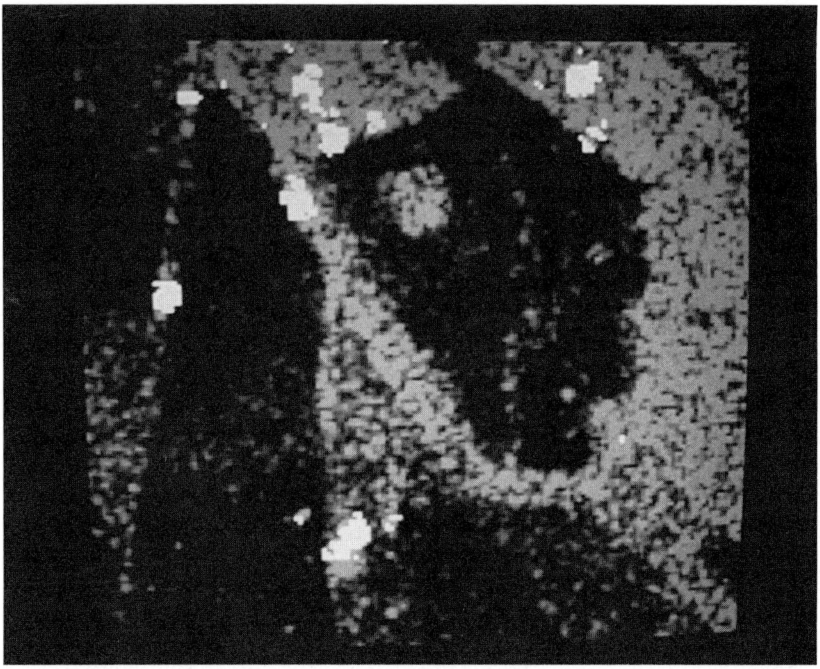

Fig. 11 Light areas are objects identified using the context sensitive approach. Note that objects now identified very close to region borders.

5. DISCUSSION

The example presented illustrates the utility of contextual processing. Although it is felt that the initial approach to the problem has been encouraging, the improvement in performance has still to be assessed. One of the difficulties in processing this type of data is relating the results to ground truth and unfortunately that is outwith the scope of this paper. A subjective analysis of the results do suggest that the algorithm is indeed more sensitive in border regions of the image but this may simply be due to more false alarms in these regions. A proper study of the correspondence between these results and ground truth is required.

A current topic of interest in this project is the interaction between signals and symbols, (Neii et al. (1982)), and the potential for using 'knowledge based' techniques in image processing. Some systems have already used a combination of A.I and signal processing, but the combination is often more accidental than intentional, (Davis et al. (1983)). The motivation for attempting to use knowledge based systems stems from the assertion that "a small amount of planned signal processing can substitute for a lot of symbolic processing", - a useful feature if ultimately we wish to interpret the image.

It is still unclear how best to map from a range of numerical values to a single symbol. For example in remote sensing, a 'high' temperature in one country may be equivalent numerically to a 'cool' temperature in another. It is seen again here that context is all important. The inverse of this problem is even more difficult. This is one problem we hope to tackle in the near future.

This paper does not deal with any real world knowledge such as the relationship between types of vegetation and likely target locations. It would be a straight-forward step to include this type of knowledge in the approach outlined in this paper although the experiments conducted here were not to detect targets but to monitor the effects of the knowledge rules. As with any rule based system there is a continual updating of the knowledge base. It is thus an important feature of this system that one is able to introduce new rules as they become available and also update the probabilities in the rules if it is felt that performance is not as desired.

A practical system would contain many more rules than the example given. This would enable the use of more complex control strategies that could ultimately explain the action taken at a point in the image and may also give greater insight to the low level algorithm itself.

It is felt that the concept of contextual processing and the ideas presented here have potential in many other areas of image processing. This is as a result of the recent advances in knowledge representation and the great number of situations where available knowledge is not used in the processing model.

6. REFERENCES

Neii, Penny, Feigenbaum, Anton, Rockmore "Signal to Symbol Transformation: HASP/SIAP Case Study", The AI magazine, pp. 23-35, Spring 1982.

Davis, Kopec, Oppenheim, "Artificial Intelligence and Signal Processing: Overview and Experiments", Technical Report, MIT, Cambridge, Mass, January 1983.

Quatieri, T.F., "Object Detection by Two Dimensional Linear Prediction", Technical Report, Lincoln Lab, MIT, January 1983.

Milios, E., Nawab, S., "Interpretation-Guided Signal Processing via Protocol Analysis", Proc. IEEE ICASSP 1985, Vol. 4, pp. 1660-63.

Sharman, D. and Durrani, T., "Knowledge Based Object Detection", Proc. IEEE ICASSP 1985, Vol 2, pp. 909-912.

Forsyth, R., "Expert Systems: Principles and Case Studies", Chapman and Hall, 1982.

Robinson, G.S., "Detection and Coding of Edges Using Directional Edge Masks", Opt. Eng., Vol. 16, No. 6, Nov-Dec, 1977.

April, G., Arsenault, H., "Properties of Speckle Integrated with a Finite Aperture and Logarithmically Transformed", J. Opt. Soc. Am., Vol. 66, No. 11, Nov. 1976.

V NUMERICAL ALGORITHMS AND ARCHITECTURES

THE NUMERICAL SOLUTION OF THE GENERAL GAUSS-MARKOV LINEAR MODEL

S. Hammarling
(Numerical Algorithms Group Ltd., Oxford)

1. INTRODUCTION

In this contribution we give an overview of the numerical solution of the general Gauss-Markov linear model given by

$$y = X\beta + e, \quad e \subset N(0,\sigma^2 W), \tag{1.1}$$

where the notation $e \subset N(0,\sigma^2 W)$ indicates that the noise vector e is assumed to come from a normal distribution with mean zero and variance-covariance matrix $\sigma^2 W$, W being a symmetric non-negative definite matrix.

We start by considering the standard case where W = I, then consider the more general case where W is any non-singular positive definite matrix and finally we consider the case where W is allowed to be singular. The material in this contribution is not new, but deserves an airing since it provides us with the basic tools for the reliable implementation of least-squares problems and the Kalman filter, and on a variety of architectures.

2. THE STANDARD GAUSS-MARKOV LINEAR MODEL

In this section we look at the standard model

$$y = X\beta + e, \quad e \subset N(0,\sigma^2 I). \tag{2.1}$$

The associated linear least-squares problem for this model is given by

$$\min_b \hat{e}^T\hat{e}, \quad \hat{e} = \hat{y} - \hat{X}b, \tag{2.2}$$

where $\hat{X}$ is the n by p matrix of observations, $\hat{y}$ the n element vector of dependent observations, b is the p element vector of regression coefficients and $\hat{e}$ is the n element error vector. Throughout this article we shall assume that there are at least as many observations as variables so that $n \geq p$, although much of the discussion is readily extended to the underdetermined case.

Tools for the reliable solution of (2.2) are the QR factorization and the singular value decomposition (SVD). In the next section we describe these factorizations. We shall not give the algorithmic detail of computing these factorizations, but instead give appropriate references.

3. THE QR FACTORIZATION AND THE SVD

The <u>QR factorization</u> of an n by p matrix X is given by

$$X = Q \begin{bmatrix} R \\ O \end{bmatrix}, \tag{3.1}$$

where Q is an n by n orthogonal matrix and R is a p by p upper triangular matrix. The SVD of X is given by

$$X = U \begin{bmatrix} D \\ O \end{bmatrix} V^T, \tag{3.2}$$

where U is an n by n orthogonal matrix, V is a p by p orthogonal matrix and D is a diagonal matrix with non-negative diagonal elements called the <u>singular values</u> of X. The factorization can be chosen so that the singular values are in descending order down D. The first p columns of U are the left singular vectors of X and the columns of V are the right singular vectors of X.

If we perform the QR factorization (3.1) and then perform an SVD of R as

$$R = \tilde{U}\tilde{D}\tilde{V}^T \tag{3.3}$$

then

$$X = Q \begin{bmatrix} \tilde{U}\tilde{D}\tilde{V}^T \\ O \end{bmatrix} = Q \begin{bmatrix} \tilde{U} & O \\ O & I \end{bmatrix} \begin{bmatrix} \tilde{D} \\ O \end{bmatrix} \tilde{V}^T$$

$$= U \begin{bmatrix} D \\ C \end{bmatrix} V^T, \quad U = Q \begin{bmatrix} \tilde{U} & O \\ O & I \end{bmatrix}, \quad D = \tilde{D}, \quad V = \tilde{V} \tag{3.4}$$

and hence X and R have the same singular values and right singular vectors.

From (3.1) and (3.2) we find that

$$X^TX = R^TR \tag{3.5}$$

$$= VD^2V^T,$$

which are respectively the Cholesky and spectral factorizations of X^TX. Thus R and (V,D) give us alternative compact representations of X^TX without having to take the numerically damaging step of explicitly forming X^TX. (Golub, 1965; Hammarling, 1985.)

For many applications we can use the QR factorization when X is not close to being rank deficient, but carry on to the SVD otherwise. The SVD is a powerful tool for the analysis of rank deficient and near rank deficient problems. For example, from (3.2)

$$Xv_j = d_ju_j,$$

where d_j, u_j and v_j are respectively the jth singular value, left singular vector and right singular vector of X. Thus

$$||Xv_j||_2 = d_j$$

so that right singular vectors that correspond to small singular values give information on the near linear dependencies in the columns of X.

Both the QR factorization and the SVD can be obtained by numerically stable methods, based upon orthogonal transformations, and there are a number of sources of quality software that implement these methods. For background on both factorizations, the algorithms to compute them and many further references see Golub and Van Loan (1983).

For signal processing applications, a useful transformation is the plane rotation. This transformation is flexible and, unlike the hyperbolic rotation, is orthogonal and therefore has the desired numerical stability. Plane rotations are described in the above reference and in Figure 3.1 we illustrate the progress from X to R using the standard sequence of plane rotations for the case n=4, p=3. (Givens, 1958.)

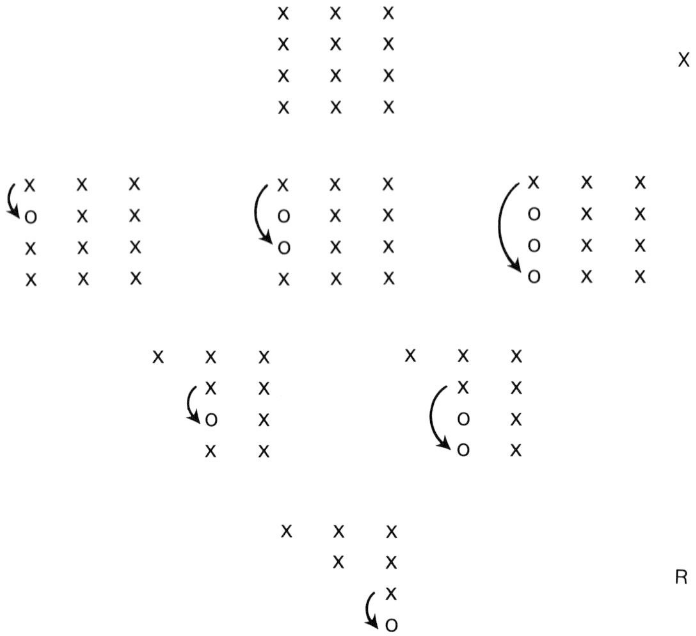

Fig. 3.1

A number of other sequences are possible, based upon introducing zeros into a column in different orders and two such alternative orderings are illustrated in Figure 3.2.

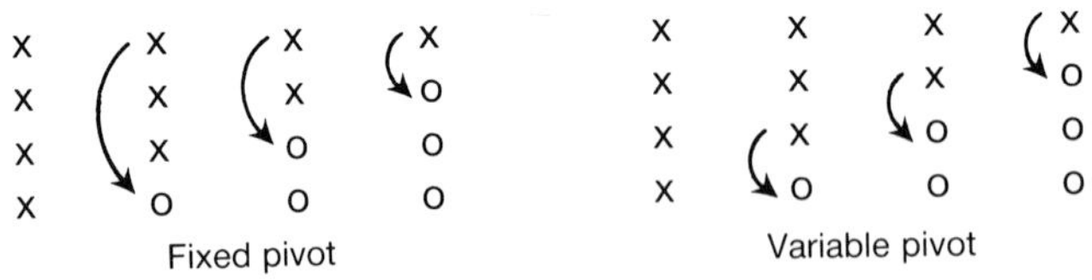

Fig. 3.2

Similarly we can perform plane rotations in disjoint planes in parallel, an idea that originates from a paper on error analysis for the QR factorization! (Gentleman, 1975; see also Modi and Clarke, 1984.) One such sequence for the case n=8, p=6 is illustrated in Figure 3.3, where rotations with the same index can be performed in parallel.

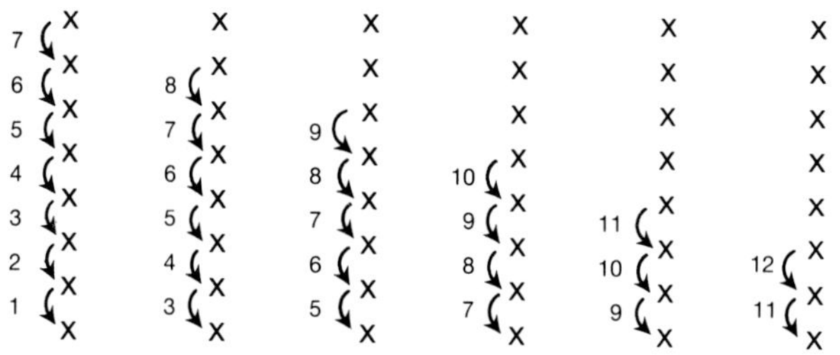

Fig. 3.3

The number of parallel operations for the illustrated sequence is (n+p-2) as opposed to ½p(2n-p-1) serial operations.

We can also readily update an existing QR factorization with additional observations as they become available, a particularly important feature for real time computing. This is illustrated in Figure 3.4 for the case p=3.

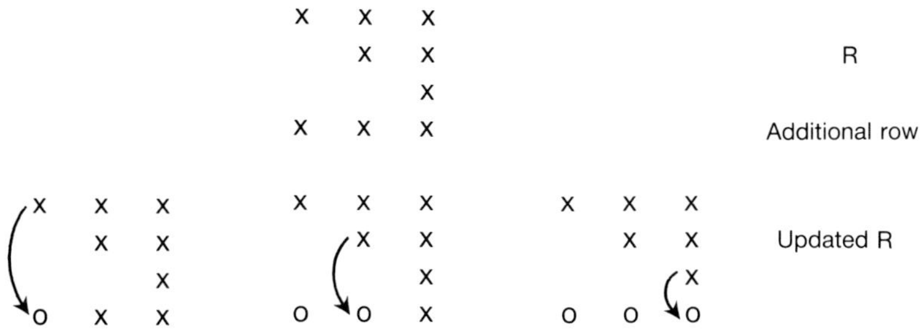

Fig. 3.4

These ideas form the basis of the systolic array implementation of the QR factorization. (Gentleman and Kung, 1981.)

For the SVD, as already mentioned, we can first apply the QR factorization and then find the SVD of R. The classical method of computing the SVD is to use the method originating with Golub and Kahan (1965) and realized as an Algol 60 algorithm by Golub and Reinsch (1970), in which R is reduced to bi-diagonal form and then the implicit QR algorithm is used to further reduce this to diagonal form. We note that R may be reduced to bi-diagonal form by applying plane rotations from both sides. This computation of the SVD is illustrated in Figure 3.5. The method corresponds to the reduction of R^TR to tri-diagonal form, followed by the QR algorithm applied to the tri-diagonal form.

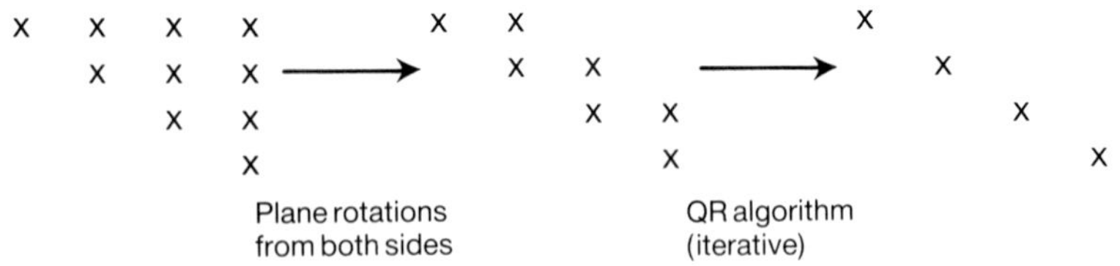

Fig. 3.5

More recently a method due to Kogbetliantz (1955) has been revived and in this method R is iteratively reduced to diagonal form, corresponding to Jacobi's method applied to R^TR. In Jacobi's method each plane rotation is chosen to solve the two by two symmetric eigenvalue problem so that $c=\cos\theta$ and $s=\sin\theta$ are such that

$$\begin{bmatrix} c & s \\ -s & c \end{bmatrix} \begin{bmatrix} x & y \\ y & z \end{bmatrix} \begin{bmatrix} c & -s \\ s & c \end{bmatrix} = \begin{bmatrix} x' & 0 \\ 0 & z' \end{bmatrix}$$

and a sweep of Jacobi's method applies $\frac{1}{2}n(n-1)$ such transformations with

$$\begin{bmatrix} x & y \\ y & z \end{bmatrix} = \begin{bmatrix} a_{ii} & a_{ij} \\ a_{ij} & a_{jj} \end{bmatrix}$$

for each i and j (j>l). The usual method chooses i and j cyclically, but other orderings are possible and, as with the QR algorithm we can apply transformations in parallel. (Modi, 1982; Modi and Parkinson, 1982.)

In the Kogbetliantz method (Figure 3.6) each transformation solves the two by two SVD problem, so that the left and right rotations are such that

$$\begin{bmatrix} c_1 & s_1 \\ -s_1 & c_1 \end{bmatrix} \begin{bmatrix} x & y \\ y' & z \end{bmatrix} \begin{bmatrix} c_2 & -s_2 \\ s_2 & c_2 \end{bmatrix} = \begin{bmatrix} x' & 0 \\ 0 & z' \end{bmatrix}$$

and one sweep of the method corresponds to one sweep of Jacobi. In the cyclic ordering one sweep makes R lower triangular and a second sweep restores the upper triangle. The Kogbetliantz approach is the basis of a systolic array implementation of the SVD. (Brent and Luk, 1985.)

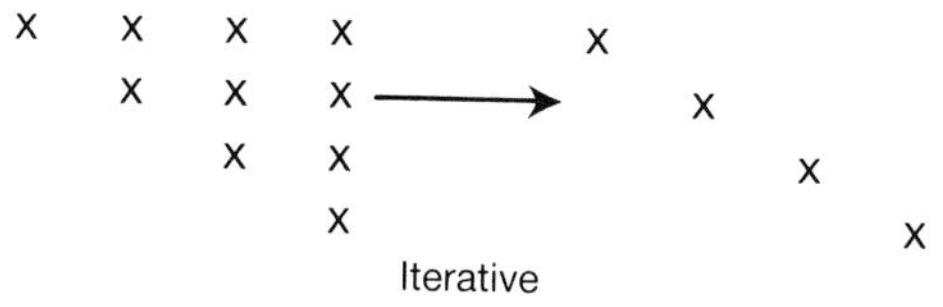

Fig. 3.6

4. SOLVING THE STANDARD LINEAR LEAST-SQUARES PROBLEM

In this section we consider the solution of the least-squares problem

$$\min \hat{e}^T\hat{e}, \quad \hat{e} = \hat{y} - \hat{X}b \tag{4.1}$$

using the QR factorization of (3.1) and the SVD. If we put

$$Q^T\hat{e} = f \equiv \begin{bmatrix} f_1 \\ f_2 \end{bmatrix}, \quad Q^T\hat{y} = z \equiv \begin{bmatrix} z_1 \\ z_2 \end{bmatrix}$$

then, since $f^Tf = \hat{e}^T\hat{e}$, (4.1) becomes

$$\min f^Tf, \quad f = z - \begin{bmatrix} R \\ 0 \end{bmatrix} b = \begin{bmatrix} z_1 - Rb \\ z_2 \end{bmatrix} \tag{4.2}$$

If $\hat{X}$ has full rank so that R is non-singular, we can choose b as the solution of the upper triangular equations

$$Rb = z_1 \tag{4.3}$$

and hence this solves (4.2). Thus

$$\hat{e}^T\hat{e} = z_2^Tz_2. \tag{4.4}$$

If $\hat{X}$ does not have full rank we can proceed with the SVD of R

$$R = \tilde{U}DV^T \tag{4.5}$$

and (4.2) is equivalent to

$$\min f_1^T f_1, \quad f_1 = z_1 - \tilde{U}DV^T b \tag{4.6}$$

Putting

$$\tilde{U}^T f_1 = g \equiv \begin{bmatrix} g_1 \\ g_2 \end{bmatrix}, \quad \tilde{U}^T z_1 = w \equiv \begin{bmatrix} w_1 \\ w_2 \end{bmatrix}, \quad V^T b = c \equiv \begin{bmatrix} c_1 \\ c_2 \end{bmatrix} \tag{4.7}$$

and

$$D = \begin{bmatrix} S & 0 \\ 0 & 0 \end{bmatrix}, \tag{4.8}$$

where S is non-singular, then (4.6) becomes

$$\min g^T g, \quad g = w - \begin{bmatrix} S & 0 \\ 0 & 0 \end{bmatrix} c, \tag{4.9}$$

and this is solved for any c such that

$$Sc_1 = w_1 \tag{4.10}$$

for which

$$\hat{e}^T \hat{e} = z_2^T z_2 + w_2^T w_2. \tag{4.11}$$

c_2 is arbitrary and the solution for which $b^T b$ is minimized is called the <u>minimal length</u> solution and is a best linear unbiased estimate of β. Since $b^T b = c^T c$, this solution is given by

$$c_2 = 0 \tag{4.12}$$

so that

$$b = V \begin{bmatrix} S^{-1} w_1 \\ 0 \end{bmatrix}, \quad S^{-1} = \text{diag } (1/d_i). \tag{4.13}$$

Other solutions, such as the reduced parameter solution, can readily be obtained via the SVD and this allows the decision on the rank of X to be made in the most favourable circumstance. (See for example, Hammarling, 1985, and the references given there.)

5. THE GENERAL GAUSS-MARKOV LINEAR MODEL

In this section we look at the general model

$$y = X\beta + e, \quad e \subset N(0, \sigma^2 W). \tag{5.1}$$

When W is non-singular this is usually associated with the weighted least-squares problem

$$\min_b \hat{e}^T W^{-1} \hat{e}, \quad \hat{e} = \hat{y} - \hat{X}b \tag{5.2}$$

If we let F be any matrix for which

$$W = FF^T \tag{5.3}$$

and put

$$r = F^{-1}\hat{e} \quad \text{so that} \quad r \subset N(0, \sigma^2 I) \tag{5.4}$$

then (5.2) becomes

$$\min r^T r, \quad r = F^{-1}\hat{y} - F^{-1}\hat{X}b \tag{5.5}$$

which is in the form of the standard problem (2.2), and so we can solve (5.5) as before using the QR factorization and the SVD. <u>But</u> this approach is numerically unstable if W is ill-conditioned and (5.2) and (5.5) are not even defined if W is singular. Adding in artificial noise is a completely unsatisfactory numerical approach for circumventing this.

In place of (5.4) we can let r be any vector such that

$$Fr = \hat{e} \tag{5.6}$$

and rearrange (5.5) to give

$$\min r^T r, \text{ subject to: } \hat{y} = \hat{X}b + Fr. \tag{5.7}$$

While being superficially a trivial rearrangement this has the vital difference of not now requiring non-singularity of F and hence of W. Notice that if

$$r \subset N(0, \sigma^2 I) \quad \text{we still have} \quad \hat{e} \subset N(0, \sigma^2 W)$$

(5.7) is called the <u>generalized linear least-squares problem</u>. (Paige, 1978.) Tools for the reliable solution of (5.7) are the QR factorization, the SVD and the generalized singular value decomposition (GSVD). In the next section we indicate how the problem may be solved with the QR factorization and the SVD and in the following section we define the GSVD and give references to its computation and its use in solving (5.7).

6. SOLVING THE GENERALIZED LINEAR LEAST-SQUARES PROBLEM

Here we consider the solution of the problem

$$\min r^T r, \text{ subject to: } \hat{y} = \hat{X}b + Fr \tag{6.1}$$

using the QR factorization of (3.1). Denote the QR factorization of $\hat{X}$ by

$$\hat{X} = Q_x \begin{bmatrix} R_x \\ O \end{bmatrix} \tag{6.2}$$

and put

$$Q_x^T \hat{y} = z \equiv \begin{bmatrix} z_1 \\ z_2 \end{bmatrix}, \quad Q_x = (Q_1 \; Q_2) \tag{6.3}$$

so that (6.1) becomes

$$\min r^T r, \text{ subject to: } z = \begin{bmatrix} R_x \\ O \end{bmatrix} b + \begin{bmatrix} Q_1^T F \\ Q_2^T F \end{bmatrix} \tag{6.4}$$

If $\hat{X}$ does not have full rank we can proceed to compute the SVD of R_x, otherwise for any r we can determine b as the solution of the upper triangular equations

$$R_x b = z_1 - (Q_1^T F) r \tag{6.5}$$

so that (6.4) becomes

$$\min r^T r, \text{ subject to: } z_2 = (Q_2^T F) r \tag{6.6}$$

If we now denote the QR factorization of $F^T Q_2$ as

$$F^T Q_2 = Q_F \begin{bmatrix} R_F \\ O \end{bmatrix} \tag{6.7}$$

and put

$$Q_F^T r = p \equiv \begin{bmatrix} p_1 \\ p_2 \end{bmatrix} \qquad (6.8)$$

then (6.6) becomes

$$\min p^T p, \quad \text{subject to:} \quad z_2 = R_F^T p_1. \qquad (6.9)$$

If R_F does not have full rank we can proceed to compute its SVD, otherwise we can solve the lower triangular system $R_F^T p_1 = z_2$ for p_1 and hence p is given by

$$R_F^T p_1 = z_2, \quad p_2 = 0 \qquad (6.10)$$

so that

$$r^T r = p_1^T p_1, \; r = Q_F \begin{bmatrix} p_1 \\ 0 \end{bmatrix}$$

and b is given from (6.5). Details on this type of approach to the solution of (6.1) are given in Paige (1978, 1979a), error and perturbation analysis in Paige (1979b) and some statistical analysis in Kourouklis and Paige (1981). Further details on the SVD approach are given in Hammarling, Long and Martin (1983).

7. THE GENERALIZED SINGULAR VALUE DECOMPOSITION

To give a gentle introduction to the GSVD we suggest a generalization of the QR factorization of the single matrix to the matrix pair $(\hat{X},F)$. We define the generalized QR factorization as

$$\hat{X} = Q_X \begin{bmatrix} R_X \\ 0 \end{bmatrix}, \; Q_X^T F = R_F Q_F^T, \qquad (7.1)$$

where Q_X and Q_F are orthogonal and R_X and R_F are upper triangular. If F is non-singular we find that

$$F^{-1}\hat{X} = Q_F \begin{bmatrix} R_F^{-1} R_X \\ 0 \end{bmatrix} \quad , \qquad (7.2)$$

which is the QR factorization of $F^{-1}\hat{X}$, the data matrix in (5.5). (7.1) gives all the information in (7.2) without relying on invertibility of F and without requiring the numerically damaging step of computing $F^{-1}\hat{X}$. Applying this to (6.1) and putting

$$Q_X^T\hat{y} = z \equiv \begin{bmatrix} z_1 \\ z_2 \end{bmatrix}, \quad Q_F^T r = p \equiv \begin{bmatrix} p_1 \\ p_2 \end{bmatrix}, \quad R_F = \begin{bmatrix} R_{11} & R_{12} \\ 0 & R_{22} \end{bmatrix} \tag{7.3}$$

we get

$$\min p^T p, \text{ subject to: } z = \begin{bmatrix} R_X \\ 0 \end{bmatrix} b + R_F p \tag{7.4}$$

and if R_X and R_{22} are of full rank we solve (7.4) as

$$R_{22}p_2 = z_2, \; p_1 = 0, \; R_X b = z_1 - R_{12}p_{22}, \; r^T r = p_2^T p_2. \tag{7.5}$$

There are variations in the way one can define the GSVD, but the appropriate form for the generalized linear least-squares problem (Paige, 1985) is to define the generalized SVD for the matrix pair $(\hat{X},F)$ as

$$\hat{X} = Q \begin{bmatrix} R \\ 0 \end{bmatrix} \begin{bmatrix} I & 0 & 0 \\ 0 & C & 0 \\ 0 & 0 & 0 \end{bmatrix} V^T, \quad F = Q \begin{bmatrix} R \\ 0 \end{bmatrix} \begin{bmatrix} 0 & 0 & 0 \\ 0 & S & 0 \\ 0 & 0 & I \end{bmatrix} U^T, \tag{7.6}$$

where Q, U and V are orthogonal, R is a non-singular upper triangular matrix and C and S are diagonal with

$$C^2 + S^2 = I, \quad c_i > 0, \quad s_i > 0.$$

We can choose the diagonal elements of C, the c_i, to be in descending order, in which case the diagonal elements of S are of course in ascending order. The pairs (c_i, s_i) are called the generalized singular values of $(\hat{X},F)$.

If F is non-singular then

$$F^{-1}\hat{X} = U \begin{bmatrix} S^{-1}C & O \\ O & O \end{bmatrix} V^T, \tag{7.7}$$

which is the SVD of $F^{-1}\hat{X}$, and

$$\hat{X}\hat{X}^T = (QR) \begin{bmatrix} C^2 & O \\ O & O \end{bmatrix} (QR)^T, \quad FF^T = (QR) \begin{bmatrix} S^2 & O \\ O & I \end{bmatrix} (QR)^T, \tag{7.8}$$

so that (QR) is the congruence matrix that simultaneously diagonalizes $\hat{X}\hat{X}^T$ and FF^T. The values (c_i^2/s_i^2) are the non-zero eigenvalues of the generalized symmetric eigenvalue problem

$$\hat{X}\hat{X}^T Z = \lambda FF^T Z.$$

For background on the GSVD the reader is strongly recommended to consult Van Loan (1976) and Paige and Saunders (1981).

THE GSVD can readily be applied to (6.1) to yield the solution of the generalized linear least-squares problem and just as the SVD is a valuable tool for the solution and analysis of the standard linear least-squares problem, so the GSVD plays the same role for the generalized problem as is deomstrated in Paige (1985). Stable numerical methods are emerging for computing the GSVD, (Stewart, 1983; Van Loan, 1984; Paige, 1984), the last of these, being based on implicit Kogbetliantz approach, has potential for systolic implementation (Brent, Luk, and Van Loan, 1983.)

8. CONCLUSION

We have discussed some of the numerical tools that are necessary for the reliable solution of the general Gauss-Markov linear model and have given references to the details necessary to implement this method. The crux of such methods is the use of orthogonal transformations, which are numerically stable (for example, Wilkinson, 1965), and the avoidance of the numerically damaging steps of forming normal matrices of the form $\hat{X}^T\hat{X}$ and of matrix inversion.

The Kalman filter can be viewed as a generalized least-squares problem and the ideas here form the basis of reliable and efficient methods for the numerical solution of the Kalman filter problem. (Hammarling, 1985.)

9. REFERENCES

Brent, R.P. and Luk, F.T., (1985) The solution of singular value and symmetric eigenvalue problems on multiprocessor arrays. *SIAM J. Sci. Stat. Comput.*, **6**, 69-84.

Brent, R.P., Luk, F.T. and Van Loan, C., (1983) Computation of the generalized singular value decomposition using mesh-connected processors. *Proc. SPIE*, Vol. 431, Real-Time Signal Processing, VI.

Gentleman, W.M., (1975) Error analysis of QR decompositions by Givens transformations. *Linear Algebra Applic.*, **10**, 189-197.

Gentleman, W.M. and Kung, H.T., (1981) Matrix triangularization by systolic arrays. *Proc. SPIE*, Vol. 298, Real-Time Signal Processing IV.

Givens, W., (1958) Computation of plane unitary rotations transforming a general matrix to triangular form. *SIAM J. Appl. Math.*, **6**, 26-50.

Golub, G.H., (1965) Numerical methods for solving linear least-squares problems. *Num. Math.*, **7**, 206-216.

Golub, G.H. and Kahan, W., (1965) Calculating the singular values and pseudo-inverse of a matrix. *SIAM J. Num. Anal.*, **2**, 202-224.

Golub, G.H. and Reinsch, C., (1970) Singular value decomposition and least-squares solutions. *Num. Math.*, **14**, 403-420.

Golub, G.H. and Van Loan, C.F., (1983) Matrix Computations. North Oxford Academic, Oxford.

Hammarling, S.J., (1985) The singular value decomposition in multivariate statistics, Signum Newsletter, No. 3.

Hammarling, S.J., (1985) The Numerical solution of the Kalman filtering problem. NAG Technical Report, TR1/85, NAG Central Office, 256 Banbury Road, Oxford, OX2 7DE, UK.

Hammarling, S.J., Long, E.M.R. and Martin, D.W., (1983) A generalized linear least-squares algorithm for correlated observations with special reference to degenerate data. NPL Report DITC 33/83, National Physical Laboratory, Teddington, Middlesex, TW11 OLW, UK.

Kogbetliantz, E.G., (1955) Solution of linear equations by diagonalization of coefficients matrix. *Quart. Appl. Math.*, **13**, 123-132.

Kourouklis, S. and Paige, C.C., (1981) A constrained least-squares approach to the general Gauss-Markov linear model. *J. Amer. Statist. Assoc.*, **76**, 620-625.

Modi, J.J., (1982) Jacobi methods for eigenvalue and related problems - in a parallel computing environment. Ph.D. Thesis, University of London.

Modi, J.J. and Clarke, M.R.B., (1984) An alternative Givens Ordering. *Numer. Math.*, **43**, 83-90.

Modi, J.J. and Parkinson, D., (1982) Study of Jacobi methods for eigenvalues and singular value decomposition on DAP. Computer Physics Communications, 26, 317-320.

Paige, C.C., (1978) Numerically stable computations for general univariate linear models. Commun. Statist.-Simula. Computa., B7(5), 437-453.

Paige, C.C., (1979a) Fast stable computations for generalized linear least-squares problems. *SIAM J. Num. Anal.*, **16**, 165-171.

Paige, C.C., (1979b) Computer solution and perturbation analysis of generalized linear least-squares problems. *Maths Comp.*, **33**, 171-183.

Paige, C.C., (1985) The general linear model and the generalized singular value decomposition. *Linear Algebra Applic.*, **70**, 269-284.

Paige, C.C., (1984) Computing the generalized singular value decomposition. School of Computer Science, McGill University, Montreal, Quebec, Canada, H3A 2K6. (Submitted to *SIAM J. Sci. Stat. Comput.*)

Paige, C.C. and Saunders, M.A., (1981) Towards a generalized singular value decomposition. *SIAM J. Num. Anal.*, **18**, 398-405.

Stewart, G.W., (1983) A method for computing the generalized singular value decomposition. In "Matrix Pencils". Eds. Kagstrom, B. and Ruhe, A., Springer-Verlag, Berlin.

Van Loan, C.F., (1976) Generalizing the singular value decomposition, *SIAM J. Num. Anal.*, **13**, 76-83.

Van Loan, C.F., (1984) Computing the CS and generalized singular value decomposition. Technical Report CS-604, Dept. Computer Science, Cornell University, Ithaca, New York.

Wilkinson, J.H., (1965) The Algebraic Eigenvalue Problem. Oxford University Press, London.

A PIPELINED ARRAY FOR LINEARLY CONSTRAINED LEAST-SQUARES OPTIMISATION

T.J. Shepherd and J.G. McWhirter
(Royal Signals and Radar Establishment, Malvern, Worcs)

1. INTRODUCTION

A particularly important aspect in the design of advanced signal processing systems is the need to match algorithms and processor architectures. It is obviously important to ensure that the algorithm solves the required problem in a numerically stable and well-conditioned manner. However, it is also important to ensure that the algorithm can be implemented at the required data rate using an efficient, cost effective processor. There is no point in having a very sophisticated algorithm if it cannot be implemented in practice and conversely there is little point in designing a very efficient processor if the associated algorithm is unreliable. The need to take both aspects into account simultaneously provides the signal processing designer with a very considerable challenge. However, the different requirements need not be mutually opposed and, indeed, the extra discipline imposed by architectural considerations may well lead the designer to a better algorithm.

In this paper some of these points will be illustrated with reference to the specific problem of linearly constrained least-squares minimisation. This is a very important problem in many aspects of science and engineering [1] with particular application to signal processing in the context of adaptive beamforming. The application to adaptive beamforming has been discussed in previous publications [2,3] and will not be described in detail here. We will concentrate instead on presenting a solution to the basic mathematical problem and relating it to an associated processor architecture. Particular attention will be devoted to a novel technique whereby the linear constraints may be incorporated neatly into the least-squares processor.

In Section 2 we present a conventional formulation of the linearly constrained least-squares problem and discuss its relationship to adaptive beamforming. The "canonical" least-squares problem is also introduced. Section 3 then describes how the canonical problem can be solved by the method of Q-R decomposition using complex Givens rotations on a pipelined triangular processor array. In Section 4 we go on to explain how the least-squares problem with one or more linear constraints may be reduced to canonical form using a simple "pre-processor" array and how the function of the pre-processor can actually be incorporated within the basic triangular array. Finally, in Section 5 we point out how the triangular processor may be implemented as a systolic array and comment on its significance.

2. LEAST-SQUARES FORMULATION

2.1 *General Least-Squares Problem with Linear Constraints*

The general linearly constrained least-squares problem may be expressed as follows. Given an (n x p) "design" matrix $\underline{\tilde{X}}(n)$ (n>p), find the p-element vector $\underline{\tilde{w}}(n)$ which minimises $||\underline{e}(n)||$, where

$$\underline{e}(n) = \underline{\tilde{X}}(n)\underline{\tilde{w}}(n) \qquad (2.1.1)$$

subject to the N (<p) constraints

$$\underline{\tilde{s}}_i^T\underline{\tilde{w}}(n) = \mu_i \quad (i=1,2,\ldots\ldots N). \qquad (2.1.2)$$

It will be assumed throughout this paper that the design matrix $\underline{\tilde{X}}(n)$ and the weight vector $\underline{\tilde{w}}(n)$ are complex, and so

$$||\underline{e}(n)||^2 = \underline{e}^H(n)\underline{e}(n) \qquad (2.1.3)$$

(where the superscript H denotes Hermitian conjugation).

For a p-element adaptive antenna array of the type depicted in Fig. 1, the design matrix $\underline{\tilde{X}}(n)$ takes the form

$$\underline{\tilde{X}}(n) = \begin{bmatrix} x_1(1) & x_2(1) & \ldots\ldots & x_p(1) \\ x_1(2) & x_2(2) & \ldots\ldots & x_p(2) \\ \cdot & \cdot & & \cdot \\ \cdot & \cdot & & \cdot \\ \cdot & \cdot & & \cdot \\ x_1(n) & x_2(n) & \ldots\ldots & x_p(n) \end{bmatrix} \qquad (2.1.4)$$

where $x_j(t)$ denotes the complex signal received by the jth antenna element at the tth sample time. Each row of the matrix $\underline{\tilde{X}}(n)$ (henceforth referred to as the "data" matrix) constitutes a single "snapshot" of data $\underline{\tilde{x}}^T(t)$ i.e. the p-element vector of complex data samples received at the corresponding sample time t. Thus the data matrix, which constitutes all snapshots of data collected from time t = 1 to t = n, may be expressed in the form

$$\underline{\tilde{X}}^T(n) = [\,\underline{\tilde{x}}(1),\ \underline{\tilde{x}}(2),\ \ldots\ \underline{\tilde{x}}(n)\,]. \qquad (2.1.5)$$

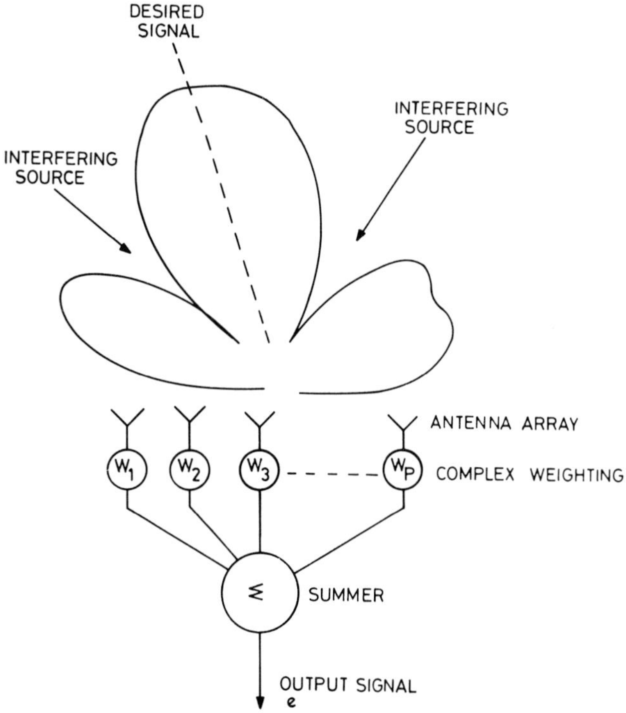

Fig. 1 Adaptive antenna array

Each column of the data matrix constitutes the vector of signal samples $\underline{x}_j(n)$ received by the corresponding antenna element j, and so it may also be written as

$$\underline{\tilde{X}}(n) = [\,\underline{x}_1(n),\ \underline{x}_2(n),\ \ldots\ \underline{x}_p(n)\,]. \qquad (2.1.6)$$

The output from an adaptive array is formed as the inner product between the latest data snapshot $\underline{\tilde{x}}(n)$ and the corresponding estimate $\underline{\tilde{w}}(n)$ of the complex weight vector.

This is known as the "*a posteriori*" residual and denoted by e(n,n), where

$$e(t,n) = \underline{\tilde{x}}^T(t)\underline{\tilde{w}}(n). \tag{2.1.7}$$

The scalar e(t,n) is simply referred to as an "output residual" and we note from equation 2.1.1 that

$$\underline{e}^T(n) = [\, e(1,n), e(2,n), \, . \, . \, . \, e(n,n) \,]. \tag{2.1.8}$$

$\underline{e}(n)$ is thus referred to as the residual vector.

In an adaptive array the complex weight vector $\underline{\tilde{w}}(n)$ determines the relative phase and amplitude of the signal in each receiver channel and so the effect of adjusting $\underline{\tilde{w}}(n)$ is to alter the spatial response characteristics of the combined antennae. Each of the N constraint vectors $\underline{\tilde{s}}_i$ ensures that the gain of the array is maintained at the constant value μ_i in a given "look direction" from which the desired signal is assumed to arrive. Selecting the weight vector $\underline{\tilde{w}}(n)$ which minimises the norm of the residual vector $\underline{e}(n)$ is thus equivalent to maximising the signal to noise ratio at the output of the array.

For a single constraint, standard minimisation techniques supply $\underline{\tilde{w}}(n)$ as a solution of the normal equations [4]

$$(\underline{\tilde{X}}^H(n)\underline{\tilde{X}}(n))\underline{\tilde{w}}(n) = \lambda\underline{\tilde{s}}^* , \tag{2.1.9}$$

where λ is a complex Lagrange multiplier given by

$$\lambda = \frac{\mu}{\underline{\tilde{s}}^T(\underline{\tilde{X}}^H(n)\underline{\tilde{X}}(n))^{-1}\underline{\tilde{s}}^*} \tag{2.1.10}$$

The complexity of forming the estimated covariance matrix $(\underline{\tilde{X}}^H(n)\underline{\tilde{X}}(n))$ and then solving equation 2.1.9 renders this direct approach operationally expensive, and so, when the direct approach is used, the weight vector is not normally updated every sample time but computed once only for each block of data $\underline{\tilde{X}}(n)$[5]. The corresponding processor architecture is also quite complicated, requiring separate units to store the data matrix, form the covariance matrix estimate, solve equation (2.1.9) and form the residual vector $\underline{e}(n)$. In addition, the system requires a number of high speed buses to transfer data between individual units and a sophisticated controller to ensure the correct sequence and timing of operations.

Since the condition number of the estimated covariance matrix $(\underline{\tilde{X}}^H(n)\underline{\tilde{X}}(n))$ is the square of the condition number of the original data matrix $\underline{\tilde{X}}(n)$, the direct approach is also inefficient from the numerical point of view and a method which does not involve squaring the data in this way would be clearly preferable.

2.2 Canonical Least-Squares Problem

We shall seek an alternative method of least-squares minimisation applied at first to a more restricted problem than that described above. We define the "canonical problem" as that in which there is a single specific constraint - namely, that the complex weight $w_p(n)$ is fixed at unity, while the weights $w_1(n), w_2(n), \ldots w_{p-1}(n)$ remain unrestricted. This is equivalent to choosing the following parameters for a single constraint,

$$\underline{\tilde{s}}^T = [0, 0, 0, \ldots 1] \quad , \quad \mu = 1 \qquad (2.2.1)$$

in the general problem. Rationalising notation by defining the reduced (n x [p-1]) data matrix $\underline{X}(n)$ and the reduced (p-1)-element vector $\underline{w}(n)$ through the relations

$$\underline{\tilde{X}}(n) = [\underline{X}(n), \underline{x}_p(n)] = [\underline{X}(n), \underline{y}(n)] \qquad (2.2.2)$$

and

$$\underline{\tilde{w}}^T(n) = [\underline{w}^T(n), 1] \qquad (2.2.3)$$

the residual vector may be written

$$\underline{e}(n) = \underline{X}(n)\underline{w}(n) + \underline{y}(n) \ . \qquad (2.2.4)$$

This is the form of the residual vector for the canonical least-squares problem. We term the vector $\underline{y}(n)$ the "primary" channel, and the columns of the reduced matrix $\underline{X}(n)$ the "auxiliary" channels.

3. THE CANONICAL LEAST-SQUARES PROCESSOR

3.1 Q-R Decomposition

A well-attested means of solving the least-squares problem without the need to form the covariance matrix estimate is provided by Q-R decomposition of the data matrix $\underline{X}(n)$[6].

The decomposition will be employed here in the form of a transformation upon the data matrix,

$$\underline{Q}(n)\underline{X}(n) = \begin{bmatrix} \underline{R}(n) \\ \underline{O} \end{bmatrix} , \tag{3.1.1}$$

where $\underline{Q}(n)$ is an (n x n) unitary matrix and $\underline{R}(n)$ is ([p-1] x [p-1]) upper triangular. Left-multiplication of the residual vector $\underline{e}(n)$ in equation 2.2.4 by the unitary matrix $\underline{Q}(n)$ then gives

$$\underline{Q}(n)\underline{e}(n) = \begin{bmatrix} \underline{R}(n) \\ \underline{O} \end{bmatrix} \underline{w}(n) + \begin{bmatrix} \underline{u}(n) \\ \underline{v}(n) \end{bmatrix} , \tag{3.1.2}$$

where the vectors $\underline{u}(n)$ and $\underline{v}(n)$ are formed from the upper (p-1) and lower (n-p+1) elements respectively of the vector $\underline{Q}(n)\underline{y}(n)$. As $\underline{Q}(n)$ is unitary, the norm of $\underline{Q}(n)\underline{e}(n)$ is equal to the norm of $\underline{e}(n)$, and so least-squares minimisation may be executed using the right-hand side of equation 3.1.2. The norm of this vector is clearly a minimum with respect to the complex weights $\{w_j(n)\}$ when the following condition is satisfied,

$$\underline{R}(n)\underline{w}(n) + \underline{u}(n) = \underline{O} . \tag{3.1.3}$$

Equation 3.1.3 is a triangularised set of linear equations which can be simply solved by means of back-substitution. Moreover, if the Q-R decomposition is exact, equation 3.1.3 is better conditioned than the corresponding normal equations, since the condition number of the matrix $\underline{R}(n)$ is equal to that of the original data matrix $\underline{X}(n)$.

3.2 Givens Rotation Algorithm and Triangular Processor

The method by which the Q-R decomposition of $\underline{X}(n)$ is to be performed is that of elementary two-dimensional plane rotations, or Givens rotations [7]. This method has the virtue that triangularisation may be carried out recursively in time (that is, as n increases), and that the elementary operations constituting the triangularisation procedure involve only two vector components at a time.

The elementary Givens rotation will be defined here to perform the following unitary transformation upon a pair of complex row vectors,

$$\begin{bmatrix} c & s^* \\ -s & c \end{bmatrix} \begin{bmatrix} 0 & 0 & \dots & r_i & r_{i+1} & \dots & r_k & \dots \\ 0 & 0 & \dots & z_i & z_{i+1} & \dots & z_k & \dots \end{bmatrix} \quad (3.2.1)$$

$$= \begin{bmatrix} 0 & 0 & \dots & r'_i & r'_{i+1} & \dots & r'_k & \dots \\ 0 & 0 & \dots & 0 & z'_{i+1} & \dots & z'_k & \dots \end{bmatrix}$$

where c and r_i are defined to be real, and

$$c^2 + |s|^2 = 1 \quad . \quad (3.2.2)$$

The values of c and s which effect this transformation are obtained from

$$r'_i = (r_i^2 + |z_i|^2)^{\frac{1}{2}} \quad (3.2.3)$$

$$c = r_i/r'_i \quad (3.2.4)$$

$$s = z_i/r'_i \quad . \quad (3.2.5)$$

(Note that r'_i is also real)

The iterative procedure by which successive triangularisations of $\underline{X}(n)$ are performed is best illustrated using the structure in Fig. 2, which represents a Q-R least-squares processor. Each processor node stores an element $r_{ij}(n)$ of the matrix $\underline{R}(n)$, or an element $u_i(n)$ of the vector $\underline{u}(n)$, as shown. Diagonal elements of $\underline{R}(n)$ are stored in the circular nodes. Assume that an $([n-1] \times [n-1])$ unitary matrix $\underline{Q}(n-1)$ has already been found which triangularises the data matrix $\underline{X}(n-1)$, and that the elements of $\underline{R}(n-1)$ and $\underline{u}(n-1)$ are stored in the appropriate nodes. A partial triangularisation of the updated data matrix $\underline{X}(n)$ will thus have been accomplished already, for, defining the (n x n) unitary matrix $\overline{\underline{Q}}(n-1)$ by

$$\overline{\underline{Q}}(n-1) = \begin{bmatrix} \underline{Q}(n-1) & \underline{0} \\ \underline{0} & 1 \end{bmatrix} \quad , \quad (3.2.6)$$

we find, from equation 3.1.1, that

$$\bar{\underline{Q}}(n-1)\underline{X}(n) = \bar{\underline{Q}}(n-1)\begin{bmatrix}\underline{X}(n-1)\\ \underline{x}^T(n)\end{bmatrix}$$

$$= \begin{bmatrix}\underline{R}(n-1)\\ \underline{O}\\ \underline{x}^T(n)\end{bmatrix} \quad . \qquad (3.2.7)$$

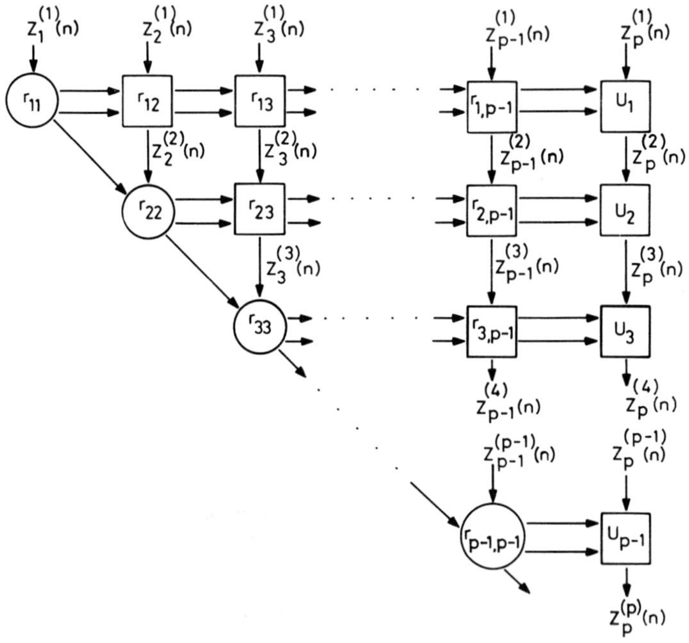

Fig. 2 Q-R Processor

It thus remains to construct a unitary (n x n) matrix $\hat{\underline{Q}}(n)$ which successively eliminates the elements of $\underline{x}^T(n)$. The first of these elements is removed using the top row of $\underline{R}(n-1)$ as the upper vector, and $\underline{x}^T(n)$ as the lower vector to be transformed in equation 3.2.1. The resulting vectors are, respectively, the new top row of $\underline{R}(n)$, and a transformed data row vector which can then be rotated in a similar way with the second row of $\underline{R}(n-1)$. The procedure is repeated with all rows of $\underline{R}(n-1)$ until all transformed data vector elements have been eliminated. It then follows that

$$\underline{Q}(n) = \hat{\underline{Q}}(n)\bar{\underline{Q}}(n-1) \quad . \qquad (3.2.8)$$

Within the structure depicted in Fig. 2, these operations are carried out row-by-row in a corresponding fashion. The new data vector $\underline{x}^T(n)$, relabelled as $[\, z_1^{(1)}(n),\ z_2^{(1)}(n)\ \ldots\ z_{p-1}^{(1)}(n)\,]$, enters the top row of the processor, and the first Givens rotation parameters are computed in the top (circular) boundary node as explained in Fig. 3a. The subsequent rotations are performed within each (rectangular) internal node of the upper row in accordance with Fig. 3b, and the transformed data vector $[\, z_2^{(2)}(n),\ z_3^{(2)}(n)\ \ldots\ z_{p-1}^{(2)}(n)\,]$ passed on to the second row, and so on.

Boundary Node

(a) 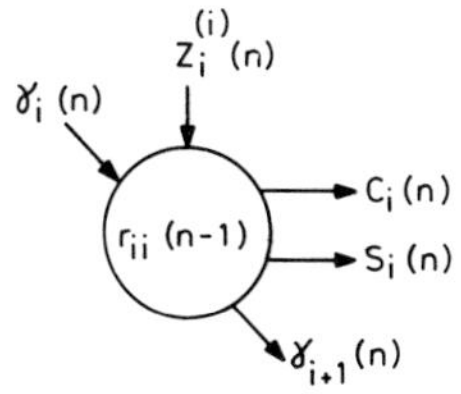

$z_i(n) \leftarrow z_i^{(i)}(n)$

If $z_i(n) = 0$, then $s_i(n) \leftarrow 0$, $c_i(n) \leftarrow 1$,
$\gamma_{i+1}(n) \leftarrow \gamma_i(n)$

else

$r_{ii}(n) \leftarrow [\, r_{ii}^2(n-1) + |z_i(n)|^2 \,]^{\frac{1}{2}}$

$c_i(n) \leftarrow r_{ii}(n-1)/r_{ii}(n)$

$s_i(n) \leftarrow z_i(n)/r_{ii}(n)$

$\gamma_{i+1}(n) \leftarrow c_i(n)\ \gamma_i(n)$

Internal Node

(b) 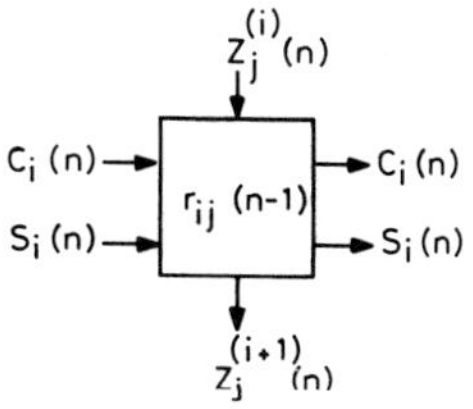

$(r_{ip}(n) = u_i(n))$

$r_{ij}(n) \leftarrow c_i(n) r_{ij}(n-1) + s_i^*(n) z_j^{(i)}(n)$

$z_j^{(i+1)}(n) \leftarrow -\ s_i(n) r_{ij}(n-1) + c_i(n) z_j^{(i)}(n)$

Fig. 3 (a) i^{th} Boundary Node (b) $(i,j)^{th}$ Internal Node

From equation 3.1.2 it is clear that the effect of the partially triangularising matrix $\bar{\underline{Q}}(n-1)$ upon the primary channel vector is as follows,

$$\bar{\underline{Q}}(n-1)\,\underline{y}(n) = \begin{bmatrix} \underline{Q}(n-1) & \underline{O} \\ & \\ \underline{O} & 1 \end{bmatrix} \begin{bmatrix} \underline{y}(n-1) \\ \\ y(n) \end{bmatrix} = \begin{bmatrix} \underline{u}(n-1) \\ \underline{v}(n-1) \\ y(n) \end{bmatrix} . \quad (3.2.9)$$

By applying identical rotations to primary channel data y(n) (relabelled as $z_p^{(1)}(n)$) in the rightmost column, the vector $\underline{u}(n-1)$ can be updated simultaneously. That is,

$$\begin{bmatrix} \underline{u}(n) \\ \\ \underline{v}(n) \end{bmatrix} = \hat{\underset{\sim}{Q}}(n) \begin{bmatrix} \underline{u}(n-1) \\ \underline{v}(n-1) \\ y(n) \end{bmatrix} . \tag{3.2.10}$$

In principle, the weight vector $\underline{w}(n)$ may now be found by stopping the processor, extracting the triangular matrix $\underline{R}(n)$ together with the corresponding vector $\underline{u}(n)$, and solving equation (3.1.3) by back-substitution using a separate processor [8]. However, this procedure is unnecessary, as will be shown in Sections 3.3 and 3.5.

3.3 Direct Residual Extraction

In adaptive beamforming, the least-squares weight vector $\underline{w}(n)$ is not the principal object of interest. Of more direct concern is the corresponding *a posteriori* residual e(n,n) since this constitutes the noise-reduced output signal from a fully adaptive array [3]. The output e(n,n) is required at each instant n and we demonstrate here how it may be obtained directly from the structure described above without prior computation of the weight vector $\underline{w}(n)$ [9]. For the canonical least-squares problem the *a posteriori* residual takes the form

$$e(n,n) = \underline{x}^T(n)\underline{w}(n) + y(n) , \tag{3.3.1}$$

in which the values of the latest data $[\underline{x}^T(n), y(n)]$ have been used in the estimation of $\underline{w}(n)$. From 2.1.8 this scalar is seen to be the nth element of the residual vector $\underline{e}(n)$. Equations 3.1.2 and 3.1.3 show that $\underline{e}(n)$ satisfies

$$\underset{\sim}{Q}(n)\underline{e}(n) = \begin{bmatrix} \underline{0} \\ \\ \underline{v}(n) \end{bmatrix} . \tag{3.3.2}$$

Defining $\underline{P}(n)$ and $\underline{S}(n)$ as the ([p-1] x n) and ([n-p+1] x n) matrix partitions of $\underset{\sim}{Q}(n)$,

$$\underset{\sim}{Q}(n) = \begin{bmatrix} \underline{P}(n) \\ \\ \underline{S}(n) \end{bmatrix} , \tag{3.3.3}$$

the residual vector can be expressed explicitly using the unitarity of $\underline{Q}(n)$,

$$\underline{e}(n) = \underline{S}^{H}(n)\underline{v}(n) \quad . \qquad (3.3.4)$$

Now, by explicit construction from the elementary rotation matrices, it is possible to show that the matrix $\hat{\underline{Q}}(n)$ takes the following form [3] ,

$$\hat{\underline{Q}}(n) = \begin{bmatrix} \underline{A}(n) & O & \underline{a}(n) \\ \underline{O} & \underline{I} & \underline{O} \\ \underline{b}^{H}(n) & \underline{O} & \gamma(n) \end{bmatrix} , \qquad (3.3.5)$$

where the constituent matrices $\underline{A}(n)$, $\underline{a}(n)$, $\underline{b}^{H}(n)$ and $\gamma(n)$ are order ([p-1] x [p-1]), ([p-1] x 1), (1 x [p-1]), and (1 x 1), respectively, and $\underline{I}$ is an ([n-p] x [n-p]) unit matrix. The scalar $\gamma(n)$ is the product of all $\{c_i(n)\}$ computed at a given time t=n, and is real. It then follows from equations 3.2.10 and 3.3.5 that

$$\underline{v}^{T}(n) = [\, \underline{v}^{T}(n-1),\ z_p(n) \,] \quad , \qquad (3.3.6)$$

where $z_p(n)$ is given by

$$z_p(n) = \underline{b}^{H}(n)\underline{u}(n-1) + \gamma(n)y(n), \qquad (3.3.7)$$

and corresponds to the fully rotated element $z_p^{(p)}(n)$ from the primary channel as illustrated in Fig. 2. From equations 3.2.6, 3.2.8, 3.3.5 and 3.3.6 the residual vector 3.3.4 may be rewritten as

$$\underline{e}(n) = \begin{bmatrix} \underline{S}^{H}(n-1) & \underline{P}^{H}(n-1)\underline{b}(n) \\ \underline{O} & \gamma(n) \end{bmatrix} \begin{bmatrix} \underline{v}(n-1) \\ z_p(n) \end{bmatrix} . \qquad (3.3.8)$$

The nth element of $\underline{e}(n)$ is read off from equation 3.3.8 as

$$e(n,n) = \gamma(n)z_p(n) \quad . \qquad (3.3.9)$$

The factor $\gamma(n)$, which we shall term the "auxiliary scalar", is simply computed within the structure of Fig. 2 by accumulating the product of the $\{c_i(n)\}$ parameters at each time n from each boundary node (Fig. 3a), while $z_p(n)$ is delivered naturally by the final internal node. The *a posteriori* residual is thus readily available at each time recursion, upon inclusion of a single extra multiplier. Note that the scalar residual is well-defined by the processor computations, even when the data matrix is not of full rank and the weight vector cannot be computed uniquely.

Another form of residual, which occurs naturally in other least-squares algorithms such as the least-squares lattice [10] and fast Kalman algorithms [11] , is the "*a priori*" residual e(n,n-1). For the canonical least-squares problem this takes the form

$$e(n,n-1) = \underline{x}^T(n)\underline{w}(n-1) + y(n) \quad . \qquad (3.3.10)$$

The *a priori* residual may also be obtained directly from the triangular processor as we shall now show.

The explicit time update of $\underline{R}(n)$ at time t=n is expressed as follows,

$$\begin{bmatrix} \underline{R}(n) \\ \underline{O} \\ \underline{O} \end{bmatrix} = \hat{\underline{Q}}(n) \begin{bmatrix} \underline{R}(n-1) \\ \underline{O} \\ \underline{x}^T(n) \end{bmatrix} \qquad (3.3.11)$$

In terms of the constituent components of $\hat{\underline{Q}}(n)$ given in equation 3.3.5, the nth row of this vector equation reads

$$\underline{O} = \underline{b}^H(n)\underline{R}(n-1) + \gamma(n)\underline{x}^T(n) . \qquad (3.3.12)$$

Substituting $\underline{x}^T(n)$ from 3.3.12 into 3.3.10 and using 3.1.3 for time t=n-1 gives

$$e(n,n-1) = \frac{1}{\gamma(n)} \underline{b}^H(n)\underline{u}(n-1) + y(n) . \qquad (3.3.13)$$

Finally, 3.3.7 allows e(n,n-1) to be expressed in terms of $z_p(n)$,

$$e(n,n-1) = z_p(n)/\gamma(n) \quad . \qquad (3.3.14)$$

Thus the *a priori* residual is computed from the same quantities as the *a posteriori* residual. They are related by the expression

$$e(n,n) = \gamma^2(n)e(n,n-1) \quad , \qquad (3.3.15)$$

and $z_p(n)$ may be written

$$z_p(n) = \{e(n,n)e(n,n-1)\}^{\frac{1}{2}}. \qquad (3.3.16)$$

In the context of least-squares lattice and fast Kalman algorithms, a scalar of the form of $z_p(n)$ has been termed a "rationalised" residual [10], and the quantity $(1-\gamma^2(n))$ a "log-likelihood" variable [12], owing to its occurrence in the description of multivariate Gaussian processes.

So far, we have only described how the triangular array may be used to carry out the sequence of Givens rotations $\hat{\underline{Q}}(n)$ on a single data snapshot $\underline{x}(n)$. However, since the procedure only utilises one row of the array during each rotation stage, it is clear that the entire process could be pipelined by inputting a new data snapshot to the array on every rotation cycle. Thus there would (in general) be p data snapshots stored within the array at any instant, each at a different stage in the reduction process. The triangular array is then capable of producing one fully computed *a posteriori* residual every rotation cycle.

3.4 Square-Root-Free Algorithm

Gentleman [13] and Hammarling [14] have derived modified versions of the Givens rotation Q-R algorithm which require no square-root operation. These can also be implemented in a highly pipelined manner on a triangular array of the type depicted in Fig. 1 and, unless otherwise stated, all of the discussion in this paper applies equally well to both versions. The essence of the square-root-free algorithm is the following factorisation of $\underline{R}(n)$,

$$\underline{R}(n) = \underline{D}^{\frac{1}{2}}(n)\underline{K}(n) \quad , \qquad (3.4.1)$$

where

$$\underline{D}^{\frac{1}{2}}(n) = \mathrm{diag}[\, r_{11}(n),\ r_{22}(n)\ \ldots\ r_{p-1,p-1}(n)\,] \ , \qquad (3.4.2)$$

and $\underline{K}(n)$ is unit upper triangular, containing elements which are rational functions of the elements of $\underline{X}(n)$. (This latter property can be proved by construction.) The squares of the elements of $\underline{D}^{\frac{1}{2}}(n)$ become stored in the boundary nodes.

For convenience we shall choose to employ the "two-multiply" square-root-free (TMSRF) algorithm described by Gentleman [13]. This can be derived after defining the following quantities,

$$k_{ij}(n) = \frac{r_{ij}(n)}{r_{ii}(n)} \quad , \quad k_{ip}(n) = \frac{u_i(n)}{r_{ii}(n)} \tag{3.4.3}$$

$$d_i(n) = r_{ii}^2(n) \tag{3.4.4}$$

$$\alpha_i(n) = \gamma_i^2(n) \tag{3.4.5}$$

$$z_j^{(i)}(n,n-1) = z_j^{(i)}(n)/\gamma_i(n) \tag{3.4.6}$$

$$C_i(n) = (c_i(n))^2, \quad S_i(n) = \frac{\gamma_i(n) z_i(n)}{r_{ii}^2(n)} \tag{3.4.7}$$

Note that $z_p^{(p)}(n,n-1)$ is the *a priori* residual e(n,n-1), which justifies the use of the notation in 3.4.6. Using the recursion relations in Fig. 3, the TMSRF algorithm may be executed according the the steps in Fig. 4, with initialisation

$$[\, z_1^{(1)}(n,n-1),\ z_2^{(1)}(n,n-1)\ \ldots\ z_{p-1}^{(1)}(n,n-1),\ z_p^{(1)}(n,n-1)\,]$$

$$\leftarrow [\, \underline{x}^T(n),\ y(n)\,] \quad ,$$

$$\alpha_1(n) \leftarrow 1 \quad . \tag{3.4.8}$$

The squares of the elements of $\underline{D}^{\frac{1}{2}}(n)$ are updated recursively and stored in the boundary nodes while the (non-diagonal) elements of the matrix $\underline{K}(n)$ are updated and recursively stored in the internal cells. When the square-root-free algorithm is used, the *a priori* residual e(n,n-1) emerges directly from the lowest internal cell in Fig. 2. The *a posteriori* residual may then be obtained by a final multiplier which computes

$$e(n,n) = \alpha_p(n)\, e(n,n-1) \quad . \tag{3.4.9}$$

(An algorithm of very similar form to this has been derived by Ling and Proakis [15], using a recursive form of the Modified Gram-Schmidt procedure [16].)

Boundary Node

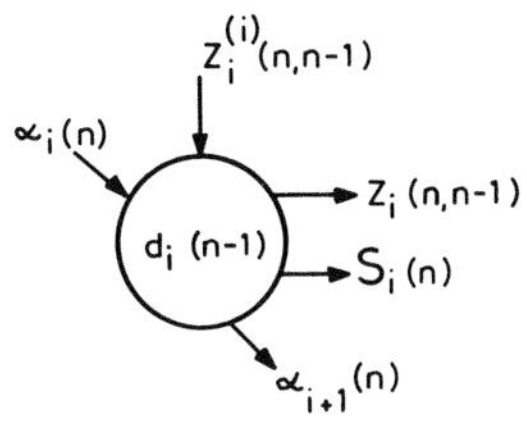

$(\alpha_1(n) = 1)$

$z_i(n,n-1) \leftarrow z_i^{(i)}(n,n-1)$

If $z_i(n,n-1) = 0$, then $S_i(n) \leftarrow 0$, $C_i(n) \leftarrow 1$,
$\alpha_{i+1}(n) \leftarrow \alpha_i(n)$

else

$d_i(n) \leftarrow d_i(n-1) + \alpha_i(n)|z_i(n,n-1)|^2$

$C_i(n) \leftarrow d_i(n-1)/d_i(n)$

$S_i(n) \leftarrow \alpha_i(n)z_i(n,n-1)/d_i(n)$

$\alpha_{i+1}(n) \leftarrow C_i(n)\alpha_i(n)$

Internal Node

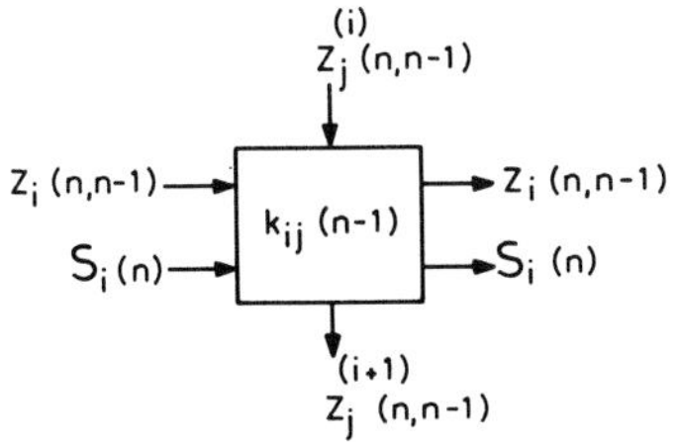

$z_j^{(i+1)}(n,n-1) \leftarrow z_j^{(i)}(n,n-1) - k_{ij}(n-1)z_i(n,n-1)$

$k_{ij}(n) \leftarrow k_{ij}(n-1) + S_i^*(n)z_j^{(i+1)}(n,n-1)$

Fig. 4 TMSRF Algorithm

3.5 Weight Flushing

Although direct computation of the least-squares weight vector may be avoided for the purposes of evaluating the *a priori* or *a posteriori* residual, the weight vector itself may well be required for other purposes. However, it may be extracted from the triangular array at any time t=n without using additional processors [3]. The adaptation is simply "frozen" by suppressing any update of the triangular matrix $\underline{R}(n)$ (or $\underline{D}^{\frac{1}{2}}(n)\underline{K}(n)$). If a data vector of the form $[\underline{x}^T, y]$ is then input to the array it will produce as output the inner product $(\underline{x}^T\underline{w}(n) + y)$ in place of the usual updated residual. It follows that by choosing $\underline{x}$ such that $x_j = \delta_{jk}$ with $y = 0$ the value of $w_k(n)$ will be output from the array. This procedure provides a powerful method for obtaining the weight vector $\underline{w}(n)$ explicitly without extracting the triangular matrix $\underline{R}(n)$ or using any additional circuitry.

It is worth noting that suppression of the adaptation can be effected most conveniently when the square-root-free algorithm is employed. In this case, the update of $\underline{D}^{\frac{1}{2}}(n)\underline{K}(n)$ is simply suppressed by setting $\alpha_1 = 0$ for any data vector input to the array during the non-adaptive period and no additional control of the array is required. On the other hand, when the basic Givens rotation algorithm is used it is necessary to suppress the update of each element of $\underline{R}(n)$ explicitly and, as a result, some additional control is required within every cell of the array.

4. PROCESSOR WITH GENERAL LINEAR CONSTRAINTS

4.1 Constraint Pre-processor

Any triangular least-squares processor, such as either of those described in Section 3, which minimises the norm of a residual vector $\underline{e}(n)$ in the canonical form of 2.2.4 will henceforth be referred to as a "canonical processor of order p", where p-1 is the order of $\underline{w}(n)$. This section will describe how the canonical processor can be employed in the more general least-squares problems outlined in Subsection 2.1.

Let us consider the most general case where there are N (< p) linear constraints of the form given in equation 2.1.2. Defining the constraint matrix $\tilde{\underline{S}}$ and gain vector $\underline{m}$ as

$$\tilde{\underline{S}} = \begin{bmatrix} \tilde{\underline{s}}_1^T \\ \tilde{\underline{s}}_2^T \\ \cdot \\ \cdot \\ \cdot \\ \tilde{\underline{s}}_N^T \end{bmatrix} \quad ; \quad \underline{m} = - \begin{bmatrix} \mu_1 \\ \mu_2 \\ \cdot \\ \cdot \\ \cdot \\ \mu_N \end{bmatrix} , \qquad (4.1.1)$$

the constraint equation then reads

$$\tilde{\underline{S}}\, \tilde{\underline{w}}(n) + \underline{m} = \underline{O} \ . \qquad (4.1.2)$$

The constraint matrix $\tilde{\underline{S}}$ is first reduced to an upper trapezoidal matrix $\tilde{\underline{S}}'$ by Q-R decomposition. Denoting the orthogonal transformation matrix by $\underline{Q}_o$, the constraints then take the form

$$\tilde{\underline{S}}'\tilde{\underline{w}}(n) + \underline{m}' = \underline{O} , \qquad (4.1.3)$$

where

$$[\tilde{\underline{S}}', \underline{m}'] = \underline{Q}_o[\tilde{\underline{S}}, \underline{m}] . \qquad (4.1.4)$$

Consider the first transformed constraint equation

$$\underline{\tilde{s}}_1'^T \underline{\tilde{w}}(n) = \mu_1' . \qquad (4.1.5)$$

Partitioning the first elements of $\underline{\tilde{s}}'_1$ and $\underline{\tilde{w}}(n)$ and the first column of the data matrix $\underline{\tilde{X}}(n)$ we define

$$\left.\begin{aligned} \underline{\tilde{s}}_1'^T &= [\, s_{11}', \ \underline{\hat{s}}_1'^T \,] \\ \underline{\tilde{w}}^T(n) &= [\, w_1(n), \ \underline{\hat{w}}^T(n)] \\ \underline{\tilde{X}}(n) &= [\, \underline{x}_1(n), \ \underline{\hat{X}}(n) \,] \end{aligned}\right\} . \qquad (4.1.6)$$

The residual vector in 2.1.7 may now be written as

$$\underline{e}(n) = \underline{x}_1(n) w_1(n) + \underline{\hat{X}}(n)\underline{\hat{w}}(n) \qquad (4.1.7)$$

and the constraint as

$$s_{11}' w_1(n) + \underline{\hat{s}}'^T_1 \underline{\hat{w}}(n) = \mu_1' . \qquad (4.1.8)$$

Thus, assuming $s'_{11} \neq 0$, $w_1(n)$ may be eliminated to express $\underline{e}(n)$ in the canonical form

$$\underline{e}(n) = \underline{X}'(n)\underline{\hat{w}}(n) + \underline{y}'(n) \qquad (4.1.9)$$

where

$$\underline{X}'(n) = \underline{\hat{X}}(n) - \frac{1}{s'_{11}} \underline{x}_1(n) \underline{\hat{s}}_1'^T \qquad (4.1.10)$$

and

$$\underline{y}'(n) = \frac{\mu'_1}{s_{11}'} \underline{x}_1(n) . \qquad (4.1.11)$$

These transformed data channels may be obtained from the input data using the fixed network shown in Fig. 5a with

$$q_{1j} = s_{1j}' (j = 1, 2 \ldots p) \quad ; \quad q_1^y = - \mu_1' . \qquad (4.1.12)$$

Thus the network in Fig. 5a serves as a pre-processor which converts the problem from one with the general constraint in equation 2.1.2 into one with the canonical constraint 2.2.1.

If there were no further constraints to consider, the norm of $\underline{e}(n)$ could now be minimised using a canonical least-squares processor of order p with inputs $[\underline{X}'(n), \underline{y}'(n)]$.

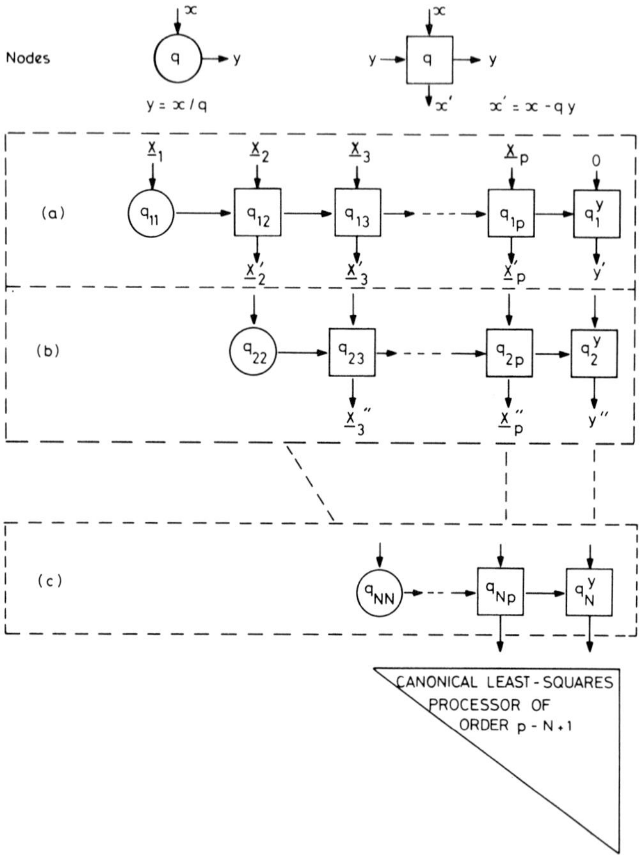

Fig. 5 Constraints Pre-processor (a) First constraint (b) Second constraint (c) N^{th} constraint

Now consider the second constraint, which may be written in the form

$$\underline{\tilde{s}}_2'^T \underline{\hat{w}}(n) = \mu_2' . \tag{4.1.13}$$

This may be used, in turn, to eliminate $w_2(n)$ from the problem so that the residual vector takes the form

$$\underline{e}(n) = \underline{X}''(n)\underline{\hat{\hat{w}}}(n) + \underline{y}''(n) \tag{4.1.14}$$

where, assuming $s'_{22} \neq 0$,

$$\underline{X}''(n) = \hat{\underline{X}}'(n) - \frac{1}{s'_{22}} \underline{x}'_2(n)\hat{\underline{s}}'^{T}_2 \qquad (4.1.15)$$

and

$$\underline{y}''(n) = \underline{y}'(n) + \frac{\mu'_2}{s'_{22}} \underline{x}'_2(n) , \qquad (4.1.16)$$

with a new partition defined by

$$\left.\begin{aligned} \underline{X}'(n) &= [\,\underline{x}'_2(n),\ \hat{\underline{X}}'(n)\,] \\ \underline{s}'^{T}_2 &= [\,s'_{22},\ \hat{\underline{s}}'^{T}_2\,] \\ \hat{\underline{w}}^{T}(n) &= [\,w_2(n),\ \hat{\hat{\underline{w}}}^{T}(n)\,] \end{aligned}\right\} . \qquad (4.1.17)$$

The transformed data $[\,\underline{X}''(n),\ \underline{y}''(n)\,]$ can be obtained by passing the data $[\,\underline{X}'(n),\ \underline{y}'(n)\,]$ through the fixed pre-processor network shown in Fig. 5b with

$$q_{2j} = s'_{2j}\,(j = 2,\ 3\ \ldots\ p) \quad ; \quad q^{y}_{2} = -\,\mu'_2 \ . \qquad (4.1.18)$$

Thus, two pre-processor networks may be used in cascade to take account of two linear constraints. The residual vector $\underline{e}(n)$ is again transformed into canonical form and, if there were no further constraints to consider, its norm could be minimised by inputting the data $[\,\underline{X}''(n),\ \underline{y}''(n)\,]$ to a canonical least-squares processor of order p-1.

The above procedure can obviously be extended to incorporate all N of the constraints in equation 4.1.2 and leads to the use of N pre-processor networks in cascade, their transformed output being input to a canonical least-squares processor of order p-N+1.

4.2 Application of Constraints by Canonical Processor

We conclude this section by pointing out how the constraint pre-processor function may readily be incorporated within a canonical least-squares processor of order p+1.

Consider first a canonical least-squares processor which employs the basic Givens rotation algorithm. The top N rows of such a processor could clearly be used to transform the initial constraint matrix $\tilde{\underline{S}}$ into the trapezoidal form $\tilde{\underline{S}}'$ (assuming that $s'_{ii} \neq 0$ for i=1,2...N). The individual elements of $\underline{\tilde{S}}'$ would then be stored as the elements of $\underline{R}$ within the corresponding processor cells. Assume that all adaptation within the top N rows is subsequently "frozen" by suppressing any update of the stored values. It follows from inspection of the cell functions in Fig. 3 that the function of these rows would then be identical to that of the pre-processor network in Fig. 5, the stored elements s'_{ij} corresponding directly to the pre-processor parameters q_{ij}. The entire minimisation procedure may then be completed by allowing the remaining rows to adapt and act as a canonical least-squares processor of order p-N+1.

The constraint pre-processor function may be incorporated similarly into a canonical least-squares processor of order p+1 which utilises the TMSRF algorithm. The top N rows are used to reduce the constraint matrix to unit upper trapezoidal form, the elements k_{ij} within the internal cells corresponding to the normalised elements s'_{ij}/s'_{ii} of the trapezoidal constraint matrix. Inspection of the cell functions specified in Fig. 4 reveals that if the adaptation within these rows is subsequently frozen by setting $\alpha_i = 0$ (and thus suppressing the update of d_i and k_{ij}) they will then perform the same function as that of the pre-processor network in Fig. 5. Allowing the remaining rows to adapt freely completes the entire constrained least-squares computation.

It should be noted once again that the suppression of adaptation required to apply the "captured" constraints can be effected much more conveniently when the TMSRF algorithm is being used since it is only necessary then to set the parameter $\alpha_i = 0$ in the boundary cell, and this automatically controls the entire row of cells.

4.3 Constraints with Leading Zeros

The constraint pre-processor technique described in section 4.1 will obviously break down if the leading element s'_{ii} of any reduced constraint vector takes zero value. The problem does not arise of course if all N leading principal sub-matrices of the constraint matrix $\underline{\tilde{S}}$ are non-singular and this condition is certainly fulfilled for a linear antenna array of uniformly spaced elements [4] . In this case each leading sub-matrix $\underline{S}[j]$ (j = 1, 2 ... N) of $\underline{\tilde{S}}$ is proportional to a matrix of the Vandermonde type,

$$\underline{S}^{[j]} \propto \begin{bmatrix} 1 & \lambda_1 & \lambda_1^2 & \cdots & \lambda_1^j \\ 1 & \lambda_2 & \lambda_2^2 & \cdots & \lambda_2^j \\ \vdots & \vdots & \vdots & \vdots & \vdots \\ 1 & \lambda_j & \lambda_j^2 & \cdots & \lambda_j^j \end{bmatrix} , \qquad (4.3.1)$$

where

$$\lambda_j = \exp(i\theta_j) \ . \qquad (4.3.2)$$

Provided that no two θ_j are equal (i.e they correspond to different look directions) these matrices are known to be non-singular. However, the problem could arise in more general circumstances and so this sub-section is devoted to showing how it may readily be overcome. The first point to note is that during the initial trapezoidal reduction of the constraint matrix $\underline{\tilde{S}}$, any vector whose first k elements are zero will not be stored within the first k rows of the triangular processor array. The Givens rotation operates in such a way that it will simply be passed down through the array until it reaches the $k+1^{th}$ row where it is either captured and stored (if the $k+1^{th}$ row is zero) or undergoes further Givens rotations. Clearly, if $k \geq N$ for any reduced constraint vector, then the trapezoidal matrix $\underline{\tilde{S}}'$ will not be captured within the first N rows of the triangular processor array. However, the reduced vectors may still be used *in situ* to apply the required constraints.

In order to demonstrate the principle, we will consider first the case of a single constraint vector $\underline{\tilde{s}}$ whose first k elements are zero. When this vector is input to the array, it will simply pass downwards until it is captured and stored within the $k+1^{th}$ row. Assume that this row of the triangular processor is subsequently frozen so that it operates in a constraint application mode as illustrated in Fig. 6. The array may now be considered in two sections, the first being an adaptive trapezoidal network which has k rows and the second being a canonical adaptive processor with a single constraint pre-processor. It will now be shown how this configuration implements the constrained least-squares minimisation defined in equations 2.1.1 and 2.1.2.

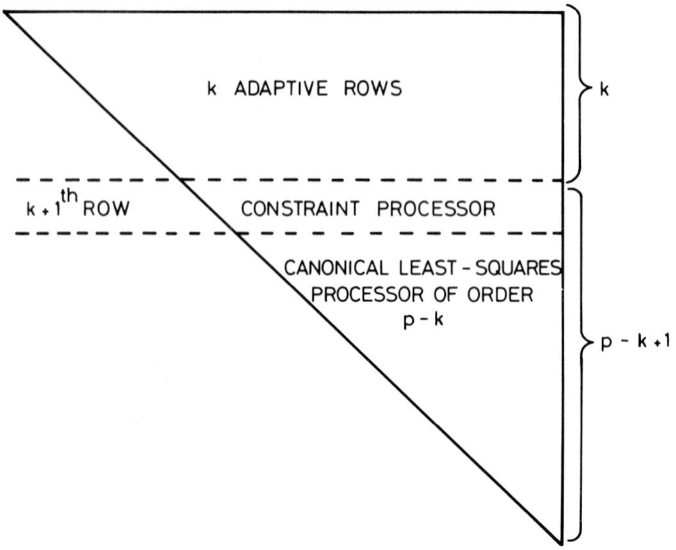

Fig. 6 Application of constraint with leading zeros

Since the first k elements of the constraint vector $\underline{\tilde{s}}$ are zero, we may express the constraint as

$$\underline{\tilde{s}}'^{T}\underline{\hat{w}}(n) = \mu \qquad (4.3.3)$$

where $\underline{\tilde{s}}'$ and $\underline{\hat{w}}(n)$ are p-k element vectors. $\underline{\tilde{s}}'$ denotes the non-zero elements of $\underline{\tilde{s}}$ and the weight vector $\underline{\tilde{w}}(n)$ has been partitioned according to

$$\underline{\tilde{w}}^{T}(n) = [\, \underline{w}^{T}(n), \ \underline{\hat{w}}^{T}(n) \,] \qquad (4.3.4)$$

The residual vector $\underline{e}(n)$ is now expressed in the form

$$\underline{e}(n) = \underline{X}(n)\underline{w}(n) + \hat{w}_1\underline{y}_1(n) + \ldots\ldots + \hat{w}_{p-k}\underline{y}_{p-k}(n) \qquad (4.3.5)$$

where

$$\underline{X}(n) = [\, \underline{x}_1(n), \underline{x}_2(n) \ \ldots\ldots\ldots \ \underline{x}_k(n) \,] \qquad (4.3.6)$$

denotes the first k columns of the data matrix and the remaining columns are denoted by

$$\underline{y}_i(n) = \underline{x}_{k+i}(n) \qquad (i=1,2,\ldots\ldots p-k) \qquad (4.3.7)$$

Consider first the problem of minimising $\|\underline{e}(n)\|$ with respect to the k element weight vector $\underline{w}(n)$, assuming that $\hat{\underline{w}}$ is fixed. Using the same notation here, for Q-R decomposition of the n x k matrix $\underline{X}(n)$, as that which was adopted for the (n x [p-1]) matrix in section 3, it follows by analogy with equation 3.1.3 that the optimum weight vector is given by

$$\underline{R}(n)\underline{w}(n) + \hat{w}_1\underline{u}_1(n) + \ldots\ldots + \hat{w}_{p-k}\underline{u}_{p-k}(n) = 0 \tag{4.3.8}$$

where $\underline{u}_i$ denotes the upper k elements of $\underline{Q}(n)\underline{y}_i(n)$. Denoting the lower n-k elements of $\underline{Q}(n)\underline{y}_1(n)$ by $\underline{v}_i(n)$, it also follows by comparison with equation 3.3.4 that

$$\underline{e}(n) = \underline{S}^H(n)(\hat{w}_1\underline{v}_1(n) + \ldots\ldots + \hat{w}_{p-k}\underline{v}_{p-k}(n)) \tag{4.3.9}$$

and, by analogy with equation 3.3.9, it can be shown that the *a-posteriori* residual is given by

$$e(n,n) = \gamma_o(n)(\hat{w}_1 z_{k+1}^{(k+1)}(n) + \ldots\ldots + \hat{w}_{p-k} z_p^{(k+1)}(n)) \tag{4.3.10}$$

The parameters $z_i^{(k+1)}(n)$ in equation 4.3.10 correspond to those in Fig. 2 while the scalar $\gamma_o(n)$ denotes the product of all cosine parameters computed in the top k boundary processors at time t=n.

In order to complete the overall least-squares minimisation problem, it remains to determine the p-k element vector of weights $\hat{\underline{w}}(n)$ which minimises the norm of the residual vector $\underline{e}(n)$ as expressed in equation 4.3.9 subject to the constraint in equation 4.3.3. Now since the matrix $\underline{S}(n)$ denotes the lower n-k rows of the unitary matrix $Q(n)$, it follows that $\underline{S}(n)\underline{S}^H(n) = \underline{I}$ and so, from equation 4.3.9 we have

$$\|\underline{e}(n)\| = (\hat{w}_1\underline{v}_1(n) + \ldots\ldots + \hat{w}_{p-k}\underline{v}_{p-k}(n))\| \tag{4.3.11}$$

The minimisation with respect to $\hat{\underline{w}}$ can therefore be achieved by supplying the vectors $\underline{v}_1(n), \underline{v}_2(n) \ldots\ldots \underline{v}_{p-k}(n)$ as input data to a canonical least-squares processor of order p-k with a single pre-processor to apply the constraint vector $\tilde{\underline{s}}$. But this is precisely the function of the bottom p-k+1 rows in

Fig. 6 and, furthermore, from equation 3.3.6 it can be seen that the elements of each vector $\underline{v}_i(n)$ are precisely the scalars $z_{k+1}^{(k+1)}(n)$ which emerge from the top k rows and enter the constraint pre-processor. It follows that the bottom p-k+1 rows in Fig. 6, treated as a separate least-squares processor, produce the output residual

$$e_1(n,n) = \hat{w}_1(n) z_{k+1}^{(k+1)}(n) + \ldots\ldots + \hat{w}_{p-k}(n) z_p^{(k+1)}(n) \qquad (4.3.12)$$

where $\hat{\underline{w}}(n)$ is the vector which minimises $\|\underline{e}(n)\|$. It can therefore be deduced from equation 4.3.10 that the *a-posteriori* residual for the overall minimisation problem is given by

$$e(n,n) = \gamma_0(n) e_1(n,n) \qquad (4.3.13)$$

This output will be produced directly by the complete array provided that the product of cosine parameters $\gamma_0(n)$ produced in the top k rows is supplied as input to the boundary processor in the $k+1^{th}$ row in the normal way. If the TMSRF algorithm is employed, then it is necessary to ensure that the value of $\alpha_k(n)$ which enters the $k+1^{th}$ boundary processor is passed on as input to the $k+2^{th}$, even though its value is set to zero within the $k+1^{th}$ boundary computation in order to freeze the adaptation.

The discussion above, which relates to a single constraint with k leading zeros, can readily be generalised to apply to a constraint matrix $\underline{\tilde{S}}$ for which the reduced trapezoidal matrix $\underline{\tilde{S}}'$ is captured on arbitrary rows within the canonical least-squares processor array. The general philosophy is then as follows. First input the constraint matrix to the processor array with all rows operating in the adaptive mode. Then freeze those rows on which the reduced constraint matrix has been captured leaving all others free to adapt. Then input the data matrix in the usual way ensuring that the appropriate product of rotation parameters is generated by the scalar which propagates along the boundary of the array.

5. DISCUSSION

In this paper we have described a triangular processor array which can be used to carry out linearly constrained least-squares optimisation in a highly efficient pipelined manner. The processor array is highly regular and has many desirable architectural features. However, it has some limitations which may readily be avoided. We note that in the array of Fig. 2, all internal cells are idle while the rotation parameters are being computed in the boundary cells and that the boundary cells, in turn, are idle while the rotations are being applied in the internal cells. We also note that the computed rotation parameters must be broadcast to all of the internal cells simultaneously - an undesirable feature for circuit and system design. These limitations can be overcome by configuring the processor to operate as a systolic array [8]. The rotation parameters are simply delayed by one rotation cycle per column as they propagate from left to right across the array and a corresponding delay is imposed on the input data vectors $\underline{x}(n)$, i.e. each sample $x_j(n)$ is input to the array one rotation cycle later than the sample $x_{j-1}(n)$ in the previous column. This data skew ensures that the algorithm is still executed correctly. Since the boundary cells are no longer required to wait for one rotation to be applied before starting to compute the next, the systolic array is fully efficient. Every cell may be used on each rotation cycle, the cycle time being dictated by the processing rate of the boundary or internal cell (whichever is slower). Furthermore, since the vertical flow of data within the array also takes place at the rate of one row per cycle, it can be seen that each processor is only required to transfer information to or from its nearest neighbours. This is a very desirable feature which renders the array circuitry much easier to design. Overall, the triangular systolic array constitutes a much more suitable architecture for real-time signal processing than the more conventional type associated with direct solution of the Gauss normal equations as discussed in Section 2. The triangular array is also based on a superior least-squares algorithm. It operates by applying Q-R decomposition directly to the data matrix using Givens rotations - an approach which is known to be very stable and efficient from the numerical point of view. The stability is further enhanced by avoiding explicit computation of the least-squares weight vector for the purposes of evaluating the residual. The weight vector solution may be ill-conditioned even though the corresponding residual is well defined and so, for example, the triangular array produces the correct (zero) residual during early stages of the recursive procedure when the data matrix $\underline{X}(n)$ contains less than p non-zero rows and is not of full rank.

ACKNOWLEDGEMENTS

The authors wish to thank Drs. C.R. Ward and P.J. Hargrave of the Standard Telecommunications Laboratories, Harlow for support in the work reported here.

REFERENCES

[1] Lawson, C.L. and Hanson, R.J., (1974) "Solving Least Squares Problems", Prentice-Hall.

[2] Ward, C.R., Robson, A.J., Hargrave, P.J. and McWhirter, J.G., (1984) "The application of a systolic least squares processing array to adaptive beamforming" in *Proc. Int. Conf. in Acoustics, Speech, and Sig. Proc.* (ICASSP 1984), San Diego, Ca, 1984.

[3] Ward, C.R., Hargrave, P.J. and McWhirter, J.G., (1986) "A novel algorithm and architecture for adaptive digital beamforming", *IEEE Trans. Antennas and Propagation,* Vol. AP-34, No. 3.

[4] Hudson, J.E., (1981) "Adaptive Array Principles", Peter Peregrinus, Ltd., Stevenage, UK and New York.

[5] Reed, I.S., Mallett, J.D. and Brennan, L.E., (1974) "Rapid convergence rate in adaptive arrays" *IEEE Trans.* AES-10, 853-863.

[6] Golub, G.H., (1965) "Numerical methods for solving linear least-squares problems", *Num. Math.* **7**, 206-216.

[7] Givens, W., (1958) "Computation of plane unitary rotations transforming a general matrix to triangular form", *J. Soc. Ind. Appl. Math.* **6**, *26-50.*

[8] Kung, H.T. and Gentleman, W.M., (1981) "Matrix triangularization by systolic arrays", in Proc. SPIE, Real-Time Signal Processing IV, San Diego, Ca., **298**, 19-26.

[9] McWhirter, J.G., (1983) "Recursive least-squares minimization using a systolic array", in Proc. SPIE Real-Time Signal Processing VI, San Diego, Ca. **431**, 105-112.

[10] McWhirter, J.G. and Shepherd, T.J., (1983) "Least-squares lattice algorithm for adaptive channel equalisation", IEE Proc. F-130, 532-542.

[11] Falconer, D.D. and Ljung, L., (1978) "Application of fast Kalman estimation to adaptive equalization", *IEEE Trans.* COM-26, 1439-1446.

[12] Lee, D.T.L., Morf, M. and Friedlander, B., (1981) "Recursive least squares ladder estimation algorithms', *IEEE Trans* CAS-28, 467-481.

[13] Gentleman, W.M., (1973) "Least-squares computations by Givens transformations without square roots", *J. Inst. Maths. Applics.* **12**, 329-336.

[14] Hammarling, S., (1974) "A note on modifications to the Givens plane rotation". *J. Inst. Maths. Applics.* **13**, 215-218.

[15] Ling, F. and Proakis, J.G., (1984) "A recursive modified Gram-Schmidt algorithm with application to least squares estimation and adaptive filtering", in Proc. Int. Symp. on Circuits and Systems (ISCAS '84), Montreal, Quebec, Canada, 781-784.

[16] Björk, Å., (1967) "Solving linear least squares problems by Gram-Schmidt orthogonalization" *BIT* **7**, 1-21.

ESTIMATION OF SPECTRAL PARAMETERS OF CORRELATED SIGNALS IN WAVEFIELDS

J.F. Böhme

(Lehrstuhl für Signaltheorie, Ruhr-Universität, Bochum, Germany)

ABSTRACT

Statistical properties of a known method to estimate the spectral matrix of signals in a wavefield are investigated. Because the waveparameters, for example bearings and ranges, of the sources are unknown in general, a suitable technique for estimating them separately is reported. Assuming the waveparameters, conditions for the identifiability of the signal spectral parameters are derived. Optimal properties of the estimate for the spectral matrix of signals are shown, for example minimum variance unbiasedness for normally distributed data. A numerical experiment demonstrates some properties of the estimates of all parameters. For example, the estimates of signal spectral powers and bearings are significantly correlated for sources close together.

1. INTRODUCTION

This paper deals with the frequently investigated problem of estimating certain parameters of sources in a wavefield which generate signals arriving at an array of sensors. Unknown parameters are the waveparameters of the sources, for example wavenumbers or bearings and ranges, signal spectral parameters and spectral powers of sensor noise. In applications such as underwater acoustics, both the location and the source-spectral-parameter estimation problems are difficult because signals can be coherent owing to multipath propagation. We apply a procedure which allows the estimation problem of the waveparameters to be separated from that of the signal spectral parameters. The method was introduced in Böhme (1984) and is discussed in Böhme (1985), Wax (1985). Having determined the waveparameters, the problem of estimating both the spectral matrix of the signals and the spectral powers of noise is then that which, in statistical literature, is called the problem of estimation of variance components, cf. Rao-Kleffe (1980), Böhme (1983). We demonstrate that a simple, nonparametrically motivated estimate of the signal spectral matrix, as introduced in Wax et al. (1982), estimates the variance components in an optimal sense. An essential assumption is that the number of sources is smaller than the number of sensors. Otherwise, we cannot uniquely parameterize the spectral matrix of the sensor outputs. This case was thoroughly investigated in d'Assumpcao (1980), where possible coherences between signals had to be neglected.

2. MODEL AND DATA

A conventional model is used: sources $m=1,\ldots,M$ generate signals which are transmitted via the wavefield. The wavefield has known properties of propagation except for some parameters. We assume to observe successive pieces of the outputs of the sensors $n=1,\ldots,N$ each of length T. The pieces are Fourier transformed with a smooth and normalized window. If $X_n^k(\omega)$ denotes the Fourier transform of the k-th piece of the n-th sensor for a fixed frequency ω of interest, we have data $\underline{X}^k(\omega) = (X_1^k(\omega),\ldots,X_N^k(\omega))'$ $(k=1,\ldots,K)$. Correspondingly, $\underline{S}^k(\omega) = (S_1^k(\omega),\ldots,S_M^k(\omega))'$ $(k=1,\ldots,K)$ are the Fourier transformed signals striking the origin. The array output is assumed to be a zero-mean stationary vector process. The geometry and the propagation-reception conditions for the signals are described by a linear system with a (NxM)-matrix $H(\omega)$ as frequency response. The m-th column of $H(\omega)$ can be interpreted as the steering vector for the m-th source which is a vector of length 1. For example, if farfield sources generate coherent waves, the (n,m)-th element of $H(\omega)$ is $N^{-1/2} \times \exp(-j\omega\tau_{nm})$, where τ_{nm} is the time delay of the m-th signal in the n-th sensor. The array output consists of filtered signals and is additively disturbed by independent sensor noise. The spectral density matrix $C_X(\omega)$ (NxN) of the sensor outputs is

$$C_X(\omega) = H(\omega)C_S(\omega)H(\omega)^* + \nu(\omega)I. \tag{2.1}$$

Here, $C_S(\omega)$ (MxM) is the spectral density matrix of the signals, the * denotes hermitian operation, $\nu(\omega)$ is the spectral power of noise and I is the unit matrix. Since we have fixed the frequency ω, we omit its notation.

Let us apply some well known asymptotic properties of Fourier transformed stationary sensor outputs. If the window length T is large, the pieces $\underline{X}^1,\ldots,\underline{X}^K$ are independent random vectors with zero mean and covariance matrix C_X as in (2.1). Asymptotically, we can write

$$\underline{X}^k = H\underline{S}^k + \underline{U}^k \quad (k=1,\ \ldots,\ K), \tag{2.2}$$

where $\underline{U}^k$ denotes Fourier transformed noise. $\underline{S}^1, \ldots, \underline{S}^K$ are independent random vectors with zero mean and covariance matrix C_S. The $\underline{S}^k$ and $\underline{U}^k$ are not observable. Unknown parameters are in general the vector $\underline{\xi}$ of waveparameters in $H = H(\underline{\xi})$, the matrix C_S and spectral noise power ν. The number M of sources is also an unknown parameter in practice. We shall indicate a simple method to estimate it, but this general problem is beyond the scope of this paper.

3. ESTIMATION OF WAVEPARAMETERS

Since in (2.2) the noise components $\underline{U}^1, \ldots, \underline{U}^K$ are asymptotically independent random vectors with zero mean and covariance matrix νI, the parameters $\underline{\xi}$ and ν can be estimated by least squares as follows. We assume $M < N$ signals and that the $\underline{S}^k$ are random parameters. We minimize the sum of squares

$$\frac{1}{K}\sum_{k=1}^{K} |\underline{X}^k - H\underline{S}^k|^2$$

over $\underline{S}^1, \ldots, \underline{S}^K$ and $\underline{\xi}$. That means we minimize the mean power of the noise vectors $\underline{U}^k$. The solution is, cf. Böhme (1984), Nickel (1983)

$$\bar{\underline{S}}^k = \hat{\underline{S}}^k(\bar{\underline{\xi}}),\ q(\bar{\underline{\xi}}) = \min_{\underline{\xi}} q(\underline{\xi}),$$

where for nonsingular H

$$\hat{\underline{S}}^k(\xi) = (H^*H)^{-1} H^*|_{\underline{\xi}}\underline{X}^k \quad (k=1,\ \ldots,\ K), \tag{3.1}$$

$$q(\underline{\xi}) = \mathrm{tr}((I-P(\underline{\xi}))\ \hat{C}_X), \tag{3.2}$$

$$P(\underline{\xi}) = H(H^*H)^{-1}H^*|_{\underline{\xi}}, \tag{3.3}$$

$$\hat{C}_X = \frac{1}{K}\sum_{k=1}^{K} \underline{X}^k\underline{X}^{k*}. \tag{3.4}$$

$\hat{C}_X$ is the well known nonparametric estimate of C_X. Since for given $\underline{\xi}$, $q(\underline{\xi})/(N-M)$ is a reasonable estimate of the spectral noise power, the optimum $\bar{\underline{\xi}}$ minimizes the estimated spectral noise power. We can use the minimum,

$$\bar{\nu} = q(\bar{\underline{\xi}})/(N-M) \tag{3.5}$$

as an estimate of ν for unknown $\underline{\xi}$.

To increase the stability of the estimate of $\underline{\xi}$, we can use the fact that Fourier transformed stationary processes are asymptotically independent random variables for different discrete frequencies $\omega_l = 2\pi n_l/T$ $(l=1, \ldots, L)$. If we write $\underline{X}_l^k = \underline{X}^k(\omega_l)$ etc., the sum of squares

$$\frac{1}{KL} \sum_{l=1}^{L} \sum_{k=1}^{K} |\underline{X}_l^k - H_l \underline{S}_l^k|^2$$

is minimized over the $\underline{S}_l^k$ to

$$Q(\underline{\xi}) = \frac{1}{L} \sum_{l=1}^{L} q_l(\underline{\xi}).$$

An optimum $\bar{\underline{\xi}}$ then minimizes $Q(\underline{\xi})$, i.e. the arithmetic mean of the estimated spectral noise powers.

We remark that in Böhme (1985) the asymptotic normality of data in the frequency domain is applied in addition. The maximum likelihood estimate of $\underline{\xi}$ requires minimizing the geometric mean of the estimated spectral noise powers and not the arithmetic one. Starting with maximum likelihood estimates, one can formulate Akaike's or Rissanen's information criterion to estimate the number M of signals if M is less than the number N of sensors, cf. Wax (1986) and Böhme (1985) for a more detailed discussion. In Böhme (1985) and in Sandkühler and Böhme (1985) finally, a relatively simple numerical procedure for minimization of $q(\underline{\xi})$ is discussed which first generates an initial estimate by a decomposition of $\hat{C}_X$ similar to a singular value decomposition and then corrects by some iterations related to Fisher's scoring. The resulting approximation of $\bar{\underline{\xi}}$ compare favourably with classic high resolution estimation methods, for example with that in Liggett (1973).

4. IDENTIFIABILITY OF SPECTRAL PARAMETERS

We assume in the sequel that the waveparameters are known or estimated before. The problem now is the estimation of C_S and ν in model (2.1). We ask for unique parameterizations of C_X. If we note that the columns of H are the known steering vectors $\underline{d}_1, \ldots, \underline{d}_M$ and denote the elements of C_S by C_{Sim}, we can write

$$C_X = \nu I + HC_S H^* = \nu I + \sum_m C_{Smm} \underline{d}_m \underline{d}_m^* \tag{4.1}$$

$$+ \sum_{i<m} [\, 2\mathrm{Re}C_{Sim} (\underline{d}_i \underline{d}_m^* + \underline{d}_m \underline{d}_i^*) + 2\mathrm{Im}C_{Sim} j(\underline{d}_i \underline{d}_m^* - \underline{d}_m \underline{d}_i^*) \,] ,$$

or simply

$$C_X = \sum_{i=o}^{M^2} \zeta_i V_i, \quad \zeta_o = \nu, \; V_o = I. \tag{4.2}$$

The ζ_i are real parameters and the V_i known hermitian matrices. That means that our problem is that of estimation of variance-covariance components cf. Rao and Kleffe (1980).

Representations (4.1) and (4.2) are unique if and only if the matrices $I, V_1, \ldots, V_{M^2}$ are linearly independent. The latter is true if and only if rank(H) = M<N. That means that the steering vectors are linearly independent and the number of signals is smaller than the number of sensors. The proof is simple: Let M<N and $\underline{d}_1, \ldots, \underline{d}_M$ be independent. Then, H^*H is positively definite and $V_1, \ldots, V_{M^2}$ are independent. If we assume in addition that $I, V_1, \ldots, V_{M^2}$ are not independent, then there are $A=A^*$ and $a \neq 0$ with $HAH^* + aI = 0$. Consequently, $A = -a(H^*H)^{-1}$ and $H(H^*H)^{-1}H = I$ which is not possible for M<N. Let us now assume that the (NxN) matrices $I, V_1, \ldots, V_{M^2}$ are independent. M^2+1 such matrices can be independent only for M<N. The independence of $V_1, \ldots, V_{M^2}$ then implies rank (H)=M.

In d'Assumpcao (1980) the matrix C_S was estimated by a diagonal matrix for a large number of sources. We now conclude that we cannot do much better in that situation. In the following, we assume rank (H) = M < N.

5. ESTIMATION OF SPECTRAL PARAMETERS

Heuristically, we can construct an estimate of C_S by use of the estimated signals $\hat{\underline{S}}^k$ from (3.1) since we assume the waveparameters. We simply apply the mean of periodograms and remove the noise part:

$$\bar{C}_S = \frac{1}{K}\sum_{k=1}^{K} \hat{\underline{S}}^k \hat{\underline{S}}^{k*} - \bar{\nu}(H^* H)^{-1}. \tag{5.1}$$

Herein, $\bar{\nu}$ corresponds to (3.5), but computed with the known wave parameters,

$$\bar{\nu} = \frac{1}{N-M} \operatorname{tr}\,((I-P)\hat{C}_X), \tag{5.2}$$

cf. (3.2, 3.3, 3.4). Replacing $\hat{\underline{S}}^k$ in (5.1) by (3.1) we find

$$\bar{C}_S = (H^* H)^{-1} H^* \hat{C}_X H (H^* H)^{-1} - \bar{\nu}(H^* H)^{-1}. \tag{5.3}$$

This estimate is known from Wax <u>et</u> <u>al</u>. (1982).

We show in this section, that the estimates (5.2) and (5.3) are well motivated parametric estimates. First, they are solutions of a least squares fit. If we minimize the quadratic function

$$\operatorname{tr}((\hat{C}_X - C_X)(\hat{C}_X - C_X))$$

over the ζ_i in (4.2) we obtain necessary conditions for the optimum $\tilde{\zeta}_i$ by partial derivations,

$$\sum_{k=0}^{M^2} \tilde{\zeta}_k \operatorname{tr}(V_k V_i) = \operatorname{tr}(V_i \hat{C}_X) \qquad (i = 0, \ldots, M^2).$$

Since this equation system is not well arranged, we use C_S, ν and (4.1). The corresponding conditions can be written, cf. Rao (1973),

$$\operatorname{tr}(\hat{C}_X - \tilde{C}_X) = 0, \tag{5.4}$$

$$\operatorname{tr}\,((\hat{C}_X - \tilde{C}_X)\, HBH^*) = 0, \text{ arbitrary } B = B^*, \tag{5.5}$$

where the optimal model is

$$\tilde{C}_X = H\tilde{C}_S H^* + \tilde{\nu} I.$$

Formula (5.5) means that we have

$$H^*(H\tilde{C}_S H^* + \tilde{\nu} I)H = H^*\hat{C}_X H.$$

Solving for $\tilde{C}_S$, gives (5.3) if we put $\tilde{\nu} = \bar{\nu}$. By use of this expression, we determine from (5.4)

$$\mathrm{tr}(H\tilde{C}_S H^* + \tilde{\nu} I_N) = \mathrm{tr}(H^*\hat{C}_X H(H^*H)^{-1} - \tilde{\nu} I_M) + \tilde{\nu}\,\mathrm{tr} I_N = \mathrm{tr}\hat{C}_X$$

and conclude with (5.2) that $\tilde{\nu} = \bar{\nu}$.

Let us discuss the statistical properties of the estimates. It is easy to see that the estimates are unbiased. If a complex-normal distribution of the $\underline{X}^k$ is assumed, $K\hat{C}_X$ is complex-Wishart distributed with K degrees of freedom and parameter matrix C_X, cf. Brillinger (1981), section 4.2. We then apply

$$\mathrm{Cov}(\mathrm{tr}(A\hat{C}_X),\ \mathrm{tr}(B\hat{C}_X)) = \frac{1}{K}\,\mathrm{tr}(AC_X B\ C_X)$$

for hermitian matrices A and B and find variances and covariances of the estimates

$$\mathrm{Var}\ \bar{\nu} = \frac{\nu^2}{K(N-M)},$$

$$\mathrm{Cov}(\mathrm{tr}(\bar{C}_S A), \bar{\nu}) = -\frac{\nu^2}{K(N-M)}\ \mathrm{tr}((H^*H)^{-1}A), \qquad (5.6)$$

$$\mathrm{Cov}(\mathrm{tr}(\bar{C}_S A),\ \mathrm{tr}(\bar{C}_S B)) =$$

$$\frac{1}{K}(\mathrm{tr}(A(C_S+\nu(H^*H)^{-1})B(C_S+\nu(H^*H)^{-1})) + \frac{\nu^2}{N-M}\,\mathrm{tr}((H^*H)^{-1}A)\mathrm{tr}(H^*H)^{-1}B)).$$

The normal distribution also implies simple distributional properties of the estimates. Inspection of (5.1), (5.2), and (5.6) yields that $W = K(\bar{C}_S + \bar{\nu}(H^*H)^{-1})$ is complex-Wishart distributed with K degrees of freedom and parameter matrix

$C_S + \nu(H^*H)^{-1}$, $w = \bar{\nu}K(N-M)/\nu$ is chisquare distributed with $2K(N-M)$ degrees of freedom, and W and w are independent random variables.

Without going into details, we indicate some more interesting properties of the estimates $\bar{C}_S$ and $\bar{\nu}$. Let us first look at a quadratic estimate of a linear combination of the parameters,

$$\gamma = tr(C_S B) + b\nu, \text{ arbitrary } B = B^* \text{ and } b,$$

in the sense of Rao and Kleffe (1980), Rao (1973). We suppose a priori knowledge about C_X by a positively definite matrix $C_o = HC_{So}H^* + \nu_o I$. Such an estimate is $\hat{\gamma} = tr(C_O^{-1}(HGH^* + gI) \times C_X^{-1}\hat{C}_X)$, where $G = G^*$ and g is chosen so that $\hat{\gamma}$ is unbiased. By some matrix algebra, we find

$$\hat{\gamma} = tr(H(H^*H)^{-1}B(H^*H)^{-1}H^*\hat{C}_X) + \frac{1}{N-M}(b-tr((H^*H)^{-1}B)\ tr((I-P)\hat{C}_X)).$$

Thus, a priori knowledge by C_o does not help to estimate γ and so C_S and ν because the last formula means to use $\bar{C}_S$ and $\bar{\nu}$. If we assume normally distributed data in addition, $\hat{\gamma}$ is also a locally minimum-variance unbiased estimate of γ given C_o, cf. Rao and Kleffe (1980). The independence of C_o implies the minimum variance unbiasedness of $\hat{\gamma}.\bar{C}_S$ and $\bar{\nu}$ consequently have similar properties.

For normally distributed data $\underline{X}^k$, $\bar{C}_S$ and $\bar{\nu}$ are maximum likelihood estimates. For, the log-likelihood function

$$L = -\log\det C_X - tr(\hat{C}_X C_X^{-1})$$

has to be maximized. The optimum parameters satisfy

$$tr(\tilde{C}_X^{-1}) = tr(\tilde{C}_X^{-1}\hat{C}_X\tilde{C}_X^{-1}),$$

$$tr(\tilde{C}_X^{-1}HBH^*) = tr(\tilde{C}_X^{-1}\hat{C}\tilde{C}^{-1}HBH^*), \text{ arbitrary } B = B^*,$$

where $\tilde{C}_X = H\tilde{C}_S H^* + \tilde{\nu}I$. We can solve for $\tilde{C}_S$ and find again that $\tilde{C}_S = \bar{C}_S$ and $\tilde{\nu} = \bar{\nu}$.

6. NUMERICAL EXPERIMENTS

A desirable property of the method discussed in this paper is the separation of the estimation of the waveparameters from that of the spectral parameters. In Böhme (1985), the asymptotic distributional properties of the waveparameter estimates for large K were discussed. In this paper, we describe the distributional properties of the spectral parameter estimates. The common behaviour of all estimates for moderate K is of interest in practice. Some answers can be found by numerical experiments with pseudorandom numbers. Such experiments have been executed on an array processor. A statistical analysis of these and other results is part of a forthcoming PhD dissertation by U. Sandkühler. Some details can be found in Sandkühler and Böhme (1985). In this paper, we describe only one representative experiment.

Model (2.1) is used for a line array of N=15 sensors spaced by a half wavelength in a plane wavefield consisting of spheric coherent waves. Three sources located approximately broadside generate uncorrelated signals. Unknown waveparameters are bearings and ranges. Ranges are measured with respect to the centre element of the line array. The parametrization of the time delays τ_{nm} is straight-forward, cf. Sandkühler and Böhme (1985). Two equally weak sources separated in bearing by 0.41 of beamwidth and a stronger source well separated from the others are assumed. Signal to noise ratios, $10 \log(C_{Sii}/\nu)$ $(i=1,2,3)$ are 3, 3 and 9dB and ranges, 35, 35 and 18 times antenna length, respectively. The matrices $K\hat{C}_X$ are simulated by complex Wishart matrices with K degrees of freedom and parameter matrix C_X resulting from the model supposed. K = 20 degrees of freedom were used for the experiment. 2048 matrices $\hat{C}_X$ and all estimates of bearings, ranges and spectral powers were computed which required approximately 35 minutes of array processor time (Floating Point Systems AP 100).

The results are indicated in the scatter diagrams in Fig. 1 and Fig. 2. Estimated SNR (with respect to exact spectral noise power) over estimated bearing are depicted in Fig. 1. Exact values given by the parameters of the model are denoted by crosses. Estimated range in multiples of array length over estimated bearing is depicted in Fig. 2. A quick analysis of these results using Spearman's rank correlation coefficient indicates the relationships between pairs of estimates. If we count the sources from left to right and denote the SNRs, bearings and ranges by SNR_i, α_i, and r_i, respectively, the following coefficients were computed:

i	1	2	3
SNR_i, α_i	.742	-.751	-.016
r_i, α_i	.012	.059	.002
SNR_1, SNR_2	.157		

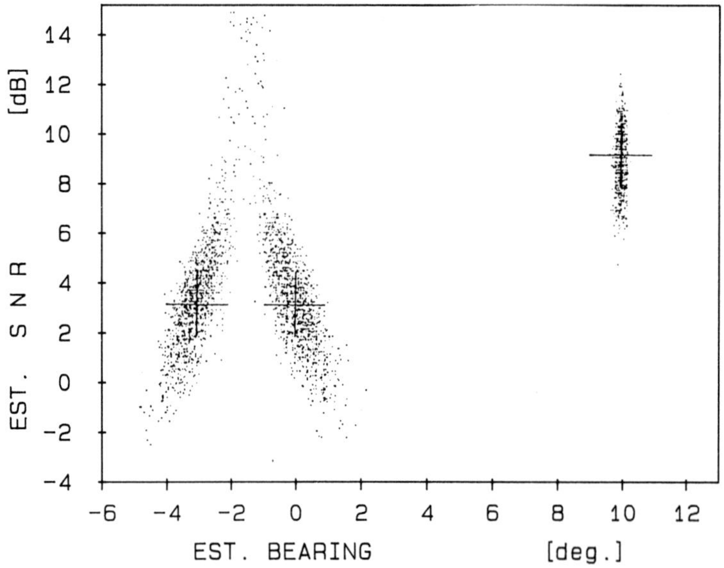

Fig. 1 Scatter diagram of 2048 estimates of bearings and signal to noise ratios for three sources. Model parameters are indicated by crosses.

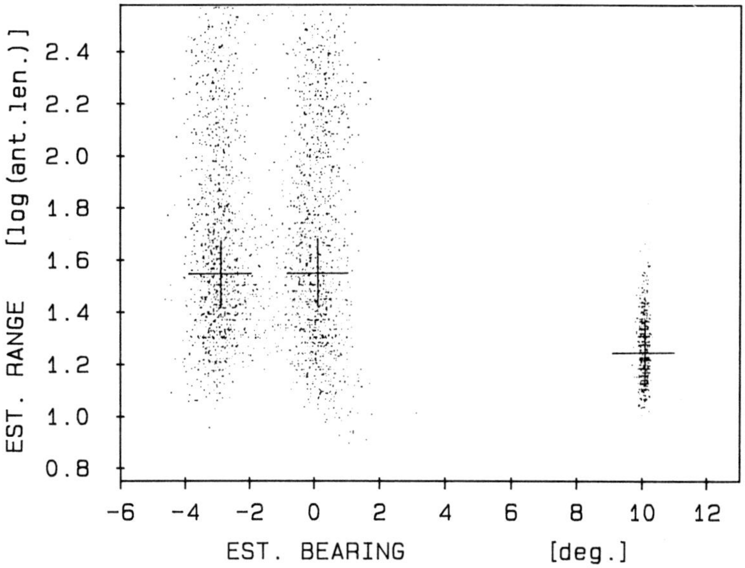

Fig. 2 Scatter diagram of estimates of bearings and ranges for the same experiment as in Fig. 1.

Let us summarize some results of this and of similar experiments. If a source is well separated from other sources, the estimates of the waveparameters and of the signal spectral power are precise and stable. The estimates tend to be uncorrelated and not affected by the other sources. A normal distribution could be used as a first approximation. This property contrasts with estimation for sources close together, for example less than a half of a beamwidth (7.3° in the experiment). Estimations of range and bearing are nearly uncorrelated, but signal spectral powers and bearings are not. We find significant influence by the neighbour source. Also an hypothesis of normal distribution should be rejected.

7. CONCLUDING REMARKS

The paper reexamines a method to estimate the waveparameters and the signal spectral parameters of waves striking over an array of sensors. First, this paper derives conditions for the identifiability of the spectral parameters. Some optimum properties of their estimates are shown. Since in this method the estimation of the waveparameters can be separated from that of the signal spectral parameters, one could suspect the estimates to be uncorrelated. A numerical experiment showed that this prediction is wrong for sources close together.

8. ACKNOWLEDGEMENT

U. Sandkühler's assistance in designing, programming and executing the numerical experiments and analysing the results, is acknowledged.

9. REFERENCES

Böhme, J.F., (1983) "On parametric methods for array processing". In "Signal Processing II", (Schüßler, H.W. ed.) North Holland, Amsterdam, 637-644.

Böhme, J.F., (1984) "Estimation of source parameters by maximum likelihood". Proc. IEEE-ICASSP, San Diego, CA, 7.3. 1-4.

Böhme, J.F., (1985) "Source Parameter estimation by approximate maximum likelihood". IEEE J. OE-10, 206-212.

Brillinger, D.R., (1981) "Time Series: Data Analysis and Theory". Expanded edition, Holden Day, San Francisco.

d'Assumpcao, H.A., (1980) "Some new signal processors for arrays of sensors". IEEE Trans. IT-26, 441-453.

Liggett, W.S., (1973) "Passive sonar: fitting models to multiple time series". In "Signal Processing" (J.W.R. Griffith et al. eds.) Academic Press, New York, 327-345.

Nickel, U., (1983) "Super-resolution by spectral line fitting" In "Signal Processing II" (Schüßler, H.W. ed.) North Holland, Amsterdam, 645-648.

Rao, C.R., (1973) "Linear Statistical Inference and its Applications". J. Wiley, New York.

Rao, C.R., and Kleffe, J., (1980) "Estimation of variance components". In "Handbook of Statistics, Vol. I" (Krishnaiah, P.R., ed.) North Holland, Amsterdam, 1-40.

Sandkühler, U. and Böhme, J.F., (1985) "Location and power estimation of multiple sources by nonlinear regression". Proc. Inst. of Acoustics Intl. Conf. Array Signal Processing, Loughborough, U.K.

Wax, M., Shan, T.J. and Kailath, T., (1982) "Location and the spectral density estimation of multiple sources". Proc. IEEE 16th Asimolar Conference, 322-326.

Wax, M., (1985) "Detection and estimation of superimposed signals". PhD dissertation, Dept. of Electrical Engineering, Stanford University.

NEW ALGORITHMS FOR INVERSION OF TOEPLITZ-LIKE PATTERNED MATRICES

M.J.C. Gover and S. Barnett
(School of Mathematical Sciences, University of Bradford)

ABSTRACT

Matrices having a Toeplitz or similar type of patterned form arise frequently in signal processing applications. This paper summarises, in a unified fashion, some recent work by the authors, its objectives being (i) to define, characterise and give some properties of a class of patterned matrices, of which both Toeplitz and block Toeplitz matrices are special cases, (ii) to present an efficient algorithm for inverting such matrices and (iii) to apply this algorithm to standard Toeplitz matrices which are not strongly nonsingular, when existing algorithms break down.

1. INTRODUCTION

If the elements t_{ij} of a matrix T of order n satisfy the relationship

$$t_{i+1,j+1} = t_{ij} , \quad i, j = 1, 2, ..., n - 1 \qquad (1.1)$$

then T is called a Toeplitz matrix. This matrix occurs frequently in applications, particularly in the field of signal processing (see Bunch (1985) for a list of references).

When the inversion of a Toeplitz matrix is required, advantage can be taken of the pattern of its elements so that the result is obtained in $O(n^2)$ or even $O(n\ \ell n^2\ n)$ flops, instead of $O(n^3)$ flops using a standard method. The same order of efficiency can also be achieved when solving a Toeplitz system of linear equations. In fact the first $O(n^2)$ algorithm for the latter problem was given in Levinson (1947) where it was used to solve equations arising in linear filtering theory. Many further algorithms have been put forward (see Roebuck and Barnett (1978)), most of which can only be used when the Toeplitz matrix involved is strongly nonsingular, i.e. its principal minors of all orders are nonzero. In Levinson (1947) the coefficient matrix is positive definite, which is, of course, strongly nonsingular. A recent computer package by Arushanian et al. (1983) gives programmes in Fortran for solving Toeplitz, block Toeplitz and other related systems of equations based on the algorithms of Trench (1967) and Zohar (1974). The stability of these and other algorithms is discussed in Cybenko (1980) and Bunch (1985).

In this paper we show that both Toeplitz and block Toeplitz matrices are special cases of a new class of matrices called r-Toeplitz (rT) matrices. Some of the known properties of Toeplitz matrices are then extended to rT matrices. These include a characterisation of nonsingular rT matrices in terms of a simple linear matrix equation, and the important result that the elements of the inverse of an rT matrix are completely determined by those elements in the first and last r rows and columns (for the Toeplitz case see Barnett and Gover (1983)). The well known Levinson-Trench algorithm for inverting a Toeplitz matrix in $O(n^2)$ flops is then generalised to this new class of matrices, and a modification is given for solving an rT system of linear equations. In Section 4 the inversion algorithm for rT matrices is used in conjunction with the Levinson-Trench algorithm to invert a Toeplitz matrix, which is not strongly nonsingular, in $O(n^2)$ flops. Another method which weakens the strong nonsingularity condition can be found in Heinig and Rost (1984).

2. r-TOEPLITZ MATRICES

We begin by defining this new class of matrices.

Definition 2.1 A matrix $A = [a_{ij}]$ of order n is an <u>r-Toeplitz</u> (rT) matrix if

$$a_{i+r,j+r} = a_{ij}, \quad i, j = 1, 2, \ldots, n - r. \qquad (2.1)$$

It is clear that a 1-Toeplitz matrix is Toeplitz and that if $n = mr$ an rT matrix can be partitioned into $r \times r$ blocks and seen to be a block Toeplitz matrix, i.e. $A = [A_{ij}]$, $i, j = 1, 2, \ldots, m$ and $A_{i+1,j+1} = A_{ij}$, $i, j = 1,2, \ldots, m - 1$.

Example 2.1 The matrices A and B below illustrate Definition 2.1, for r = 1 and r = 2 respectively, when n = 5.

$$A = \begin{bmatrix} a_0 & a_{-1} & a_{-2} & a_{-3} & a_{-4} \\ a_1 & a_0 & a_{-1} & a_{-2} & a_{-3} \\ a_2 & a_1 & a_0 & a_{-1} & a_{-2} \\ a_3 & a_2 & a_1 & a_0 & a_{-1} \\ a_4 & a_3 & a_2 & a_1 & a_0 \end{bmatrix} \qquad B = \begin{bmatrix} b_{11} & b_{12} & b_{13} & b_{14} & b_{15} \\ b_{21} & b_{22} & b_{23} & b_{24} & b_{25} \\ b_{31} & b_{32} & b_{11} & b_{12} & b_{13} \\ b_{41} & b_{42} & b_{21} & b_{22} & b_{23} \\ b_{51} & b_{52} & b_{31} & b_{32} & b_{11} \end{bmatrix}$$

1-Toeplitz (Toeplitz) 2-Toeplitz

An algorithm for the inversion of a block Toeplitz matrix is given in Akaike (1973). A special case of a 2T matrix called conjugate-Toeplitz, where $a_{i+1,j+1} = \bar{a}_{ij}$, has been introduced in Barnett and Gover (1983) and an algorithm for its inversion can be found in Gover and Barnett (1984).

We now define two matrices which are important in the following results.

Definition 2.2 A matrix J_n of order n is the reverse unit matrix if $[J_n]_{ij} = \delta_{i,n-j+1}$ where δ_{ij} is the Kronecker delta. The reverse unit matrix therefore has ones in the secondary diagonal and zeros elsewhere.

Definition 2.3 An r-companion matrix of order n has the form

$$C = \begin{bmatrix} O & I_{n-r} \\ & C^1 \end{bmatrix} \tag{2.2}$$

where C^1 is an arbitrary r × n matrix.

A 1-companion matrix is a standard companion matrix.

Some properties of rT matrices are now given.

Theorem 2.1 (Characterisation Theorem) A nonsingular matrix A is r-Toeplitz if and only if there exist r-companion matrices C_1 and C_2 such that

$$C_1A = AJC_2^TJ \tag{2.3}$$

A 1-Toeplitz matrix is persymmetric, i.e. symmetric about the secondary diagonal, and it was shown in Barnett and Gover (1983) that in this case $C_1 = C_2$.

If $Y = A^{-1}$ then from (2.3) $YC_1 = JC_2^TJY$ and equating elements gives:

Theorem 2.2 If $Y = [y_{ij}]$ is the inverse of an rT matrix, then for i, j = 1, 2, ..., n - r

$$y_{i+r,j+r} = y_{ij} + [y_{1,j+r}, \ldots, y_{r,j+r}]\, Y_{1r}^{-T}[y_{i+r,1}, \ldots, y_{i+r,r}]^T$$

$$- [y_{i,n-r+1}, \ldots, y_{in}]\, Y_{nr}^{-1}[y_{n-r+1,j}, \ldots, y_{nj}]^T \qquad (2.4)$$

provided that the first and last principal submatrices of Y of order r, denoted by Y_{1r} and Y_{nr} respectively, are nonsingular.

This theorem shows that the elements of Y are completely determined by its first and last r rows and columns. For 1-Toeplitz matrices this result is well known (Greville and Trench (1979)).

Since the inverse of a triangular matrix is triangular an immediate consequence of Theorem 2.2 is:

Theorem 2.3 If A is a nonsingular upper (lower) triangular rT matrix, then so is its inverse.

Further details and proofs of the preceding Theorems can be found in Gover and Barnett (1983a). This report also contains, for rT matrices, a characterisation in terms of displacement rank, being a generalisation of Kailath, Kung and Morf (1979), and a representation of the inverse in terms of block triangular rT matrices which for Toeplitz matrices is given in Gohberg and Fel'dman (1974).

3. INVERSION OF AN rT MATRIX

The method used for inversion is a modification of Trench's algorithm (Trench (1964), Zohar (1969)). However, since rT matrices are not in general persymmetric, it is necessary to apply the algorithm simultaneously to both A_n and a related matrix $B_n = JA_n^TJ$. Consider the nonsingular rT matrices A_n and B_n, of order n, and their inverses in the partitioned forms

$$A_n = \begin{array}{c} \begin{array}{cc} \;\; r \;\; & \;\; n-r \;\; \end{array} \\ \left[\begin{array}{cc} A_r & R^T_{n-r} \\ & \\ S_{n-r} & A_{n-r} \end{array}\right] \begin{array}{l} r \\ \\ n-r \end{array} \end{array}, \qquad A_n^{-1} = \begin{array}{c} \begin{array}{cc} \;\; r \;\; & \;\; n-r \;\; \end{array} \\ \left[\begin{array}{cc} Y_{n-r} & E^T_{n-r} \\ & \\ F_{n-r} & M_{n-r} \end{array}\right] \begin{array}{l} r \\ \\ n-r \end{array} \end{array} \tag{3.1}$$

$$B_n = \begin{array}{c} \begin{array}{cc} \;\; r \;\; & \;\; n-r \;\; \end{array} \\ \left[\begin{array}{cc} B_r & U^T_{n-r} \\ & \\ V_{n-r} & B_{n-r} \end{array}\right] \begin{array}{l} r \\ \\ n-r \end{array} \end{array}, \qquad B_n^{-1} = \begin{array}{c} \begin{array}{cc} \;\; r \;\; & \;\; n-r \;\; \end{array} \\ \left[\begin{array}{cc} Z_{n-r} & G^T_{n-r} \\ & \\ H_{n-r} & P_{n-r} \end{array}\right] \begin{array}{l} r \\ \\ n-r \end{array} \end{array} \tag{3.2}$$

If $n = mr + p$, assume that the leading and trailing principal minors of A_n of order $\ell r + p$, $\ell = 1, 2, \ldots, m - 1$ are non-zero. Note that the trailing principal minors of A_n are the leading principal minors of B_n. The steps of the algorithm are now as follows.

<u>Step 1</u> (Calculation of starting values Y_p, Z_p, E_p, F_p, G_p, H_p)

Find A^{-1}_{r+p}, using a standard inversion procedure, and hence B^{-1}_{r+p} $(=JA^{-T}_{r+p}J)$.

<u>Step 2</u> (Recursion procedure for finding the r-width border of A_n^{-1})

(a) For $\ell = 0, 1, \ldots, m - 2$ and $k = \ell r + p$ determine

$$\alpha_k = Z_k J_r R^1_{k+r} + G^T_k J_k R_k, \quad \beta_k = Z^T_k J_r S^1_{k+r} + H^T_k J_k S_k \tag{3.3}$$

$$\gamma_k = Y_k J_r U^1_{k+r} + E^T_k J_k U_k, \quad \delta_k = Y^T_k J_r V^1_{k+r} + F^T_k J_k V_k \tag{3.4}$$

where R^1_{k+r} is the matrix of the last r rows of R_{k+r} and similarly for S^1_{k+r}, U^1_{k+r} and V^1_{k+r} .

(b) Solve for Y_{k+r} and Z_{k+r} the equations

$$Y_{k+r} - \gamma_k \beta_k Y_{k+r} \alpha_k^T \delta_k^T = Y_k + \gamma_k Z_k^T \delta_k^T \tag{3.5}$$

$$Z_{k+r} - \alpha_k \delta_k Z_{k+r} \gamma_k^T \beta_k^T = Z_k + \alpha_k Y_k^T \beta_k^T \tag{3.6}$$

(c) Determine

$$E_{k+r} = \begin{bmatrix} E_k \\ O \end{bmatrix} Y_k^{-T} Y_{k+r}^T - \begin{bmatrix} J_k H_k Z_k^{-1} \\ J_r \end{bmatrix} \alpha_k Y_{k+r}^T \tag{3.7}$$

$$F_{k+r} = \begin{bmatrix} F_k \\ O \end{bmatrix} Y_k^{-1} Y_{k+r} - \begin{bmatrix} J_k G_k Z_k^{-T} \\ J_r \end{bmatrix} \beta_k Y_{k+r} \tag{3.8}$$

$$G_{k+r} = \begin{bmatrix} G_k \\ O \end{bmatrix} Z_k^{-T} Z_{k+r}^T - \begin{bmatrix} J_k F_k Y_k^{-1} \\ J_r \end{bmatrix} \gamma_k Z_{k+r}^T \tag{3.9}$$

$$H_{k+r} = \begin{bmatrix} H_k \\ O \end{bmatrix} Z_k^{-1} Z_{k+r} - \begin{bmatrix} J_k E_k Y_k^{-T} \\ J_r \end{bmatrix} \delta_k Z_{k+r} \tag{3.10}$$

<u>Step 3</u> (Completion of $A_n^{-1} = Y_n$)

For i, j = 1, 2, ..., n - 2r

$$[A_n^{-1}]_{i+r,j+r} = [A_n^{-1}]_{ij} + [F_{n-r} Y_{n-r}^{-1} E_{n-r}^T - J_{n-r} G_{n-r} Z_{n-r}^{-T} H_{n-r}^T J_{n-r}]_{ij} \tag{3.11}$$

<u>Notes</u> (i) If p = 0, Step 1 gives $Y_0 = A_r^{-1}$, $Z_0 = B_r^{-1}$ and the terms involving E_0, F_0, G_0 and H_0 are deleted. In this case the whole algorithm is equivalent to that for block Toeplitz matrices in Akaike (1973).

(ii) Step 2(b) can be solved using for instance Golub, Nash and Van Loan (1979).

(iii) The algorithm is carried out in $O(8rn)^2$ flops. When $r = 1$ persymmetry reduces this to $O(4n^2)$. Also $A = B$, $Y = Z$ and so only (3.3), (3.5), (3.7), (3.8) and (3.11) are required. The algorithm is then identical to that in Zohar (1969).

The following modifications are needed to solve a set of linear equations

$$A_n X_n = C_n$$

where X_n and C_n are column vectors, and A_n is an rT matrix of order n.

Step 1 We need the additional calculation of $X_{r+p} = A_{r+p}^{-1} C_{r+p}$.

Step 2 Include the additional calculations θ_k and X_{k+r} where

$$\theta_k = Z_k^{-T} H_k^T J_k C_k + J_r C_{k+r}^1 \ , \ X_{k+r} = \begin{bmatrix} X_k + J_k G_k \theta_k \\ J_r Z_k^T \theta_k \end{bmatrix} \tag{3.12}$$

and C_{k+r}^1 consists of the last r elements of C_{k+r}.

Step 3 is not required.

Notes (iv) When $r = 1$ this algorithm is identical to that in Zohar (1974).

(v) An operation count for the solution of X gives $O(6rn^2)$ flops.

(vi) It is not necessary to restrict X and C to a single column, so that several sets of equations can be solved simultaneously.

Finally it should be noted that if $|A_n|$ denotes the determinant of A_n then

$$|A_n| = \frac{|A_{r+p}|}{\prod_{\ell=1}^{m-1} |Y_{\ell r+p}|} \tag{3.13}$$

Proofs of the algorithms and other results in this section, together with further details, are given in Gover and Barnett (1985).

4. INVERSION OF TOEPLITZ MATRICES WHICH ARE NOT STRONGLY NONSINGULAR

The algorithm of Trench, as derived from the previous section with $r = 1$ is:

Step T1 $Y_0 = \frac{1}{a_0}$.

Step T2 For $k = 0, 1, \ldots, n - 2$

$$\alpha_k = Y_k A_{-k-1} + E_k^T J_k R_k, \quad \beta_k = Y_k a_{k+1} + F_k^T J_k S_k, \quad Y_{k+1} = \frac{Y_k}{1 - \alpha_k \beta_k} \tag{4.1}$$

$$E_{k+1} = \frac{Y_{k+1}}{Y_k} \begin{bmatrix} E_k - \alpha_k J_k F_k \\ -\alpha_k Y_k \end{bmatrix}, \quad F_{k+1} = \frac{Y_{k+1}}{Y_k} \begin{bmatrix} F_k - \beta_k J_k E_k \\ -\beta_k Y_k \end{bmatrix} \tag{4.2}$$

Note: For $k = 0$, E_0 and F_0 are deleted.

Step T3 For $i, j = 1, 2, \ldots, n - 1$, $i + j \leqslant n - 1$

$$[A_n^{-1}]_{i+1,j+1} = [A_n^{-1}]_{ij} + \frac{1}{Y_{n-1}} [(F_{n-1})_i (E_{n-1})_j - (F_{n-1})_{n-j} (E_{n-1})_{n-i}] \tag{4.3}$$

where $A = [a_{ij}] = [a_{i-j}]$ is a Toeplitz matrix and $(E_{n-1})_i$ and $(F_{n-1})_i$ denote the ith element of the vectors E and F respectively. Note that α_k, β_k and Y_k are scalars.

It can be seen from (4.1) that the algorithm breaks down when $1 - \alpha_k \beta_k = 0$ and it is shown in Gover and Barnett (1983b) that this is equivalent to the principal minor of A of order $k + 2$ being zero. However, since a Toeplitz matrix is also r-Toeplitz for any $r < n$, the algorithm in Section 3 may be used for some suitable r, to "jump over" the singularity. If there are t consecutive zero principal minors of increasing order then, in general, $r = t + 2$. It is also necessary for at least two nonzero principle minors to follow the sequence of zero principal minors in order that the r-Toeplitz algorithm can be used. After "jumping over" the singularity we return to Step T2. The method outlined here will clearly only be efficient if the number of zero principal minors is small in comparison to n.

More details and analysis of the above procedure can be found in Gover and Barnett (1983b, 1985).

The example below illustrates the procedure in the case of a single zero principal minor.

Consider the Toeplitz matrix of order 8 given by

$$A_8 = \begin{bmatrix} 2 & 1 & -4 & -1 & 6 & -5 & 104 & -65 \\ -1 & 2 & 1 & -4 & -1 & 6 & -5 & 104 \\ 0 & -1 & 2 & 1 & -4 & -1 & 6 & -5 \\ -2 & 0 & -1 & 2 & 1 & -4 & -1 & 6 \\ 7 & -2 & 0 & -1 & 2 & 1 & -4 & -1 \\ -18 & 7 & -2 & 0 & -1 & 2 & 1 & -4 \\ 47 & -18 & 7 & -2 & 0 & -1 & 2 & 1 \\ -124 & 47 & -18 & 7 & -2 & 0 & -1 & 2 \end{bmatrix} \qquad (4.4)$$

<u>Step T1</u> $\gamma_0 = \frac{1}{2}$.

<u>Step T2</u> $k = 0$: $\alpha_0 = \frac{1}{2}$, $\beta_0 = -\frac{1}{2}$, $\gamma_1 = \frac{2}{2}$, $E_1 = \left[-\frac{1}{5}\right]$, $F_1 = \left[\frac{1}{5}\right]$

$$k = 1: \ \alpha_1 = -\frac{9}{5}, \ \beta_1 = -\frac{1}{5}, \ \gamma_2 = \frac{5}{8}, \ E_2 = \begin{bmatrix} \frac{1}{4} \\ \frac{9}{8} \end{bmatrix}, \ F_2 = \begin{bmatrix} \frac{1}{4} \\ \frac{1}{8} \end{bmatrix}$$

$$k = 2: \ \alpha_2 = -\frac{1}{2}, \ \beta_2 = -\frac{11}{8}, \ \gamma_3 = 2, \ E_3 = \begin{bmatrix} 1 \\ 4 \\ 1 \end{bmatrix}, \ F_3 = \begin{bmatrix} \frac{23}{4} \\ \frac{3}{2} \\ \frac{11}{4} \end{bmatrix} \qquad (4.5)$$

$$k = 3: \ \alpha_3 = -4, \ \beta_3 = -\frac{1}{4}, \ 1 - \alpha_3\beta_3 = 0.$$

This indicates that $|A_5| = 0$ and so the Toeplitz algorithm breaks down. Step T3 is now used to determine the first three rows and columns of A_4^{-1} so that the algorithm in Section 3 can be used with r = 3.

<u>Step T3</u> (n = 4) $[A_4^{-1}]_{22} = \frac{7}{2}$, $[A_4^{-1}]_{23} = \frac{47}{4}$, $[A_4^{-1}]_{32} = 1.$

(4.6)

Using (4.5) and (4.6), together with the persymmetry of A_4^{-1}, we have

$$A_4^{-1} = \left[\begin{array}{ccc|c} 2 & 1 & 4 & 1 \\ \frac{23}{4} & \frac{7}{2} & \frac{47}{4} & 4 \\ \frac{3}{2} & 1 & \frac{7}{2} & 1 \\ \hline \frac{11}{4} & \frac{3}{2} & \frac{23}{4} & \cdot \end{array}\right] = \begin{bmatrix} Y_1 & E_1^T \\ F_1 & \cdot \end{bmatrix}$$

where Y_1, E_1 and F_1 are now the submatrices in (3.1).

<u>Step 2</u> of the rT algorithm gives

(a)

$$\alpha_1 = \begin{bmatrix} 226 & -12 & -4 \\ 647 & -35.5 & -12 \\ 171 & -9 & -5 \end{bmatrix}, \quad \beta_1 = \begin{bmatrix} -4.5 & 1.25 & -0.25 \\ -12 & 4.5 & -1.5 \\ -10.5 & 3.25 & -1.25 \end{bmatrix}$$

(b)

$$Y_4 = \frac{-1}{32} \begin{bmatrix} 18 & 76 & 870 \\ 90 & 380 & 4318 \\ 130 & 556 & 6166 \end{bmatrix}$$

(c)

$$E_4 = \frac{-1}{32}\begin{bmatrix} 1028 & 4948 & 6852 \\ 1956 & 9620 & 13604 \\ 1492 & 7076 & 9620 \\ 324 & 1492 & 1956 \end{bmatrix}, \quad F_4 = \frac{-1}{32}\begin{bmatrix} 66 & 300 & 3094 \\ 59 & 258 & 2809 \\ 6 & 36 & 258 \\ 1 & 6 & 59 \end{bmatrix}$$

Return to the Toeplitz algorithm using the first row and column of Y_4 and the first columns of E_4 and F_4 for Y_6 in (4.1) and E_6 and F_6 in (4.2).

<u>Step T2</u> $k = 6$: $\alpha_6 = -\frac{119}{4}$, $\beta_6 = -\frac{1}{32}$, $\gamma_7 = -8$

$$E_7 = -[47, 466, 1237, 1742, 2382, 1334, 238]^T$$

$$F_7 = -\frac{1}{4}[178, 314, 226, 162, 59, 6, 1]^T$$

Finally using Step T3 with $n = 8$ gives

$$A_8^{-1} = -\frac{1}{4}\begin{bmatrix} 32 & 188 & 1864 & 4948 & 6968 & 9528 & 5336 & 952 \\ 178 & 1048 & 10378 & 27632 & 38888 & 53244 & 29868 & 5336 \\ 314 & 1856 & 18338 & 49092 & 68992 & 94696 & 53244 & 9528 \\ 226 & 1344 & 13234 & 35716 & 50068 & 68992 & 38888 & 6968 \\ 162 & 960 & 9474 & 25436 & 35716 & 4092 & 27632 & 4948 \\ 59 & 354 & 3469 & 9474 & 13234 & 18338 & 10378 & 1864 \\ 6 & 36 & 354 & 960 & 1344 & 1856 & 1048 & 188 \\ 1 & 6 & 59 & 162 & 226 & 314 & 178 & 32 \end{bmatrix}$$

The calculations in this example were carried out using MATLAB (Moler (1981)).

REFERENCES

Akaike, H., (1973) "Block Toeplitz matrix inversion". *SIAM J. Appl. Math.* **24**, 234-241.

Arushanian, O.B., et al. (1983) "The Toeplitz Package Users Guide". ANL-83-16, Argonne Nat. Lab., Argonne, Illinois.

Barnett, S. and Gover, M.J.C., (1983) "Some extensions of Hankel and Toeplitz matrices". *Linear and Multilinear Algebra* **14**, 45-65.

Bunch, J.R., (1985) "Stability of methods for solving Toeplitz systems of equations". *SIAM J. Sci. Stat. Comput.* **6**, 349-364.

Cybenko, G., (1980) "The numerical stability of the Levinson-Durbin algorithm for Toeplitz systems of equations". *SIAM J. Sci. Stat. Comput.* **1**, 303-319.

Gohberg, I.C. and Fel'dman, I.A., (1974) "Convolution equations and projection methods for their solution". Amer. Math. Soc. Providence, R.I.

Golub, G.H., Nash, S. and Van Loan, C., (1979) "A Hessenberg-Schur method for the problem AX + XB = C". *IEEE Trans. Auto. Contr.* AC-24, 909-913.

Gover, M.J.C. and Barnett, S., (1983a) "An extension of block Toeplitz matrices I: Characterisation and displacement rank". Mathematical Sciences Reports on Theory and Application of Matrices, TAM 83-13, University of Bradford.

Gover, M.J.C. and Barnett, S., (1983b) "An extension of block Toeplitz matrices II: Inversion and solution of sets of linear equations". Mathematical Sciences Reports on Theory and Application of Matrices, TAM 83-14, University of Bradford.

Gover, M.J.C. and Barnett, S., (1984) "Inversion of certain extensions of Toeplitz matrices". *J. Math. Anal. and Appl.* **100**, 339-353.

Gover, M.J.C. and Barnett, S., (1985) "Inversion of Toeplitz matrices which are not strongly nonsingular". *IMA J. Num. Anal.* **5**, 101-110.

Greville, T.N.E. and Trench, W.F., (1979) "Band matrices with Toeplitz inverses". *Linear Algebra Appl.* **27** 199-209.

Heinig, G. and Rost, K., (1984) "Algebraic methods for Toeplitz-like matrices and operators". Akademie-Verlag, Berlin, DDR.

Kailath, T., Kung, S-Y. and Morf, M., (1979) "Displacement ranks of matrices and linear equations". *J. Math. Anal. Appl.* **68**, 395-407.

Levinson, N., (1947) "The Wiener RMS error criterion in filter design and prediction". *J. Math. and Phys.* **25**, 261-278.

Moler, C.B., (1981) "MATLAB users guide". Techn. Report CS81-1, Dept. of Comput. Sci., Univ. of New Mexico.

Roebuck, P.A. and Barnett, S., (1978) "A survey of Toeplitz and related matrices". *Int. J. Systems Sci.*, **9**, 921-934.

Trench, W.F., (1964) "An algorithm for the inversion of finite Toeplitz matrices". *J. SIAM* **12**, 515-522.

Trench, W.F., (1967) "Weighting coefficients for the prediction of stationary time series from the finite past". *SIAM J. Appl. Math.* **15**, 1502-1510.

Zohar, S., (1969) "Toeplitz matrix inversion: The algorithm of W.F. Trench". *J. Assoc. Comput. Mach.* **16**, 592-601.

Zohar, S., (1974) "The Solution of a Toeplitz set of linear equations". *J. Assoc. Comput. Mach.* **21**, 272-276.

STATISTICAL TESTS FOR STREAM CIPHERS

M. Kimberley
(Goldsmiths' College, London)

and

F. Piper
(Royal Holloway and Bedford New College, London)

ABSTRACT

Many modern methods of encipherment use stream ciphers. An important element of a stream cipher system is the keystream sequence which is an infinite (but normally periodic) binary sequence. The desirable mathematical and statistical properties of sequences which are used in this way are discussed. In particular the sequences need to have good randomness properties and methods of testing these properties are described.

1. INTRODUCTION

There are many situations in which it is desirable or essential to keep messages secret to authorised people. These situations may occur in a commercial, diplomatic, military or personal context. The problem is that most means of transmitting messages - post, courier, telegraph, telephone, radio, data link - can be intercepted, albeit with varying degrees of difficulty. One safeguard is to disguise the message by changing it before transmission and, of course, by changing it back after reception; this disguise is the aim of a *cipher system*.

The theory of designing cipher systems is called *cryptography*, whilst the theory of breaking cipher systems is called *cryptanalysis*. The branch of knowledge which encompasses both cryptography and cryptanalysis is called *cryptology*. There is a sense in which the separation into cryptography and cryptanalysis is misleading since, in order to decide whether a system is 'good' or not, the cryptographer assesses what is needed to break it, whilst the cryptanalyst clearly needs a knowledge of cryptography to be effective.

In a cipher system the information before it is disguised is called the *plaintext* (or message) which is transformed into the *ciphertext* (or *cryptogram*) by the process of *enciphering*. The process of transforming the ciphertext back into the plaintext is called *deciphering* and clearly this process must give unique plaintext for given ciphertext. Enciphering and deciphering both depend on an algorithm and a key (or keys) which controls the algorithm. It is the key which provides the security and so the deciphering key must be kept secret - except from the legitimate receiver of course. If the deciphering key can be easily deduced from the enciphering key then both keys must be kept secret.

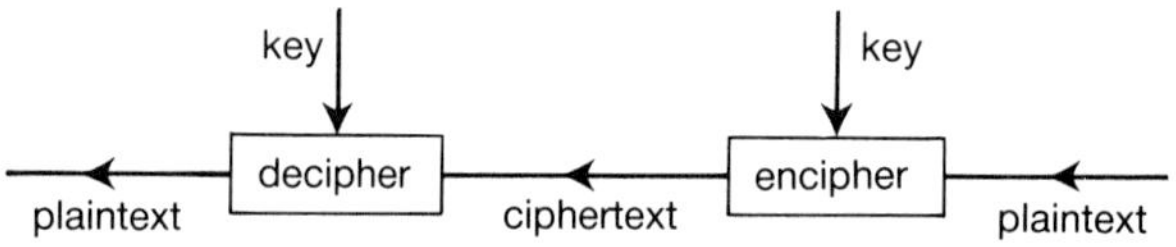

Fig. 1 Enciphering and deciphering

Modern cryptography is often used to protect data transmitted over high-speed data links or stored in computer systems. Data may need to be cryptographically protected in two ways: against unauthorised interception of messages, and against unauthorised modification or introduction of false data. The problem is particularly serious because successful interception or tampering can go unnoticed. Thus the widespread use of the computer has caused a rapid increase in the number of cryptographic problems to be solved.

There is another important way in which computers have changed cryptology: the cryptanalyst will be able to use computer power to break systems so that the strength of a cipher system has to be judged against the time needed for a fast computer to break it - the higher the stakes the more the cryptanalyst can afford to invest. For this reason modern cipher systems have to be complex.

2. PERFECT SECRECY

In order to decide whether or not a cipher system is 'good' we must have some idea - or make some assumptions - about how much information might be available to a cryptanalyst who attempts to break it. It is common to assume the following worst case conditions as stated by Beker and Piper (1982):

1. The cryptanalyst has a complete knowledge of the cipher system.

2. The cryptanalyst has obtained a considerable amount of ciphertext.

3. The cryptanalyst knows the plaintext equivalent of a certain amount of the ciphertext.

Of course the words 'considerable' and 'certain' are ill-defined and will vary according to the situation. It is clear that with these worst case conditions the security of the system relies on the secrecy of the deciphering key.

The enciphering key of a cipher system determines a transformation from the set of all possible messages, called the *message space*, to the set of all possible cryptograms, called the *cryptogram space*. A cipher system is formally defined as a finite set K of reversible transformations (or enciphering keys) of a message space M into a cryptogram space C. Associated with each message m in M is the a priori probability p(m) of m being transmitted, where $\sum_{m \in M} p(m) = 1$. Let $p(m|c)$ be the probability that message m was transmitted given that cryptogram c was received. The cipher system is said to have *perfect secrecy* if $p(m|c) = p(m)$ for every m in M and every c in C. If perfect secrecy can be achieved then the frustrated cryptanalyst obtains no information about the transmitted plaintext from the intercepted ciphertext.

Shannon (1948, 1949) discussed the idea of perfect secrecy, and established that it was achieved by the one-time pad. The one-time pad is important because it has motivated the design of many modern cipher systems, in particular stream ciphers. We shall describe the one-time pad using binary digits but the system can use other characters. Let the message to be enciphered be $m = m_o m_1 \ldots m_{n-1}$ where each $m_i = 0$ or 1. The key is a sequence $k_o k_1 \ldots k_{n-1}$ where each k_i is 0 or 1 and k_i has been chosen randomly from 0 and 1 independently of the other k_j's. The message and key are then mixed modulo-2 and bit by bit to give $c = c_o c_1 \ldots c_{n-1}$ i.e. c is a sequence of 0's and 1's given by $c_i = m_i \oplus k_i$. The system is shown in Figure 2. It is called the one-time pad because in one form of the system the sender uses a pad with a key sequence written on each page, informs the receiver (who has a similar pad) which page to use and then destroys the page after use.

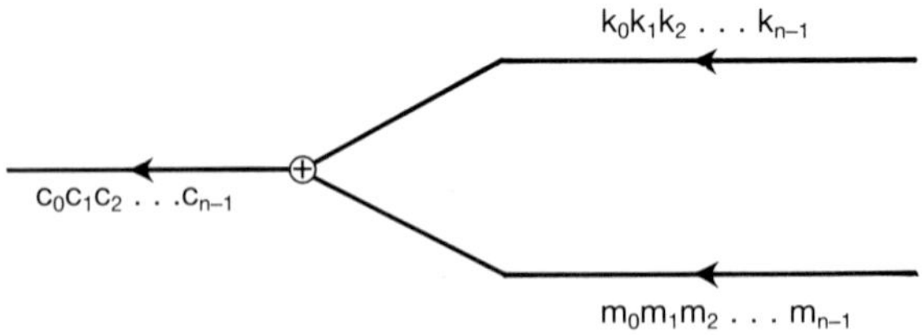

Fig. 2 The one-time pad

3. STREAM CIPHERS

Since most modern cipher systems are implemented electronically we shall assume from now on that plaintext is always a sequence of bits, 0 and 1.

A stream cipher is a system in which the plaintext is converted into ciphertext by combining it bit by bit with an infinite binary sequence called the keystream sequence. One advantage of the system is that there is no error propagation because the mixing is bit by bit. From Figure 3 the similarity to the one-time pad is clear.

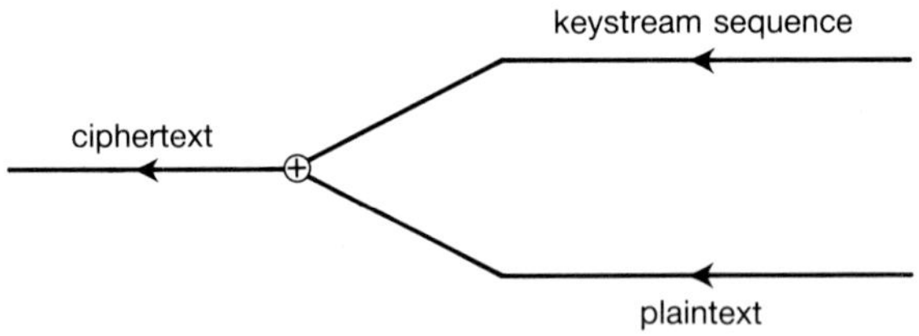

Fig. 3 Stream Cipher

The keystream sequence is generated by feeding the key to an algorithm (usually called the keystream generator). The mixer used is usually an XOR-gate denoted by $\oplus$. The ciphertext is then converted to plaintext by the receiver by mixing the keystream sequence and the received ciphertext.

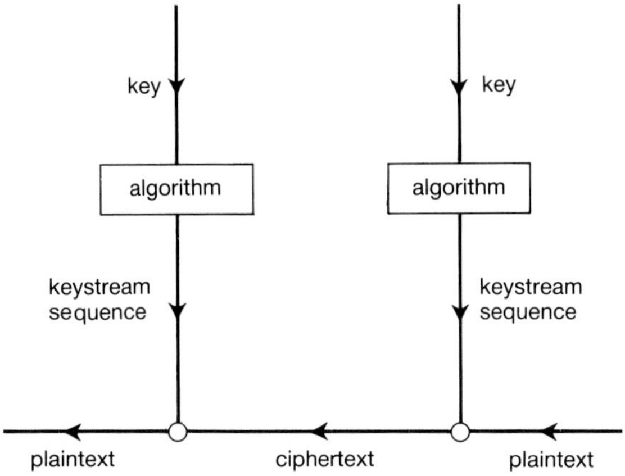

Fig. 4 Stream cipher system

An assumption of our worst case conditions is that an interceptor knows the plaintext equivalent of a certain amount of the ciphertext, which for a stream cipher implies that a certain amount of the keystream sequence is known since $k_i = m_i \oplus c_i$. It is clearly important that this should not allow the complete keystream sequence to be determined by the cryptanalyst. However in most practical situations the keystream sequence will be *periodic*, that is there exists an integer $p \neq 0$ such that $u_{t+p} = u_t$ for all t; in other words the sequence has the form $u_o u_1 u_2 \ldots u_{p-1} u_o u_1 \ldots u_{p-1} u_o \ldots$. The least value of p is called the *period* of the sequence and if p consecutive bits of the sequence are known then the sequence is determined apart from the starting position. A keystream sequence then must have large period to prevent the cryptanalyst gaining complete knowledge of it. At the very least the period needs to be greater than the length of any message sent using that particular keystream sequence.

In addition it may be that knowledge of the algorithm will allow the cryptanalyst to construct the keystream sequence from a part of the sequence even though that part is small in comparison with the period. Another requirement of the keystream sequence then is that knowledge of a part of the sequence (less than the period) should not give knowledge of the rest of the sequence.

The plaintext language of a cipher system possesses natural statistics related to the frequencies with which single letters, double letters etc. normally occur in the language. If these natural statistics are not disguised in encipherment then the cryptanalyst will use them to help break the system. A method of disguise is to make the keystream sequence as unpredictable and as unpatterned as possible. Ideally we want to ensure that the cryptanalyst cannot predict the plaintext from the ciphertext alone and also cannot predict one bit of the keystream sequence from a knowledge of other sections of the keystream sequence. The latter requirement is impossible for a periodic sequence but it is worth attempting for subsequences of length much shorter than the period. Intuitively we want sections of the keystream sequence to appear to behave randomly; of course the keystream sequence is deterministic and cannot be random, but what is required is that it should be what is usually called *pseudo-random*. We explore what is meant by this in more detail.

4. PSEUDO-RANDOM SEQUENCES

In this section we shall discuss the model of a perfect (binary) sequence and then use the following definition. An $\alpha\%$ *pseudo-random (binary) sequence* is a finite (binary) sequence of length n which, when considered as data, fits the model of a perfect sequence when tested at the $\alpha\%$ level by the set of statistical tests described below.

A *perfect sequence* A is a finite sequence of independent random variables $A_o A_1 \ldots A_{n-1}$ such that the probability that $A_j = 0$ is $\frac{1}{2}$ and the probability that $A_j = 1$ is $\frac{1}{2}$. Associated with A is the cyclic sequence $\bar{A}$ defined by regarding the first entry A_o of A as also following the last entry. When discussing $\bar{A}$ we shall use symbols obtained by placing the bar over the appropriate symbol relating to A.

Let t be an integer such that $1 \leq t \leq \frac{1}{2}(n+1)$ then a sequence of t bits is called a *t-sequence*. A t-sequence $r = r_o r_1 \ldots r_{t-1}$ is said to occur at position j in A if $r_o = A_j, r_1 = A_{j+1} \ldots r_{t-1} = A_{j+t-1}$ for some j, $0 \leq j \leq n-t$ and to occur at position j in $\bar{A}$ if the same condition holds for $0 \leq j \leq n-1$ and A_{n+i} is identified with A_i for all i. Let N_r be the number of occurrences of the sequence $r = r_o r_1 \ldots r_{t-1}$ in A and let $\bar{N}_r$ be similarly defined for $\bar{A}$; of course the occurrences of r are not independent events.

We now consider $\bar{\psi}_t^2 = \frac{2^t}{n} \Sigma(\bar{N}_r - \frac{n}{2^t})^2$ where the sum is over all 2^t possible t-sequences r. Good (1953) showed that the asymptotic distribution of $\bar{\psi}_t^2 - \bar{\psi}_{t-1}^2$ (where $\bar{\psi}_o^2 = 0$) is a χ^2-distribution with 2^{t-1} degrees of freedom.

This result leads to the *generalised serial test:* Let $k = k_o k_1 \ldots k_{n-1}$ be a binary sequence and choose $t \geq 1$ such that $\frac{n}{2^t} \geq 5$. The test statistic

$$X_t^2 = \frac{2^t}{n} \sum_r (\bar{n}_r - \frac{n}{2^t})^2 - \frac{2^{t-1}}{n} \sum_s (\bar{n}_s - \frac{n}{2^{t-1}})^2$$ where $\bar{n}_r$, $\bar{n}_s$ are the observed values of $\bar{N}_r$ and $\bar{N}_s$ respectively, r ranges over all $\underline{t}$-sequences and s ranges over all (t-1) - sequences (with $\bar{n}_s = n$ if $t = 1$). A binary sequence $k = k_o k_1 \ldots k_{n-1}$ is said to have *passed the t-serial test at the α% level* if $X_t^2 \leq P_{2^{t-1},\alpha}$ where $P_{2^{t-1},\alpha}$ is the (100-α) percentage point of the χ^2 distribution with 2^{t-1} degrees of freedom.

If n is large there will be many choices of t which satisfy the conditions. If the sequence is intended as a keystream sequence there may be certain values of t that are particularly relevant; for instance if we wish to encipher data that has been converted to binary by using the 7-bit code ITA5, the case t=7 would be important. The special cases t=1 and t=2 are often called the *frequency test* and the *serial test* respectively.

We return now to our model of a perfect sequence A. Let t be an integer such that $1 \leq t \leq n$. We partition A into consecutive hands of length t where a hand is a subsequence of the form $A_{wt} A_{wt+1} \ldots A_{(w+1)t-1}$ $(0 \leq w \leq \left[\frac{n}{t}\right] - 1)$. There are 2^t distinct possibilities for a hand (as above). However in this case we only count occurrences of $r = r_o r_1 \ldots r_{t-1}$ at positions wt $(0 \leq w \leq \left[\frac{n}{t}\right] - 1)$. Since A is a perfect sequence and the hands do not overlap, these occurrences of r are independent events and the probability of r occurring at position wt is $\frac{1}{2^t}$. Let $f = \left[\frac{n}{t}\right]$ and let F_r be the number of occurrences of r at position wt then clearly $E[F_r] = \frac{f}{2^t}$ and $\sum_r F_r = f$.

Now let k be a binary sequence as above with f_r the observed value of F_r so that $\sum_r f_r = f$. We can apply a χ^2 goodness of fit test to decide whether k could be an α% pseudo-random sequence by comparing the test statistic $\frac{2^t}{f} \sum_r (f_r - \frac{f}{2^t})^2$ and the χ^2 distribution with $2^t - 1$ degrees of freedom. Thus k is said to have *passed the poker test (for some t) at the α% level* if $\frac{2^t}{f} \sum_r (f_r - \frac{f}{2^t})^2 \leqslant P_{2^t - 1,\alpha}$ where $P_{2^t - 1,\alpha}$ is the $(100 - \alpha)\%$ point of the χ^2 distribution with $2^t - 1$ degrees of freedom where $t \geqslant 1$ is chosen so that $\frac{f}{2^t} \geqslant 5$.

As above there will be many permissible choices of t for large n and certain values of t may be appropriate. We note that when t = 1 the poker test and the frequency test are identical, but that for t > 1 the poker test and the generalised serial test differ. (The original version of the poker test was due to Kendall and Babington-Smith (1938)).

There is also a more complicated test called the *runs test*. If $k = k_o k_1 \ldots k_{n-1}$ is a binary sequence then a run is defined to be a subsequence of consecutive identical bits which is neither preceded nor succeeded by the same bit. The runs test is then based on the numbers of runs of zeroes and ones of different lengths in k. The underlying distribution theory is fully described in Mood (1940).

When considering the randomness properties of a periodic sequence it is often possible to obtain results theoretically when dealing with a complete period. However when a message is enciphered the part of the keystream sequence used will usually be much shorter than the period. It is to test small sections of the potential keystream sequence that the statistical tests above are needed. In practice they are usually implemented as part of a computer package.

5. CONCLUSION

It is important to realise that if a sequence possesses one or some of the desirable properties described above, it may still not be suitable as a keystream sequence. Also assessing the security of a cipher system involves many more considerations than the keystream sequence.

6. REFERENCES

Beker, H. and Piper, F., (1982) "Cipher Systems". Van Nostrand, Reinhold, UK.

Good, I.J., (1953) "The serial test for sampling numbers and other tests for randomness". *Proc. Camb. Phil. Soc.* **49**, pp. 276-284.

Kendall, M.G. and Babington-Smith, B., (1938) "Randomness and random sampling numbers". *Jour. Royal Stat. Soc.* (Series A) 101, pp. 147-166.

Mood, A.M., (1940) "The distribution theory of runs". *Ann. Math. Stats.* **11**, pp. 367-392.

Shannon, C.E., (1948) "A Mathematical Theory of Communication", *Bell Syst. Tech. J.*, pp. 379-423 and 623-656.

Shannon, C.E., (1949) "Communication Theory of Secrecy Systems". *Bell Syst. Tech. J.* **28**, pp. 656-715.

FAST APPROXIMATE REALIZATION WITH NO MULTIPLICATIONS OF LINEAR FILTERS

J. Shen and S. Castan
(CERFIA, Université Paul Sabatier, Toulouse, France)

ABSTRACT

Linear filters are widely used in signal processing and in image processing. The fast realization of these filters has been an important research subject.

In the present paper, by use of the spline functions, we propose a new technique which realizes approximately a given linear filter with a factor difference by the sum of the translated outputs of some sum-box filters requiring no multiplications. This method is highly efficient in image processing, in particular, for large filter kernels.

1. INTRODUCTION

Linear filters are widely used in signal processing and image processing. It is well known that the output of a linear filter can be expressed as the convolution of the input signal with the filter kernel (i.e., the impulse response of the filter). So filtering the input signal by a finite impulse response filter (FIR) of a large kernel involves many multiplications and the fast realization of such a filter has been an important research subject.

Many methods based on the decomposition of a large filter kernel into small kernels have been proposed in order to reduce the number of multiplications in the convolution or to simplify the hardware realization of the filter. The SVD (Singular Value Decomposition) and the SGK (Small Generating Kernel) techniques [1,2] decompose a linear filter kernel into a sequence of small convolution kernels and the HDC (Hierarchical Discrete Convolution) technique [3,4] calculates the convolution of the input signal with a large kernel by a weighted sum of convolutions with smaller kernels, and these in turn can be computed as weighted sums of convolutions with still smaller kernels. These methods reduce effectively the computation complexity or make easier the realization of the filter by a specialized hardware. However, the convolution of the input signal with the small kernels always requires some multiplications.

A very simple linear filter is the box filter [5] which takes as the output the average value of the input signal in a window and the fast algorithm of an N-dimensional box filter involves only 2N additions and one multiplication for one output point. Unfortunately a single box filter is of little use in signal processing because it produces a stopband characteristic with many sidelobes. Some authors have proposed the technique realizing the Gaussian filters and their derivatives by a cascade of box filters [5,6] but the decomposition of a given linear filter into box filters has not yet been solved.

In a previous paper [8], we proposed to realize a given linear filter by so-called sum-box filters which take as the output the sum of the input signal in a window. An N-dimensional sum-box filter requires only 2N additions for one output point and is therefore still simpler than a box filter. We show that a given linear filter can be decomposed into the weighted sum of some translating cascading sum-box filters. But with this method, some multiplications are always necessary because of the weight coefficients.

In this paper, we propose an efficient realization of a finite impulse response (FIR) digital filter based on the decomposition of the impulse response in terms of B-Spline functions. Under some conditions, the B-Spline functions are approximated as the Gaussian functions. On the other hand, a weighted Gaussian function can be realized as a cascade of sum-boxes, which do not require multiplications in its realization. Overall, a linear filter can be approximately realized with a factor difference by the parallel/cascade combination of sum-boxes, thereby requiring no multiplications.

2. CASCADING SUM-BOX FUNCTION

At first, we review briefly the cascading sum-box function and its relation with the B-Spline functions.

2.1 B-Spline Functions

A k-order B-Spline function is defined as:

$$\Omega_k(x) = \sum_{j=0}^{k+1} \frac{(-1)^j C_{k+1}^j (x + \frac{k+1}{2} - j)_+^k}{k!}$$

where k is the order of the B-Spline function $\Omega_k(x)$, C means the combination number and:

$$x_+^k = \begin{cases} x^k & \text{for } x \geq 0, \\ 0 & \text{otherwise.} \end{cases}$$

$\Omega_k(x)$ has the following important property [7] :

$$\Omega_k(x) = \Omega_{k-1}(x) * \Omega_0(x) \tag{1}$$

where * means the convolution.

2.2 Cascading Sum-box Function

A sum-box filter takes as the output the sum of the input signal values in a window. A 1-D sum-box filter is defined by:

$$S_0(m) = \sum_{j=-l_1}^{l_1} S_i(m-j) \tag{2}$$

where S_0 is the output, S_i is the input, and the window size is $(2\,l_1 + 1)$.

Evidently, a 1-D sum-box filter requires only 2 additions for one output point [8] and an N-dimensional sum-box filter is separable and can be decomposed as a cascade of N 1-D sum-box filters respectively in the corresponding dimensions, i.e.:

$$S_0(m_1,m_2,\ldots,m_N) = \sum_{j_1=-l_1}^{l_1} \sum_{j_2=-l_2}^{l_2} \cdots\cdots \sum_{j_N=-l_N}^{l_N} S_i(m_1-j_1,\ldots,m_N-j_N)$$

where the window size is $(2l_1 + 1) \times \ldots \times (2l_N + 1)$. We have immediately that an N-dimensional sum-box filter requires 2N additions for one output point.

Obviously, the effect of a 1-D sum-box can be described by its impulse response $W_1(x,a)$, (Fig. 1) :

$$W_1(x,a) = \Omega_O(\frac{x}{a})$$

where a is the window size of the sum-box filter.

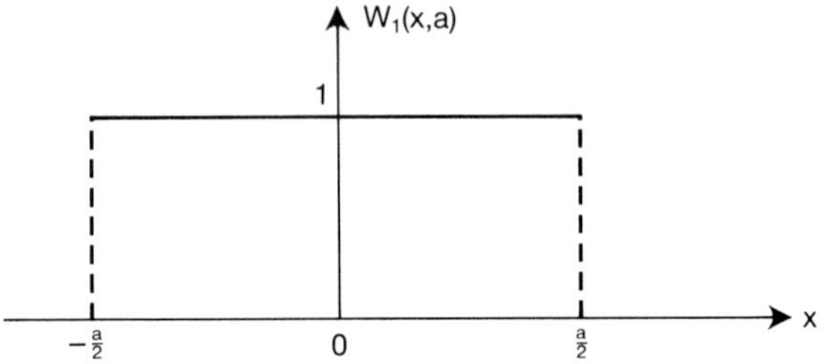

Fig. 1 The impulse response of a sum-box filter

Consider the input signal $S_i(x) = 1$, we have the output signal of the sum-box filter as the following:

$$S_O(x) = \int_{-\infty}^{\infty} S_i(t).W_1(x-t,a)\,dt$$

$$= a$$

That is, the input signal is amplified and we say that the gain of the sum-box is a.

The impulse response of a cascade of N 1-D sum-box filters in the same dimension is therefore:

$$W_N(x,a) = \underbrace{\Omega_O(\frac{x}{a}) * \ldots * \Omega_O(\frac{x}{a})}_{N \text{ times}}$$

which gives [8] :

$$W_N(x,a) = a^{N-1}\,\Omega_{N-1}(\frac{x}{a}) \qquad (3)$$

we call the function $W_N(x,a)$ the cascading sum-box function.

3. SUM-BOX REALIZATION OF WEIGHTED B-SPLINE FUNCTION

We show in this section how to realize a weighted B-Spline function by a cascade of sum-box filters.

Let $C_i . \Omega_k(\frac{x}{a})$ be the impulse response of the filter to be realized, where $\Omega_k(x)$ is the k-order B-Spline function of x; a > 1, a positive integer called the scale factor; and C_i, the weight coefficient, $C_i > 0$.

As is mentioned in a previous paper [6] , a k-order B-Spline function $\Omega_k(x)$ approaches to a Gaussian function $G(x,\sigma)$ when k is large:

$$\Omega_k(x) \approx G(x,\sigma) \equiv \frac{1}{\sqrt{2\pi}.\sigma} \exp\left(-\frac{x^2}{2\cdot\sigma^2}\right)$$

with

$$\sigma = \sqrt{\frac{k+1}{12}}$$

which gives:

$$\Omega_k(\frac{x}{a}) \approx a \cdot G(x, \sqrt{\frac{k+1}{12}} \cdot a) \qquad (4)$$

Multiplying Eq. (4) by C_i, we have:

$$C_i \cdot \Omega_k(\frac{x}{a}) \approx C_i .\ a.\ G(x, \sqrt{\frac{k+1}{12}} \cdot a) \qquad (5)$$

On the other hand, from Eqs. (3) and (4), the impulse response of a cascade of N_i sum-box filters with the window size b_i can be expressed as :

$$W_{N_i}(x, b_i) \approx b_i^{N_i} .G(x, \sqrt{\frac{N_i}{12}} b_i) \qquad (6)$$

Comparing Eq. (5) with Eq. (6), we see that: if

$$\begin{cases} C_i \cdot a = b_i^{N_i} & (7) \\ \sqrt{\frac{N_i}{12}} \cdot b_i = \sqrt{\frac{k+1}{12}} \cdot a & (8) \end{cases}$$

then the impulse response $C_i . \Omega_k(\frac{x}{a})$ can be realized by the cascade of N_i sum-box filters of window size b_i. So the essential problem for realizing the filter kernel $C_i . \Omega_k(\frac{x}{a})$ is to find the solutions N_i and b_i of the equation system (7) and (8).

3.1 Solution Existence Condition

From Eqs. (7) and (8), we have:

$$\begin{cases} N_i = \dfrac{\ln(C_i . a)}{\ln b_i} & (9) \\ N_i = (k+1) . (\dfrac{a}{b_i})^2 & (10) \end{cases}$$

Eqs. (9) and (10) give:

$$\ln b_i = \frac{b_i^2 \ln(C_i . a)}{(k+1) \cdot a^2} \qquad (11)$$

The graphs of the left and the right sides of Eq. (11) are shown in Fig. 2, where b_i is taken as the variable.

From Fig. 2, we see that there exist four possible cases:

Case 1 : $C_i = C_{mc}$, we have $b_i = b_o$ as the unique solution of Eq. (11);

Case 2 : $C_i > C_{mc}$, because the value of the right side of Eq. (11) increases with C_i, no solutions of Eq. (11) exist;

Case 3 : $\frac{1}{a} < C_i < C_{mc}$. In this case, there exist two solutions $b_i = b_{i1}$ and $b_i = b_{i2}$ for the equation (11), of which one is larger than b_o (b_{i2}) and another is smaller than b_o (b_{i1}) ;

Case 4 : $0 < C_i \leq \frac{1}{a}$, we have the unique solution for b_i in Eq. (11) and this solution $b_i \leq 1$.

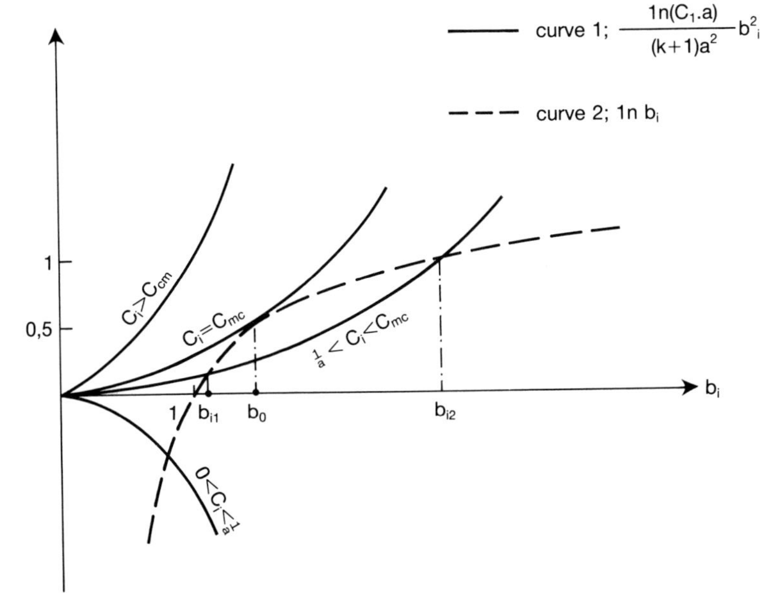

Fig. 2 Graphs of $\frac{\ln(C_i.a)}{(k+1)a^2} b_i^2$ and $\ln b_i$

The values C_{mc} and b_o can be calculated as the following. From Fig. 2, it is obvious that when $C_i = C_{mc}$, the curves 1 and 2 have the same tangent line at $b_i = b_o$, i.e. :

$$\left\{\begin{array}{l} \left.\dfrac{d \ln b_i}{d b_i}\right|_{b_i = b_o} = \left.\dfrac{d\left\{\dfrac{b_i^2 \ln(C_{mc} a)}{(k+1) a^2}\right\}}{d b_i}\right|_{b_i = b_o} \qquad (12) \\[2em] \ln b_o = \dfrac{b_o^2 \ln(C_{mc} a)}{(k+1) a^2} \qquad (13) \end{array}\right.$$

Eq. (12) gives :

$$\frac{b_o^2 \, . \ln(C_{mc} a)}{(k+1) a^2} = \frac{1}{2} \tag{14}$$

Substituting Eq. (14) in Eq. (13), we have :

$$b_o = \sqrt{e} \simeq 1.649 \tag{15}$$

which gives from Eq. (14) :

$$C_{mc} = \frac{1}{a} \cdot e^{\frac{1}{2e}(k+1) a^2} \tag{16}$$

In order that the solution b_i of the equation system (7) and (8) exists, C_i should satisfy the following condition:

$$0 < C_i \leq C_{mc} \equiv \frac{1}{a} \cdot e^{\frac{1}{2e}(k+1) a^2} \tag{17}$$

3.2 Some Practical Constraints

Because b_i is the window size of the sum-box filters in the cascade, in practice, it should satisfy :

$$b_i \geq b_{min} \tag{18}$$

where b_{min} is the minimal window size permitted of a sum-box. Evidently, the window size of a realizable sum-box is always larger than or equal to 2, we have:

$$b_{min} \geq 2 > \sqrt{e} \tag{19}$$

This means that only the solution $b_{i2} > b_o$ in the case 3 is realizable. In this case, we see from Eq. (11) that the solution $b_{i2} \geq b_o$ decreases when C_i increases. In order to assure this solution $b_{i2} \geq b_{min}$, C_i should satisfy:

$$\frac{\ln b_{min}}{b_{min}^2} \geq \frac{\ln(C_i \cdot a)}{(k+1)a^2}$$

i.e.,

$$C_i \leq \frac{1}{a} \cdot \left[b_{min}^{\frac{(k+1)a^2}{b_{min}^2}} \right] \equiv C_{min} \tag{20}$$

Considering of the constraints proposed for the case 3 (section 3.1), C_i should satisfy also:

$$\frac{1}{a} < C_i < C_{mc} \tag{21}$$

Another practical constraint is that for the solution N_i. Because N_i is the number of the sum-boxes in the cascade that we use to approximate the impulse response of a Gaussian filter (Eq. (6)), N_i should satisfy :

$$N_i \geq N_{min} \tag{22}$$

where N_{min} is the minimal number of sum-boxes in a cascade in order that the impulse response of the cascade approaches well to that of a Gaussian filter. From Eqs. (9) and (10), we have:

$$N_i \ln \frac{(k+1)a^2}{N_i} = 2 \ln(C_i . a) \tag{23}$$

Differentiating the left side of Eq. (23) with respect to N_i, we have:

$$p \equiv \frac{d}{dN_i} \left\{ N_i \ln \frac{(k+1)a^2}{N_i} \right\}$$

$$= \ln \frac{(k+1)a^2}{N_i} - 1$$

$$= \frac{2 \ln(C_i \cdot a)}{N_i} - 1$$

which gives from Eq. (9):

$$p = 2 \ln b_i - 1$$

When $C_i = C_{mc}$, we have from Eq. (15) $b_i = b_o = \sqrt{e}$, which gives $p_{C_i = C_{mc}} = 0$. And when $\frac{1}{a} < C_i < C_{mc}$, for the solution $b_{i2} > b_o$, we have $p_{C_i < C_{mc}} > 0$. Considering $\frac{d}{dC_i}\{2 \ln(C_i.a)\} > 0$, we have that if $C_i \leq C_{mc}$, the solution N_i corresponding to $b_{i2} \geq b_o$ increases with the increment of C_i. In order that the solution N_i is larger than or equal to N_{min}, C_i should satisfy the following condition from Eq. (23) :

$$C_i \geq \frac{1}{a} \left[\frac{(k+1)a^2}{N_{min}} \right]^{\frac{1}{2} N_{min}} \equiv C_{ml} \tag{24}$$

Considering of the constraint $b_i \geq b_{min}$, i.e., $C_i \leq C_{mm}$, we have from Eq. (10) that when $C_i = C_{mm}$, N_i takes its maximum:

$$(k+1)(\frac{a}{b_{min}})^2$$

and when $C_i = C_{ml}$, N_i takes its minimum N_{min}. We have therefore:

$$N_{min} \leq N_i \leq (k+1)(\frac{a}{b_{min}})^2 \tag{25}$$

Eq. (25) implies:

$$(k+1)\left(\frac{a}{b_{min}}\right)^2 \geq N_{min}$$

i.e., N_{min} and b_{min} should satisfy:

$$N_{min} \cdot b_{min}^2 \leq (k+1) \cdot a^2 \qquad (26)$$

3.3 Concluding Remarks

Combining the analyses in the sections 3.1 and 3.2, i.e., Eqs. (17), (20), (21), (24) and (26), we conclude that the solutions N_i and b_i of the equation system (7) and (8) satisfying $N_{min} \leq N_i \leq (k+1)\left(\frac{a}{b_{min}}\right)^2$ and $b_i \geq b_{min}$ exist and are unique if the following conditions are satisfied:

$$\frac{1}{a} < C_i < C_{mc} = \frac{1}{a} \cdot e^{\frac{1}{2e}(k+1)a^2} \qquad (27)$$

and
$$C_i \leq C_{mm} = \frac{1}{a}\left[b_{min}^{\frac{1}{b_{min}^2}(k+1)a^2}\right] \qquad (28)$$

and
$$C_i \geq C_{ml} = \frac{1}{a}\left[\frac{(k+1)a^2}{N_{min}}\right]^{\frac{1}{2}N_{min}} \qquad (29)$$

and
$$N_{min} \cdot b_{min}^2 \leq (k+1)a^2 \qquad (30)$$

Because $b_{min} \geq 2$, we have always $C_{mm} < C_{mc}$. And because a is a positive integer and $a > 1$, Eq. (30) implies $C_{ml} > \frac{1}{a}$. The conditions (27) - (30) are equivalent to:

$$C_{ml} \leq C_i \leq C_{mm} \qquad (31)$$

and
$$N_{min} \cdot b_{min}^2 \leq (k+1)a^2 \qquad (32)$$

We have therefore the following conclusion: if the conditions (31) and (32) are satisfied, a filter with the impulse response $C_i.\Omega_k(\frac{x}{a})$ can be approximately realized by a cascade of N_i sum-box filters of window size b_i with $N_i > N_{min}$ and $b_i \geq b_{min}$. N_i and b_i are solved from Eqs. (7) and (8).

In practical digital filtering, because N_i is the number of sum-box filters and b_i is the filter window size, only the integer solutions N_i and b_i are realizable. So if the solutions N_i and b_i that we find from Eqs. (7) and (8) are integers, they can be directly used in the sum-box realization of the weighted B-spline impulse response. If the solutions N_i and b_i that we find are not integers, we should adjust appropriately the solutions, i.e., choose N_{di} and b_{dj} (j = 1, ..., N_{di}) such that:

$$\begin{cases} \sum_{j=1}^{N_{di}} b_{dj}^2 \approx (k+1)a^2 \\ \prod_{j=1}^{N_{di}} b_{dj} \approx C_i.a \end{cases} \tag{33}$$

According to the central limit theorem, when the b_{dj}'s $(j=1,...,N_{di})$ are not too different from one another, the weighted B-Spline impulse response $C_i.\Omega_k(\frac{x}{a})$ can be realized by the cascade of N_{di} sum-box filters with respectively the window sizes $b_{d1},..., b_{dN_{di}}$. In this process of adjustment, the solutions N_i and b_i can serve us as a guide for the selection of the initial values of N_{di} and b_{dj} to find the approximate integer solutions of the equation system (33) with a sufficient precision.

4. SUM-BOX REALIZATION OF LINEAR FILTERS

In this section, we propose the technique which can realize approximately a given FIR linear filter by the cascades of sum-box filters requiring no multiplications.

4.1 B-Spline Decomposition of Linear Filters

Given a linear filter whose impulse response is f(x), we take a positive integer $a > 1$ and generate a function:

$$g(x) = f(ax)$$

Without loss of generality, suppose that g(x) is non-zero only in the interval [O,M], where M is a positive integer. (A function which is non-zero in an interval [P,Q] can always be obtained by a simple translation if we take M = Q-P).

g(x) can be expressed by its spline function [7], i.e. :

$$g(x) = \sum_{i=0}^{M+k-1} \beta_i \, \Omega_k\left(x - i + \frac{k+1}{2}\right) \quad (34)$$

where k is the order of the spline function.

Eq. (34) gives:

$$f(x) = \sum_{i=0}^{M+k-1} \beta_i \, \Omega_k\left(\frac{x}{a} - i + \frac{k+1}{2}\right) \quad (35)$$

i.e.

$$f(x) = \sum_{i=0}^{M+k-1} \operatorname{sign}(\beta_i) . |\beta_i| . \Omega_k\left(\frac{x}{a} - i + \frac{k+1}{2}\right) \quad (36)$$

Choose b_{min} and N_{min} such that:

$$N_{min} . b_{min}^2 < (k+1)a^2$$

and calculate C_{mm} and C_{ml} from Eqs. (20) and (24).

Take two thresholds C_{max} and C_{min}, such that:

$$C_{ml} \leq C_{min} \ll C_{max} \leq C_{mm} \quad (37)$$

Let $\beta = \max_{i=0}^{M+k-1} |\beta_i|$. Multiply the two sides of Eq. (36) by

$R = \frac{C_{max}}{\beta}$, we have:

$$R.f(x) = \sum_{i=0}^{M+k-1} \text{sign}\ (C_i).|C_i|.\Omega_k(\frac{x}{a} - i + \frac{k+1}{2}) \qquad (38)$$

where $C_i = R.\beta_i$.

Because $C_{max} >> C_{min}$, all the terms in Eq. (38) with $|C_i| < C_{min}$ can be neglected, and Eq. (38) becomes:

$$\left.\begin{array}{ll} & R.f(x) \simeq \sum_{i=0}^{M+k-1} \text{sign}(C_i)\ E^{(\frac{k+1}{2} - i)a} \{|C_i|.\Omega_k(\frac{x}{a})\} \\ \text{with} & \forall\, i \in \{0, \ldots, M+k-1\}, \\ & C_{min} \leq |C_i| \leq C_{max} \\ \text{or} & |C_i| = 0 \end{array}\right\} \qquad (39)$$

where E is a translation operator, $E^c\ \{f(x)\} \equiv f(x+c)$.

From Eqs. (37) and (39), all the non-zero terms in the B-Spline expression (39) of R.f(x) satisfy:

$$C_{ml} \leq |C_i| \leq C_{mm}.$$

According to the conclusion in the section 3.3, each non-zero term $|C_i|.\ \Omega_k(\frac{x}{a})$ in Eq. (39) can be realized by a cascade of sum-box filters requiring no multiplications. Overall, the filter with the impulse response R.f(x) can be realized by the sum (or difference, depending on sign (C_i)) of the outputs of the cascades of sum-box filters with respectively a translation $(\frac{k+1}{2} - i)a$, thereby requiring no multiplications.

Evidently, in order that we can find C_{max} and C_{min} such that $C_{ml} < C_{min} << C_{max} \leq C_{mm}$ (eq. (37)), it is necessary that $C_{ml} << C_{mm}$. It should be noted that this condition can be easily satisfied in practice, for example, if we take a = 7, k = 3, $b_{min} = 3$ and $N_{min} = 3$, we have:

$$C_{mc} = \frac{1}{7}\left[e^{\frac{1}{2e}(3+1).7^2}\right] = 6.49 \times 10^{14}$$

$$C_{mm} = \frac{1}{7}\left[{}_3\frac{(3+1).7^2}{3}\right] = 3.51 \times 10^{9}$$

$$C_{ml} = \frac{1}{7}\left[\frac{(3+1)\ 7^2}{3}\right]^{\frac{3}{2}} = 75.44$$

We see that $C_{mm} >> C_{ml}$, and there is no problem for selecting C_{max} and C_{min} such that $C_{ml} \leq C_{min} << C_{max} \leq C_{mm}$, so the elimination of all the terms of which $|c_i| < C_{min}$ brings about no important errors.

$$N_{min} . b^2_{min} = 27$$

and

$$(k+1)a^2 = 196$$

the constraint $N_{min} . b^2_{min} < (k+1)a^2$ is easily satisfied.

As a summary, the translating cascading sum-box technique to realize a given filter is the following:

(1) According to the given filter kernel f(x) to be realized, create the function g(x) = f(ax), where a is a positive integer, $a \geq 2$;

(2) Develop g(x) as its spline function (Eq. (34)) ;

(3) The kernel f(x) is expressed in the form of Eq. (36);

(4) Determine the minimal window size b_{min} of the sum-box filters that we shall use in the cascade, and the minimal number of iterations N_{min} in the cascade, such that $N_{min}.b_{min}^2 < (k+1)a^2$. In general, $b_{min} \geq 2$ and $N_{min} \geq 3$;

(5) Calculate C_{mm} and C_{ml} from Eqs. (20) and (24);

(6) Take two thresholds C_{max} and C_{min}, such that:

$$C_{ml} \leq C_{min} << C_{max} \leq C_{mm}$$

(7) Take $R = \dfrac{C_{max}}{\underset{i=0}{\overset{M+k-1}{\text{Max}}} |\beta_i|}$, multiply the expression (36) of f(x) by R and let $\forall\, i \in \{0,\ldots, M+k-1\}$, $C_i = R.\beta_i$, neglect all the terms in Eq. (38) whose $|C_i| < C_{min}$. The approximate expression (39) of R.f(x) is thus obtained;

(8) For each B-Spline function term in Eq. (39); calculate the solutions N_i and b_i of the equation system (7) and (8).

(i) If the solutions N_i and b_i are integer, the weighted B-Spline filter $|C_i|.\Omega_k(\frac{x}{a})$ is realized by the cascade of N_i sum-box filters of window size b_i;

(ii) If the solutions N_i and b_i are not integer, adjust the solutions by use of the equation system (33), and the weighted B-Spline filter is realized by the cascade of N_{di} sum-box filters with respectively the window sizes b_{dj} ($j= 1,\ldots, N_{di}$);

(9) Take the sum (or difference, according to sign (C_i)) of the outputs of the cascading sum-box filters with respectively a translation $(\frac{k+1}{2} - i)a$. This sum will be equivalent to the signal filtered by the linear filter R.f(x). The filter with kernel R.f(x) is therefore realized by the sum-box technique requiring no multiplications.

5. CONCLUSION

In the present paper, we propose the translating cascading sum-box technique which realizes approximately a given linear filter by the sum-box filters requiring no multiplications. The given filter kernel is at first decomposed into the sum of some translated weighted B-Spline filters and each weighted B-Spline filter is then realized by a cascade of sum-box filters. Of course, our method gives an approximate realization, but the forms of many filters used in signal processing and in image processing are rarely strict, as is well known. Our analysis is based on 1-D filters. Evidently, this highly efficient method can be extended to separable 2-D filters widely used in image processing. In particular, for large filter kernels, this method is potentially one of the fastest filter algorithms. Another advantage of the translating cascading sum box technique is that one can easily calculate the derivatives of a cascade of sum-box filters [6], so with a simple sum-box hardware, one can realize a given linear filter and its derivatives for different purposes. Further research will be done in the analysis of the errors introduced by this approximated realization and the application of the sum-box technique to some large kernel FIR filters actually used in image processing.

REFERENCES

[1] Abramatic, J.F. and Faugeras, O.D., (1982) Sequential Convolution Techniques for Image Filtering. *IEEE TASSP*, Vol. ASSP-30, N° 1.

[2] Lee, S.U., (1980) Design of SVD/SGK Convolution Filters for Image Processing, Technical Report 950, Univ. of Southern California.

[3] Burt, P.J., (1981) Fast Filter Transforms for Image Processing, *CGIP*, Vol. 16, N° 1, pp. 20-51.

[4] Castan, S. and Shen, J., (1985) Fast Filter Transform Theory and Design for Image Processing, Proceedings of CVPR'85, pp. 420-422, San Francisco.

[5] McDonnell, M.J., (1981) Box-Filtering Techniques, *CGIP*, Vol. 17, n° 1, pp. 65-70.

[6] Castan, S. and Shen, J., (1985) Box Filtering for Gaussian-Type Filters by Use of the B-Spline Functions, Proceedings of the 4th Scandinavian Conference on Image Analysis, pp. 235-243, Norway.

[7] Li, Y.S. and Qi, T.X., (1979) Spline Function Methods, Scientific Press, Peking.

[8] Castan, S. and Shen, J., (1985) Fast Approximate Realization of Linear Filters by Translating Cascading Sum-box Technique, Proceedings of CVPR'85, pp. 678-680, San Francisco.

VI ADAPTIVE TECHNIQUES

NEW RECURSIVE ALGORITHM FOR MAXIMUM LIKELIHOOD TOEPLITZ COVARIANCE MATRIX ESTIMATION

M.K. Ibrahim
(Department of Electrical and Electronic Engineering, University of Nottingham)

and

C.E. Goutis
(School of Electrical Engineering, University of Patras, Greece)

ABSTRACT

A new maximum likelihood algorithm for Toeplitz covariance matrix estimation is introduced. The new algorithm solves for the reflection coefficients rather than the covariance lags. It calculates these coefficients successively starting from filter length one - the Kth reflection coefficient, a_{kk}, is computed by maximizing the associated likelihood function, and assuming that the reflection coefficients a_{nn}, n = 1, .. , k-1, have been previously calculated. The algorithm does not assume that the noisy data is expressed by an AR filter. This assumption is incorrect since noise is always present in real signals and usually has serious consequences in low signal to noise ratio where the approximation due to the assumption is significant and the true spectra is required for reliable detection.

The optimisation associated with the Kth order recursion specifying the Kth reflection coefficient, a_{kk}, produces a cubic equation whose root with absolute value less than one is taken equal to a_{kk}; if more than one root satisfies this condition, the one with higher likelihood is chosen. The rest of the Kth order filter coefficients are computed using the Levinson recursion. The stability of the algorithm is guaranteed as the reflection coefficients are constrained to have absolute value less than one during the recursion. Implementation of the algorithm gave the expected performance showing improved spectra.

1. INTRODUCTION

Spectrum estimation using either the conventional methods or the maximum entropy method requires an estimate of the autocorrelation function of the given data record. Two common methods of calculating the autocorrelation function are the conventional lag product method and the Burg technique (1968). The latter generates the autocorrelation function using prediction error operator whose coefficients are constrained to obey the Levinson recursion. The reflection coefficients are calculated in succession by minimizing the sum of the power of the forward and backward linear prediction error.

A generalisation of the Burg technique for the Multichannel and Multidimensional spectral estimation was presented by by Burg et al. (1982). It is based on choosing the covariance matrix of specified structure which maximizes the likelihood function of the probability density function of the given data. Assuming that the data is drawn from a zero mean multivariable Gaussian process, an iterative algorithm was presented in Burg et al. (1982) to give the nearest Toeplitz covariance matrix to the non-Toeplitz one computed from the data; this was achieved by decomposing the covariance into a series of Toeplitz basis functions and estimating the coefficients of this decomposition iteratively. The algorithm is terminated when the likelihood function reaches its maximum value. The final solution is ensured to be positive definite by calculating the determinant at each iteration and reducing the step size whenever the updated matrix is non positive definite. The performance of the algorithm based on the generalized procedure is superior to that of the Burg technique, but the number of computations for the former far exceeds that of the latter. In addition, the generalized procedure does not ensure uniqueness of the final solution.

In this paper, a new recursive algorithm is presented that finds an efficient estimate of the exact maximum likelihood solution. The computational requirements of the new algorithm are comparable to that of the Burg technique. The cost function used in the new algorithm is the same as that presented in Burg et al. (1982). However, it is maximized recursively with respect to the reflection coefficients rather than the correlation lags. It uses the fact that a set of reflection coefficients has a one-to -one relationship with a symmetric Toeplitz matrix whose main diagonal element is unity, see Levinson (1947). Positive definiteness of the resulting covariance matrix is ensured by simply constraining the reflection coefficients to have a magnitude of less than one.

The reflection coefficients are estimated in a similar fashion to that of the Burg technique but do not minimize the sum of the forward and backward prediction errors. As in the algorithm presented by Kay (1983), one requires to solve a cubic equation for each reflection coefficient. The primary difference is that we do not assume that the data belong to an autogressive process, an assumption which does not hold in presence of white noise. The cubic equation is guaranteed to have at least one real root within the range (-1, +1) which is adopted. In cases where more than one real root lies between (-1, +1) the one which maximizes the likelihood function is chosen.

The derivation of the algorithm is presented in the next section while the computed results and conclusions are presented in sections 3 and 4 respectively.

2. DERIVATION OF THE ALGORITHM

Given M independent vectors X_i, i = 1,.., M drawn from an (N + 1) - dimensional Gaussian process with zero mean and covariance matrix R_{N+1}, the corresponding joint probability density function is given by

$$P(X_1,..,X_M/R_{N+1}) = (2\pi)^{-\frac{M(N+1)}{2}} |R_{N+1}|^{-\frac{M}{2}} \exp\{-\frac{1}{2}\sum_{i=1}^{M} X_i^T R_{N+1}^{-1} X_i\}$$

where T denotes transpose and $|R|$ is the determinant of R. It is shown by Burg et al. (1982) that maximizing the above equation is equivalent to maximizing

$$g(S_{N+1}/R_{N+1}) = -\log|R_{N+1}| -tr(R_{N+1}^{-1} S_{N+1}) \qquad (2.1)$$

where S_{N+1} is given by

$$S_{N+1} = \frac{1}{M}\sum_{i=1}^{M} X_i X_i^T \qquad (2.2)$$

In the following, we will maximise equation (2.2) with respect to the variance of the process, σ_o. Letting

$$U_{N+1} = \frac{1}{\sigma_o} R_{N+1}$$

then

$$R_{N+1}^{-1} = \frac{1}{\sigma_o} U_{N+1}^{-1}$$

and

$$|R_{N+1}| = \sigma_o^{N+1} \; |U_{N+1}|$$

Substituting for R_{N+1} and $|R_{N+1}|$ in equation (2.1) their values above, we have

$$g(S_{N+1}/U_{N+1}, \sigma_o) = - \log \sigma_o^{N+1}|U_{N+1}| - \frac{1}{\sigma_o} \operatorname{tr}(U_{N+1}^{-1} S_{N+1}) \tag{2.3}$$

Setting the derivative of the above equation with respect to σ_o equal to zero and rearranging, we obtain

$$\sigma_o = \frac{1}{N+1} \operatorname{tr}(U_{N+1}^{-1} S_{N+1})$$

which is the maximum likelihood estimate of σ_o. Using this value of σ_o in equation (2.3), it results in

$$g(S_{N+1}/U_{N+1}) = - (N+1) \log [\operatorname{tr}(U_{N+1}^{-1} S_{N+1})]$$

$$- \log |U_{N+1}| + (N+1)\log(N+1) - (N+1) \tag{2.4}$$

From the Levinson algorithm (1947), it can be shown that U_{N+1}^{-1} and $|U_{N+1}|$ have a one to one relationship with a set of reflection coefficients. Therefore $g(S_{N+1}/U_{N+1})$ can be maximized iteratively with respect to the reflection coefficients rather than the elements of U_{N+1} directly. However, this will result in a computationally expensive algorithm. The approach adopted here is to maximize $g(S_{N+1}/U_{N+1})$ recursively starting with N = 1. In what follows, it is assumed that the reflection coefficients, a_{nn}, n = 1, .., N - 1, have been previously calculated and that they have a magnitude less than one. Also, we maximise $g(S_{N+1}/U_{N+1})$ with respect to the Nth reflection coefficient only. The final result is then generalized for all N.

The inverse of U_{N+1} is found using the matrix inversion lemma for partitioned matrices which, for symmetric Toeplitz matrices, is given by (see Carayanis et al. 1982),

$$\alpha_N U_{N+1}^{-1} = \begin{bmatrix} \alpha_N U_N^{-1} + W_N W_N^T & W_N \\ W_N^T & 1 \end{bmatrix} \tag{2.5a}$$

where

$$\alpha_N = \frac{|U_{N+1}|}{|U_N|} = \alpha_{N-1} (1 - a_{NN}^2) \tag{2.5b}$$

$$\alpha_o = 1.0$$

$$W_N = [a_{NN}, \ldots, a_{N1}]^T$$

a_{mi}, i = 1, .., m, m= 1, .., N, are the prediction error filter coefficients at order m, see Durbin (1960), satisfying the Levinson recursion (Levinson (1947)),

$$a_{mi} = a_{m-1,i} + a_{mm} a_{m-1,m-i}$$

and a_{ii} is the ith reflection coefficient. It is important to note that equations (2.5a and b) imply that U_N and $|U_N|$ are independent of a_{NN}. Writing equation (2.5a) in a different form,

$$U_{N+1}^{-1} = [\sum_{n=O}^{N-1} \frac{1}{\alpha_n} {}_{n+1}G_{N+1} + \frac{1}{\alpha_N} A_{N+1}] \tag{2.6}$$

where

$${}_{n+1}G_{N+1} = \begin{vmatrix} A_{n+1} & {}_{n+1}O_{N-n} \\ {}_{n+1}O_{N-n}^T & {}_{N-n}O_{N-n} \end{vmatrix}$$

$$[A_{n+1}]_{ij} = a_{n,n-i+1} \quad a_{n,n-j+1} \quad i,j = 1,..,n+1 \tag{2.7}$$

${}_iO_j$ is an i x j zero matrix and a_{no} = 1.0, n=0, .., N. Using equations (2.5 and 2.6) in equation (2.4), one obtains,

$$g(S_{N+1}/U_N, a_{NN}) = -(N+1)\log\{E_N + \frac{1}{\alpha_N}\operatorname{tr}(A_{N+1}S_{N+1})$$

$$-\log\alpha_N - \log|U_N| + (N+1)\log(N+1) - (N+1) \tag{2.8}$$

where

$$E_N = \operatorname{tr}(\sum_{n=0}^{N-1}\frac{1}{\alpha_n}\,{}_{n+1}G_{N+1}S_{N+1})$$

or

$$E_N = \sum_{n=0}^{N-1}\frac{1}{\alpha_n}\operatorname{tr}(A_{n+1}S_{n+1}) \tag{2.9}$$

Note that E_N and α_n, $n = 0, .., N-1$, do not depend on a_{NN}. Using the Levinson recursion and equation (2.7), it is shown that

$$\operatorname{tr}(A_{n+1}S_{n+1}) = \sum_{k=1}^{M}\{\sum_{i=0}^{n}[X_k]_i\, a_{ni}\}^2 \tag{2.10}$$

for n = 0, .., N, which are positive and they are equal to zero only when the data consist of pure sinusoids which have a low possibility of occurring. Since it is assumed that $|a_{NN}|<1$, $n = 1, .., N-1$, α_n, $n = 0, .., N-1$, and hence E_N are positive as seen from equations (2.5b and 2.9) and that they have a finite value.

From the Levinson recursion and equation (2.7), it is shown that

$$\operatorname{tr}[A_{N+1}S_{N+1}] = B_N + a_{NN}C_N + a^2_{NN}D_N \tag{2.11}$$

where

$$B_N = \sum_{i=1}^{N+1}\sum_{j=1}^{N+1}[S_{N+1}]_{ij}\, a_{N-1,N-i+1}\, a_{N-1,N-j+1}$$

$$C_N = \sum_{i=1}^{N+1}\sum_{j=1}^{N+1}[S_{N+1}]_{ij}\,(a_{N-1,N-j+1}\, a_{N-1,i-1}$$

$$+ a_{N-1,N-i+1}\, a_{N-1,j-1})$$

$$D_N = \sum_{i=1}^{N+1} \sum_{j=1}^{N+1} [S_{N+1}]_{ij} \; a_{N-1,j-1} \; a_{N-1,i-1}$$

and $a_{mi} = 0$ for $m < i$.

Substituting equations (2.11 and 2.5b) for $tr(A_{N+1} S_{N+1})$ and α_N in equation (2.8) and after some manipulation, it follows that,

$$\begin{aligned} g(S_{N+1}/U_N, a_{NN}) = &- (N+1)\log [\, \alpha_{N-1} E_N (1-a_{NN}^2) \\ &+ B_N + a_{NN} C_N + a_{NN}^2 D_N \,] + N \log \alpha_{N-1} \\ &+ N \log (1-a_{NN}^2) - \log |U_N| \\ &+ (N+1) \log(N+1) - (N+1) \end{aligned} \qquad (2.12)$$

Since it is assumed that a_{ii}, $i = 1, .., N-1$, have already been estimated, E_N, B_N, C_N, D_N, α_{N-1} and $\log |U_N|$ can be computed and so are treated as constants. Setting the derivative of $g(S_{N+1}/U_N, a_{NN})$ with respect to a_{NN} equal to zero, rearranging and regrouping in terms of powers of a_{NN}, we have

$$T_N + Q_N a_{NN} + V_N a_{NN}^2 + Z_N a_{NN}^3 = 0 \qquad (2.13)$$

where

$$\begin{aligned} T_N &= - (N+1) C_N \\ Q_N &= -2(N+1) D_N - 2N B_N + 2 \alpha_{N-1} E_N \qquad (2.14) \\ V_N &= - (N-1) C_N \end{aligned}$$

and

$$Z_N = 2 D_N - 2 \alpha_{N-1} E_N$$

The solution of the above cubic equation produces the estimate of a_{NN}. Since equation (2.13) was derived for a general N, it is used to calculate the reflection coefficients a_{ii}, i = 1, .., N, recursively starting with order one. Although the solution obtained is suboptimal, the resulting algorithm has computational requirement comparable to that of the Burg technique.

In order to ensure that the resulting matrix, U_{N+1} is positive definite, it is shown below that the above cubic equation has at least one real root within the range (-1, +1). Assuming that $|a_{ii}|<1$, i=1, .., N-1, we will first prove it for a_{NN}, the result is then generalized for the rest of the reflection coefficient by induction. Maximizing $g(S_{N+1}/U_N, a_{NN})$ is equivalent to maximizing

$$g_{,}(S_{N+1}/U_N, a_{NN}) = \alpha_N^{N-1} \{\alpha_N \sum_{n=0}^{N-1} \frac{1}{\alpha_n} \text{tr}\,(A_{n+1} S_{n+1})$$

$$+ \text{tr}\,(A_{N+1} S_{N+1}) \}^{-N} |U_N|^{-\frac{1}{2}}$$

As mentioned earlier tr $(A_{n+1}S_{n+1})$ n = 1, .., N+1, are equal to zero only when X consists solely of pure sinusoids and that this case is neglected since it has a low possibility of occurring. If a_{NN} tends towards unity, α_N and hence $g(S_{N+1}/U_N, a_{NN})$ will tend to zero. For variations of a_{NN} within (-1,+1) $g(S_{N+1}/U_N, a_{NN})$ is (i) positive since α_i and tr $(A_{i+1} S_{i+1})$, i = 0,..,N, are positive and (ii) continuous and differentiable with respect to a_{NN}.

Therefore by Rolle's theorem, see Grossman (1981), $g(S_{N+1}/U_N, a_{NN})$ has at least one maximum within the interval (-1,+1). Since the above results hold for any N, the proof is extended to a_{ii}, i=1, .., N-1, by induction.

In order to make maximum use of the data $[X_i]_j$, i=1,..M,j=1, ..,N+1, a second method for calculating the covariance is presented. It is based on recomputing the covariance for each new stage as follows,

$$[S_{n+1}]_{kl} = \frac{1}{M(N-n)} \sum_{i=1}^{M} \sum_{t=1}^{N-n} [x_i]_{t+k-1} [x_i]_{t+l-1} \tag{2.15}$$

for n=1,..N and k,l = 1,..,n-1. This does not change the structure of the algorithm or the characteristic of the cubic equation since each independent vector can be decomposed into N-n overlapping vectors when computing the nth reflection coefficient.

The steps of the algorithm are:

1. set n = 1 and repeat steps 2-4 until n is equal to the desired filter length which should not be greater than N.

2. calculate S using equation (2.15) and then compute T_n, Q_n, V_n and Z_n using equation (2.14).

3. Find the roots of equation (2.13) and the root which has a magnitude less than 1 is set equal to a_{nn}. If more than one root lies between the range (1, -1), the one with the maximum $g(S_{n+1}/R_{n+1})$ is set equal to a_{nn}.

4. Calculate the filter coefficients $a_{n,i}$ i = 2, n-1 using the Levinson recursion and then increment n.

Other methods of calculating S can also be used in the above algorithm.

3. SIMULATION RESULTS

In this section, the new algorithm and the Burg technique are compared using one vector sample of sinusoidal data in white noise. The results shown in this section were found using 20 samples of a sine wave frequency = 20Hz, sampling frequency = 100Hz, filter order = 10 and SNR = -0.5dB. To demonstrate the performance in detecting sinusoids, the maximum entropy spectra of the corresponding solutions was plotted.

Simulation results have shown that in general the new algorithm has higher resolution and less spurious peaks than than of the Burg technique. This is indicated in Figs. 1 and 2 where the spectrum shown in Fig. 1 is that of the Burg technique, while the spectrum of the new algorithm is shown in Fig. 2.

Tests have also shown that the performance of the algorithm in estimating the location of frequencies depends very much on selecting the appropriate order. Simulation results have also shown that the new algorithm exhibits line splitting.

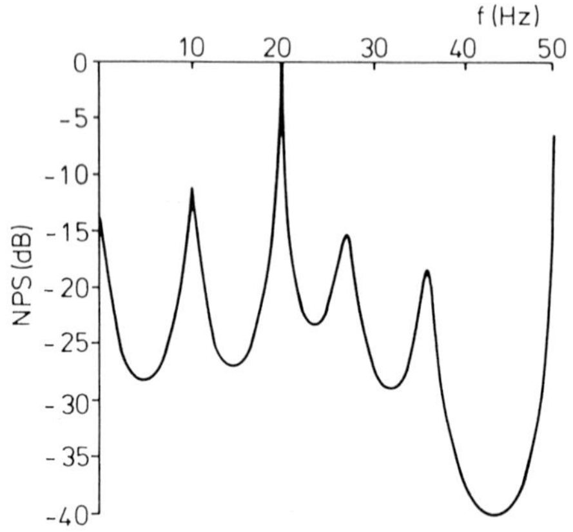

Fig. 1 Normalised power spectra of the Burg technique for a sinusoid with f=20 Hz, sampling frequency = 100 Hz, filter order =10, number of samples = 20 and SNR = -0.5 dB.

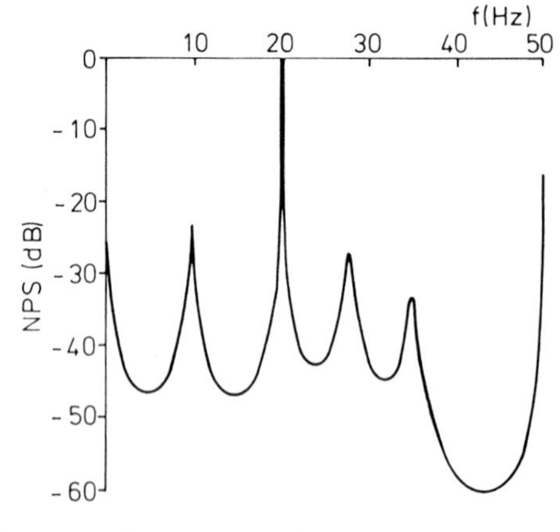

Fig. 2 Spectrum of the new algorithm for the same data as in Fig. 1.

CONCLUSION

A new recursive algorithm for estimating the reflection coefficient is presented. It is based on maximizing the likelihood function of the p.d.f. of the given data. It does not minimize the prediction error as in the Burg technique and does not assume an AR process as in the algorithm presented by Kay (1983), both of which are undesirable in detecting sinusoids in white noise.

The method guarantees the positive definitives of the resulting correlation matrix since the reflection coefficients are ensured to have magnitude less than one. It is computationally more efficient than that of the iterative algorithm presented in Burg et al. (1982). Simulation results have shown improved resolution, and lower side lobes over that of the Burg techique.

REFERENCES

Burg, J.P., (1968) "A New Analysis Technique for Time Series Data". Adv. Study Inst. on Signal Processing, NATO, Enschede, Holland.

Burg, J.P., Luenberger, D.G., and Wenger, D.L., (1982) "Estimation of Structured Covariance Matrices". IEEE Proc. Vol. 70, No. 9, pp. 963-974.

Durbin, J., (1960) "The Fitting of Time-Series Models". Rev. Inst Inst. Statist., Vol. 28, pp. 233-244.

Grossman, I., (1981) "Calculus". Academic Press, New York and London.

Kay, S.M., (1983) "Recursive Maximum Likelihood Estimation of Autogressive Processes". IEE Trans. on Acoustic Speech and Signal Processing, Vol. ASSP-31, No. 1, pp. 56-65.

Levinson, N., (1947) "The Wiener RMS Error in Filter Design and Prediction". *J. Math. Phys.* Vol. 25, pp. 261-278.

Carayannis, G., Kalouptsidis, N., and Manolakis, D.G., (1982) "Fast Recursive Algorithms for a Class of Linear Equations". IEEE Trans. Acoustics, Speech and Signal Processing, Vol. 30, pp. 227-239.

SELF-TUNING FILTERS AND PREDICTORS FOR TWO-DIMENSIONAL SYSTEMS

P.E. Wellstead
(Control Systems Centre, UMIST)

and

J.R. Caldas Pinto
(Departamento de Engenharia Mecanica, Instituto Superior Technico, Portugal)

SUMMARY

Self-tuning algorithms as developed in the control literature are discussed in the context of two-dimensional signal processing. In particular, it is shown that one-dimensional predictors and smoothers based upon ARMA models can be extended to handle two-dimensional (2-D) raster scanned data.

1. INTRODUCTION

In the control literature the technique known as self-tuning control is widely used and researched (see for example Harris and Billings (1981)). The basic idea in self-tuning is to combine in one on-line, recursive sequence the following steps:

(i) Estimation - in which a recursive estimator is used to update the parameters of a model of the data generating mechanism.

(ii) Synthesis - in which the estimated model parameters are transformed into the coefficients of an appropriate signal processing algorithm (e.g. a controller, predictor or filter).

(iii) Implementation - in which the signal processing algorithm is applied to the data.

As the data is processed the above self-tuning sequence is applied at each step. The algorithm commences with little or no knowledge of the data generating mechanism. However, as the number of processed data points increases, an adaptive process takes place whereby the estimator coefficients, and hence the signal processing algorithm, self-tunes to the configuration which would have been obtained using off-line design tools.

The motivation for using a self-tuning approach to filter design is to short-cut the off-line design cycle and provide a semi-automatic design method for certain parametric signal processing applications. In this connection we note that Bellman's 'curse of dimensionality' applies such that even the simplest two-dimensional (2-D) filter can involve the selection of eight or more filter coefficients. Clearly an automatic procedure for filter set-up can save significant design effort. Moreover, since many two-dimensional data fields are non-stationary, the inherent adaptive capability of self-tuning algorithms can allow filters to track changes in the 2-D data generator.

In one dimension, self-tuning predictors and filters based upon input/output models have been developed (Wittenmark (1974); Hagander and Wittenmark (1977)). The contribution of this paper lies in the extension of the one-dimensional self-tuners to the two-dimensional case. Historically, the work described here has developed through an interest in signal processing of data which can be arranged in the form of a two-dimensional data field. Basic work in this context is contained in the thesis work of Sym (1982), with explicit two-dimensional self-tuners appearing first in Caldas Pinto (1983). The material presented in this paper is abstracted from the technical reports (Wellstead and Caldas Pinto (1984), Caldas Pinto and Wellstead (1984a, 1984b)).

The scope of the paper is constrained by space to a review of two-dimensional self-tuning algorithms, together with a discussion of the problems associated with the background theory. In this way we hope to attract the interest of signal processing theorists to this area of research.

2. STATEMENT OF THE PROBLEM

A two-dimensional data source can be digitized such that it is defined over a discrete 2-D region consisting of a lattice of regularly spaced points. Once in this form the data can be scanned in any one of a number of ways. A conventional raster scan will be assumed in which the data is analysed row by row, from the bottom up and from left to right, such that the data points $y(m,n)$ are used as indicated in Figure 1. The problem considered is that of filtering and prediction using the data $y(m,n)$, under the assumption that the data generation mechanism is that of a two-dimensional ARMA (Auto-Regressive/ Moving Average) process s(m,n), corrupted by a zero mean, white Gaussian process e(m,n) with covariance R.

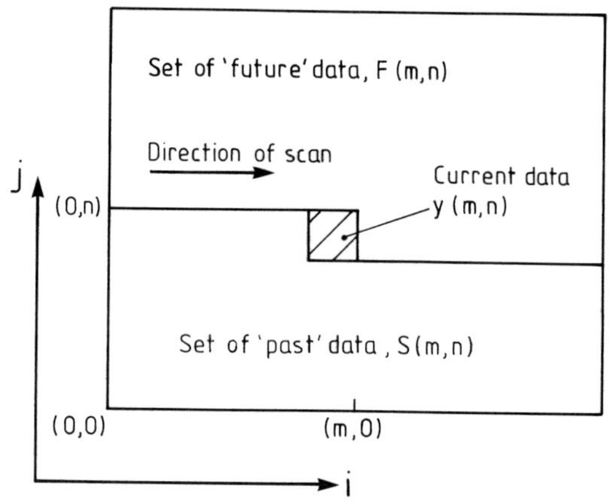

Fig. 1. Showing a raster scan, two dimensional (2-D) data field

It is assumed that s(m,n) is related to a subset of "past" data by a recursion of the form:

$$s(m,n) + \sum_{(i,j)\in\beta-(0,0)} a(i,j)\,s(m-i,n-j) = \sum_{(i,j)\in\alpha} c(i,j)\,v(m-i,n-j) \quad (2.1)$$

where i,j are integer indices and v(m,n) is a zero mean white Gaussian 2-D process with variance Q, (v(m,n) and e(m,n) are independent processes). The measured data y(m,n) is then defined as

$$y(m,n) = s(m,n) + e(m,n) \quad (2.2)$$

Physically, v(m,n) and equation (2.1) represent the data generation mechanism while e(m,n) represents output measurement or sensor noise. We cannot at the moment deal with the case of coloured output noise, unless the dynamics of colouring are known, in which case the solution is a trivial prefiltering operation on the data.

The sets β,α in equation (2.1) indicate the so-called "support" of the recursion. Conventionally, these supports include the origin (0,0); when we wish to exclude the origin (as in the first summation in equation (2.1)) from the support we write e.g. $\beta-(0,0)$ or $\alpha-(0,0)$. Note also that the support α for $c(i,j)$ is coincident, or included in β the support of $a(i,j)$. We assume that the support β and α for the 2-D filter is the usual one for raster scan data whereby the signal is assumed to depend upon data from a set of "past" points in the picture. This kind of non-symmetric half-plane support (Huang (1981)) is illustrated in Figure 2. With reference to Figure 2, the area $U(m,n)$ is often termed the global state vector (Woods (1981)). The smaller region $R(M,N)$ is that used to form the data point $y(m,n)$ and is termed the local state.

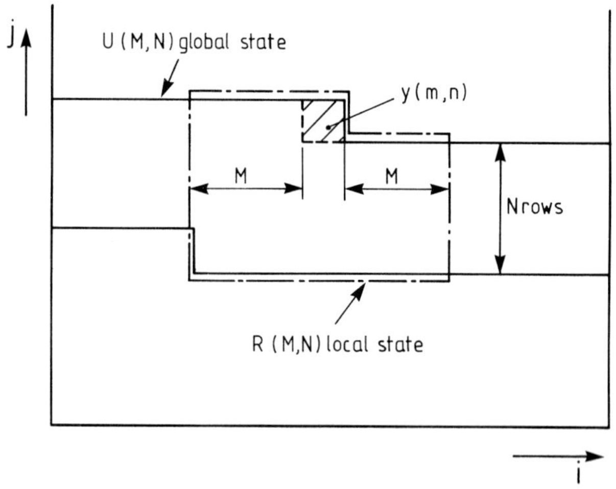

Fig. 2. Illustrating the concepts of non symmetric half plane support, global state vector and local state vector

The problem is to formulate self-tuning algorithms for fixed k-lag smoothers, filters and k-step ahead predictors based on the 2-D data field $y(m,n)$. The filters/predictors are to be based upon 2-D ARMA models recursing on a sub-set of $U(m,n)$. The fixed k-lag smoother for $s(m,n)$ is ${}_k\hat{s}(m,n)$ and is the best least squares estimate of $s(m,n)$ given the measured data up until $y(m+k,n)$. Thus,

$$ {}_k\hat{s}(m,n) = E\{s(m,n)\,|\,y(i,j),\ (i,j)\in S(m+k+1,n)\} \qquad (2.3)$$

where $S(m+k+1,n)$ is the set of "past" data defined in Figure 1. Note that the filter for $s(m,n)$ is obtained as the fixed lag smoother with $k=0$.

The k-step predictor $\hat{y}(m+k,n)$ will be taken to be the best least squares estimate of $y(m+k,n)$ given data up until $y(m,n)$. In equation form this becomes

$$\hat{y}(m+k,n) = E\{y(m+k,n)\,|\,y(i,j),\ (i,j) \in S(m+1,n)\} \qquad (2.4)$$

Again $S(m+1,n)$ is the set of past data as defined in Figure 1.

3. 2-D SELF-TUNING SMOOTHERS AND FILTERS

3.1 Model Formulation

Given a 2-D data field defined by equations (2.1) and (2.2), it is possible, as in the 1-D case (Anderson and Moore (1979)) to write

$$y(m,n) = \tilde{y}(m,n) + \sum_{(i,j)\in S(m,n)} \ell(i,j)\tilde{y}(m-i,n-j) \qquad (3.1)$$

where the innovations process $\{\tilde{y}(m,n)\}$ is a zero mean white Gaussian random process with variance W, and $\{\ell(i,j)\}$ is the "impulse response" of the process. Note that the key assumption here is that a 2-D innovations model of the data exists (see Strintzis (1978); Caldas Pinto (1983)).

Correspondingly, an innovations model may be proposed:

$$y(m,n) + \sum_{(i,j)\in\beta-(0,0)} a(i,j)y(m-i,n-j) = \sum_{(i,j)\,\varepsilon\gamma} d(i,j)\tilde{y}(m-i,n-j) \qquad (3.2)$$

where $\{a(i,j)\}$, $\{d(i,j)\}$ correspond to stable invertible processes and γ is the support for $d(i,j)$. Assuming that $y(m,n)$ is a stationary process, then equation (3.2) will be the innovations model of equations (2.1) and (2.2) if

$$A(w,z)A(w^{-1}z^{-1})R+C(w,z)C(w^{-1},z^{-1})Q = D(w,z)D(w^{-1}z^{-1})W \qquad (3.3)$$

This is a 2-D form of the usual spectral factorization relationship (Anderson and Moore (1979)), where

A(w,z), C(w,z), D(w,z) are respectively the 2-D z-transforms (Dudgeon and Mersereau (1984)) of {a(i,j), c(i,j), d(i,j)}.

Note that in equation (3.3) D(w,z) has infinite order. This implies that the corresponding sequence {d(i,j)} has support γ equal to U(m,n). In practice the support of {d(i,j)} can be truncated. The minimum practically permissible support being that of {a(i,j)}, with p extra columns added either side of R(M,N) (Figure 3). This form of support for {d(i,j)} is termed EXS(p). In practice, most of the arrays a(i,j) are defined in a R(M,N) such that $M \leq 3$ and $N \leq 3$. In such cases p=2 or even p=1 is sufficient to ensure that the innovations model and the data generating mechanism (equations (2.1) and (2.2)) have power spectra which are indistinguishable (Caldas Pinto (1983).

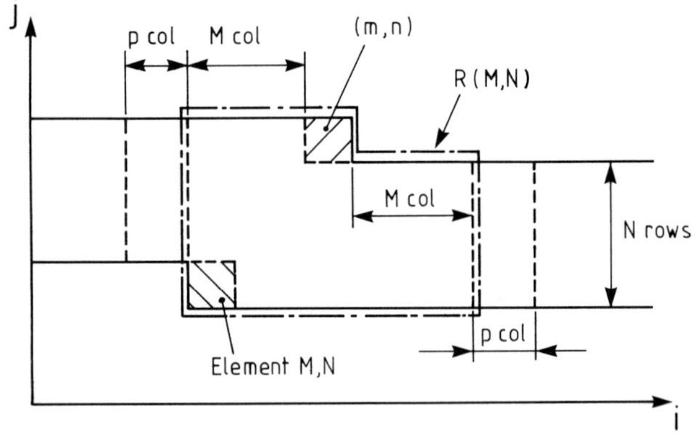

Fig. 3. Illustrating the extended support provided for [d(i,j)] by adding an extra p data columns to R(m,n) and termed EXS(p)

3.2 The Algorithm

From section 3.1 it is clear that a finite dimension innovations model in the ARMA form of equation (3.2) will be an approximation to the true innovations representation. As a result any ARMA smoothing algorithms will also be approximate, the quality of approximation increasing as the support index p is increased. With this point in mind it is possible to show (Caldas Pinto, 1983; Wellstead and Caldas Pinto, 1984) that the k-lag smoother ${}_{k}\tilde{s}(m,n)$ is given by:

$$ {}_k\hat{s}(m,n) = y(m,n) - \frac{R}{W} \sum_{i=0}^{k} h(i,0)\tilde{y}(m+i,n) \tag{3.4} $$

where R, W are respectively the variances of $e(m,n)$ and $\tilde{y}(m,n)$.

In addition, the ratio of variances R/W can be calculated, analogously to the 1-D case, as

$$ \frac{R}{W} = \frac{d(M,N)}{a(M,N)} \tag{3.5} $$

where $d(M,N)$, $a(M,N)$ correspond to the (M,N)th elements (Figure 3) taken from $\{d(i,j), a(i,j)\}$ respectively.

Appropriate coefficients of $\{h(i,j)\}$ are evaluated from the transform relationship:

$$ H(w,z) = \frac{A(w,z)}{D(w,z)} \tag{3.6} $$

where $H(w,z)$, $A(w,z)$, $D(w,z)$ are the w,z transforms of the sequence $\{h(i,j), a(i,j)$ and $d(i,j)\}$ respectively. Only the first row $\{h(i,0)\}$ of the $h(i,j)$ bi-sequence is required to compute ${}_k\hat{s}(m,n)$. Hence it is possible to use

$$ H'(w) = \frac{A'(w)}{D'(w)} \tag{3.7} $$

where $A'(w)$, $D'(w)$, $H'(w)$ are respectively the first rows of $A(w,z)$, $D(w,z)$ and $H(w,z)$.

The self-tuning version of the 2-D smoother/filters operates in an analogous way to the 1-D version developed by Hagander and Wittenmark (1977). At each point $y(m,n)$ in the raster scan of the data the following steps are executed:

STEP (i) Update the estimates $\{\hat{a}(i,j), \hat{d}(i,j)\}$ of the RML (Ljung and Soderstrom, (1983)) model of the 2-D process given by:

$$ y(\bar{m},n) = -\sum_{(i,j)\in\beta-(0,0)} \hat{a}(i,j)\,y(\bar{m}-i,n-j) + \sum_{(i,j)\in\gamma_t} \hat{d}(i,j)\,\varepsilon(\bar{m}-i,n-j) + \varepsilon(\bar{m},n) \tag{3.8} $$

where $\varepsilon(\bar{m},n)$ is the current prediction error in the RML recursion and $\bar{m} = m+k$.

STEP (ii) Determine $\hat{h}(p,0)$ $0 \leq p \leq k$ and $\hat{R}/\hat{W}$ from equations (3.7) and (3.5).

STEP (iii) Calculate the current estimate of the smoothed/filtered signal as:

$$_{k}\hat{s}(m,n) = y(m,n) - \frac{\hat{R}}{\hat{W}} \sum_{p=0}^{k} \hat{h}(p,0)\varepsilon(m+p,n) \tag{3.9}$$

In this self-tuning algorithm we exploit the fact that, provided the order of the estimated polynomials is correct, then the process $\{\varepsilon(m,n)\}$ can converge to the innovations process $\{\tilde{y}(m,n)\}$ (Ljung and Soderstrom (1983)). The assumption here is that the extended support EXS(p) is sufficient to cover the dynamics of d(i,j) and allow convergence of the RML recursion. For further discussion of this point see Caldas Pinto (1983).

4. 2-D SELF-TUNING PREDICTORS

Consider a general 2-D recursive ARMA model for y(m,n) given by:

$$y(m,n) + \sum_{(i,j)\in\beta-(0,0)} a(i,j)y(m-i,n-j) = \sum_{(i,j)\in\alpha} c(i,j)e(m-i,n-j) \tag{4.1}$$

in which a(i,j) and c(i,j) are stable arrays, c(0,0) = 1, and e(m,n) is a bi-variate white noise process with variance R. It is also assumed that $\alpha = \beta = R(M,M)$ (see Figure 2).

The k-step ahead predictor is defined by:

$$\hat{y}_k(m,n) = E\{y(m+k,n)\,|\,y(i,j),\ (i,j) \in S(m+1,n)\} \tag{4.2}$$

where S(m,n) represents "past" data, (see Figure 1).

Again using techniques drawn from the 1-D case, it can be shown that the 2-D predictor can be written as:

$$\hat{y}_k(m,n) = -\sum_{(i,j)\in\beta-(0,0)} a(i,j)\hat{y}_k(m-i,n-j) + \sum_{(i,j)\in\Theta} g(i,j)e(m-i,n-j) \tag{4.3}$$

The corresponding prediction error $\Phi_k(m,n)$ is given by:

$$\Phi_k(m,n) = y(m+k,n)-\hat{y}_k(m,n) = \sum_{p=0}^{k=1} f(p,0)e(m+k-p,n) \tag{4.4}$$

The coefficients of $\{g(i,j)\}$ and $\{f(i,0)\}$ are given by the identity:

$$C(w,z) = A(w,z)F(w,1) + w^k G(w,z) \tag{4.5}$$

The support Θ (illustrated in Figure 4) is specific to each value of k and assures a unique solution to the identity (3.14) provided the support of β is R(M,M).

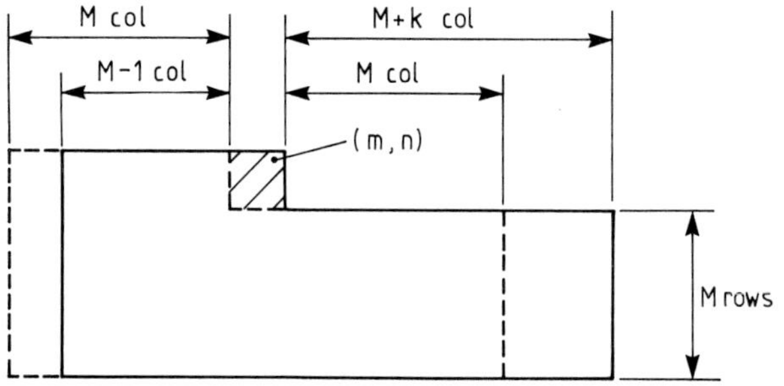

Fig. 4. Illustrating the support Θ for [g(i,j)] in the 2-D self tuning k-step ahead predictor

In an implicit self-tuning algorithm, however, we wish (if possible) to avoid solving an identity for the predictor parameters. This can be done by rewriting the predictor in the alternative form:

$$\hat{y}_k(m,n) = -\sum_{(i,j)\in\gamma-(0,0)} q(i,j)\hat{y}_k(m-i,n-j) + \sum_{(i,j)\in\Theta} g(i,j)\Phi_k(m-k-i,n-j) \tag{4.6}$$

where $\{q(i,j)\}$ is essentially the product of $\{a(i,j)\}$ and $\{f(i,1)\}$. That is to say,

$$Q(w,z) = A(w,z)F(w,1) \tag{4.7}$$

and the support γ for $\{q(i,j)\}$ is given by (Figure 5):

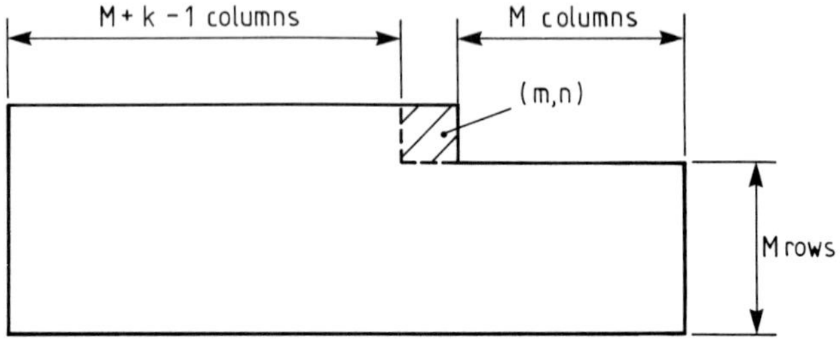

Fig. 5. Illustrating the support γ for $[q(i,j)]$ in the self tuning k-step ahead predictor recursion

$$\gamma = \{(i,j): 0\le i\le M+k-1, j=0\} \cup \{(i,j): -M\le i\le M+k-1, 1\le j\le M\}$$

The self-tuning prediction algorithm then runs as follows. Scanning the data in the normal raster procedure, at point (m,n), the following steps are computed:

STEP (i) Update the estimates $\{\hat{q}(i,j), \hat{g}(i,j)\}$ of the recursive least squares model of the process given by:

$$y(m,n) = -\sum_{(i,j)\in\gamma-(0,0)} \hat{q}(i,j)\hat{y}_k(m-k-i,n-j) + \sum_{(i,j)\in\Theta} \hat{g}(i,j)\hat{\Phi}_k(m-2k-i,n-j) + \hat{\Phi}_k(m-k,n) \tag{4.8}$$

STEP (ii) Use the estimates $\{\hat{q}(i,j)\}$, $\{\hat{g}(i,j)\}$ to form the k-step ahead prediction using:

$$\hat{y}_k(m,n) = -\sum_{(i,j)\in\gamma-(0,0)} \hat{q}(i,j)\hat{y}_k(m-i,n-j) + \sum_{(i,j)\in\Phi} \hat{g}(i,j)\hat{\Phi}_k(m-k-i,n-j) \tag{4.9}$$

It is possible to invoke the self-tuning property (Wittenmark (1974)) to assert that if the recursion converges then $\hat{\Phi}_k(m,n)$ can converge to the true prediction error sequence, and $\hat{y}_k(m,n)$ becomes the true prediction. The problem of parameterisation (encountered in the smoothing case) does not occur in this (prediction) case since a direct minimum phase ARMA representation of the process y(m,n) is used.

5. DISCUSSION

The algorithms presented here have been tested extensively using simulated data. In all cases these tests (Caldas Pinto (1983)) show that the self-tuning algorithms converge and give comparable filtering and prediction errors to those encountered with off-line designs. In particular, after about six lines of sixty four pixel per line, the performance of the self-tuning and off-line designs were indistinguishable. From a practical viewpoint the filtering techniques have been used as image enhancement algorithms on digitized photographs (Caldas Pinto (1983)). Again, the algorithms worked well and produced a significant reduction in the level of noise corruption. It seems therefore that the self-tuning concepts discussed have a useful part to play in two-dimensional signal processing. However, before they can be generally used a number of issues, theoretical and computational, require further research.

From a theoretical viewpoint, the question of support truncation in the self-tuning smoother needs further study to establish the limits of the approximations involved. The most troublesome computational areas concern the speed of computation. It is possible that fast regression algorithms or special VLSI architectures may assist here, although a more useful objective would be to develop an implicit form of the smoothing algorithm such that the synthesis step (equation (3.7)) need not be solved at each iteration. The most enjoyable area of further work is, however, in applications, since it is our experience that two-dimensional filters and predictors have a wide range of uses and often yield dramatically improved results when compared with simple 1-D algorithms.

5. REFERENCES

Anderson, B.D.O. and Moore, J.B. (1979), "Optimal Filtering". Prentice-Hall.

Caldas Pinto, J.R. (1983), "Self-Tuning Filters for Two-Dimensional Systems". *Ph.D. Thesis,* Control Systems Centre, UMIST, Manchester.

Caldas Pinto, J.R. and Wellstead, P.E. (1984a), "Self-Tuning Filters and Predictors for Two-Dimensional Systems: 2 - Smoothing Applications". Control Systems Centre Report 614, UMIST, Manchester.

Caldas Pinto, J.R. and Wellstead, P.E. (1984b), "Self-Tuning Filters and Predictors for Two-Dimensional Systems; 3 - Prediction Applications". Control Systems Centre Report 616, UMIST, Manchester.

Dudgeon, D.E. and Mersereau, R.M. (1984), "Multidimensional Digital Signal Processing". Prentice-Hall.

Harris, C.J. and Billings, S.A. (1981), "Self-Tuning and Adaptive Control". P. Peregrinus.

Ljung, L. and Soderstrom, T. (1983), "Theory and Practice of Recursive Identification". MIT Press.

Strintzis, M.G. (1978), "Dynamic Representation and Recursive Estimation of Cyclic and Two-Dimensional Processes". *IEEE Trans. AC-23,* **5**.

Sym, N. (1982), "Applications of Multivariable Self-Tuning Prediction, Adaptive Filtering and Identification". *Ph.D Thesis,* Control Systems Centre, UMIST, Manchester.

Wellstead, P.E. and Caldas Pinto (1984), "Self-Tuning Filters and Predictors for Two-Dimensional Systems; 1 - Algorithms". Control Systems Centre Report 612, UMIST, Manchester.

Wellstead, P.E. and Zarrop, M.B. (1983), "Non-Parametric Self-Tuning Regulators". *Int. J. Control,* **37**, 4, pp. 787-807.

Wittenmark, B. (1974), "A Self-Tuning Predictor". *IEEE Trans. AC-19,* **6**, pp. 484-851.

Wittenmark, B. and Hagander, P. (1977), "A Self-Tuning Filter for Fixed-Lag Smoothing". *IEEE Trans. IT-23,* **3**, pp. 377-384.

Woods, J.W. (1981), In "Two Dimensional Digital Signal Processing - Linear Filters". (Ed. Huang), Springer-Verlag.

CONVERGENCE ANALYSIS OF NON-LINEAR ADAPTIVE FILTERS USING TRANSPOSE DISTRIBUTED ARITHMETIC

M.J. Smith
(Currently with the Radar Systems Department, Ferranti Defence Systems Ltd., Edinburgh)

and

C.F.N. Cowan
(Department of Electrical Engineering, University of Edinburgh, Edinburgh)

ABSTRACT

This paper describes the convergence properties of transpose distributed arithmetic adaptive filters used for non-linear system modelling. Classical analysis presented in the form of an evaluation of the elements of the filter autocorrelation matrix is shown to provide an insufficient description of the convergence behaviour of the filter. A novel state variable analysis technique is then introduced and developed for the case of a first order adaptive system. This is shown to completely describe the convergence characteristics of the filter and is subsequently shown to be sufficiently general to encompass the description of arbitrarily higher orders of distributed arithmetic adaptive filters.

1. INTRODUCTION

The specific problem which motivated this work was that of describing the convergence properties of a non-linear adaptive echo canceller for the type of multilevel digital baseband transmission system discussed by Adams et al. (1984) and illustrated in Fig. 1.1.

The non-linearities in such a system are predominantly caused by component tolerancing in a Digital to Analogue Converter (DAC). See Agazzi et al (1982). A non-linear system modelling technique developed to represent these non-linearities has been described by Cowan and Adams (1984). The physical realisation of the non-linear system was shown to take the form of a transpose distributed arithmetic filter. The filter taps in a system whose architecture is based on the table look-up method take the form of random access memories (RAMs) as illustrated in Fig. 1.2. In a 4 level digital transmission system each RAM can be shown to contain 4 locations. Each location represents some combination of non-linear coefficients and linear dispersion coefficients. In other words each location contains a specific transmission level acted upon by a non-linear factor and an appropriate weighting factor in a linear dispersion.

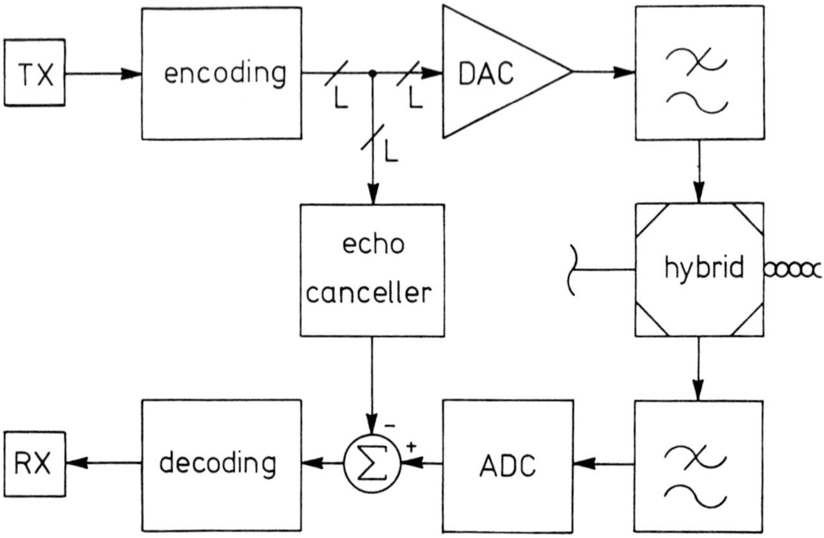

Fig. 1.1 System block diagram

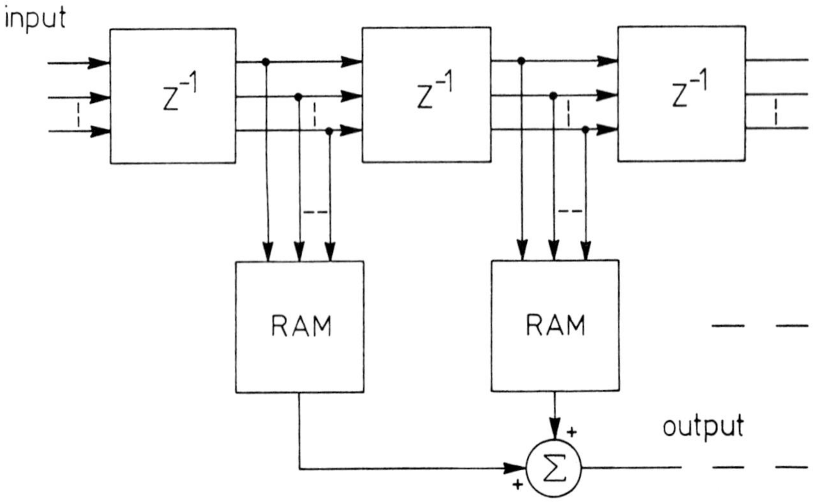

Fig. 1.2 General nonlinear filter

In many transmission systems it is desired to provide some digital filtering before the DAC. This produces an additional complexity by adding a linear dispersion before the non-linearity. It is still possible to use a transpose distributed arithmetic filter as the basis of an echo canceller for such a system, with an appropriate increase in the complexity of the adaptive filter, e.g. a general two stage dispersion before the DAC necessitates a factor of 4 increase in the number of locations in each RAM of the filter. Again it is necessary to describe and understand the performance of such filters.

2. DISTRIBUTED ARITHMETIC ADAPTIVE FILTERS

A general first order distributed arithmetic filter is shown in Fig. 1.2 The signal supplied by the transmitter of Fig. 1.1 is a binary data stream at rate B bits/sec which is encoded K bits at a time into a multi-level analogue signal with $L = 2^K$ levels. The output of this filter may be written as:

$$y(t) = \underline{H}_A^T(t)\underline{X}_A(t) \tag{2.1}$$

where the tth set of augmented input signals and tap weights are respectively:

$$\underline{X}_A^T(t) = [x_1(t), x_2(t), .., x_L(t), x_1(t-1), .., x_L(t-N+1)]$$

$$\underline{H}_A^T(t) = [h_{11}(t), h_{21}(t), .., h_{L1}(t), h_{12}(t), .., h_{LN}(t)]$$

such that the individual elements $x_i(t-k)$ have value 1 or 0 and represent the presence or absence of a particular signal level (level i) at time t-k, where t and k are integer multiples of the signalling interval K/B. The augmented vector $\underline{H}_A(t)$ is the weighting applied to each of these signals where each term is the product of the linear weight required to restore the true analogue level, the non-linear weight applied to that level and the linear impulse value at time t-k.

Applying the LMS stochastic gradient search algorithm, see Cowan and Grant (1985), it is easily shown that the following relation is obtained:

$$\underline{H}_A(t+1) = \underline{H}_A(t) + 2\mu e(t)\underline{X}_A(t) \tag{2.2}$$

where e(t) is the error signal at time t, μ is the convergence factor and N-1 is the order of the filter i.e. N is the number of taps in the filter. The update product in 2.2 above is automatically formed by the RAM address procedure since for any given values of t and k only one of the $x_i(t-k)$ values can be non-zero. That value of $x_i(t-k)$ is arbitrarily taken to be unity and therefore the only weight vector to be updated is the one currently addressed by the RAM. The update is simply achieved by adding a scaled portion of e(t) to the current RAM contents.

If the following two important assumptions are made an expression for the predicted behaviour of the filter can be obtained.

Assumption #1 The individual input vectors come from a mutually independent sequence i.e. $\underline{X}(t)$ and $\underline{X}(t-1)$ are uncorrelated.

Assumption #2 The covariance or interdependence between $\underline{X}(t)\underline{X}^T(t)$ and $\underline{H}(t)$ is zero.

These are common assumptions in adaptive filtering theory. Assumption #2 is particularly valid when the convergence factor μ is small enough to make the gross temporal variation in $\underline{H}(t)$ far slower that that of $\underline{X}(t)$. Using these assumptions:

$$E[\underline{H}_A(t)] = E[\underline{H}_A(t)] - 2\mu E[\underline{X}_A(t)\underline{X}_A^T(t)]\,E[\underline{H}_A(t)] + 2\mu E[\underline{X}_A(t)d(t)] \tag{2.3}$$

Thus if $E[\underline{X}_A(t)\underline{X}_A(t)]$ and $E[\underline{X}_A(t)d(t)]$ are calculated, equation 2.3 can be used to predict the convergence behaviour of the first order distributed arithmetic adaptive filter.

3. CLASSICAL ANALYSIS OF CONVERGENCE BEHAVIOUR

Using a probabilistic analysis it is possible to evaluate the elements of the auto and crosscorrelation matrices and thus obtain an expression for the convergence behaviour of distributed arithmetic adaptive filters.

To aid understanding the autocorrelation of $\underline{X}_A(t)$ is represented in block matrix form. Accordingly:

$$E[\underline{X}_A(t)\underline{X}_A^T(t)] = E\begin{bmatrix} \underline{X}(t)\underline{X}^T(t) & \underline{X}(t)\underline{X}^T(t-1) & . & . \\ \underline{X}(t-1)\underline{X}^T(t) & \underline{X}(t-1)\underline{X}^T(t-1) & . & . \\ . & . & . & . \\ . & . & . & \underline{X}(t-N+1)\underline{X}^T(t-N+1) \end{bmatrix}$$

where:

$$\underline{X}^T(t-k) = [\, x_1(t-k), x_2(t-k), .., x_L(t-k) \,]$$

Consider first the subset D of diagonal block correlation matrices where:

$$D \in \{E[\underline{X}(t)\underline{X}^T(t)], E[\underline{X}(t-1)\underline{X}^T(t-1)], .., E[\underline{X}(t-N+1)\underline{X}^T(t-N+1)]\}$$

Investigating the first element of D gives:

$$E[\underline{X}(t)\underline{X}^T(t)] = E\begin{bmatrix} x_1(t)x_1(t) & x_2(t)x_1(t) & . & . \\ x_2(t)x_1(t) & x_2(t)x_2(t) & . & . \\ . & . & . & . \\ . & . & . & x_L(t)x_L(t) \end{bmatrix} \qquad (3.1)$$

The task is now reduced to the calculation of the expected values of scalar elements. To do this we first note that when an individual address line $x_u(t)$ (level u) within a RAM is activated at time t $x_u(t) = 1$. Alternatively when the individual address line is not activated $x_u(t) = 0$. Assuming a stochastic ergodic signal source with a uniform (rectangular) probability density function each address line is activated with the same relative frequency $f = 1/L$. For example, a 2 bit 4-level signal activates each address line on average once in every four clock cycles. This gives the a priori probability that $x_u(t)=1$, $P(x_u(t)=1) = 1/4 = 0.25$, and $P(x_u(t)=0) = 1 - 1/L = 0.75$. Analysing the individual correlative elements in equation 3.1:

$$E[x_u(t)x_v(t)] = \sum_{i=0}^{1}\sum_{j=0}^{1} P(x_u(t)=i, (x_v(t)=j) \times (x_u(t)=i) \times (x_v(t)=j)$$

where $P(x_u(t)=i, x_v(t)=j)$ is the probability that x_u takes the value i at time t and x_v takes the value j at time t. As this is ensemble averaging the summation is over all the states which can theoretically occur, i.e. $i \in \{0,1\}$ and $j \in \{0,1\}$.

There are two cases of specific interest:

1) $u = v$

$$\begin{aligned} E[x_u(t)x_u(t)] = {} & P(x_u(t)=0, x_u(t)=0) \times (x_u(t)=0) \times (x_u(t)=0) \\ & + P(x_u(t)=0, x_u(t)=1) \times (x_u(t)=0) \times (x_u(t)=1) \\ & + P(x_u(t)=1, x_u(t)=0) \times (x_u(t)=1) \times (x_u(t)=0) \\ & + P(x_u(t)=1, x_u(t)=1) \times (x_u(t)=1) \times (x_u(t)=1) \end{aligned}$$

Although all the joint probabilities can be readily calculated this is unnecessary owing to the frequent occurrence of zero as a multiplying factor. Thus the only probability of direct relevance is $P(x_u(t)=1, x_v(t)=1)$. This fact is used throughout the analysis. In this case:

$$P(x_u(t)=1, x_v(t)=1) = P(x_u(t)=1) = \frac{1}{L}$$

which gives:

$$E[x_u(t)x_u(t)] = 0 + 0 + 0 + \frac{1}{L} = \frac{1}{L}$$

2) $u \neq v$

$$\begin{aligned} E[x_u(t)x_v(t)] = {} & P(x_u(t)=0, x_v(t)=0) \times (x_u(t)=0) \times (x_v(t)=0) \\ & + P(x_u(t)=0, x_v(t)=1) \times (x_u(t)=0) \times (x_v(t)=1) \\ & + P(x_u(t)=1, x_v(t)=0) \times (x_u(t)=1) \times (x_v(t)=0) \\ & + P(x_u(t)=1, x_v(t)=1) \times (x_u(t)=1) \times (x_v(t)=1) \end{aligned}$$

Noting that:

$$P(x_u(t)=1, x_v(t)=1) = 0$$

due to the fact that two different signal levels cannot occur at the same instant of time, gives:

$$E[x_u(t)x_v(t)] = 0$$

Thus the leading internal block correlation matrix of $E[\underline{X}_A(t)\underline{X}_A^T(t)]$ is:

$$E[\underline{X}(t)\underline{X}^T(t)] = \frac{1}{L}\begin{bmatrix} 1 & 0 & . & 0 \\ 0 & 1 & . & 0 \\ . & . & . & . \\ 0 & 0 & . & 1 \end{bmatrix} \tag{3.2}$$

If the signal vector is wide sense stationary all associated probabilities are time-invariant thus:

$$E[\underline{X}(t)\underline{X}^T(t)] = E[\underline{X}(t-1)\underline{X}^T(t-1)] = E[\underline{X}(t-k)\underline{X}^T(t-k)] \tag{3.3}$$

$$\left\{ k \in \{0,1,2,..N-1\} \right\}$$

Equations 3.2 and 3.3 define all the diagonal block matrices of the complete autocorrelation matrix. Through the use of the analysis technique employed above it may be shown that all the remaining block matrices are identical and are given by:

$$E[\underline{X}(t-r)\underline{X}^T(t-k)] = \frac{1}{L^2}\begin{bmatrix} 1 & 1 & . & 1 \\ 1 & 1 & . & 1 \\ . & . & . & . \\ 1 & 1 & . & 1 \end{bmatrix} \tag{3.4}$$

$$\left\{ \begin{matrix} k \in \{0,1,2,..,N-1\} \\ r \in \{0,1,..,N-1\} \\ r \neq k \end{matrix} \right\}$$

3.1 Computer Simulations of Classical Analysis

The normal method of investigating the validity and accuracy of the preceding probabilistic analysis would be to diagonalise the autocorrelation matrix derived in §3 and express the convergence transients in terms of the asymptotic behaviour of the natural modes of the system. However, given the fixed form of the autocorrelation matrix this is unneccessarily complicated. It is simpler to use a form of equation 2.3 directly as shown below.

$$E[\underline{H}_A(t+1)] = E[H_A(t)] - 2\mu E[\underline{X}_A(t)\underline{X}_A^T(t)]\, E[(\underline{H}_A(t) - \underline{H}^{opt})] \tag{3.1.1}$$

where $\underline{H}^{opt}$ is the optimum weight vector corresponding to the least mean square error. It is possible to directly compare the behaviour of equation 3.1.1, using the theoretically predicted form of the autocorrelation matrix, against the performance indicator.

The relative tap weight norm is defined as:

$$r_e = 10\log_{10}\left\{\frac{[h_{11}^{opt} - h_{11}(t)]^2 + [h_{21}^{opt} - h_{21}(t)]^2 + .. + [h_{LN}^{opt} - h_{LN}(t)]^2}{[h_{11}^{opt}]^2 + [h_{21}^{opt}]^2 + .. + [h_{LN}^{opt}]^2}\right\}$$

A comparison between the theoretical convergence and actual convergence of a 2 RAM 4-level filter with $\mu = 0.25$ is shown in Fig. 3.1.1 The disparity between the two traces can be clearly seen. This disparity is the subject of the following section.

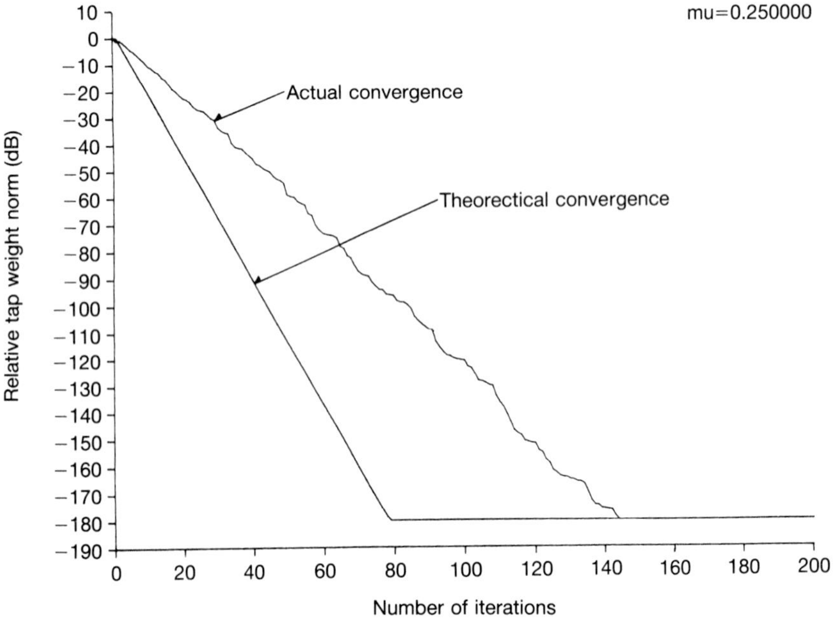

Fig. 3.1.1

4. STATE VARIABLE ANALYSIS OF CONVERGENCE BEHAVIOUR

The reasons for the disparity between the theoretical and actual performance of the adaptive filter lie in assumptions #1 and #2 introduced in §2. In reality assumption #1 is not valid as successive input vectors are correlated due to the fact that a tapped-delay line filter is being used: each input vector consists partly of a time shifted version of the previous input vector. However removal of this effect produces only a marginal improvement in the disparity between the predicted and actual performance of the filter. Therefore assumption #2 also contributes to the error in the classical analysis.

A solution to the shortcomings of the preceding analysis can be obtained by re-considering equation 3.1.1. Initially, to ease notational requirements, note that:

$$[(\underline{H}_A(t+1) - \underline{H}^{opt})] = [(\underline{H}_A(t) - \underline{H}^{opt})] + 2\mu \underline{X}_A(t)\underline{X}_A^T(t)[\underline{H}_A(t) - \underline{H}_A^{opt}]$$

Letting:

$$\underline{H}_A'(t) = \underline{H}_A(t) - \underline{H}^{opt}$$

gives:

$$\underline{H}_A'(t+1) = \underline{H}_A'(t) + 2\mu \underline{X}_A(t)\underline{X}_A^T(t)\underline{H}_A'(t)$$

Taking expected values:

$$E[\underline{H}_A'(t+1)] = E[\underline{H}_A'(t)] + 2\mu E[\underline{X}_A(t)\underline{X}_A^T(t)\underline{H}_A'(t)]$$

The important step is the use of:

$$E[\underline{X}_A(t)\underline{X}_A^T(t)\underline{H}_A'(t)]$$

rather than:

$$E[\underline{X}_A(t)\underline{X}_A^T(t)] \times E[\underline{H}_A'(t)]$$

This step is equivalent to calculating every "state" through which the filter can pass: a state being a potential set of actual tap values which can be attained on going from one iteration to the next, as illustrated schematically in Fig. 4.1. Mathematically this can be expressed as some performance indicator acting on an expanding set of attainable weight vectors. This set of weight vectors is written as:

$$\underline{H}_{L^N i+j}(t+1) = \underline{H}_i(t) + 2\mu \underline{A}_j \underline{H}'(t) \quad \left\{ \begin{array}{l} j \in \{0, 1, \ldots L^N-1\} \\ i \in \{0, 1, \ldots, (L^N)^t\} \end{array} \right\} \tag{4.1}$$

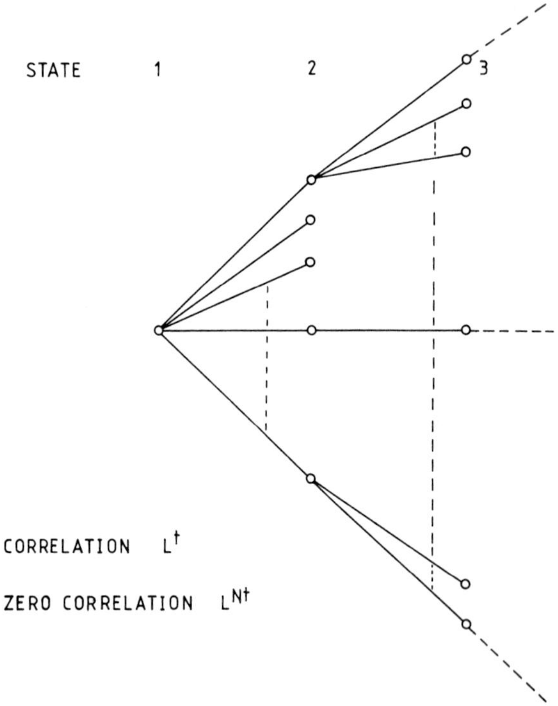

Fig. 4.1

This relation says that the $(L^N i + j)^{th}$ possible weight vector at time t+1 is given by the sum of the ith weight vector at time t plus one of L^N vectors determined by $\underline{A}_j$, $\underline{H}_i(t)$ and $\underline{H}^{opt}$. Every possible weight vector at time t+1 is calculated by inserting all possible values of i and j. By operating on the total set of these resultant weight vectors, a true performance indicator can be obtained which gives a description of the average performance of the adaptive filter at time t+1. The matrix $\underline{A}_j$ is a "controlling" matrix and is one of a set of matrices representing every possible instantaneous correlation of $\underline{X}_A(t)$. These autocorrelation matrices are easily derived by considering all the possible input signals into the RAMs.

The introduction of coding prior to Digital to Analogue conversion only requires the derivation of a different set of controlling matrices $\underline{A}$. It should also be noted that when crosscorrelation between successive input vectors is introduced ie. by considering a tapped-delay line, this effectively reduces the number of states which can be attained by the filter in "moving" from one iteration to the next. The only difference this makes is the imposition of a (known) constraint on the derivation of the set of controlling matrices $\underline{A}$. These two points indicate that the state variable technique is completely general in nature and completely describes the behaviour of FIR distributed arithmetic adaptive filters.

4.1 Computer Simulations of State Variable Analysis

The major drawback of the state variable theorem is the "explosion" in the number of elements in the set of possible weight vectors with increasing iteration number. These vectors increase as $(L^N)^t$, where t is the iteration number. The possible weight vectors are required to calculate an average performance indicator of the adaptive filter at each iteration. Although still analytically sound, this rapidly makes explicit calculation impracticable, if not impossible.

Fortunately it is possible to circumvent this problem by linearly extrapolating from the first few (easily calculated) performance values. Fig. 4.1.1 shows the excellent agreement between theoretical and actual convergence for a 2 RAM 4-level filter, with $\mu = 0.25$ adapting to the same channel described in §3.1 using the relative tap norm as a performance indicator (cf. Fig. 3.1.1).

5. CONCLUSIONS

Conventional analysis has been shown to provide an insufficient description of the convergence characteristics of distributed arithmetic adaptive filters. In response to this a powerful analysis technique was presented and shown to be accurate both for arbitrary orders of adaptive filters and for the case of successive vector correlation.

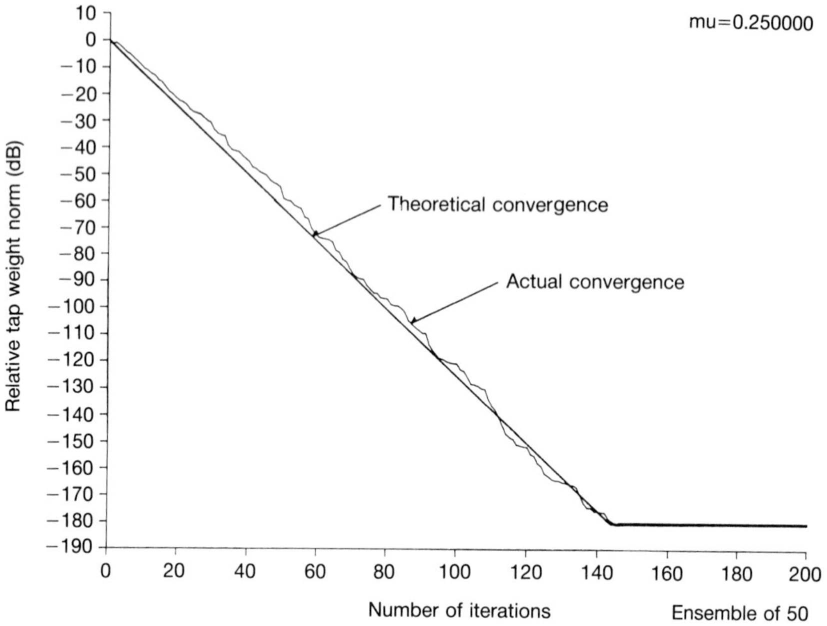

Fig. 4.1.1

ACKNOWLEDGEMENTS

Acknowledgement is made to the Director of Research of the British Telecom Research Laboratories for permission to publish this paper. The funding from the Science and Engineering Research Council is also gratefully acknowledged. Finally the authors would like to thank B. Mulgrew and P.F. Adams for their help and constructive criticisms.

REFERENCES

Adams, P.F., Cox, S.A. and Glen, P.J., (1984) "Long-reach duplex transmission systems for ISDN access", BTTJ, Vol. 2, No. 2.

Agazzi, O., Hodges, D.A. and Messerschmitt, D.G., (1982) "Large-scale integration of hybrid-method digital subscriberloops", IEEE Trans. Comm., 30, (9), pp. 2095-2108.

Cowan, C.F.N. and Adams, P.F., (1984) "Non-linear system modelling: concept and application", Proc. ICASSP84, pp. 45.6.1-45.6.4.

Cowan, C.F.N. and Grant, P.M., (1985) "Adaptive filters", Prentice-Hall, Englewood Cliffs, N.Y.

STABILISATION OF FAST RLS ALGORITHMS

G. Demoment and A. Houacine
(Laboratoire des Signaux et Systèmes (CNRS/ESE) France)

ABSTRACT

Divergence problems often appear with the usual fast recursive least-squares (FRLS) algorithms. This is due to the ill-posedness of the problem which occurs in presence of poorly exciting inputs and when an exponential weighting of the past observations is used with a growing memory filter. We study the limitations of this approach and we propose a regularisation method incorporating a priori information about the solution and using a finite memory filter in order to develop new fast adaptive algorithms. Simulation examples are used to make a comparison between our method and other fast algorithms, either recursive or not.

1. INTRODUCTION

During the last few years, many fast algorithms for transversal filter estimation, or for finite impulse response (FIR) system identification, were developed [e.g. Carayannis et al. (1984), Cioffi (1984)] . They are essentially fixed-order, fast recursive least-squares (FRLS) filters, improperly called "Fast Kalman Estimators" (FKE). They yield substantial improvement in initial convergence, if compared with Least-Mean-Squares (LMS) algorithms, with a modest increase in computational requirement. Adaptive filtering is usually performed with an exponential weighting of the past observations.

But the validity of these methods is limited to the case where the least-squares (LS) solution is acceptable, i.e. when the normal matrix of the problem is well-conditioned. This condition is not satisfied when the input signal has a poor frequency content and when the reference signal is highly corrupted by noise.

In this paper we address the problem of stabilising FRLS algorithms. For that purpose, we shall use regularisation techniques which must be introduced with the unusual constraints of an adaptive context and low computational requirements.

2. STATEMENT OF THE PROBLEM

2.1 Adaptive estimation

The problem of the adaptive estimation of time-varying transversal filters or regression coefficients can be described by the block-diagram of figure 1.

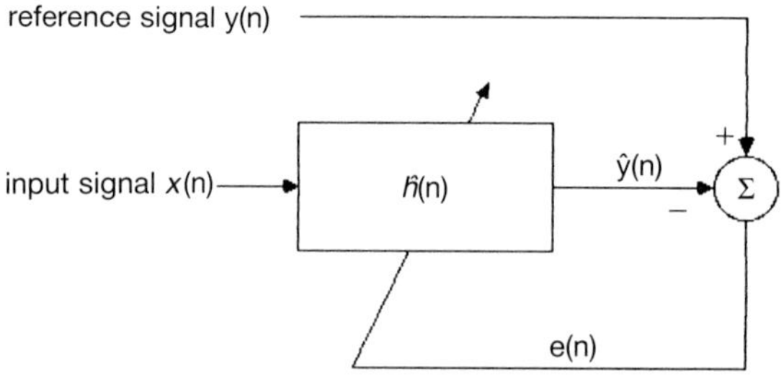

Fig. 1 Adaptive filter diagram

The discrete-time reference signal y(n) is assumed to depend linearly on the unknown m-order parameter vector $\underline{h}(n)$, according to the observation equation

$$y(n) = \underline{x}(n)^t \underline{h}(n) + b(n)$$

where

$$\underline{h}(n) = [h_o\ h_1\ \dots\ h_{m-1}]_n^t$$

$$\underline{x}(n) = [x(n)\, x(n-1)\ \dots\ x(n-m+1)]^t$$

$\underline{x}(n)$ is a vector of delayed samples of an input signal x(n), and b(n) is an independent gaussian white noise with distribution N(0,1). Adaptive estimation of the parameter vector is based on the innovation or residual error

$$e(n) = y(n) - \hat{y}(n)$$

The parameter vector is assumed to be slowly time-varying. This is a situation often encountered in various engineering fields such as digital communications or system analysis. For obvious reasons of real-time implementation, the problem is to estimate these time-varying parameters with fast recursive adaptive algorithms, i.e. with a complexity O(m) per recursion.

2.2 Current approaches

Two main families of recursive algorithms are commonly used to estimate adaptive transversal filters. The first one is the LMS family which minimises a stochastic version of the LS criterion.

$$J(n) = E[e^2]$$

through gradient techniques. They provide a suboptimal solution with the advantage of a very low complexity (2m or 3m arithmetic operations per recursion). But these algorithms suffer from a slow initial convergence.

This is why a lot of FRLS algorithms were developed during the last few years. They yield substantial improvement in initial convergence with a modest increase in complexity. They minimise a weighted LS criterion

$$J(n) = \sum_{i=1}^{n} \lambda^{n-i} e^2(i) \quad \text{with} \quad 0 < \lambda \leq 1$$

where the exponential data weighting is used to make these algorithms adaptive. But they often present stability problems.

3. PITFALLS OF THE EXPONENTIAL DATA WEIGHTING

3.1 Stability and numerical problems

The divergence and instability problems that appear in the common FRLS algorithms are a direct consequence of the ill-posedness of the corresponding least-squares problem. Computation of the estimate requires the inversion of the weighted normal matrix $[X(n)^t W(n) X(n)]$, where

$$X(n) = [\underline{x}(1), \underline{x}(2), \ldots, \underline{x}(n)]^t$$

and

$$W(n) = \text{diag}[\lambda^{n-1}, \ldots, \lambda, 1]$$

This weighted normal matrix has to be well-conditioned to get a stable numerical solution. In the stationary case, the weighting factor is chosen equal to one, and this corresponds to the well-known persistence of excitation condition which can be roughly stated as

$$\lim_{n \to \infty} (1/n) X(n)^t X(n) = \text{constant invertible matrix}$$

But when $\lambda < 1$, the problem is complicated by the introduction of the matrix $W(n)$ and becomes more and more ill-conditioned as the time goes along. The covariance matrix of the estimate is bounded by the inverses of the normal and weighted normal matrices

$$[X(n)^t X(n)]^{-1} \leq \text{cov}\{\underline{\hat{h}}(n)\} \leq [X(n)^t W(n) X(n)]^{-1}$$

and, obviously, the usual persistence of excitation condition is necessary but not sufficient.

A sufficient condition is difficult to establish. The additional constraint

$$\lambda = \lambda(n) \neq \text{constant}, \quad \prod_{i=1}^{\infty} \lambda(i) \neq 0$$

would imply

$$X(n)^t W(n) X(n) > [\prod_{i=1}^{n} \lambda(i)] X(n)^t X(n)$$

and therefore provides a sufficient condition. But this corresponds to the variable forgetting factor technique used in automatic control, [Fortescue et al. (1981) and Zarrop (1982)], which cannot be used to track slow parameter variations.

The main defect of the exponential weighting technique is then the following: although the weighted normal matrix may formally be a full rank one, modification of the scaling of its elements introduced by the exponential forgetting modifies its conditioning and leads to instability when finite precision computations are performed.

3.2 Regularisation

It is well-known that a convenient way to get an acceptable solution of an inverse problem is to perform a regularisation of the initial problem by combining the information content of the data with a priori information about the solution. This may correspond to the minimisation of a criterion

$$J(\underline{h}) = J_1(\underline{h}) + J_2(\underline{h})$$

where J_1 is the output criterion measuring the infidelity to the observations, and J_2 is a regularising function measuring the infidelity to the a-priori.

But the difficulty lies in the adaptive context of the problem, since attention must be paid to maintain a balance between the two functionals J_1 and J_2 in order to get the regularisation effective all along the time. As shown below, this is difficult to obtain under the additional constraint of a low computational load, and when the exponential forgetting is used.

3.3 Appropriate regularisation of growing memory filters

Growing memory filters take into account an increasing number of data that are exponentially weighted to get an adaptive algorithm. In a LS framework, and for example with a smoothness priors, an appropriate regularisation consists of taking J_1 and J_2 such as

$$J_1(\underline{h}) = \left(\sum_{i=1}^{n} \lambda^{n-i}\right)^{-1} \sum_{i=1}^{n} \lambda^{n-i} e(i)^2$$

$$J_2(\underline{h}) = \mu \, ||\underline{h}(n) - \underline{h}_o||^2_{P_o^{-1}}$$

μ is the regularising parameter which adjusts the balance between the two functionals. J_1 is the usual weighted least-squares criterion normalised in order to maintain the requested balance. This leads to the minimisation of the balanced criterion:

$$J(\underline{h}) = \sum_{i=1}^{n} \lambda^{n-i} e(i)^2 + \mu\left(\sum_{i=1}^{n} \lambda^{n-i}\right) ||\underline{h} - \underline{h}_o||^2_{P_o^{-1}}$$

We can see that the main difference from previous approaches lies in time-dependence of the a priori weight

$$\mu(n,\lambda) = \mu\left(\sum_{i=1}^{n} \lambda^{n-i}\right)$$

In the usual approaches, this weight is taken constant or even time decreasing. This is a fundamental difference in the "esprit" in which the a priori is introduced. In the usual FRLS techniques, it is used only to facilitate the initialisation and is quickly forgotten as the time increases. In our technique, it is used to stabilise the solution and must be propagated all along the time. The optimal solution minimising $J(\underline{h})$ is given by:

$$\underline{h}(n) = [X(n)^t W(n) X(n) + \mu(n,\lambda) P_o^{-1}]^{-1} [X(n)^t W(n) \underline{y}(n) + \mu(n,\lambda) P_o^{-1} \underline{h}_o]$$

and it can be computed recursively by

$$\underline{h}(n+1) = \underline{h}(n) + P(n+1)\,[e(n+1)\underline{x}(n+1) - \mu P_o^{-1}(\underline{h}(n) - h_o)]$$

$$P(n+1)^{-1} = \lambda P(n)^{-1} + \underline{x}(n+1)\underline{x}(n+1)^t + \mu P_o^{-1}$$

This algorithm is formally similar to the one proposed by Ljung (1985). But, unlike Ljung's work, the a priori mean is also taken into account and the perturbating matrix μP_o^{-1} is not introduced in a tricky way, but is directly derived from the LS smoothing norm chosen to perform the regularisation. Therefore, the user can utilise all the results published in the mathematics literature on the necessary link between the smoothing norm and the spectral content of the input signal.

There are two correction terms at each recursion, one proportional to the innovation, the other propagating the a priori. But the inverse of P(n) is modified at each recursion by the matrix P_o which, unlike the dyadic matrix $\underline{x}(n)\underline{x}(n)^t$, has no interesting shift property and does not lead to fast algorithms.

Computation of an exact solution minimising the balanced criterion is impossible with fast algorithms, and non-exact or suboptimal ways can be investigated. To avoid updating and inversion of the error covariance matrix at each time instant, this matrix is approximated by the identity matrix with a scalar coefficient performing a scaling with respect to the input signal level, as in the LMS algorithms, (see Segalen et al. (1982))

$$P(n+1) \dashrightarrow \alpha\ (\underline{x}(n+1)^t \underline{x}(n+1))^{-1} I_m$$

the recursive estimator is then computed as

$$\underline{h}(n+1) = \underline{h}(n) + \frac{\alpha}{\underline{x}(n+1)^t \underline{x}(n+1)} [e(n+1)\underline{x}(n+1) - \mu P_o^{-1}\underline{h}(n)]$$

If P_o is diagonal, (this corresponds to the euclidean norm constraint about the solution), the complexity is only 4m. If P_o^{-1} is the second differences band-diagonal matrix (smoothness priors about the solution), the complexity becomes 9m. The stability is improved but, in both cases, the initial convergence remains slow. Therefore, fast initial convergence, stability and adaptivity requirements imply a basic change.

4. FIXED MEMORY FILTERS

4.1 The general concept

Since the propagation of the a priori cannot be done with fast algorithms implementing growing (exponentially weighted) memory filters, an alternative way consists of using fixed-memory filters which use at each recursion a fixed number of data and which, consequently, allow a constant weight of the a priori. In a LS framework, this corresponds to the minimisation of:

$$J(\underline{h}) = \sum_{n-\ell+1}^{n} \lambda^{n-i} e(i)^2 + \mu(1) \| \underline{h}(n) - \underline{h}_o \|^2_{P_o^{-1}}$$

where the regularising parameter is no longer a function of time-index n. Two kinds of solution may be used to perform a fixed-memory filtering. The most natural one consists of using a sliding window (fig. 2). The data are uniformly processed but, at the expense of a greater computational burden due to the computation of two corrections at each recursion (see details below).

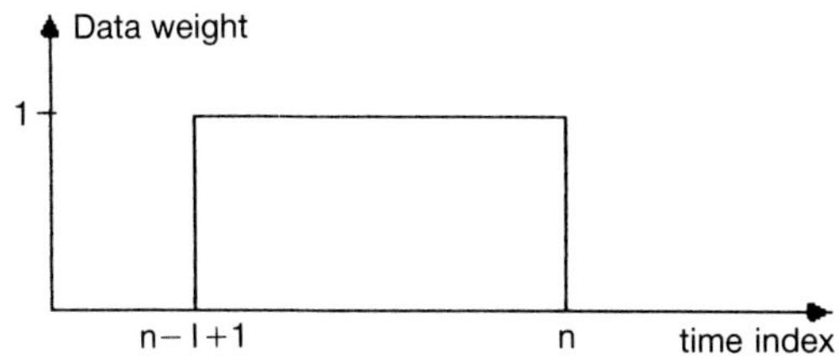

Fig. 2 Sliding window

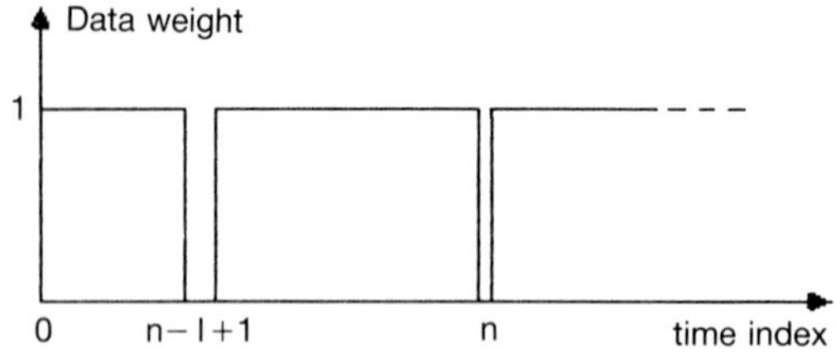

Fig. 3 Block processing

Another way consists of dividing the data into blocks, having the same length or not, and being adjacent or not, and performing the minimisation for each block (fig. 3). The estimate computed at the end of each block is then used as the a priori mean for the following one. The computational load is then reduced, but the data are not uniformly processed. Some recently proposed modifications of existing algorithms such as periodical reinitialisations [e.g Lin (1984)], correspond indeed to a block processing of the data.

4.2 Derivation of fast algorithms

In order to derive fast algorithms, the shift property of the successive input vectors must be exploited. We present here two different ways of using this important property.

4.2.1 State space model

First, let us consider a block-processing of the data. Define the extended parameter vector as

$$\underline{H}^t(n) = [0 \ldots 0 \; h_{m-1} \; h_{m-2} \ldots h_o \; 0 \ldots 0] \quad \text{N-dim.,}$$

and the global input vector corresponding to the block to be processed as

$$\underline{X}^t = [x_{-m+1} \; x_{-m+2} \ldots x_o \ldots x_{N-m}] .$$

Then, the observation equation may be written as

$$y(n) = \underline{X}^t \underline{H}(n) + b(n),$$

and a state equation may be added to translate the shift property

$$\underline{H}(n+1) = D \, \underline{H}(n), \; D \text{ being the shift operator.}$$

Now, the optimal estimate may be recursively computed with a Kalman filter. But the state-space model is invariant, since $\underline{X}$ and D are constant. We can replace the Riccati equations by Chandrasekhar ones, using factorisation techniques. The complete derivation of the corresponding Block-processing Chandrasekhar Adaptive Regularizer (B-CAR) algorithm can be found in Houacine and Demoment (1985a).

When a sliding-window filter is used, two correction gains $\underline{k}^+$ and $\underline{k}^-$ have to be computed at each recursion. The estimate is then given by

$$\underline{h}(n+1) = \underline{h}(n) + \underline{k}^+(n+1)\ e^+(n+1) - \underline{k}^-(n+1)e^-(n+1)$$

where $e^+(n+1)$ and $e^-(n+1)$ are the innovations based respectively on the present data and the old one to be discarded. The resulting Sliding Chandrasekhar Adaptive Regularizer (S-CAR) algorithm has been given in Houacine and Demoment (1985b). Initialisation is done using the B-CAR algorithm during the ℓ first time instants.

The resulting complexity is 11m to 17m for a B-CAR and 34m to 54m for a S-CAR, depending on the displacement rank of the a priori matrix P_o. This complexity is at its lowest level when the parameters are assumed a priori uncorrelated.

4.2.2 Partitioned-matrix inversion

This is a well-known technique popularised by Carayannis et al. (1984). Let us define the m-dimension input vector

$$\underline{x}_m^t(n) = [\, x_n\ x_{n-1}\ \dots\ x_{n-m}\,]$$

the shift property is represented by the order relation

$$\underline{X}_{m+1}(n) = \begin{bmatrix} x(n) \\ \hline \underline{x}_m(n-1) \end{bmatrix} = \begin{bmatrix} \underline{x}_m(n) \\ \hline x(n-m-1) \end{bmatrix}$$

Defining the regularised covariance matrix corresponding to the fixed memory balanced criterion by

$$R_m(n) = \sum_{n-\ell+1}^{n} \underline{x}_m(p)\underline{x}_m(p)^t + \mu P_o^{-1}$$

we have the partitioned matrices relations

$$R_m(n) = \left[\begin{array}{c|c} R_m(n) & \underline{r}_b(n) \\ \hline \underline{r}_b^t(n) & r_{bo}^t(n) \end{array}\right] = \left[\begin{array}{c|c} r_{fo}(n) & \underline{r}_f^t(n) \\ \hline \underline{r}_f(n) & R_m(n-1) \end{array}\right]$$

and using the partitioned-matrix inversion lemma and the shift property, we get the Sliding-window Partitioned-matrix Adaptive Regularizer (S-PAR) algorithm which was proposed in Houacine and Demoment (1985c). The resulting complexity is only 24m, due to the fact that P_o must be chosen proportional to identity matrix in order to apply the partitioning technique.

5. SIMULATED EXAMPLES

We present here simulation results for the B-CAR algorithm. In the stationary case we estimate the 7-point impulse response of Marple and Rabiner (1983) for a white noise input (fig. 4.A), a filtered input (fig. 4.B) (by the 25-point filter of Marple and Rabiner), and a non-stationary input (fig. 4.C) (Chirp $x(i) = \sin(0.04\,\pi\, i^2/m)$).

In the non-stationary case we consider two situations corresponding to a drift in the parameter vector (fig. 4.D) ($\underline{h}_{i+1} = a\,\underline{h}_i$ with $a = 0.99$), and a random perturbation (fig. 4.E) ($\underline{h}_{i+1} = \underline{h}_i + b\,\underline{u}_i$ with $b = .001$ and $\underline{u}_i$ white noise $N(0,1)$).

The performance of the estimation is measured at each recursion by

$$Q_{i\,(dB)} = 10 \log_{10}[(\underline{h}_i - \hat{\underline{h}}_i)^t(\underline{h}_i - \hat{\underline{h}}_i)]/(\underline{h}_i^t\underline{h}_i)$$

In all cases the B-CAR exhibits a fast convergence, and we can see obvious improvement in the stability and the estimation quality compared with LS or LMS algorithms.

The influence of the a priori appears clearly. The performance index is represented for various weights of the a priori in fig. 4.F. We can see that there is an optimal a priori weight factor μ which gives a good compromise between the estimation quality (low performance index) and the stability of the solution (represented by the variance of the performance index).

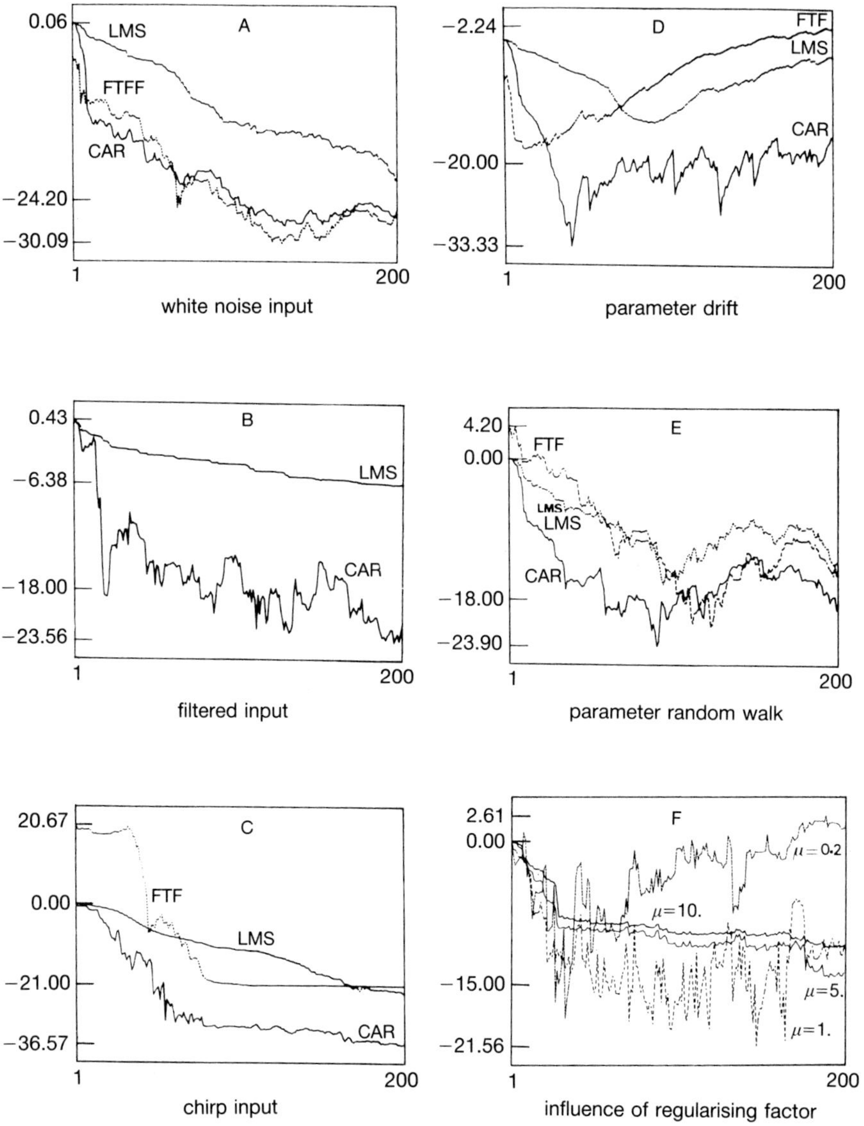

Fig. 4 Simulation results

6. CONCLUSION

We have explained why the use of a growing memory filter with an exponential data weighting is not really appropriate to derive fast, stable and adaptive algorithms.

We propose, as an alternative, an interesting way to solve the problem of stabilising FRLS algorithms. This way consists of using a finite-memory filter combined with an a priori information about the solution, which can be obtained by a block-processing of the data or by the use of a sliding window.

7. REFERENCES

Cioffi, J.M., (1984) "Fast transversal filters for communications applications", *PhD. Thesis,* Standford University.

Fortescue, T.R., Kershenbaum, L.S., Ydstie, B.E., (1981) "Implementation of self-tuning regulators with variable forgetting factors", *Automatica,* vol. 17, no 6, pp. 831-835.

Houacine, A. and Demoment, G., (1985a) "Identification de réponses impulsionnelles finies par filtrage de Kalman rapide", Proc. 10th GRETSI Symp., Nice, 20-24 May, pp. 483-388.

Houacine, A. and Demoment, G., (1985b) "Application of Chandrasekhar factorisation techniques to fixed memory filtering", L2S. Int. Rep. no. 85/062.

Houacine, A. and Demoment, G., (1985c) "Fast, recursive, fixed memory least squares algorithm", L2S. Int. Rep. no 85/063.

Kalouptsidis, N., Carayannis, G. and Manolakis, D., (1984) "A fast covariance algorithm for sequential least-squares filtering and prediciton", *IEEE Trans.,* vol. AC-29, pp. 752-755.

Lin, D.W., (1984) "On digital implementation of the fast Kalman algorithms", *IEEE, vol. ASSP-32,* no. 5, pp. 998-1005.

Ljung, L., (1985) "Error propagation in adaptation algorithms with poorly exciting signals", CNRS-GRETSI Workshop on "Quantization and adaptivity", Nice, 24-25 May, pp. 150-160.

Marple, S.L., (1981) "Efficient least squares FIR system identification, *IEEE Trans.,* vol. ASSP-29, pp. 62-73.

Marple, S.L. and Rabiner, L.R., (1983) "Performance of a fast algorithm for FIR system identification using least squares analysis", *Bell Tech. J.*, vol. 62, pp. 717-742.

Ségalen, A. and Demoment, G., (1982) "Constrained LMS algorithm", *Electron. Lett.*, vol. 18, pp. 226-227.

Zarrop, M.B., (1982) "Variable forgetting factors in parameter estimation", 6th IFAC Identification and System Parameter Estimation Symp., vol. 2, pp. 994-998.

ANALYSIS OF THE LMS ADAPTIVE FILTER APPLIED TO PERIODIC SIGNALS

P.M. Clarkson
*(Institute of Sound and Vibration Research,
University of Southampton)*

ABSTRACT

Convergence results for the LMS adaptive filter applied to the cancellation of periodic interferences are developed. It is demonstrated that in general for deterministic inputs the LMS filter can be approximated by a discrete transfer function relating the desired input to either the filter output or to the error.

For periodic inputs this transfer function relation is often exact. Particular results are obtained for several example simple inputs. For more general inputs methods for inferring stability are demonstrated.

1. INTRODUCTION

Although the LMS adaptive algorithm (for example [1]), has found widespread use in numerous applications and practically is well known as a simple and robust algorithm, its properties have proved far more difficult to demonstrate analytically, (for example [1]).

For random imputs most work has been based on the somewhat artificial case of successive input vectors being uncorrelated. In the case of deterministic imputs, the only reported work has been that of Glover [2], who has demonstrated for sinusoidal inputs that the LMS adaptive system can be approximately represented by a discrete transfer function relation. In this paper the transfer function approximation is first generalised to include all deterministic input signals. A special form of this relation, valid for periodic inputs is derived and it is shown that when the filter length is an integer multiple of the period, the transfer function is exact.

A number of simple examples are developed (including recovery of the sinusoidal input case of Glover as a special case). For more general inputs it is demonstrated that stability conditions for the adaptive filter can be obtained using well known stability tests from discrete control theory. The accuracy of the approximation is also examined by reconsidering the sinusoidal input case in the light of a more general relation than the simple transfer function.

2. TRANSFER FUNCTION FOR DETERMINISTIC INPUTS

Consider the adaptive filtering configuration as displayed in Figure 1.

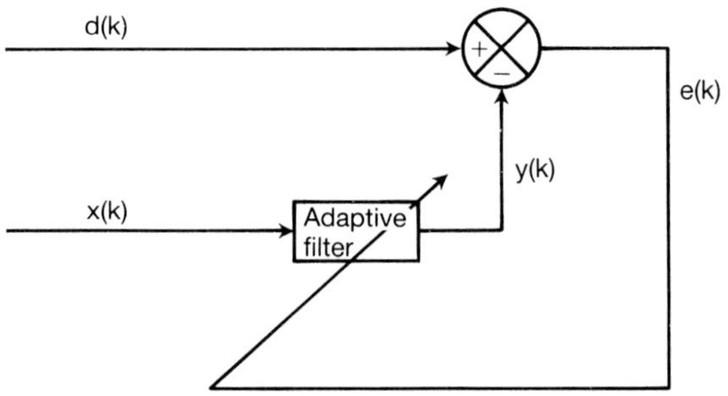

Fig. 1 Adaptive Filtering System

The LMS update equation [1], can be written:

$$\underline{f}_n = \underline{f}_n + \alpha e_{n-1}\underline{x}_{n-1} \qquad (1)$$

where

$$\underline{f}_n^{\,t} = (f_n(o),\ f_n(1)\ \ldots\ f_n(L-1))$$

$$\underline{x}_n = (x(n),\ x(n-1),\ \ldots\ x(n-L+1))$$

$e_n = d_n - y_n$ is the error and α is the adaptation constant.

Repeated application of (1) gives

$$\underline{f}_n = \alpha \sum_{i=0}^{n-1} e_i\, \underline{x}_i \qquad (2)$$

(assuming $\underline{f}_o \equiv 0$)

The filter output is

$$y_n = \underline{f}_n^t \underline{x}_n = \alpha \sum_{i=0}^{n-1} e_i \, \underline{x}_i^t \underline{x}_n \tag{3}$$

also $e_n = y_n - d_n$, so that

$$e_n + \alpha \sum_{i=0}^{n-1} e_i r_{in} = d_n \tag{4}$$

where

$$r_{in} = \sum_{j=0}^{L-1} x_{i-j} x_{n-j}$$

This equation is a relation between the error, e, and the desired signal, d, in terms of a purely recursive difference equation (of nth order).

In general, the coefficients r_{in} are time-varying and thus it is difficult to reach an explicit solution for e_n. However, for deterministic inputs

$r_{in} \approx r_{n-i}$ as $L \longrightarrow \infty$ (assuming x has finite power),

so that

$$e_n + \alpha \sum_{i=0}^{n-1} r_{n-i} e_i = d_n \tag{5}$$

This equation may be z-transformed to give:

$$\frac{E(z)}{D(z)} = \frac{1}{1+\alpha R(z)} = H(z) \tag{6}$$

where $R(z) = z^{-1} r_1 + z^{-2} r_2 + \ldots..$

and where we have used the one-sided z-transformation. In other words, we have replaced the adaptive filter of Figure 1 with a linear transfer function (for any given x), relating e and d (see Figure 2).

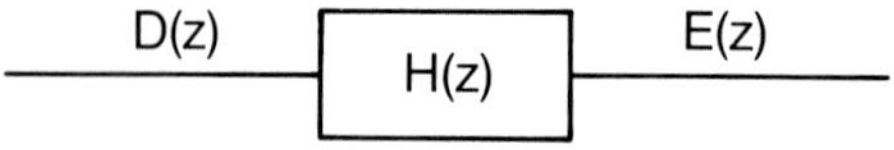

Fig. 2 Equivalent Transfer Function

3. PERIODIC INPUTS

In general, the transfer function of equation (6) is an approximate relation due to the assumption $r_{ni} = r_{n-i}$ used in the derivation. For some inputs, however, the results are exact, one such class of inputs is periodic signals. In general if x_n is a periodic sequence (with period N, say) then the r_{in} are doubly periodic, that is

$$r_{in} = r_{i+kN,n} = r_{i,kN+n}; \quad k = 0, \pm 1, \pm 2, \ldots.$$

In particular, if the filter length L is an integer multiple of N then $r_{in} \equiv r_{i-n}$ and the transfer function is exact even for finite length filters.

Also, in this case since there are only N distinct r_i's the transfer function becomes

$$\frac{E(z)}{D(z)} = \frac{1 - z^{-N}}{1+\alpha\,\hat{R}(z)-z^{-N}} \tag{7}$$

where $\hat{R}(z) = r_i z^{-1} + r_2 z^{-2} + \ldots + r_{N-1}\, z^{-N+1} + r_0 z^{-N}$

This transfer function has zeroes equispaced around the unit circle in the z plane and poles whose locations are determined by the zeroes of

$$1 + \alpha\hat{R}(z)-z^{-N} \tag{8}$$

In general it is difficult to obtain the zeroes explicitly from this Nth order polynomial however, simpler results can be obtained for some special cases. We shall consider several such examples.

(i) Sinusoidal Inputs

By considering inputs of the form

$$x_n = A\cos(\omega_0 nT+\phi)$$

it is possible to recover the transfer function described by Glover [2] , as a special case of equation (6). Using the fact that

$$r_i = \frac{A^2 L}{2} \cos(\omega_o iT) \tag{9}$$

in equation (6) gives

$$\frac{E(z)}{D(z)} = \frac{1 - 2z^{-1}\cos\omega_o T + z^{-2}}{[1 - \frac{\alpha LA^2}{2}]\, z^{-2} + [\frac{\alpha LA^2}{2} - 2]\, z^{-1}\cos\omega_o T + 1} \tag{10}$$

which is the required result.

This transfer function has zeroes on the unit circle at $z = e^{\pm i\omega_o T}$ and poles located approximately at

$$z = (1 - \frac{\alpha LA^2}{4})\, e^{\pm i\omega_o T}$$

The system is thus a notch filter centred on ω_o and whose bandwidth is controlled by α, L and A^2.

As described by Glover this result generalises to the case of M sinusoidal inputs. However, the result becomes less realistic since the requirement that the filter should span a full period becomes increasingly difficult to realise (particularly when the sinusoids are closely spaced in frequency). The nature of the solution in these cases is examined in Section 4.

(ii) Impulse Train Input

Inputs of the form

$$x(i) = \sum_{n=0}^{\infty} \delta(i - nN) \tag{11}$$

have been considered by Elliott and Darlington [3] .

In this case the transfer function can be obtained as a special case of equation (7) by noting that for this example

$$\hat{R}(z) = z^{-N} \text{ for } L = N .$$

Hence

$$\frac{E(z)}{D(z)} = \frac{1 - z^{-N}}{1-(1-\alpha)\,z^{-N}} \tag{12}$$

This transfer function has zeroes equispaced around the unit circle and poles on the same radial lines $|z| = |(1-\alpha)^{1/N}|$, (see Figure 3).

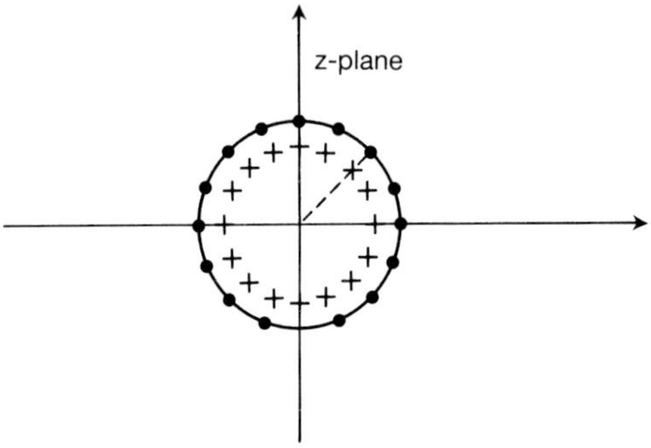

Fig. 3 Pole-zero pattern for impulse train input. The time domain response is a comb filter.

(iii) Square Wave Input

Consider a simple square wave input of the form

$$x = (1,\ 1,\ -1,\ -1,\ 1,\ 1,\ \ldots\ ) \tag{13}$$

Such an input gives rise to $R(z) = -Lz^{-2} + Lz^{-4}$ so that using equation (7), we have

$$\frac{E(z)}{D(z)} = \frac{1-z^{-4}}{1-L\alpha z^{-2} + (L\alpha-1)\ z^{-4}} \tag{14}$$

This transfer function has poles at ± 1 (cancelling zeroes there) and at $z^2 = L\alpha - 1$. For $\alpha < {}^1/L$ these poles lie on the imaginary axis, (see Figure 4), so that the result is once again a notch filter centred on $^{\pi}/2$ radians. It is interesting to note that in this respect the filter response is indistinguishable from that for a sinusoidal input of frequency $\omega_0 = {}^{\pi}/2$.

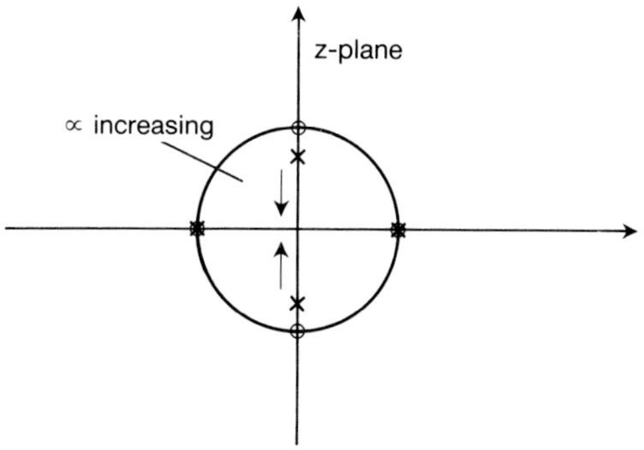

Fig. 4 Pole-zero pattern for square wave input.

(iv) General Inputs

As mentioned in a previous section it is generally difficult to determine the roots of $1 + \alpha\hat{R}(z) - z^{-N}$ explicitly, but it is often possible to examine the positions of the poles with respect to the unit circle by constructing a Jury Table or using similar or similar stability results [4].

An example of this approach is provided by considering the behaviour of the adaptive filter when the input process x has the form shown in Figure 5.

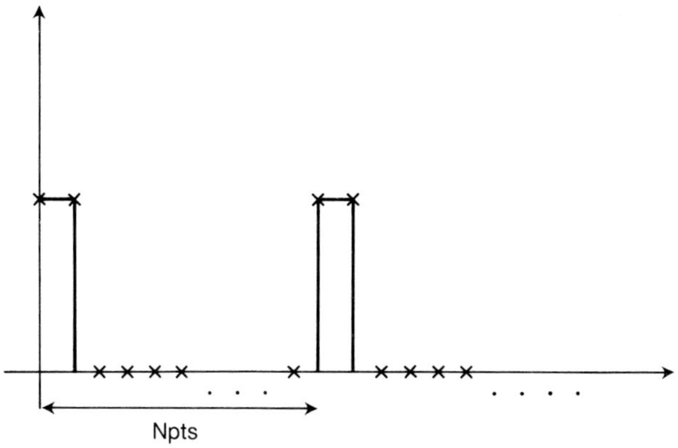

Fig. 5 2-point pulse in N point cycle.

That is, x_n has the form of a pulse, 2 samples wide which is repeated every N points. For this example

$$\hat{R}(z) = z^{-1} + z^{-N+1} + 2z^{-N}$$

Hence

$$1 + \alpha R(z) - z^{-N} = 1 + \alpha z^{-1} + \alpha z^{-N+1} + (2\alpha-1)z^{-N} \quad (15)$$

and the poles of the adaptive filter transfer function are given by the zeroes of this expression. It is difficult to obtain values for these zeroes directly, however, if we use the Jury Table we can at least determine whether they lie within $|z|=1$, and thus whether the filter is stable.

Multiplying the polynomial by z^N gives

$$P(z) = z^N + \alpha z^{N-1} + \alpha z + (2\alpha - 1) \quad (16)$$

For the general Nth order polynomial

$$P(z) = b_0 z^N + b_1 z^{N-1} + \dots\dots\dots\dots + b_{N-1} z + b_N \quad (17)$$

the Jury Table can be written as:

$$\begin{array}{llllll}
b_N & b_{N-1} & b_{N-2} & \cdots\cdots & b_2 & b_1 \quad b_0 \\
b_0 & b_1 & b_2 & \cdots\cdots & b_{N-2} & b_{N-1} \quad b_N \\
c_{N-1} & c_{N-2} & & \cdots\cdots & c_1 & c_0 \\
c_0 & c_1 & & & c_{N-2} & c_{N-1} \\
\cdot & \cdot & & & & \\
\cdot & \cdot & & & & \\
\cdot & \cdot & & & & \\
\cdot & \cdot & & & & \\
\cdot & \cdot & & & & \\
\cdot & \cdot & & & & \\
r_0 & r_1 & r_2 & r_3 & & \\
s_2 & s_1 & s_0 & & &
\end{array}$$

where $c_k = \begin{vmatrix} b_N & b_{N-k-1} \\ b_0 & b_{k+1} \end{vmatrix}$; $k = 0, 1, \ldots, N-1$

$$d_k = \begin{vmatrix} c_{N-1} & c_{N-2-k} \\ c_0 & c_{k+1} \end{vmatrix} ; \quad k = 0, \ldots, N-2, \text{ etc}$$

(see [4] for more details).

The N + 1 conditions required for all roots of P(z) to be within $|z|=1$ are

$$P(1) > 0, \; P(-1) > 0 \quad (N \text{ even})$$

$$P(-1) < 0, \quad (N \text{ odd})$$

$$|b_N| < |b_0|, \; |c_{N-1}| > |c_0|, \; |d_{N-2}| > |d_0|, \; \ldots, \; |s_2| > |s_0|$$

For the example of equation (16), $b_0 = 1$, $b_1 = \alpha$, $b_{N-1} = \alpha$ and $b_N = (2\alpha - 1)$.

This gives $P(1) = 4\alpha > 0$ if $\alpha > 0$

$P(-1) = 0$ if N is even - this implies that there is a pole at $z = -1$, however, since from equation (7) this is cancelled by a zero at the same location this has no effect on stability.

$$P(-1) = 2\alpha - 2 < 0 \text{ if } \alpha < 1 \; (N \text{ odd})$$

$$c_{N-1} = 4\alpha(\alpha-1); \; c_{N-2} = 2\alpha(\alpha-1) \; c_0 = 2\alpha(\alpha-1)$$

$$c_i \equiv 0 \; i \neq 0, N-1, N-2$$

Hence $|c_{N-1}| > |c_0|$

$$d_{N-2} = -12\alpha^2(\alpha-1)^2; \; d_{N-3} = 8\alpha^2(\alpha-1)^2, \; d_0 = -4\alpha^2(\alpha-1)^2$$

$$d_i \equiv 0 \; i \neq 0, N-1, N-2.$$

Hence $|d_{N-2}|>|d_0|$

$$e_{N-3} = 144\alpha^4(\alpha-1)^4 + 16\alpha^4(\alpha-1)^4; \quad e_0 = 32\alpha^4(\alpha-1)^4$$

$$e_i \equiv 0 \text{ otherwise}$$

Hence $|e_{N-3}|>|e_0|$

$$f_{N-4} = e_{N-3}^2 > f_0 \ (=0)$$

$$g_{N-5} = f_{N-4}^2 > g_0 \ (=0)$$

Collecting up the results, the N + 1 conditions are satisfied (and thus the system is stable) provided

$$0 < \alpha < 1$$

4. APPROXIMATE EFFECTS

Thus far it has been convenient to use the approximation $r_{ni} = r_{n-i}$ which leads to the transfer functions (6) and (7). Although, practically this approximation generally works very well it is important to examine the situation when the approximation does not hold in order that possible pathological cases can be recognised.

Consider again the case of a sinusoidal input as in the previous section. As described earlier the transfer function gradually becomes increasingly approximate as the number of sinusoids in the input increases because the requirement that the filter length should span a full period becomes increasingly difficult to achieve (particularly if the sinusoids are closely spaced in frequency). Consequently the transfer function generally becomes increasingly inaccurate. A more general form for the response can be derived to account for the effects which occur when $L \neq N$. For the single sinusoid case this has the form:

$$E(z)=H(z)[D(z)+G_1(z)H(ze^{-2i\omega_0 T})D(ze^{-2i\omega_0 T})$$

$$+ G_2(z)H(ze^{i2\omega_0 T})D(ze^{i2\omega_0 T})] \qquad (18)$$

where H(z) is the transfer function of equation (10),

$$G_1(z) = \frac{\alpha A^2}{4} \left(\sum_{i=0}^{L-1} e^{i2\phi_i} \right) U(ze^{-i\omega_0 T}) \qquad (19)$$

and

$$G_2(z) = \frac{\alpha A^2}{4} \left(\sum_{i=0}^{L-1} e^{-i2\phi_i} \right) U(ze^{i\omega_0 T}) \qquad (20)$$

$$U(z) = (z-1)^{-1}$$

where $\phi_i = \phi - \omega_0 iT$ and in general

$$X(ze^a) = z \text{ transform of } \{x(k)e^{-ak}\}.$$

This generalised transfer function is depicted in Figure 6. It is clear that the response is no longer a conventional linear transfer function between d(n) and e(n).

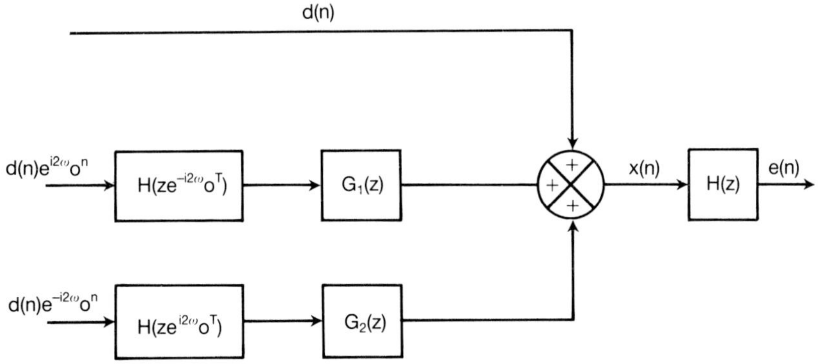

Fig. 6 Generalised Transfer Function for Sinusoidal Input

It is instead the sum of three components. The first component is the usual linear transfer function relation, the other two are obtained by heterodyning d(n) at twice the reference frequency, notch filtering using H(z) rotated in the same manner and then filtering with a first order system. These components are summed and passed through the usual notch. The effect of the non-linear components of the response is, primarily, to create an amplitude scaled and rotated (in frequency) version of the linear response. The magnitude of those components is determined from $G_1(z)$, $G_2(z)$ of equations (19) and (20) by terms of the form:

$$= \sum_{j=0}^{L-1} e^{\pm i2\phi_j} \qquad \text{where } \phi_j = \omega_0 jT$$

so that the magnitude of the non-linear terms is proportional to

$$\left| \sum_{j=0}^{L-1} e^{i2\omega_0 jT} \right| = \left| \frac{\sin 2L\omega_0 T}{\sin 2\omega_0 T} \right| \qquad (21)$$

From this it can be seen that the magnitude of the non-linear components is greatest as $\omega_0 T \to 0$ and decreases as $2\omega_0 T \to \pi/2$ (for L even) so that equation (18) reduces to equation (10).

A similar relation can also be obtained for multiple input sinusoids [5] , the results generalise in a natural way so that in addition to the usual linear components, the response also contains components rotated at twice the individual interference frequencies and at the sums and differences thereof.

The effect of these extra components can be illustrated using a simple example. Consider the adaptive filter with an input consisting of two sinusoids of equal amplitudes and frequencies 500 Hz and 833 Hz at 10 kHz sample rate. The effect of running the adaptive filter (with this data as both desired and reference inputs) using a typical value for α (0.01) and 4 weights (to make sure $L \neq N$) is to produce an error whose spectrum is as shown in Figure 7.

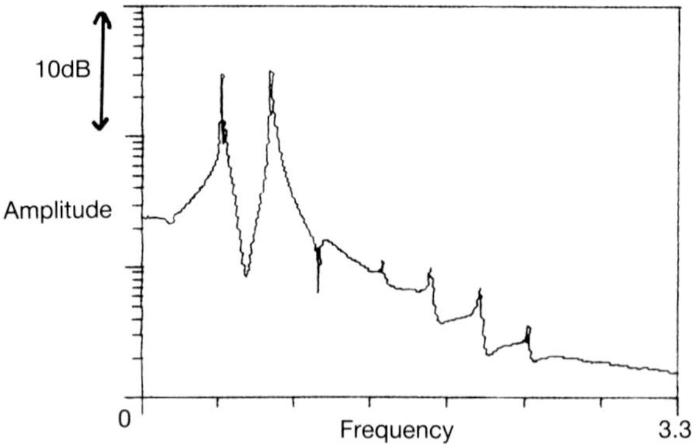

Fig. 7 Noise canceller error spectrum (2 sinusoids ω_0 = 5000 Hz, ω_1 = 833.3 Hz).

From this it can be seen that in addition to the main peaks due to the partially cancelled sinusoids (during convergence) there are a number of secondary components in the response. Careful inspection of Figure 7 shows that these peaks correspond to the non-linear (heterodyned) components predicted by equation (18) occurring as they do at $\omega_0 + 2\omega_0$, $-\omega_0 + 2\omega_1$, $\omega_1 + 2\omega_0$, etc. The important thing to note here is that the amplitude of these non-linear components is generally very small even though the deviation from the ideal L=N is significant.

5. CONCLUSIONS

It has been demonstrated that for deterministic inputs the LMS adaptive algorithm can be approximated by a linear transfer function.

For the special case of periodic inputs this relationship is exact when the filter length is an integer multiple of the period. For simple inputs the pole-zero pattern can be explicitly obtained and the stability and performance inferred. With more general inputs stability conditions can be determined using the Jury Table or similar results.

Detailed consideration of the sinusoidal example has demonstrated that the approximation is typically very accurate even when the filter length does not satisfy the required conditions.

REFERENCES

[1] Widrow, B., (1970) "Adaptive Filters", in Aspects of Network and System Theory, Kalman, R.E. and De Claris, N. (Eds), New York, Rinehart, Winston.

[2] Glover, J.R., (1977) "Adaptive Noise Cancelling applied to Sinusoidal Interferences", IEEE Transactions on Acoustics, Speech and Signal Processing, v25, pp. 484-491.

[3] Elliott, S.J. and Darlington, P., (1985) "Adaptive Cancellation of Periodic, Synchronously Sampled Interferences, IEEE Transactions on Acoustics, Speech and Signal Processing, pp. 715-717.

[4] Jury, E.I., (1964) "Theory and Application of the z-transform Method", Wiley.

[5] Clarkson, P.M., (1984) "Adaptive Approaches to Signal Enhancement and Deconvolution", Ph. D. Thesis, University of Southampton.

AN ADAPTIVE SINGULAR VALUE DECOMPOSITION ALGORITHM BASED ON GENERALIZED CHEBYSHEV RECURSIONS

B. De Moor* and J. Vandewalle
(ESAT, Electrical Engineering Department, Katholieke Universiteit Leuven, Belgium)

ABSTRACT

In this paper, two algorithms for singular value decomposition are studied in a unifying framework: the power method and the Chebyshev iteration method.

They allow to compute adaptively the partial SVD (dominant, smallest or intermediate triplets) of matrices that are slowly time varying and perturbed by noise. The only routines used are matrix vector multiplication and orthogonalization procedures. Some original convergence expressions are derived and convergence rates of both methods are compared. Some backward error theorems permit the control of the convergence level with respect to the noise level.

1. INTRODUCTION: SCOPE OF THE PAPER

Over the past few years, the singular value decomposition of matrices has become a very reliable and numerically robust tool in the analysis and design of algorithms in numerical linear algebra and is used, as an efficient matrix decomposition method, in the solution of many numerical problems (linear equations, identification and realization, model reduction, algorithm analysis, image and signal processing, ...).

In this paper the power method for the singular value decomposition is studied followed by a modification of this classical technique, which is based upon the properties of Chebyshev polynomials. It will be demonstrated that this algorithm is very well suited for a special class of applications, where other algorithms are likely to perform less efficiently. A geometric and numerical convergence analysis of both power and Chebyshev method will be presented

* Sponsored by the IWONL.

in a unifying approach, since it will be demonstrated that the Chebyshev method is nothing else than the power method, applied to an extended matrix. This leads to completely similar convergence expressions for both methods. Furthermore, some attention will be paid to a backward error analysis based upon the Rayleigh quotient. This will permit the estimation of the necessary convergence level of the approximating vector in the iteration.

Finally, it should be mentioned that the scope of this paper is not to provide a general, complete algorithm for the SVD.

This paper only summarizes some simple but original results and some new viewpoints on rather classical techniques.

Also in [De Moor, 1984] some similar results are obtained, and demonstrated in more detail, but only for the power method.

2. THE SINGULAR VALUE DECOMPOSITION

In this section, the main theoretic and well known results on the singular value decomposition are summarized for notational convenience. Also, some historical background is provided together with an overview of existing algorithms. Finally, we summarize the main specifications for the adaptive SVD strategy to be developed here.

2.1 The Autonne-Eckart-Young theorem (restricted to real matrices)

Every real rectangular mxn matrix A can be decomposed as

$$\underset{mxn}{A} = \underset{mxn}{U} \, . \, \underset{mxn}{\Sigma} \, . \, \underset{mxn}{V^t}$$

where U & V are orthogonal $U^t.U = I_m = U.U^t$ and $V^t.V = I_n = V.V^t$ and Σ is diagonal rectangular: $\Sigma = \mathrm{diag}\,(\sigma_1, \sigma_2, \ldots, \sigma_r, 0, \ldots 0)$ where by convention $\sigma_1 \geq \sigma_2 \geq \ldots \geq \sigma_r > 0$ and r = algebraic rank of A.

The columns $u_i (v_i)$ of U (V) are the left (right) singular vectors and the real positive numbers σ_i are the singular values.

From the theorem it follows immediately that:

1. Every matrix A can be written as a direct sum of r rank one matrices (dyadic decomposition)

$$A_{m\times n} = \sum_{i=1}^{r} u_i . \sigma_i . v_i^t$$

2. The singular values squared are the eigenvalues of $A.A^t$ and $(A^t.A)$ because $A.A^t = U.\Sigma . \Sigma^t . U^t$ and hence $(A.A^t). u_i = \sigma_i^2.u_i$.

These facts will be used in the development of the Chebyshev algorithm. Of course, there are many more extremely interesting properties of the SVD, that can be found in excellent textbooks e.g. [Golub (1984), Hansen (1985)].

2.2 Algorithms for the SVD

Of course, it is straight forward to compute the SVD of A via the eigendecomposition of $A.A^t$ or $A^t.A$. However, the explicit computation of $A.A^t$ & $A^t.A$ introduces numerical errors that halve the numerical precision achievable with this technique, compared to operating with the matrix A itself. Especially, the small singular values are sensitive to the implicit squaring (Klema & Laub, 1978).

One of the most reliable and efficient algorithms for the SVD of an arbitrary matrix is of course that of Golub (1965). It uses orthonormal transformations to bidiagonalize the matrix, whereafter the QR iteration is used to compute the singular values.

However, this algorithm is very costly with respect to memory and computational requirements ($O(n^3)$) and since it operates on the matrix entries itself, it deletes completely a possible structure of the matrix (Hankel, Toeplitz, Sparse...). However, no adaptive version has been developed where a priori information can be used (for instance SVD of a previous time step).

These drawbacks are not shared by the Lanczos algorithm, which was studied extensively in [Paige (1972)] and is nowadays considered as a very promising algorithm for large, structured and preferable sparse eigenvalue problems.

Another class of algorithms is the Jacobi methods, an approach that will probably lead to parallel computer implementation [Hari, 1984]. In this paper we give some promising results for another algorithm that may possibly perform better in a number of applications, that are characterized by the following specifications.

2.3 Specifications

1 Partial SVD

In many applications, only few singular values of the singular spectrum are needed. Let us mention some examples:

-In the solution of overdetermined sets of linear equations, corrupted by white noise, the solution vector is given by the right singular vector, corresponding to the smallest singular value (De Moor, 1985 b). In Pisarenko's line spectrum estimation method, the minimal eigenvector of the symmetric covariance matrix is needed (Furhmann, 1984).

-In linear system theory, a minimal realization is obtained via the SVD of a block Hankel matrix. Here only some few dominant triplets are needed, (Kung, Zeiger Mc Ewen) similarly in image compression applications and algebraic image restoration techniques (Shlien, 1982).

-In order to extract a fetal ECG, noisy measurements are projected on line onto a subspace spanned by an intermediate singular vector of the measurement matrix, that can be shown to correspond to the fetal heart, where the dominant singular values correspond to the mother heart (Vanderschoot, 1983). Other applications where intermediate eigen-and singularvalues are needed are mentioned in Jensen (1972). In all those examples, only a partial SVD is needed.

2 Large and/or structured matrices

There are a lot of applications where the matrix dimensions are large (say a few hundred) and the matrix itself can be highly structured ((block)Hankel, - Toeplitz, ...). To avoid excessive memory and computation requirements, one should not operate on the matrix entries itself, as the Golub algorithm does, hence deleting the possible structure.

For those cases, it is much more convenient to access the data indirectly, via a matrix vector multiplication subroutine, which does not alter the matrix entries.

This matrix-vector access concept will also be exploited in the following specification.

3 Adaptive computation of SVD

Two types of time varying sequences of matrices are possible:

-Adding a new column or row and omitting one (for instance, when new data become available and old data are to be neglected).

-Another possibility is time variance element-wise when all elements change slowly in each timestep.

For those cases the SVD's on two subsequent timesteps are related and hence, the SVD on timestep T could be used as initial guess for the SVD on T + 1 in an iterative algorithm.

The algorithm that will be described here is perfectly well suited for this, since it uses a matrix vector subroutine and since it is iterative. Golub's algorithm makes no use of an a priori available first initial guess.

4 Convergence level control

Measured data are always corrupted by noise. Also the matrix singular value decomposition is perturbed by the noise although the singular values are themselves perfectly well conditioned (Golub, 1983). But, since the data are noisy, a full 7 or 14 (or more) digit accuracy is of no use and in most cases only a limited number of accurate digits will suffice, especially when only few singular vectors are to be computed. In this paper, some original results are obtained that permit one to control the level of convergence. Together with the backward error analysis, this is a powerful tool in the analysis of the SVD of noisy matrices.

3. THE CLASSICAL POWER METHOD

Because the Chebyshev algorithm is related to the power method (in Wilkinson it is considered as an acceleration of the power method), we first present briefly some results on the power method for the SVD. More details can be found in [De Moor, 1985 a].

3.1 Algorithm

Suppose already p-1 singular triplets have been computed : (u_i, σ_i, v_i), i = 1, (p-1) let e_p^k, s_p^k, f_p^k be the k-th iteration approximation for u_p, σ_p, v_p; let e_p^0 be an initial guess for u_p, then step k in the iteration for the triplet is:

step k

1. Orthonormalize e_p^{k-1} with respect to $(u_1 \dots u_{p-1})$ so that

$$||e_p^{k-1}|| = 1 \text{ (Deflation step)}$$

2. Compute

$$f_p^k = A^t.e_p^k$$

$$s_p^k = ||A^t.e_p^k||$$

$$f_p^k = f_p^k/s_p^k$$

3. Test for convergence $||f_p^k - f_p^{k-1}|| < TOL$

   If so, convergence in triplet p

   If not, go to 4

4. Compute $$e_p^k = A.f_p^k \,/\, ||A.f_p^k||$$

5. Set k = k+1, go to 1.

The properties of this simple iteration scheme and possible refinements are extensively studied in De Moor (1983). (Omission of normalization, orthogonalization, acceleration algorithm, optimal initial guesses, deflation techniques, ...).

Before proving the convergence of the iteration scheme, we first need the following

Definition: The internal coordinates of an iteration vector e_p^k (f_p^k) are the coordinates of e_p^k (f_p^k) in the left (right) singular base, given by $U^t.e_p^k$ $(V^t.f_p^k)$.

3.2 Convergence

3.2.1 Infinite precision arithmetic

Theorem: If e_p^O is not orthogonal to u_p then

$$\lim_{k\to\infty} e_p^k = u_p, \quad \lim_{k\to\infty} f_p^k = v_p, \quad \lim_{k\to\infty} s_p^k = \sigma_p.$$

Proof: Suppose in the algorithm, each orthogonalization and normalization step is omitted, then

$$e_p^k = \underbrace{A.A^t \;..........\; A.A^t}_{k \text{ times } A.A^t}.e_p^O$$

and $A.A^t = U.(\Sigma.\Sigma^t)U^t$ where $U^t.U = I$

so that $(U^t e_p^k) = (\Sigma.\Sigma^t)^k . (U^t.e_p^O)$

Here, $U^t e_p^k$ and $U^t.e_p^O$ are the internal coordinates of e_p^k and e_p^O.

Since e_p^O is orthonormalized to $(u_1, \ldots u_{p-1})$, the internal coordinates have the form

$$U^t.e_p^O = (O, \ldots, O, g_p, g_{p+1}, \ldots g_n)^t$$

$$U^t.e_p^k = (O, \ldots, O, \left(\sigma_p\right)^{2k} g_p, \left(\sigma_{p+1}\right)^{2k} g_{p+1} \ldots)^t$$

and after normalizing and taking the limit and with the generic assumption that $\sigma_p > \sigma_{p+1} > \ldots > \sigma_m$ we have for the i-th component

$$\lim_{k\to\infty} \frac{\sigma_i^{2k} g_i}{\sqrt{\sigma_p^{4k} g_p^2 + \sigma_{p+1}^{4k} g_{p+1}^2 + \ldots}} = \delta_{ip}$$

where δ_{ip} is the Kronecker delta.

Hence
$$\lim_{k\to\infty} \frac{U^t.e_p^k}{||U^t e_p^k||} = (O, \ldots O, \underset{\underset{p}{\uparrow}}{1}, O \ldots O)^t$$

but these are nothing else than the internal coordinates of u_p.

The same technique can be used in proving the convergence of f_p^k to v_p and s_p^k to σ_p.

3.2.1 *Finite precision arithmetic*

Finite precision arithmetic is always characterized by the so called machine precision $\varepsilon_m = \sup \{\varepsilon | fl(1 + \varepsilon) = 1\}$ where $fl(.)$ denotes the floating point operator (Wilkinson 1965). The effect of finite machine precision on the addition of two real numbers a,b can be seen from $(a+b) = a(1 + \frac{b}{a})$. If $\frac{b}{a} < \varepsilon_m$, then $fl(1 + \frac{b}{a})=1$ and $fl(a+b) = a$. Hence, if two numbers a and b differ more than $(-\log_{10} \varepsilon_m)$ decimal digits, addition has no effect.

This finite precision causes the internal coordinates of e_p^O to be of the form $fl(U^t.e_p^k) = (\varepsilon_1, \varepsilon_2, \ldots, \varepsilon_{p-1}, g_p, g_{p+1}, \ldots g_m)$ where the ε_i are of the order of magnitude of ε_m.

Now, if we suppose that the orthogonalization keeps the first p-1 internal coordinates small ($\sim\varepsilon_m$), then it is not difficult to see that numerically, geometric convergence occurs when the p-th internal coordinate dominates all others by more than $[-\log \varepsilon_m]$ decimal digits. Since for a "good" initial guess for the p-th left singular vector u_p, the internal coordinates satisfy $g_p > g_{p+1} > g_i$, $i = p+2,m$ convergence occurs if

$$\log_{10} \frac{|\sigma_p^{2k} g_p|}{|\sigma_{p+1}^{2k} g_{p+1}|} > -\log_{10}\varepsilon_m$$

or after

$$k = \frac{-\log_{10} \varepsilon_m - \log_{10} \left[\frac{g_p}{|g_{p+1}|}\right]}{2 \log_{10} \left[\frac{\sigma_p}{\sigma_{p+1}}\right]}$$

matrix vector multiplications with $A.A^t$. under the assumption that the first (p-1) are kept to machine precision level with orthogonalization. The derived formula is of course very interesting since it shows clearly the three different effects that influence the number of iterations.

1. $\log_{10} \frac{\sigma_p}{\sigma_{p+1}}$: the larger the gap between the computed value σ_p and the subsequent one σ_{p+1}, the faster the convergence. For clustered singular values, $\sigma_p/\sigma_{p+1} \approx 1$, convergence will be very slow. For those cases of slow convergence, acceleration algorithms have been developed and described in (De Moor, 1983).

2. $\log_{10}(g_p/g_{p+1})$: the better the quality of the initial guess, the smaller the number of iterations. For a very good initial vector, the internal coordinates of e_p^0 satisfy $g_p >> g_{p+1}$.

3. $(-\log_{10} \varepsilon_m)$: This quantity measures the number of desired correct digits in the iteration vector. Hence if for instance only t correct digits are necessary, one puts $-\log_{10} 10^{-t} = t$ instead of $(-\log \varepsilon_m)$, which is equivalent to a full machine precision result.

   Hence, the numerator [-log t -log (g_p/g_{p+1})] measures the number of digits that have to be modified in order to obtain the singular vector up to the desired precision $t = -\log_{10} 10^{-t}$ decimal digits.

   It is important to mention that the factors $\log \frac{g_p}{g_{p+1}}$ and $\log \frac{\sigma_p}{\sigma_{p+1}}$ can be estimated dynamically from the convergence pattern (De Moor, 1983).

Using this numerical analysis in internal coordinates, not only the number of iterations for t digit precision can be estimated during the iteration. In [De Moor, 1984], one can find results on the determination of the optimal moment for fixing the initial guess for the next triplet and on the estimation of the critical deflation-orthogonalisation frequency.

4. THE CHEBYSHEV ALGORITHM

4.1 The Chebyshev polynomials

Orthogonal polynomials in general, and Chebyshev polynomials in particular, play an important role in numerical linear algebra (Stiefel (1955), Fox (1968), Rivlin (1976), ...).

The application of Chebyshev polynomials in the eigenvalue problem is studied for the symmetric case in Parlett (1980), Stewart (1969) and for the non symmetric case in Wilkinson (1965), Saad (1982, 1984) and Manteuffel (1977). These references show that the idea of using Chebyshev polynomials in order to accelerate iterative eigenvalue algorithms is certainly not new.

Chebyshev polynomials belong to the class of polynomials that are orthogonal with respect to a given density function (Stiefel, 1955). They can be defined in several ways (Rivlin, 1976) but we only need here some important, well known properties.

4.1.1 Recurrence relation

All orthogonal polynomials satisfy a three term recurrence relation. For the Chebyshev polynomials, this becomes

$$T_0(x) = 1$$

$$T_1(x) = x$$

$$T_{n+1}(x) = 2xT_n(x) - T_{n-1}(x)$$

4.1.2 Behaviour of the Chebyshev polynomials

An important property is $\forall\, n,\ x \in [-1,1] : |T_n(x)| \leqslant 1$

$$x \notin [-1,1] : |T_n(x)| \geqslant 1$$

This follows of course from the definition

$$T_n(x) = \cos(n\ \mathrm{Bgcos}\ x),\quad |x| \leqslant 1$$

$$T_n(x) = \cosh(n\ \mathrm{Bgcosh}\ x),\quad |x| \geqslant 1$$

This means that in $[-1,1]$, the Chebyshev polynomials show an equiripple oscillation with turning points and zeros symmetrically disposed about $x=0$.

For $x \notin [-1,1]$, the behaviour is characterized by

$$|T_{n+1}(x)| > |T_n(x)|$$

$$|T_n(x+\Delta x)| > |T_n(x)|, \quad (\Delta x>0)$$

and by the following original theorem

Theorem

For $d_1 > d_0 > 1$ the Chebyshev polynomials satisfy

$$\frac{T_{n+1}(d_1) - T_{n+1}(d_0)}{T_n(d_1) - T_n(d_0)} > 1$$

Proof

Can be easily proved by induction.

The following theorem is extremely important for the SVD algorithm that will be developed.
The proof can be found in Rivlin (1976).

Theorem

The Chebyshev polynomial $2^{1-n}.T_n(s)$ is always the largest possible polynomial of degree n with leading coefficient 1, outside $[-1,1]$. For every $p_n(x)$, polynomial of degree n with leading coeff. 1 and if $M = \max |p_n(x)|$, $x \in [-1,1]$, then for $|y| > 1$

$$|p_n(y)| \leq M.2^{1-n}|T_n(y)|$$

It will turn out further on that this theorem shows that the Chebyshev algorithm is the best possible strategy.

4.1.3 Asymptotic behaviour for large n

For large n, the following expressions hold (Parlett, 1980)

$$T_n(1+2\varepsilon) \sim \frac{1}{2}(1+2\sqrt{\varepsilon}+2\varepsilon)^n \text{ for } 0 \leq \varepsilon \leq 0.1$$

$$\sim \frac{1}{2}e^{(2n\sqrt{\varepsilon})} \text{ for } n\sqrt{\varepsilon} > 1$$

It is this rapid growth of $T_n(1+2.\varepsilon)$ for small ε that makes the Chebyshev polynomials attractive for computation of the SVD.

4.2 *The Chebyshev algorithm*

4.2.1 *Computation of the dominant triplet*

Suppose one wants to compute the dominant triplet (u_1, σ_1, v_1) of A (m<n) with $\sigma_1(A) > 1$ and $0 < \sigma_i(A) < 1$ i=2, m and the SVD of $A = U.\Sigma.V^t$.

Let e_1^0 be an initial guess for u_1 and define $e_1^k = T_k(A.A^t).e_1^0$. Since polynomials are analytic functions one has $T_k(A.A^t) = U.T_k(\Sigma.\Sigma^t).U^t$ and $U^t.e_1^k = T_k(\Sigma.\Sigma^t).(U^t.e_1^0)$

$(U^t e_1^k)$ and $(U^t e_1^0)$ are the internal coordinates of the k-th iteration vector e_1^k and the initial guess e_1^0.

Let $U^t e_1^0 = (g_1, g_2, \ldots, g_m)$ then

$$U^t e_1^k = (T_k(\sigma_1^2)g_1, T_k(\sigma_2^2)g_2, \ldots, T_k(\sigma_m^2)g_m)$$

Because of the conditions on the singular values of A and the properties of the Chebyshev polynomials, one has

$$|T_{k+1}(\sigma_1^2)| > |T_k(\sigma_1^2)| > 1$$

$$|T_k(\sigma_i^2)| \leqslant 1 \qquad i=2,m$$

Hence, the first internal coordinate grows for increasing k while all other coordinates are bounded by 1.

In the next section the convergence will be proved analytically but first, the vector iteration scheme is deduced.

By the recurrence relation of the Chebyshev polynomials

$$T_k(AA^t) = 2(A.A^t)T_{k-1}(AA^t) - T_{k-2}(AA^t)$$

so that

$$\boxed{\begin{array}{l} e_1^k = 2\,A.(A^t e_1^{k-1}) - e_1^{k-2} \\ \text{with } e_1^1 = A.(A^t.e_1^0) \end{array}}$$

This is the fundamental iteration scheme of the SVD Chebyshev iteration method.

4.2.2 Convergence

Under the assumption $\sigma_1(A) > 1$ and $0 < \sigma_i(A) < 1$, i=2,m numerical convergence of $t=(-\log \varepsilon)$ digits will be reached if the first internal coordinate dominates the other by t decimal digits.
Hence, numerical convergence occurs if

$$\log \frac{|T_k(\sigma_1^2)\ g_1|}{\max\limits_{i=2,m} |T_k(\sigma_i^2) g_i|} = -\log \varepsilon$$

Since $|T_k(\sigma_i^2) g_i| < 1$, i=2,m this becomes for the worst case

$$\log |T_k(\sigma_1^2)|$$

This equation determines the number of iterations k (multiplications with $A.A^t$) necessary for numerical convergence of $t=-\log \varepsilon$(decimal digits). Of course, σ_1 is unknown, hence $T_k\ (\sigma_1^2)$, but the number of multiplications can be estimated by considering the asymptotic convergence behaviour as follows:

In iteration step k, the number of correct decimal digits is $\log |T_k(\sigma_1^2).g_1/\max\limits_{i=2,m} g_i|$.

In iteration step k+1, this becomes $\log |T_{k+1}(\sigma_1^2)|.g_1/\max\limits_{i=2,m} g_i|$

Hence, from k to k+1 the gain in number of correct digits is

$$\log |T_{k+1}(\sigma_1^2)| - \log |T_k(\sigma_1^2)| = \log \frac{|T_{k+1}(\sigma_1^2)|}{|T_k(\sigma_1^2)|}$$

For "large" k, we now prove the following

<u>Theorem</u>

If $d > 1$, then $\lim\limits_{k\to\infty} \frac{T_{k+1}(d)}{T_k(d)} = d + \sqrt{d^2 - 1} = L$

Proof

From $T_{k+1}(d) = 2dT_k(d) - T_{k-1}(d)$ it follows that

$$\frac{T_{k+1}(d)}{T_k(d)} = 2d - \frac{T_{k-1}(d)}{T_k(d)}$$

But under very general conditions

$$L = \lim_{k\to\infty} \frac{T_{k+1}(d)}{T_k(d)} = \left[\lim_{k\to\infty} \frac{T_k(d)}{T_{k-1}(d)}\right]^{-1}$$

and hence $L = 2d - \frac{1}{L}$ or $L^2 - 2\,dL + 1 = 0$ with solutions $L = d \pm \sqrt{d^2 - 1}$

But since for $d > 1$, $|T_{k+1}(d)| > |T_k(d)|$ also $L > 1$ and hence $L = d + \sqrt{d^2 - 1}$ is the only acceptable solution.

From this theorem, it follows that for each iteration with $A.A^t$, $\log(\sigma_1^2 + \sqrt{\sigma_1^4 - 1})$ decimal digits converge.

From section 3.2, it then follows that the number of matrix multiplications with AA^t is approximated by

$$k = \frac{-\log\varepsilon - \log(g_1/\max_{i=2,m} g_i)}{\log(\sigma_1^2 + \sqrt{\sigma_1^4 - 1})}$$

The expression in the numerator is asymptotically correct. One can show numerically that it is an overestimation of the true value

$$\left[\frac{T_{k+1}(\sigma_1^2)}{T_k(\sigma_1^2)}\right]$$

which however, converges rather fast to the correct limit.

The interpretation of the convergence expression is quite similar to the case of the power method. Only the numerator is changed into the expression $(\log(\sigma_1^2 + \sqrt{\sigma_1^4 - 1})$, which is the number of left multiplications with $A.A^t$ (Chebyshev iteration) per decimal digit of convergence. As we shall derive further on, this factor can be estimated from the convergence pattern. To conclude this section, let us mention that some similar expression of the form $d+\sqrt{d^2-1}$ in the convergence behaviour can be found in a very interesting paper on Chebyshev iteration in Saad (1984).

4.2.3 Spectrum transformation

Till now, we only considered the case where $\sigma_1(A) > 1$, $0 < \sigma_i(A) < 1$, $i = 2,m$ and proved that in this case, the algorithm converges to the dominant triplet. In order to compute the dominant, smallest or some intermediate triplets, of a general matrix, some more information is needed in the form of upper- and lower bounds on the desired singular value(s).

4.2.3.1 Dominant triplet.

Suppose one knows real numbers a and b such that $\sigma_2^2 \leq a \leq \sigma_1^2$ and $0 \leq b \leq \sigma_m^2$.

Then the linear function

$$f(x) = \frac{2x - (a+b)}{a-b}$$

satisfies $f(a) = 1,\ f(b) = -1$

$$f(\sigma_1^2) > 1, \quad |f(\sigma_i^2)| \leq 1 \quad i = 2,m$$

such that

$$f(A.A^t) = \frac{2}{a-b} A.A^t - \frac{a+b}{a-b} I$$

$$= V \left[\frac{2}{a-b} \Sigma . \Sigma^t - \frac{a+b}{a-b} I\right] . V^t$$

$$= U.D.V^t$$

is a matrix satisfying the conditions of §4.2.1, with eigenvalues

$$|d_1| = \left| \frac{2\sigma_1^2 - (a+b)}{a-b} \right| > 1 \text{ and } |d_i| = \left| \frac{2\sigma_i^2 - (a+b)}{a-b} \right| \leq 1$$

i=2,m.

Only the singular values (eigenvalues of $A.A^t$) are changed but the singular vectors are conserved.

This means that the following Chebyshev matrix vector iteration scheme will converge to the dominant singular triplet.

Iteration scheme

1. Determine a,b
2. Choose e_1^0
   compute $e_1^1 = A.(A.^t.e_1^0)$

Step k

3. $e_1^{k+1} = \frac{4}{a-b} A.(A^t e_1^k) - 2\,\frac{a+b}{a-b}\, e_1^k - e_1^{k-1}$

4. Test for convergence

$$\left\| \frac{e_1^{k+1}}{\|e_1^{k+1}\|} - \frac{e_1^k}{\|e_1^k\|} \right\| < TOL$$

If so, stop.
If not, go back to 3

or alternatively

4' Estimation of quantities in convergence expression.

If conv., stop
If not go to 3

Of course, the convergence rate is now proportional to

$$1/\log\,(d_1 + \sqrt{d_1^2 - 1}\,)$$

where $d_1 = \dfrac{2\sigma_1^2 - (a+b)}{a-b} > 1$

Now, from the properties of the Chebyshev polynomials, converge is optimal if $|d_1-1|$ is as large as possible. It is not difficult to prove that this corresponds to optimal choices of a and b as $a = \sigma_2^2$ and $b = \sigma_m^2$.

Hence, the best possible convergence is determined by the ratio $K_{12} = \sigma_2^2 / \sigma_1^2$ and $K_{2m} = \sigma_m^2 / \sigma_2^2$.

The smaller K_{12} and the larger K_{2m}, the better the convergence rate. From the theorem of optimality of Chebyshev polynomials, it follows that this strategy is the best possible using only matrix polynomials.

4.3.2 Smallest triplet

The necessary bounds a and b must now satisfy : $a > \sigma_1^2$

$$\sigma_m^2 < b < \sigma_{m-1}^2$$

and the spectrum transformation function is here

$$f(A.A^t) = \frac{2.AA^t}{b-a} - \frac{a+b}{b-a} I$$

since

$$f(a) = -1 \qquad f(\sigma_m^2) > 1$$

$$f(b) = 1 \qquad |f(\sigma_i^2)| < 1 \qquad i = 1, (m-1)$$

The Chebyshev iteration now becomes

$$e_m^{k+1} = \frac{4}{b-a} A.(A^t e_m^k) - 2\frac{a+b}{b-a} e_m^k - e_m^{k-1}$$

and the convergence rate is determined by

$1/ \log (d_m + \sqrt{d_m^2 - 1})$ iterations per decimal digit with

$d_m = [\ \sigma_m^2 - (a+b)\ ] / (b-a)$

Optimal convergence condition occurs for $a = \sigma_1^2$, and $b = \sigma_{m-1}^2$.

4.2.3.3 Intermediate triplet (u_i, σ_i, v_i).

Now, 4 bounds are needed for triplet i

$A > \sigma_1^2$ (with $1 < i < m$)

$B < \sigma_m^2$

$\sigma_{i-1}^2 < b < \sigma_i^2 < a < \sigma_{i+1}^2$

and define $M_a = A-a$, $M_b = b-B$, $D = a-b$, $M = \max(M_a, M_b)$

The following strategy is <u>optimal</u> for the symmetric case where $M_a = M_b = M$.

Instead of a linear spectrum transformation spectrum, now a parabola is used

$$f(x) = px^2 + qx + t \text{ with } p = \frac{-2}{M(M+d)},\ q = -(a+b)\,p$$

$$t = 1 + ab.p$$

For more details about the deviation, omitted here for brevity, we refer to De Moor (1985a).

The Chebyshev iteration now becomes

$$e_i^{k+1} = 2\,p\,A\,[\,A^t\,(A(A^t e_i^k) + \frac{q}{p}\,e_i^k)\,] + 2te_i^k - e_i^{k-1}$$

Each iteration now demands 4 matrix vector multiplications.

Optimal convergence occurs when $\sigma_i^2 = \frac{a+b}{2}$ and then

$$d_i = f(\sigma_i^2) = 1 + \frac{1}{2(\frac{M}{D})\ (\frac{M}{D} + 1)}$$

while the optimal convergence factor is then: $1/\log\,(d_i + \sqrt{d_i^2 - 1})$ iterations per decimal digit.

From this it follows that, the smaller M/D, the faster will be the convergence. Unfortunately the M/D ratio is usually large in practical situations. A possible strategy then would consist in allowing more iteration vectors and hence perform a kind of subspace iteration (De Moor 1985a).

In the case where the interval D is located asymmetrically ($M_a \ll M_b$ or vice versa), the ratio M/D can be "preprocessed" in order to obtain attractive convergences rates (De Moor 1985a).

To conclude this section, let us illustrate the computation of an intermediate triplet by a simple example.

Let $A.A^t$ = diag (10;9;6;4;1) and given bounds B=0, A=11, a=8, b=5 then p=-0.05; q=0.65- t=-1.

Then $f(AA^t)$ = diag (0.5; 0.8; 1.1; 0.8; -0.4) and hence there are needed

$$\frac{1}{\log\left(d_1 + \sqrt{d_1^2 - 1}\right)} = \frac{1}{\log\left(1.1 + \sqrt{(1.1)^2 - 1}\right)} = 5.2$$

iterations per desired digit of precision. This is equivalent with 5.2 x 4 = 21 matrix vector multiplications per digit.

4.3 Relation Chebyshev-power iteration

4.3.1 Number of iterations

For identical initial guess vector and machine precision, it follows from the convergence expression that the ratio of matrix-multiplications of the dominant triplet for the Chebyshev method compared to the power method, is given by

$$S = \frac{n_{(\text{Chebyshev})}}{n_{(\text{power method})}} = \frac{2\log(\sigma_1/\sigma_2)}{\log\left(d_1 + \sqrt{d_1^2 - 1}\right)}$$

where $d_1 = \dfrac{2\sigma_1^2 - (a+b)}{-b + a}$ with $b<\sigma_m^2$ and $\sigma_2^2 < a < \sigma_1^2$.

When $S < 1$, then the Chebyshev method is faster than the power method.

Now take b=0, $\alpha = a/\sigma_2^2$ and $r = \sigma_1/\sigma_2$

$$\text{then } S = \frac{\log r^2}{\log\left(2(r^2/\alpha) - 1 + \sqrt{(2r^2/\alpha - 1)^2 - 1}\right)}$$

This convergence speed ratio is plotted in Fig. 1 for different values of α and r.

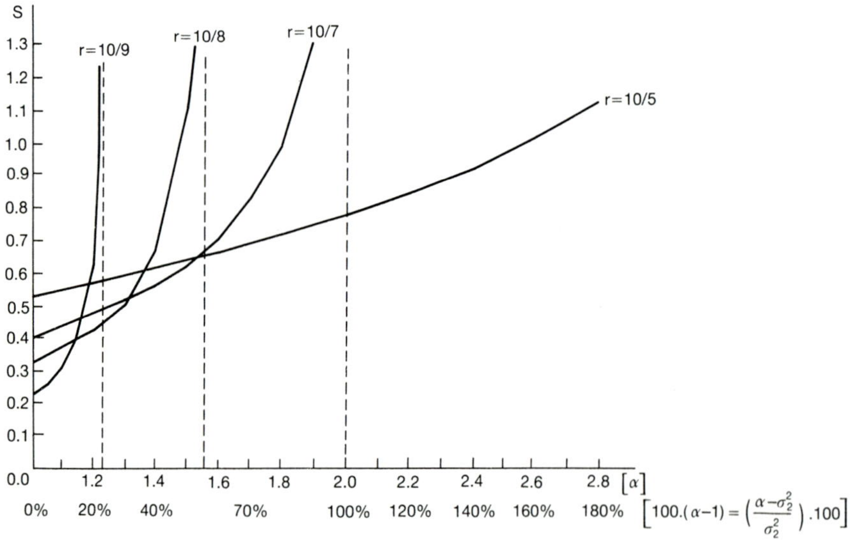

Fig. 1 r is the singular value ratio. For values of S smaller than 1, the Chebyshev method is faster than the power method. For small values of r, it is necessary to estimate the lower bound a of σ_1 accurately (optimally $a=\sigma_2$).

From this it immediately follows that the closer the singular values are ($r \longrightarrow 1$) the larger the gain in speed of the Chebyshev iteration in comparison with the power method but of course, the more difficult it will be to estimate a satisfying $\sigma_2^2 < a < \sigma_1^2$. For correct and optimal estimation of a ($= \sigma_2^2$), S, as a function of r is given in Fig. 2.

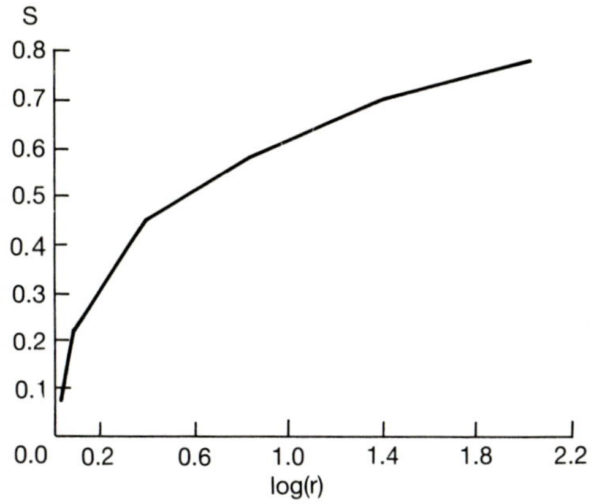

Fig. 2 If the bound a is estimated accurately, the Chebyshev method is almost always faster than the power method. The more the singular values are clustered, the larger the difference in iterations between Chebyshev and power method, but the more difficult it is to estimate a.

From this it is clear that, the more clustered σ_1 and σ_2 are, ($\log r \longrightarrow 0$), the better is the convergence rate of the Chebyshev iteration compared to the power method.

In some forthcoming work, the application of acceleration algorithms will be investigated for the case of clustered singular values (see De Moor 1985 a).

4.3.2 Relation with power method

The following statement considerably simplifies the study of the sometimes rather complicated looking Chebyshev schemes.

Theorem

The Chebyshev iteration scheme

$$e_1^{\,1} = f(A.A^t)e_1^{\,0}$$

$$e_1^{\,n+1} = 2\,f(AA^t)e_1^{\,n} - e_1^{\,n-1}$$

is completely equivalent with the eigenvalue power method applied to the matrix

$$B = \begin{bmatrix} -I & 2\,f(AA^t) \\ -2f(AA^t) & 4\,[\,f(AA^t)\,]^2 - I \end{bmatrix}$$

with initial vector $\begin{bmatrix} e_1^{\,0} \\ e_1^{\,1} \end{bmatrix}$

<u>Proof</u>

From $e_1^{\,n+1} = 2\,f(AA^t)e_1^{\,n} - e_1^{\,n-1}$

it follows that

$$e_1^{\,n+2} = 2\,f(AA^t)e_1^{\,n+1} - e_1^{\,n}$$

$$= [\,4\,[\,f(AA^t)\,]^2 - I\,]e_1^{\,n} - 2\,f(AA^t)\,e_1^{\,n-1}.$$

which can be written as

$$\begin{bmatrix} e_1^{\,n+1} \\ e_1^{\,n+2} \end{bmatrix} = \begin{bmatrix} -I & 2f(AA^t) \\ -2f(AA^t) & 4f(AA^t)^2 - I \end{bmatrix} \begin{bmatrix} e_1^{\,n-1} \\ e_1^{\,n} \end{bmatrix}$$

This is nothing else than the eigenvalue power iteration method. Moreover, the eigenvalue spectrum of this matrix is easily obtained as follows:

Since $f(AA^t) = Vf(\Sigma\,\Sigma^t)V^t$ the eigenvalues of B follow from

$$B = \begin{bmatrix} V & O \\ O & V \end{bmatrix} \cdot \begin{bmatrix} -I & 2f(\Sigma\,\Sigma^t) \\ -2f(\Sigma\,\Sigma^t) & 4[\,f(\Sigma\,\Sigma^t)\,]^2 - I \end{bmatrix} \begin{bmatrix} V^t & O \\ O & V^t \end{bmatrix}$$

$$= U.D.U^t$$

and by putting det $(D-\Lambda) = 0$ it is easily computed that

$$\Lambda = 2[f(\Sigma\Sigma^t)]^2 - I \pm 2f(\Sigma\Sigma^t)\sqrt{f(\Sigma\Sigma^t)^2 - I}$$

Now put $f(\sigma_1^2) = d_i$ then the eigenvalues are real for $d_i>1$ and imaginary for $d_i<1$.

Suppose for simplicity of explanation that only $d_1>1$ and all other $d_i<1$.

Then

- for $d_1 > 1$ $\quad \lambda_1 = 2d_1^2 - 1 + 2d_1\sqrt{d_1^2-1}$

$$\lambda_{-1} = 2d_1^2 - 1 - 2d_1\sqrt{d_1^2-1}$$

and one can prove that

$$\lambda_1 > 1 \text{ and } 0 < |\lambda_{-1}| < 1$$

and that $\lambda_{-1} = \dfrac{1}{\lambda_1}$

- for $d_i < 1$

$$\lambda_i = 2d_i^2 - 1 + 2d_i j\sqrt{1-d_i^2} \qquad \text{where } j^2 = -1$$

$$\lambda_{-1} = 2d_i^2 - 1 - 2d_i j\sqrt{1-d_i^2}$$

and $|\lambda_i| = |\lambda_{-i}| = 1$

The desired part of the spectrum ($d_i>1$) is mapped on the real axis and split into two eigenvalues $\lambda_i > 1$ and $\lambda_{-i} = \frac{1}{\lambda_i}$ for each singular value and the rest of the spectrum is mapped on the unit circle !!

The convergence pattern can now be predicted with the techniques of §3 for the power method.

The internal coordinate corresponding to the dominant eigenvalue λ_i will grow every iteration by a factor (worst case)

$$\log \frac{\lambda_1}{\max_{i>2} |\lambda_{\underline{+}i}|} = \log [\ 2d_1^{\ 2} -1+2d_1\sqrt{d_1^{\ 2} - 1} \ ]$$

But one step of the eigenvalue power method corresponds to 2 Chebyshev recursions. Hence one Chebyshev recursion step is equivalent to a growth determined by

$$[\ 2d_1^{\ 2}-1 + 2d_1\sqrt{d_1^{\ 2} - 1}) ^{\frac{1}{2}} = ((d_1 + \sqrt{d_1^{\ 2} - 1})^2)^{\frac{1}{2}}$$

$$= d_1 + \sqrt{d_1^{\ 2} - 1}$$

which is precisely the quantity computed asymptotically in §4.4.

5. BACKWARD ERROR MODIFICATION

In many practical cases, the matrix A for which the SVD is desired, is perturbed by noise and uncertainties in the data. It is a well known result that the singular values are very well conditioned [Golub, 1983] and that the condition of the singular vectors depends on the relative spread of the singular values.

The methods we discuss here, apply at every moment in the iteration, to an approximation of both left and right singular vectors.

Since obviously it is useless to require a higher accuracy in the results than can be guaranteed by the data, some tools must be constructed in order to check the relative accuracy to the data accuracy level. The theorems developed further on are straightforward generalizations of those in [Parlett 1980].

Let A be mxn, x an m-vector approximation to the left singular vector u_k, and y an n-vector approximation to the right singular vector v_k with $|| \ x \ || = || \ y \ || = 1$.

Define the generalized Rayleigh quotient as $\rho = x^t.A.y = y^t.A^t.x$. It can be proved that ρ is the best possible approximation for σ_k, using x and y only.

Define the residues $\underset{nx1}{r_x} = A^t.x - \rho.y$ and $\underset{mx1}{r_y} = A.y - \rho.x$

then the following theorems have very practical implications.

Theorem

The residues r_x and r_y are orthogonal to the approximations x and y.

$$r_x^t.y = 0 \quad r_y^t.x = 0$$

Proof : Trivial

Theorem Backward error modification.

The vectors x and y are left and right singular vectors of a modified matrix A - M with corresponding singular value s where

$$M = x.r_x^t + r_y.y^t$$

Proof it is easy to show that

$$(A - M)y = s .x$$
$$(A^t - M^t).x = s .y$$

Theorem The singular value decomposition of M is given as

$$M = \begin{bmatrix} x & \frac{r_y}{||r_y||} \end{bmatrix} \begin{bmatrix} ||r_x|| & 0 \\ 0 & ||r_y|| \end{bmatrix} \begin{bmatrix} r_x^t/||r_x|| \\ y^t \end{bmatrix}$$

Proof Trivial

From this, the norms of M are given as $|| M || = \max[||r_x||, ||r_y||]$ and $||M||_F^2 = ||r_x||^2 + ||r_y||^2$.

Now, one can prove that for two matrices A,B with singular values a_i and b_i, $|a_i - b_i| < ||A - B||$. Then put B=A-M, which has a singular value s, $|\sigma_k - s| < ||A - (A-M)|| = ||M|| = \max(||r_x||, |r_y||)$. This means that the error in the approximation of a singular value is bounded by the norm of the backward error matrix M.

If this norm $||M||$ becomes smaller than the estimated uncertainty level in one singular value, further iteration is useless.

As an example, consider the uncertainty level in a matrix A which is perturbed by white noise with standard deviation σ. Then the uncertainty level on one singular value is proportional to $\max(m,n).\sigma^2$ (De Moor 1985b). Hence iteration can stop when $||M|| < \max(m,n)\sigma^2$. However, this backward error modification expression has to be linked with the convergence expressions derived earlier.

6. AN ADAPTIVE STRATEGY AND PERSPECTIVES

In this section, we briefly summarize some ideas and perspectives to implement the Chebyshev iteration scheme in an adaptive environment. The data are not modified during computation and may be accessible for instance only in secondary memory.

The data themselves are assumed to be slowly time varying either in a way that at each time step a new line of information is added or such that minor changes occur elementwise. Moreover, some noise level characteristics are assumed to be known. The initial guess for each triplet can be obtained from the SVD of a previous time step or from a previous obtained triplet. The number of iterations for convergence is determined with respect to the desired accuracy level, that in its turn is determined by the noise level.

In the near future, the adaptive estimation of optimal singular value bounds is to be studied, for which interesting ideas can already be found in the work of Saad (1984) and Manteuffel (1977). The operation count is to be optimized by acceleration algorithms and optimal orthogonalization (deflation) strategies (De Moor 1985a).

A main problem of course is the resolution of closely spaced singular values. But also other methods such as the Lanczos method, where the relative spread of singular values equally well determines the convergence rate (Parlett, 1980), suffer from this problem.

In the case of the Lanczos algorithm some block versions have been developed for this purpose (Golub, 1981), and extensions of our Chebyshev method to the subspace iteration case are straight-forward (Parlett, 1980). Moreover, there is a large parallelism between this Chebyshev iteration and the Lanczos method. Both use only matrix vectors subroutines, need initial guesses to start up the iteration, need some deflation techniques such as selective orthogonalization, and convergence is governed by the condition and spread of the singular spectrum.

However, a direct comparison is not straight-forward and it would be very interesting to set up one.

7. CONCLUSIONS

In this paper two SVD algorithms, the power method and the Chebyshev iteration method, were studied in a unifying framework. Convergence expressions were obtained and it was proven that the Chebyshev method is also a power method. Comparisons in convergence speed were made. It was demonstrated that the typical applications for these algorithms are those where the SVD of a slowly time varying sequence of matrices is to be computed. The matrices are large and/or structured, and accessible only via matrix vector products. Knowledge of the characteristics of the noise on the data can be used to estimate the number of desired digits in the results or to follow the convergence via the backward error modification technique.

REFERENCES

1. De Moor B., Vandewalle, J., Staar, J. and Lauwers, J., (1984), An adaptive singular value decomposition algorithm and its applications to adaptive realization. Proc. of the 6th Intern. Conference on Analysis and Optimization of systems, Nice.

2. De Moor B., (1985a) Some new results for adaptive singular value decomposition algorithms. Int. report, ESAT, K.U. Leuven.

3. De Moor B., (1985b) Overdetermined sets of linear equations: trade-off between uncertainty and number of solutions. Int. report, ESAT, K.U. Leuven.

4. Fox, L. and Parker, I., (1968) Chebyshev polynomials in numerical analysis. Oxford Univ. Press, London.

5. Fuhrmann, D.R. and Liu, B., (1984) An iterative algorithm for locating the minimal eigenvector of a symmetric matrix, Proc. of the 1984 ISCAS conference.

6. Golub, G.H. and Kahan, W., (1965) Calculating the singular values and pseudo-inverse of a matrix. *Siam J. Num. Anal.* **2**, 205-24.

7. Golub, G.H., Luk, F.T. and Overton, M.L., (1981) A block Lanczos method for computing the singular values and corresponding singular vectors of a matrix. ACM. Transactions on mathematical Software, Vol. 7, No. 2.

8. Golub, G.H. and Van Loan, C.F., (1983) Matrix computations. Johns Hopkins University Press, Baltimore.

9. Hansen, P.C., (1984) SVD, Theory and applications. rep. NI-84-05, Techn. Univ. Denmark.

10. Hari, V. and Veselic, K., (1981) On Jacobi methods for singular value decomposition. Int. report of department of mathematics, University of Zagreb, Yugoslavia.

11. Jensen, P.S., (1972) The solution of large symmetric eigenproblems by sectioning. *Siam. J. Num. Anal.*, vol. 9, No. 4.

12. Klema, V.C. and Laub, A.J., (1980) The singular value decomposition: its computation and some applications. *IEEE Trans. Aut. Contr.*,

13. Kung, S.Y., (1978) A new identification and model reduction algorithm via singular value decomposition. Proc. 12th Asitomar Conf. Circ. Syst.,

14. Manteuffel, T., (1977) The Chebyshev iteration for nonsymmetric linear systems. *Num. math.* **28**, pp. 307-327.

15. Paige, C., (1972) Computational variants of the Lanczos method for the eigenproblem. *J. Inst. math. Appl.* **10**.

16. Parlett, B., (1980) The symmetric eigenvalue problem. Prentice Hall Series in Computational mathematics. Prentice Hall Inc., Englewood Cliffs, N.J. 07632.

17. Rivlin, T.J., (1976) The Chebyshev polynomials. J. Wiley and Sons, New York.

18. Saad, Y., (1984) Chebyshev acceleration techniques for solving nonsymmetric eigenvalues problems. Math. of Comput., Vol. 42, No. 166, p. 567-588.

19. Schlien, S., (1982) A method for computing the partial singular value decomposition. *IEEE Trans. on pattern analysis and machine intelligence,* vol. PAMI 4, No. 6.

20. Stiefel, E.L., (1955) Kernel polynomials in linear algebra and their numerical application. *Appl. Math. Ser. Nat. Bur. Stand,* **49**, 1-22.

21. Vanderschoot, J. and Vandewalle, J. et al (1984) Extraction of weak bioelectrical signal by means of singular value decomposition. Proc. of 6th Internat. Conf. on Analysis and Optimization of Systems, Nice.

22. Veselic, K., (1976) A convergent Jacobi method for solving the eigenproblem of arbitrary real matrices. *Num. Math.* **25**, 179-184.

23. Wilkinson, (1965) The algebraic eigenvalue problem. Clarendon Press, Oxford.

24. Zeiger, H.P. and Mc Ewen, A.J., (1974) Approximate linear realizations of given dimension via Ho's algorithm. *IEEE Trans. Aut. Contr.*, Vol. AC-19, pp. 153.

A REMARKABLE DISCRETE UNITARY TRANSFORM*

E.A. Trachtenberg
(Dept. of ECE, Drexel University, Philadelphia, USA)

ABSTRACT

This paper describes a class of discrete unitary γ-transforms with good computational qualities. They are derived as bases in which "convolution type theorems" hold for varying structure systems (VLS) over finite groups. The VLS are used as models of classical digital signal processing systems and consequently the γ-transforms approximate the corresponding unitary transforms. It is shown how to construct such γ-transforms so as to satisfy the opposing criteria of computational effectiveness and accuracy of approximation. A γ-transform can be constructed (in nontrivial cases) with computational effectiveness similar to that of a diagonal transform. As another extreme, it can be constructed to be of distance zero (in the Hilbert-Schmidt norm) from a given unitary transform. Any trade off between these two extremes can be achieved as well. These properties of γ-transform are achieved by making it dependable upon certain free parameters, whose choice determines their behaviour with respect to the two criteria. Now, part (or all) of these parameters can be chosen to affect the computational performance of a γ-transform whereas the other part is used to minimize its distance from a given transform.

1. INTRODUCTION

The following two areas of digital and logic signal processing utilize heavily orthogonal transforms:

1) Linear systems which exhibit some inherent space or time invariant properties and hence the use of spectral analysis is

* Research supported by NSF under grant ECS 85 12748.

facilitated. For example, the matrices of the DFT (Discrete Fourier Transform) are composed of eigenvectors of linear time invariant systems. That fact coupled with the existence of Fast Fourier Transform Algorithms [1,2] provides a powerful means for fast spectral analysis and synthesis of such systems. Analogously, the matrices of the Walsh-Hadamard Transform are composed of eigenvectors of dyadic-invariant systems, which find widespread application in Computer Engineering [3,4]. That fact coupled with the existence of Fast Walsh Transform Algorithms provides a powerful means for fast spectral analysis, synthesis and optimization of systems in Computer Engineering. These two are examples of use of group transforms (the cyclic group in the case of the DFT and the dyadic group in the case of the WHT) in the theory of group-invariant systems and signals.

2) Suboptimal approximations to optimal (by a given criterion) signal processing systems. Here, usually more than one, sometimes opposing, performance criteria are involved and orthogonal transforms are utilized in order to improve statistical or computational performance of systems and signals involved or to permit a trade-off between these two performance criteria [5,6,7,8]. For example, in addition to computational effectiveness a transform may also provide the accuracy of approximation by a suboptimal system in a given task. These two are opposing criteria and the problem of trade-off of one at the expense of the other has become topical [1,2,5,9].

A classical linear system like a filter, a decoder, a time invariant control system with a zero initial state is given by the following equation:

$$z(t) = (w*u)(t) = \sum_{\zeta\varepsilon N} w(t-\zeta)u(\zeta), \quad t\varepsilon N \triangleq \{0,1,\ldots,n-1\}, \tag{1}$$

where $u,z,w\colon N \to C$ are the input, the output and the impulse response functions of the system and C is the field of complex numbers. The symbol * stands for convolution of the input and the impulse response functions.

A linear group invariant system is defined as follows [9]. Let us impose a group structure on the set N, i.e., G stands for an Abelian group of order n with the carrier set N and with group operation $\circ$. Consider the convolution type linear invariant system defined on G as follows:

$$y(t) = (h \circledast u)(t) \triangleq \sum_{\zeta\varepsilon G} h(\zeta^{-1}\circ t)u(\zeta), \quad t \varepsilon G, \tag{2}$$

where u,y,h: $G \to C$ are the input, the output and the impulse response functions of the group system and $*$ stands for the operation of group convolution, and ζ^{-1} is the inverse of ζ in G. There are three reasons (at least) for approximating a classical system (1) by a group model (2):

(i) The group operation $\zeta^{-1} \circ t$ in (2) (e.g. componentwise addition of vectors mod 2) is more easily implementable than the subtraction operation $t-\zeta$ in (1).

(ii) The matrix multiplication in (2) can always be done (for non-prime n's) using fast computational algorithms [14,15].

(iii) It is desirable to achieve convergence between the outputs of the original system (1) and its group model (2) i.e. $||z-y|| \underset{n \to \infty}{\to} 0$, where $||f|| = (\sum_{t \in N} f(t)\overline{f(t)})^{1/2}$, $\overline{f(.)}$ denotes the complex conjugate of f(.).

The theory of group models (2) is well developed [9-11] and originates in the works of Frobenius [12] in 1895. The advantages of using models (2) e.g. analyticity of solutions, speed of computation, etc., has to do with the structure of the underlying group G and its dual object.

It was shown [5,9,11] that by using linear group models (2) of systems (1) we achieve goals (i), (ii), but, in general, it is very difficult to construct (for a given family of systems (1), $n \to \infty$) a family of groups G of increasing order $n \to \infty$, such that (iii) is satisfied. Therefore we will introduce (see Section 3) varying structure linear systems (VLS) which constitute an extension of linear models (2). The VLS are related to a special class of bases in $L = \{f: G \to C\}$ which are closed under group translations and in which convolution type theorems for VLS hold. These bases possess remarkable properties and they are the primary object of the paper which consists of 5 sections. The next section is devoted to some necessary facts from Fourier Analysis over finite Abelian Groups. VLS and related bases are studied in Section 3. Computational properties of these bases are studied in Section 4 and their accuracy of approximation is discussed in Section 5.

2. FOURIER ANALYSIS OVER FINITE ABELIAN GROUPS

Consider an Abelian group G with the carrier set $N=\{0,1,\ldots, n-1\}$ and with $\circ$ as the group operation. 0 is the group identity and i^{-1} is the inverse of $i \in G$ Recall [13], that a character χ_ω of a finite commutative group G is defined as a homomorphism of

G into the multiplicative group of complex numbers C. There are n characters χ_ω, $\omega = 0,1,\ldots,n-1$, which are enumerated by the elements of G. The set of all characters $\hat{G} = \{\chi_\omega\}$ is a multiplicative group isomorphic to G [13, p. 35] and is called the dual object of G. The value of character χ_ω at the point $t\varepsilon G$ will be denoted by $\chi_\omega(t)$, $\omega = 0,1,\ldots,n-1$. The characters constitute a complete and orthogonal basis in the space $L(G,C) \triangleq \{f:G \to C\}$, that is:

$$n^{-1} \sum_{t\varepsilon G} \chi_\omega(t)\ \chi_\lambda(t^{-1}) = \delta_{\omega,\lambda}\ , \tag{3}$$

where $\delta_{\omega,\lambda}$ is the Kronecker symbol and if $\sum_{t\varepsilon G} f(t)\chi_\omega(t^{-1})=0$ for all $\chi_\omega \varepsilon \hat{G}$ then $f(t)=0$ for all $t\varepsilon G$. Therefore, for $f:G \to C$ we define the Fourier transform and inverse Fourier transform over G as follows:

$$\hat{f}(\omega) = n^{-1} \sum_{t\varepsilon G} f(t)\chi_\omega(t^{-1}), \qquad \chi_\omega \varepsilon \hat{G}, \tag{4}$$

$$f(t) = \sum_{\chi_\omega \varepsilon \hat{G}} \hat{f}(\omega)\chi_\omega(t) \qquad , \qquad t\varepsilon G. \tag{5}$$

The properties of Fourier transform (4), (5) such as linearity group translation, group convolution, Plancherel, Poisson and Wiener-Khinchine theorems, etc., can be found in [9,13]. Let n be a composite number $n = \prod_{i=1}^{r} n_i$, where n_i's are powers of primes and are not necessarily distinct, we consider the group G of order n, which is the direct product of cyclic groups H_i, $i=1,2,\ldots,r$, $G=H_1 x H_2 x \ldots x H_r$. That is, G is the group of r-tuples, $g\varepsilon G$, under the componentwise addition mod n_i; $g=(h_1,h_2,\ldots,h_r)$, $h_i=0,1,\ldots,n_i-1$; $i=1,2,\ldots,r$; and $g \circ g' = (h_1,h_2,\ldots,h_r)\circ(h_1',h_2',\ldots,h_r')=(h_1+h_1' \bmod n_1,\ldots,h_r+h_r' \bmod n_r)$. The characters of G are computed by formula:

$$\chi_\omega(t) = \exp\{2\pi i \sum_{s=1}^{r} \omega_s t_s/n_s\}\ \omega,t = 0,1,\ldots,n-1; \tag{6}$$

$$\omega = (\omega_1,\omega_2,\ldots,\omega_r)\varepsilon G;\ t = (t_1,t_2,\ldots,t_r)\varepsilon G;\ i = \sqrt{-1}.$$

In this case H_i is isomorphic to $\hat{H}_i$, i=1,2,...,r and it follows from (6) (see also [14,15]) that to compute the Fourier transform (4), (5) we need at most $n \sum_{i=1}^{r} n_i$ computer operations by the corresponding algorithms of Fast Fourier transform over G. In the case of a cyclic group G, formulas (4), (5) define the well known Discrete Fourier transform (DFT). In case of a dyadic group G of binary n-vectors with componentwise addition mod 2 as its group operation, (4) and (5) define the Walsh-Hadamard Transform. If G is the group of p-ary n-vectors under the group operation of addition mod p, then the characters defined in (6) are known as Chrestenson functions and formulas (4), (5) are called the Chrestenson Transform (see [13]).

3. VLS AND BASES CLOSED WITH RESPECT TO GROUP TRANSLATIONS

We consider VLS over G with one input and one output. Let $v \in L(G,C)$ and $\beta \in L(G \times G,C)$. Then the VLS is defined for the input u and output y ($u,y \in L(G,C)$) as follows:

$$y(t) = \sum_{\zeta,\eta \in G} \beta(\zeta^{-1} \circ t, \eta^{-1} \circ t)\, v(\zeta) u(\eta) \tag{7}$$

If (7) is true for given u and y, then we say that the VLS computes the input/ output pair (u,y). We have the following advantages by utilizing systems with variable structure:

1. A pair (u,y) that is not realizable by a linear convolution type system (2), may be implemented by a VLS (7) by an appropriate choice of function β. Thus we expand the class of input/output pairs which are realizable by linear systems. Note that a linear system (3) is obtained from (7) by setting $v(\zeta)=1$ for all $\zeta \in G$ and $h(\eta^{-1} \circ t) \triangleq \sum_{\zeta \in G} \beta(\zeta^{-1} \circ t, \eta^{-1} \circ t)$.

2. By an appropriate selection of β the error in approximation problems can be decreased (see [5, Example 6]).

3. By setting v=u and $\beta(s,t) = h(s)h(t)$ for all $s,t \in G$ we obtain a class of quadratic systems, which we may investigate. Such systems (with y=const) are used for error detection [10].

4. There is a close connection between group characters and the translation invariant bases used in investigation of system (7). That permits analytical methods of solution of problems as well as use of fast computational algorithms like FFT.

We shall use orthonormal bases closed with respect to group translation to investigate VLS (7). A matrix H whose columns represent vectors of such a basis belongs to the set Fro (G) defined as follows:

$$\mathrm{Fro}\ (G) = \{H = (H_{t,\zeta}) \stackrel{\Delta}{=} ((h(\zeta^{-1} \circ t)) \,|\, h:G \to C\}. \tag{8}$$

Note that a basis matrix $T \varepsilon \mathrm{Fro}(G)$ must be unitary i.e. $TT^* = I$, where I is the (nxn)-identity matrix. These matrices were introduced by Frobenius [12] in his study of group characters. They possess fast computational algorithms for multiplication by n-vectors because such a multiplication is equivalent to group convolution. Their eigenvectors (for Abelian groups) are computed analytically and are represented by the corresponding group characters (6). We consider, therefore, a basis $T_i \varepsilon L(G,C)$, $i=0,1,\ldots,n-1$ which has a generating function $T \varepsilon L(G,C)$ such that:

$$T_i(j) = T(i \circ j) \quad \text{for all } i,j \ \varepsilon\ G. \tag{9}$$

The basis vectors $T_o, T_1, \ldots, T_{n-1}$ are arranged as rows (or columns) into a unitary matrix, denoted by the same letter T without causing ambiguity:

$$T = (T(i \circ j)), \quad i,j \varepsilon G \ ; \quad TT^* = I \quad , \quad T \varepsilon \mathrm{Fro}(G) \tag{10}$$

<u>Theorem 1:</u> Any orthonormal basis $T_o, T_1, \ldots, T_{n-1}$ in L(G,C) satisfies (9), (10) if and only if

$$T(j) = n^{-1} \sum_{\chi_\omega \varepsilon \hat{G}} \hat{\gamma}(\omega)\ \chi_\omega(j) \quad , \tag{11}$$

where $\hat{\gamma}(\omega)\ \hat{\gamma}(\omega^{-1}) = 1$ for all $\chi_\omega \varepsilon \hat{G}$. (12)

It follows now from Theorem 1 that the parameters $\hat{\gamma}(\omega)$, $\omega = 0,1,\ldots,n-1$ satisfying (12) can be chosen in such a way that T has the desired properties. For example, they can be chosen so the the multiplication of T by an n-vector requires less than the upper bound of $n \sum_{s=1}^{r} n_s$ computer operations for all group transforms. As an example of another extreme, the numbers $\hat{\gamma}(\omega)$, $\omega=0,1,\ldots,n-1$ satisfying (12) can be chosen in such a way as to make T asymptotically equivalent to a given KLT (Karhunen-Loeve Transform) which is best from the point of view of statistical performance [2,5]. More important for applications is the trade off between these two performance criteria. In that respect, in the next two sections, we shall discuss how to choose part of the numbers $\hat{\gamma}(\omega)$ to improve the computational performance of T so that the remaining part of the numbers $\hat{\gamma}(\omega)$ can be used to improve its statistical peformance with the obvious result as to the trade off between these two performance criteria: the more we improve computational performance (by selecting more of the numbers $\hat{\gamma}(\omega)$, $\omega = 0,1,\ldots, n-1$ satisfying (12)) the less we can do with respect to the statistical performance of the system (the less numbers $\hat{\gamma}(\omega)$, satisfying (12) we have to select from) and vice versa.

Let $\tilde{G}^{\gamma} = \{T_{\omega}^{\gamma}\}$ be a dual object of G for some γ chosen by (12) and (5). Then the following orthonormal γ-transform $f \leftrightarrow \tilde{f}^{\gamma}$ is used:

$$\tilde{f}^{\gamma}(\omega) = \sum_{\zeta \varepsilon G} f(\zeta) T_{\omega}^{\gamma}(\zeta^{-1}) \quad , \tag{13}$$

$$f(\zeta) = \sum_{T_{\omega}^{\gamma} \varepsilon \tilde{G}^{\gamma}} \tilde{f}^{\gamma}(\omega) T_{\omega}^{\gamma}(\zeta) \quad . \tag{14}$$

We have the following procedure for computation of $\tilde{f}^{\gamma}$:

1. Given f: $G \to C$, compute $\hat{f}$ by (4).

2. Compute $\hat{\phi}^{\gamma}(s) \stackrel{\Delta}{=} \hat{\gamma}(s) \cdot \hat{f}(s)$ for all $s \varepsilon G$ and for the chosen $\hat{\gamma}$.

3. $\tilde{f}(\omega) = \phi^{\gamma}(\omega)$, $\chi_{\omega} \varepsilon \hat{G}$; where ϕ^{γ} is computed by (5) for $\hat{\phi}^{\gamma}$ defined above.

It follows from the procedure that the γ-transform of a function $f \varepsilon L(G,C)$ may be computed by two successive applications of Fourier transform (4) (5). Since (4) and (5) may be computed in at most $n \sum_{i=1}^{r} n_i$ computer operations by the algorithms of the FFT over G, it follows that $\tilde{f}^{\gamma}$ (or f) may be computed in at most $n(2 \sum_{i=1}^{r} n_i + 1)$ computer operations, where n_i is the order of the cyclic subgroup H_i, $G = \prod_{i=1}^{r} H_i$. Thus we established a fast computational algorithm for γ-transform (13), (14) based on FFT for (4), (5). We note that the number of computer oeprations required to evaluate γ-transform may be reduced substantially (see next two sections) for special choices of γ. For the γ-transform (13), (14) the usual properties such as linearity, Plancherel and convolution theorems are valid.

4. COMPUTATIONAL PROPERTIES OF γ-TRANSFORM

We consider non-trivial choices of $\hat{\gamma}(\omega)$ satisfying (12), that is $\hat{\gamma}(\omega) \neq \chi_\nu(\omega)$ for any $\nu \varepsilon G$. Denote by $|T|^{(n)}$ the number of non-zero entries of the (nxn) matrix T. That is $|T|^{(n)}$ is the Hamming weight of T.

THEOREM 2: For any composite number $n = n_1 \cdot n_2 \cdot \ldots \cdot n_r$ we can choose $\gamma(\omega)$, $\omega = 0,1,\ldots,n-1$ satisfying (12) such that

$$n \cdot 2 < |T|^{(n)} \leq n \cdot n_m, \; n_m = \min_{1 \leq i \leq r} \{n_i > 2\} \qquad (15)$$

Note that the right hand side in (15) is achievable (and hence the left hand side follows directly) for the following construction of T. Consider the group $G=H_1 x H_2 x \ldots x H_r$ of order n, where the order of H_i is n_i, $i=1,2,\ldots,r$. The group operation is componentwise addition mod n_i, $i=1,2,\ldots,r$. Suppose H_m is a subgroup (not necessarily unique), such that $n_m = \min_{1 \leq i \leq r} \{n_i > 2\}$. In case of more than one subgroup of the same order n_m, we select any of them. Let $\hat{\gamma}(\omega_m)$, $\omega_m = 0,1,\ldots,n_m-1$ be some arbitrary complex numbers satisfying (12). For some $t^{-1} \varepsilon G$, $t^{-1} = (t_1^{-1},\ldots,t_{m-1}^{-1},0,t_{m+1}^{-1},\ldots,t_r^{-1})$ we write

$$\hat{\gamma}(\omega) = \hat{\gamma}(\omega_m) \prod_{i \neq m} \chi_{\omega_i}(t_i^{-1}), \text{ and by (11),} \qquad (16)$$

$$T(j) = \begin{cases} n_m^{-1} \sum_{\omega_m=0}^{n_m-1} \chi_{\omega_m}(j_m)\hat{\gamma}(\omega_m), & j=(t_1,\ldots,t_{m-1},j_m,t_{m+1},\ldots,t_r); \\ 0, & \text{otherwise,} \end{cases} \qquad (17)$$

where $j_m = 0,1,\ldots,n_m-1$.

The entries of the matrix T defined by (17), (10), (9) are, generally speaking, complex. Hence, to multiply an n-vector by T directly we need $n \cdot n_m$ complex multiplications and $n \cdot (n_m-1)$ complex additions. It follows from Theorem 2 that by restricting ourselves to functions $T:G\to C$ which are non-zero on some subgroups of G only, we can control the computational complexity of T by choosing subgroups in (16). A similar approach was undertaken in [9,10]. The number of non-zero entries $|T|^{(n)}$ of the (nxn)-matrix T constructed by (17), (10), (9) grows more slowly than n^2 when $n \to \infty$, the γ-transform is computationally convenient in the sense of the following Corollary.

COROLLARY: If $n_1 = \ldots = n_r = n_m$, i.e. $n = n_m^r$ and T is constructed by (17), (10), (9) then

$$\lim_{r\to\infty} (|T|^{(n)} n_m^{-2r}) = 0. \tag{18}$$

Thus, it is extremely convenient to store T in a computer, especially in recursive algorithms, when $n\to\infty$. We note, that neither Walsh nor Discrete Fourier Transforms possess that property. Like the Haar-Transform, the γ-transform (17), (10), (9) is locally sensitive in the sense that its expansion coefficients depend only on the "local" behaviour of the function i.e. on its behaviour at a few points which are "close together" as components of cosets of G by H_m. To make it globally sensitive we have only to increase the order of the chosen subgroup in (16).

Consider the optimal (computationally) case of the γ-transform, i.e. T is real, its entries have only two different non-zero values like the Walsh and Haar Transforms and the multiplication of T by an n-vector is faster than that of Theorem 2. Consider first, the case $n = 2^r$. In this case, by Theorem 2, we can take $n_m = 4$. Define $\hat{\gamma}(\omega_m)=-1$ for $\omega_m = 0$, k_m ; $k_m \varepsilon\{1,2,3\}$ and $\hat{\gamma}(\omega_m)=1$ otherwise. Then by (11)

$$T(j) = \begin{cases} (n_m-2)/n_m, & j = (t_1,\ldots,t_{m-1},0,t_{m+1},\ldots,t_r) ; \\ -2/n_m, & j = (t_1,\ldots,t_{m-1},j_m,t_{m+1},\ldots,t_r), j_m \neq 0; \\ 0 & , \text{otherwise}. \end{cases} \tag{19}$$

Substituting $n_m = 4$ we have $T_{i,j} \varepsilon\{0,+1,-1\}$; in each column and each row of T there are one -1 and three 1's (for each positive number r) and we need only 2^{r+1} additions to multiply by T.

Haar's Transform requires only $2^{r+1}-2$ additions and 2^{r-1} normalizations for its computation. Note that the direct and inverse matrices of Haar's Transform are not the same and its columns cannot be obtained by a shift operation.

In the general case, we put $\alpha = (n_m-2)/2$ and we have from (19) $T_{i,j} \varepsilon\{0,1,\alpha\}$, each column of T is obtained by a shift on G of the previous column, T is computationally convenient in the sense of the Corollary of Theorem 2 and to multiply by T directly we need $n(n_m-1) \simeq n \cdot n_m$ real additions, n real multiplications on α and n multiplications on the normalizing constant $(-\frac{2}{n_m})$. In that case T consists of n/n_m diagonal blocks, where each block is an $(n_m x n_m)$-matrix with α's on the main diagonal and with 1's for all the other entries. To multiply it by an n_m-vector $(b_1,b_2,\ldots,b_{n_m})^T$ we need n_m-2 additions to compute the 1st partial component $B_1^p = b_2 + b_3 +\ldots+ b_{n_m}$ of the result and 2 additions per each of the other components. For example, $B_2^p = B_1^p - b_2 + b_1$. That makes a total of $n_m-2 + 2(n_m-1) = 3n_m-4$. In addition, we need another n_m additions because in row i we must add $\alpha.b_i$ to B_1^p, $i=1,2,\ldots,n_m$. Hence $4n_m-4$ real additions are needed per block. We need $\frac{n}{n_m}(4n_m-4) \approx 4n$ real additions, n real multiplications by α and n multiplications on $(-\frac{2}{n_m})$ (i.e. 2n multiplications) to multiply by T. Now if n is an arbitrary (not necessarily composite) integer, then we put $\gamma(\omega) = -1$, $\omega=0,k$, $k\varepsilon\{1,2,\ldots,n\}$ and $\gamma(\omega)=1$ otherwise. As in the previous case we need $4n-n \approx 4n$ additions and 2n multiplications to multiply by T. Hence, by using all the numbers $\gamma(\omega)$, $\omega = 0,1,\ldots,n-1$ we achieve a computational performance for T similar to that of a diagonal matrix. The price for that, however, is that we have no parameters left to affect the accuracy of T. Because of that, in the next section we use $(n-n_m)$ numbers $\hat{\gamma}(\omega)$ defined by (16) in order to reduce the computational complexity of T to $n_m \cdot n$ operations, so that we have n_m unspecified numbers $\hat{\gamma}(\omega)$ to affect the statistical performance of T.

5. APPROXIMATION ACCURACY OF γ-TRANSFORM

In this section, we follow the approach developed in [8,9,11] in finding computationally "good" approximations to statistically "good" transforms. In [8,9] a measure of closeness between two transform sequences was chosen as the Hilbert-Schmidt (the weak) norm. Two sequences $A^{(n)}$, $B^{(n)}$ of unitary (nxn) matrices, $n\to\infty$ are said to be asymptotically equivalent [2,8] if the Hilbert-Schmidt norm of their differences vanishes as $n\to\infty$:

$$\lim_{n\to\infty}\|A^{(n)} - B^{(n)}\| \overset{\Delta}{=} \lim_{n\to\infty}(1/n\ \mathrm{Trace}((A^{(n)}-B^{(n)})^*(A^{(n)}-B^{(n)})))^{1/2} = 0, \tag{20}$$

THEOREM 3: Two sequences $A^{(n)}$, $B^{(n)}$ of unitary (nxm) matrices, $n\to\infty$ are asymptotically equivalent if and only if

$$\mathrm{Trace}(D^{(n)^T} D^{(n)}) = n,\ n = 1,2,\ldots\ , \tag{21}$$

where $D^{(n)} = (D^{(n)}_{s,r}) = (|<a_s, b_r>|^2)$; a_s, b_r are the row vectors of the unitary matrices $A^{(n)}$ and $B^{(n)}$ respectively for fixed n; s, r = 0,1,...,n-1; $< .,.>$ stands for the inner product. It follows from Theorem 3 that if for a given n, A is a "computationally deficient" transform, then in order to replace it by an equivalent and computationally convenient γ-Transform we have the following procedure for a composite number n:

1) Select a group G of order n so that it has subgroups of order at least 3 (see Theorem 2). Select the numbers $\hat{\gamma}(\omega)$ satisfying (12) by (16). That means that the $n-n_m$ numbers $\hat{\gamma}(\omega)$ determine the computational complexity of T as $n\cdot n_m$ computer operations.

2) Compute D and Trace D^TD using the n_m unspecified numbers $\hat{\gamma}(\omega_m)$ as unknowns.

3) Select the remaining n_m numbers $\hat{\gamma}(\omega_m)$ (if possible) so as to satisfy equation (21) subject to restriction (12).

We note that if n is a prime number then T can be chosen by specifying all of the numbers $\hat{\gamma}(\omega)$ either to achieve computational convenience (see end of section 4) or else by (20) to achieve asymptotic equivalence. For prime numbers n it is not possible to achieve the trade off between the two performance criteria because for a given prime number n, the corresponding cyclic group is unique and has no subgroups. We illustrate the use of Theorem 3 by means of the following examples.

Example 1: Let G be the cyclic group of integers 0,1...,n-1 with addition mod n as its group operation, n is a prime number, n > 2. We consider the approximation of the DFT (which is denoted by A) by the γ-transform, constructed on G. That is:

$$a_{s,t} \overset{\Delta}{=} n^{-1/2}\chi_s(t), b_{r,t} \overset{\Delta}{=} T_{r,t} = n^{-1}\sum_{\omega=0}^{n-1}\hat{\gamma}(\omega)\chi_\omega(r \circ t), s,r,t=0,1,\dots,n-1;$$

where $\chi_s(t) = \exp\{2\pi ist/n\}$, $i = \sqrt{-1}$. To solve for $\hat{\gamma}(\omega)$ satisfying (12) we assign $\hat{\gamma}(0)=1$; $\hat{\gamma}(s)=x$, $1\leq s\leq (n-1)/2$; $\hat{\gamma}(s)=x^{-1}$, $(n+1)/2\leq s\leq n-1$; $x\neq 1$ and we have from (21): $y^2 - 2(n+1)y + 1 = 0$, $y = x^4$. That is, according to the procedure for a prime number n, we have specified all the numbers $\hat{\gamma}(s)$, s = 0,1,...,n-1 to make T asymptotically equivalent to the Discrete Fourier Transform. On the other hand, if all the numbers $\hat{\gamma}(s)$, s = 0,1,...,n-1 are specified as at the end of previous section, then the best computational performance of T is achieved and it is not possible to use T as an approximation to the Discrete Fourier Transform.

Example 2: Let now $G=H_1 x H_2 x \dots x H_r$, $g\varepsilon G$, $g=(h_1,h_2,\dots,h_r)$, $g=0,\dots,n-1$; $h_s=0,\dots,n_s-1$; $s=1,\dots,r$. Suppose H_m is the subgroup of order $n_m = \min_{1\leq i\leq r}\{n_i>2\}$. We choose again the transform to be approximated as $A=(a_{s,t})=n^{-1/2}(\chi_s(t))$, s,t=0,1, ...,n-1. The γ-Transform is defined by (11) and applying the procedure following Theorem 3 in case of a composite number $n = \prod_{s=1}^{r} n_s$ we select the $n-n_m$ numbers $\hat{\gamma}(\omega)$ satisfying (12) by (16). Then the computational complexity of T is reduced to $n_m \cdot n$ computer operations. Then, at the 2nd step of the procedure, there are n_m numbers $\hat{\gamma}(\omega_m)$ which are treated as unknown x's in computing $D^{(n)}$ in the equation (21). And similarly to the previous example we have (e.g. for n_m odd) from (21): $(x^4+x^{-4})(n_m-1)/2 = n_m \cdot n-1$, where $\hat{\gamma}(\omega_m)=x$, $1\leq\omega_m\leq(n-1)/2$; $\hat{\gamma}(\omega_m)= x^{-1}$, $(n+1)/2\leq\omega_m\leq n-1$. Hence, the Fourier Transform over G which requires $n \sum_{i=1}^{r} n_i$ operations for its implementation is replaced by γ-Transform which is asymptotically equivalent to it and which requires $n_m.n$ operations only.

REFERENCES

[1] Trachtenberg, E.A. and Karpovsky, M.G., (1985) In "Spectral Techniques and Fault Detection" (M. Karpovsky, ed.) pp. 179-217, Academic Press, New York.

[2] Elliott, D.R. and Rao, K.R., (1983) "Algorithms, Analyses and Applications. Academic Press, New York.

[3] Hurst, S.L., Miller, D.M. and Muzio, J.C., (1985) "Spectral Techniques in Digital Logic". Academic Press, New York.

[4] Beauchamp, K.G., (1984) "Applications of Walsh and Related Functions". Academic Press, New York.

[5] Trachtenberg, E.A., (1985) "Construction of Group Transforms Subject to Several Performance Criteria". IEEE Trans. on ASSP, vol. ASSSP-33, No. 6, Dec. 1985, pp. 1521-1531.

[6] Karpovsky, M.G. and Trachtenberg, E.A., (1985) "Statistical and Computational Performance of a Class of Generalized Wiener Filters". IEEE Trans. on Inf. Theory, vol. IT-32, No. 2, March 1986, pp. 303-307.

[7] Trachtenberg, E.A., (1983) "Systems Over Finite Groups as Suboptimal Filters: A Comparative Study". Proc. Intern. Symp. Math. Theory of Systems and Networks, Beer-Sleva, Israel, Springer Verlag, pp. 856-863.

[8] Yemini, Y. and Pearl, J., (1979) "Asymptotic Properties of Discrete Unitary Transforms". IEEE Trans. on PAMI, Vol. PAMI-1, pp. 366-371.

[9] Karpovsky, M.G. and Trachtenberg, E.A., (1977) "Some Optimization Problems for Convolution Systems over Finite Groups". Inform and Control., 34, pp. 227-247.

[10] Karpovsky, M.G. and Trachtenberg, E.A., (1979) "Fourier Transforms over finite Groups for Error Detection and Error Correction in Computation Channels". Inform. and Control., 40, pp. 335-358.

[11] Pearl, J., (1975) "Optimal Dyadic Models of Time-Invariant Systems". IEEE Trans. on Computers, vol. C-24, pp. 598-602.

[12] Frobenius, F.G., (1895) "Uber die Primfaktoren der Gruppendenterminante" Gesammelte Abhandlungenm Band III, Springer-Verlag, pp. 38-77.

[13] Karpovsky, M.G., (1976) "Finite Orthogonal Series in the Design of Digital Devices". J. Wiley, New York.

[14] Karpovsky, M.G., (1977) "Fast Fourier Transform on Finite Non-Abelian Groups". IEEE Trans on Computers vol C-26, pp. 1028-1030.

[15] Nicholson, P.J., (1971) "Algebraic Theory of Finite Fourier Transforms". J. Comput. Syst. Sci., 5. pp. 524-547.

IMPROVEMENT OF LINKED PYRAMID SEGMENTATION SCHEME WITH VARIABLE WEIGHTING FUNCTION

A.S. Sethi and M. Ahmadi
(Department of Electrical Engineering, University of Windsor, Ontario, Canada)

1. INTRODUCTION

Machine extraction of features from the digitized representation of an image is of great interest to investigators working in fields like Pattern recognition, Scene analysis, and Robotic vision. The linked pyramid segmentation scheme has been effectively used in the segmentation of a variety of images. We present here an improvement on the technique to obtain better results on image segmentation with no increase in the processing time or the memory requirements.

2. LINKED PYRAMID SEGMENTATION SCHEME [2], [3]

This algorithm creates a pyramid data structure from the input image array. The entire data structure is then stabilized by a reiterative linking of elements at consecutive levels followed by recomputing the element values based on the links. The image array is finally segmented by translating values down the linked pyramid structure.

Tanimoto and Pavlidis [1] suggested the use of a pyramid data structure for object isolation. Subsequently, Burt, Hong and Rosenfeld [2] used the reiterative pixel linking scheme for image segmentation.

The steps involved in the algorithm can be summed up as follows:

(1) Initialize the node values of each father level array by simple block averaging each node's 16 sons. The pyramid gets created in this step. The process is carried out bottom - up from level to level.

(2) Link each son level node to that one of its four fathers whose value is closest to its own. The linked structure is created in this step top - down from level to level.

(3) Recompute the node values at the father level by averaging the pixel values of only the linked sons. This step carried out bottom - up along the pyramid.

(4) Recompute the links between the son and father elements. Due to the changed pixel values obtained in step 3, the links computed in this step would be different from the ones computed earlier.

(5) Repeat steps (3) and (4) as many times as desired. In practice there is no change after 10 iterations for most images. The pyramid data structure is said to have stabilized at this stage.

(6) Translate pixel values down the stabilized linked structure starting at a specified top level. The array at the lowest level is the segmented image. The number of segments present in it equals the number of elements in the starting top level array.

3. IMPROVEMENT IN THE TECHNIQUE

A link as defined in the basic algorithm is a boolean expression. A son node can get linked to any one of its four candidate father nodes based on a criterion that decides which one of them is 'closest' to it. Typical criteria for this decision could be the difference in the gray tones, Euclidean distance, or a combination of the above two. In case more than one father is equally 'close', an arbitrary tie-breaking rule could be used. Once the closest father element has been established, a link is defined with the son element giving no consideration to the degree of 'closeness'. The proposed enhancement assigns a weighting factor to each link defined. This weight is a function of the difference between the gray tones of the father element and its linked sons.

3.1 Links with weighting variable [4]

The linked pyramid technique basically computes reiteratively the mean value of a cluster of pixels which (finally) define the segments comprising the input image. An adjustment made to the gray tones of the son elements based on the degree of its closeness with the linked father element could therefore result in a more accurate value of the cluster mean. This correction if carried out over all the iterations that the algorithm is performed is likely to result in segments of value closer to the true average of the regions comprising them.

Based on this theory, the authors tried a number of schemes to adjust the links based on the degree of closeness. A very simple scheme that gave the best results is outlined here:

During recomputation of averages, the gray tone value of the linked son is adjusted by a factor

$$F = \text{Abs}\,[\,c(i,j,m) - c'(i',j',m-1)\,] \;*\; W$$

where c and c' are the gray tones of the son and father elements respectively; and W is the weighting variable, a parameter to control the results.

The algorithm was tried for a range of values of W. For values between 0.01 and 0.05, the results were an improvement over the ones with no weighting factor. It is interesting to note that with W = 0, we get F = 0 (unweighted links).

4. TEST RESULTS AND DISCUSSION

In order to evaluate the effectiveness of the improved linked pyramid technique, the authors studied the radiographic image of a human chest. The image was digitized to a spatial resolution of 256 x 256 pixel squared with 8 bit gray level resolution. Next, it was preprocessed using a contrast enhancement filter.

Figure 1 shows the original image of the chest X-Ray. Figures 2 and 3 show respectively the results of the pyramid linking algorithm with no weighting function and with a variable weighting function and W = 0.01.

The lower ribs not appearing distinctly in Figure 2, do so in Figure 3 thus confirming the suggested improvement.

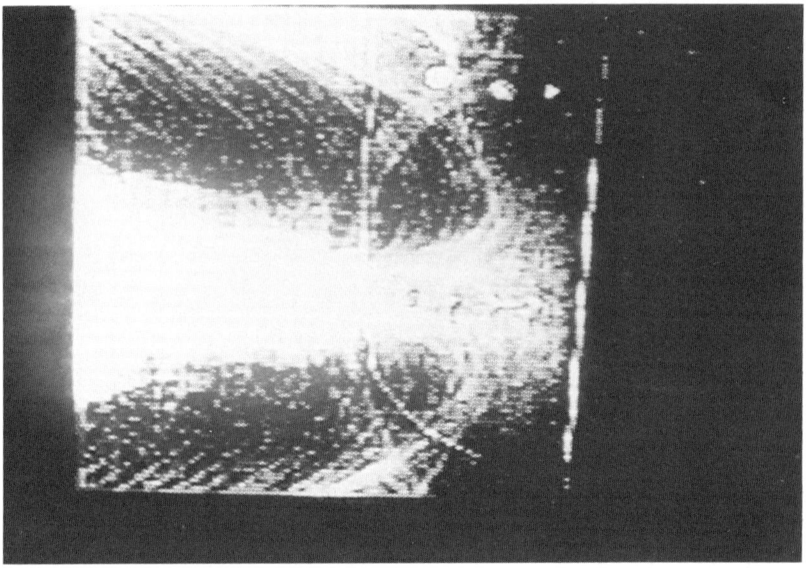

Fig. 1 Original Image of a Chest X-Ray

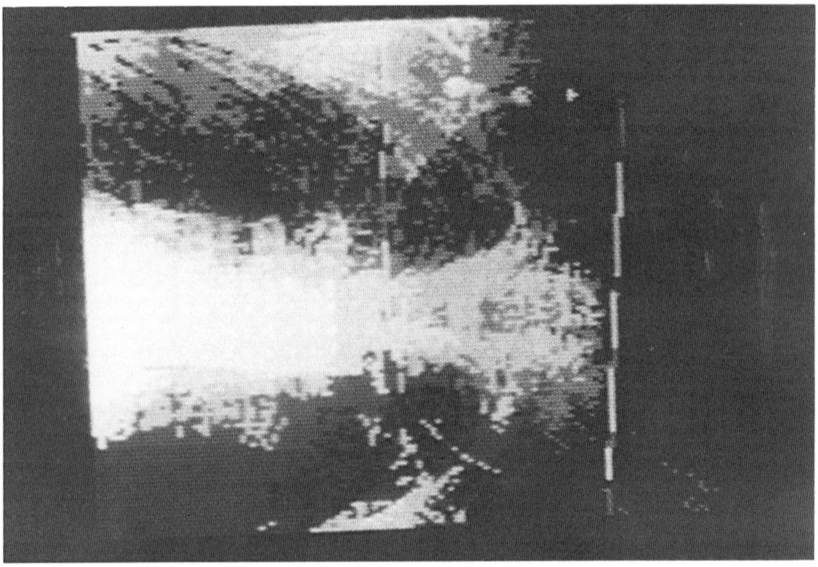

Fig. 2 Result of segmentation using the Linked Pyramid algorithm

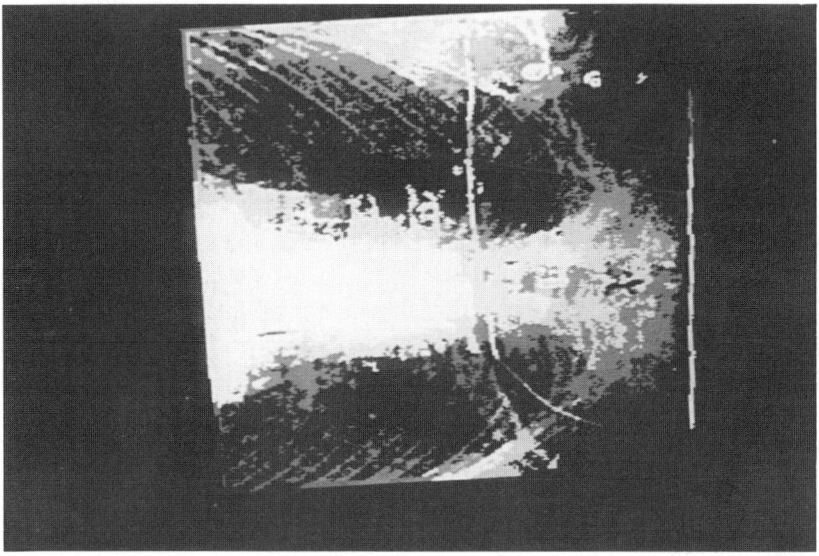

Fig. 3 Result of segmentation by the Improved Linked Pyramid technique with Weighting Variable W = 0.01.

5. REFERENCES

[1] Tanimoto, S. and Pavlidis, T., "A Hierarchical Data Structure for Picture Processing", Computer Graphics and Image Processing No. 4, pp. 104-119, 1975.

[2] Burt, P.J., Hong, T.H. and Rosenfeld, A., "Segmentation and Estimation of Image Region Properties through co-operative Hierarchical computation", IEEE Transactions on Sys, Man, and Cybernetics, Vol. SMC-11, No. 12 pp. 802-809, 1981.

[3] Hong, T.H., et al, "Image Snoothing and Segmentation by Multiresolution pixel linking: Further experiments and Extensions", IEEE Transactions on Sys, Man. and Cybernetics, Vol. SMC-12, No. 5, pp. 611-622, Sept/Oct, 1982.

[4] Sethi, A.S., "Segmentation of Biomedical Images by Multiresolution Pixel Linking", M.A.Sc. thesis, University of Windsor, Canada, 1984.

MULTIPLE SIGNAL DETECTION WITH UNKNOWN BACKGROUND NOISE

D.J. Jeffries and D.R. Farrier
(Department of Electrical Engineering, Southampton University)

ABSTRACT

A high resolution spectrum estimation is presented, which is suitable for processing with a linear array of equispaced sensors. Current high resolution methods often require large amounts of prior information, which are not always available in practice. The approach in this paper is to model the underlying noise field by two separate models, which resemble the types of model which we can expect in practice.

1. INTRODUCTION

It is often necessary to process the data from an array of equispaced sensors and achieve a bearing distribution of the incident noise power. These arrays sometimes have limited aperture, and high resolution methods are often advocated for reducing this defect. In this context, the so called Maximum Likelihood Method (MLM), the Maximum Entropy Method (MEM) and Multiple Signal Classification Method (MUSIC) are possible contenders. Of these methods, the MUSIC method proposed by Schmidt (1) has the best resolution and appears to have significant advantages over other possible methods.

Although the MUSIC method (and other eigenvector methods) has very good resolution properties, it does require that the background noise field is uncorrelated (or can be made uncorrelated if the noise statistics are known). In practice this is not always the case and in these circumstances the method gives incorrect results. The present paper proposes a technique for overcoming this particular difficulty. This method will not rely on knowing the background noise field. A number of assumptions will be made, but these encompass a large number of situations which can be expected to occur in practice. These assumptions are much less restrictive than those required by eigenvector methods, such as the MUSIC algorithm.

2. NOISE MODEL

It is helpful to consider first the type of noise field which we can expect to meet in practice. If the array is a passive acoustic device, then the noise is liable to come from a number of distinctly different sources.

Much of the noise which is incident on the array will be caused by natural phenomena in the ocean. Examples of this might be (1) wave noise (2) ocean turbulence or (3) marine life. This noise can be expected to be spectrally "smooth" about the array, and might even be symmetrical about the array.

Other components of the incident noise will be man-made noise. This noise will be primarily due to shipping, and will come from discrete directions, corresponding to the individual vessel's bearing relative to the array. It is clear that this noise will not be a smooth function of bearing, but will consist of a number of discrete line sources.

When it comes to modelling such a complex and diverse noise field we immediately have a serious problem. Methods such as the MEM are suitable for modelling smooth spectra but will not perform well on line spectra. Other methods, such as MUSIC, model the signal as a number of line spectra and will not be capable of modelling the "smooth" ocean noise. What we really require is a combination of these two models.

At this stage we are faced with a serious dilemma, for there are an infinite number of possible models which are all consistent with the data. These may range from purely discrete models to a large number of smooth spectra. It is clear that we need to supply some further constraints in order to obtain a unique solution. The authors' suggestion is that:-

1) Any non-symmetric noise is modelled by a discrete source model - since we may expect shipping noise from any bearing angle.

2) Noise which is symmetric about the array will be modelled by a continuous spectral distribution.

The techniques for implementing this particular model are outlined in the rest of the paper. It is desirable to first obtain the discrete sources (Section 3) and then the continuous spectra (Section 4).

3. DISCRETE MODEL

The imaginary part of the covariance matrix is only dependent on the non-symmetric noise, in contrast to the real part which depends on both types of noise. It is therefore necessary to model the non-symmetric case first. If $R = \{r_{\ell m}\}$ represents a (pxp) covariance matrix with k discrete sources assumed, having powers (σ_j^2) and directions (θ_j) then

$$\mathrm{Im}\left(r_{\ell m}\right) = \sum_{j=1}^{k} \sigma_j^2 \sin\left[\frac{2\pi d\,(\ell-m)\cos\theta_j}{\lambda}\right] \qquad (1)$$

The linear array of equispaced sensors has spacing d and the wave length is λ.

In order to obtain a unique solution for the unknowns σ_j and θ_j in equations (1) it is obviously necessary that 2k be less than p-1. For a deterministic covariance matrix it is possible to exactly evaluate the unknowns.

Of course in practice this is not the case and only finitely sampled covariance matrices are available. The parameters in this case may be estimated by a modification of Prony's algorithm [2]. The accuracy of these estimates depends on the number of samples used and the signal to noise ratio.

The sampled covariance matrix will be non Toeplitz, but may be made Toeplitz using the well known Burg [3] algorithm. Prony's approach is very similar to Hildebrand's [2] for treating sinusoidal signals, except additional equations are available. With a Toeplitz problem we may write

$$f_\ell = \mathrm{Im}\left(r_{\ell 1}\right) = \sum_{j=1}^{k} \sigma_j^2 \sin\left(\omega_j \ell\right) \; : \; \ell = 0, 2, \ldots p-1 \qquad (2)$$

where $\omega_j = \left(2\pi d \cos\theta_j\right)/\lambda$.

Hildebrand shows how to construct the following equation for $\left(\omega_j, j=1,\ldots k\right)$, where the $\left(\alpha_j, j=1,\ldots,k\right)$ are unknown real constants,

$$\cos\left(k\omega_j\right) - \alpha_1 \cos\left((k-1)\omega_j\right) \ldots - \alpha_{k-1}\cos\left(\omega_j\right) - \tfrac{1}{2}\alpha_k = 0 \qquad (3)$$

The unknown α's are now obtained from the following equations

$$\sum_{j-1}^{k-1} \left(f_{j+1} + f_{2k-j+i}\right)\alpha_j + f_{k+i}\,\alpha_k = f_i + f_{2k+i} \qquad (4)$$

for $(i = 0,1,\ldots, p-1-2k)$

However, unlike Hildebrand's approach, the data is derived from a Toeplitz matrix and there are additional equations that contain useful information. In order to illustrate this consider the case when k=2. Each equation in (4) takes 5 different values of f, which are shown below for $\ell = 0,1,\ldots, p-1-2k$.

$$f_0, f_1, f_2, f_3, f_4$$

$$f_1, f_2, f_3, f_4, f_5$$

$$\vdots$$

$$f_{N-6}, \ldots, f_{N-1}$$

Since the matrix is Toeplitz the sine function continues on either side of the diagonal, we have the following set of data points which represent new information.

$$f_{-1}, f_0, f_1, f_2, f_3$$

It may be shown that in the general case of k signals there are (k-1) additional equations to those given in (4), which consequently becomes valid for $(i = -k+1,\ldots, 0,\ldots, p-1-2k)$. This improves the conditioning of equations (4) and enables better estimates for $\{\alpha_j\}$ and $\{\theta_j\}$ to be obtained. These estimates may be improved again, by using them as starting values in a non-linear least squares procedure based on equation (2) for example. This method has quick convergence but is of course not optimal. Many other techniques could be applied to this area.

Having found the $\{\theta_j\}$ the $\{\sigma_j^2\}$ may be evaluated by a simple least squares technique. The background noise covariance matrix may then be estimated from

$$R_O = \mathrm{Re}\left(r_{\ell m}\right) - \sum_{j=1}^{k} \sigma_j^2 \cos\left(\omega_j(\ell-m)\right) \tag{5}$$

The imaginary part of $r_{\ell m}$ will of course contain a small residual error, but the assumption is that this is due to finite averaging only and so will be ignored, consequently R_O is purely real.

4. CONTINUOUS MODEL

R_O is always real and so will yield a symmetric spectrum. In order to model this the authors chose a band-limited maximum entropy spectral estimate [4]. This has the advantage over the usual Maximum Entropy method (MEM) of being able to reconstruct discontinuities in the spectrum. Many symmetric noise fields exhibit these discontinuities and exist only within a specified region $(-\Delta,\Delta)$ instead of over the whole Nyquist region. Examples of such symmetric noise fields are cylindrically or spherically isotropic noise fields.

The usual MEM seeks to maximise the entropy

$$H = \int_{-\frac{1}{2}}^{\frac{1}{2}} \ln f(\lambda)\, d\lambda$$

subject to the constraint

$$r_j = \int_{-\frac{1}{2}}^{\frac{1}{2}} f(\lambda)\ \exp\,(2\pi i\lambda j)\, d\lambda \ , \ j = 0,\ 1, \ldots, (p-1)$$

where $f(\lambda)$ is the desired spectrum

However in many realistic situations the following is true,

$$r_j = \int_{-\Delta}^{\Delta} f(\lambda)\ \exp(2\pi i\lambda j)\, d\lambda$$

consequently the definition of entropy must be modified to give

$$H = \int_{-\Delta}^{\Delta} \ln f(\lambda)\, d\lambda$$

It is seen that the important property of consistency is retained in the band-limited case. In order to evaluate f(λ) it is shown [4] that the following equation must be solved for the unknown γ's.

$$\left(\underline{\delta}_j^T R\right) \underline{\gamma} = \int_{-\Delta}^{\Delta} \frac{\exp(2\pi i\lambda j)\, d\lambda}{\sum_{m=0}^{p-1} \underline{\gamma}_m^* \exp(2\pi i m\lambda)} \qquad (6)$$

$$j = 0,1,..,(p-1)$$

$$\underline{\delta}_j^T = (0,\ldots,0,1,0,\ldots,0)$$

The integral in equation (6) may be evaluated analytically and an iterative method then applied to evaluate $\underline{\gamma}$. The method provides very good reconstruction of band-limited type spectra as demonstrated in the results section. The resolution of the technique is not significantly better than MEM, but in this situation this is not particularly important. Symmetric signals which are represented by a continuous model in this paper are spatially well separated (except at broadside to the array) and so good resolution can be obtained. Two symmetric signals in background noise are modelled in section 3.

5. RESULTS

This section shows the type of representation that can be obtained using this method compared to some other high resolution techniques. All simulations apply to 10 element linear equispaced arrays, with element separation of λ/5. Cylindrically isotropic noise of power 0dB is assumed with sensor noise of -10dB. Four discrete signals are modelled with wavenumbers (normalised to ±0.5) -0.1, 0.1, 0.12 and 0.16 having powers -4dB, -4dB, -1dB and -1dB respectively. The true spectrum is shown in Figure 1. Figure 2 shows the results that can be obtained by MUSIC, assuming an exact covariance matrix, 4 signals and a known background. As expected all of the signals have been correctly positioned. If the background is unknown or incorrectly estimated the performance of MUSIC is severely degraded. This is illustrated in Figure 3 which shows the MUSIC spectrum when a spherically isotropic background (which obeys a sinx/x law rather than the correct Bessel function) was assumed. In order to allow for some mismatch 6 signals were modelled, but it is clear that the two signals at 0.1 and 0.12 are not resolved, even though an exact covariance matrix was used.

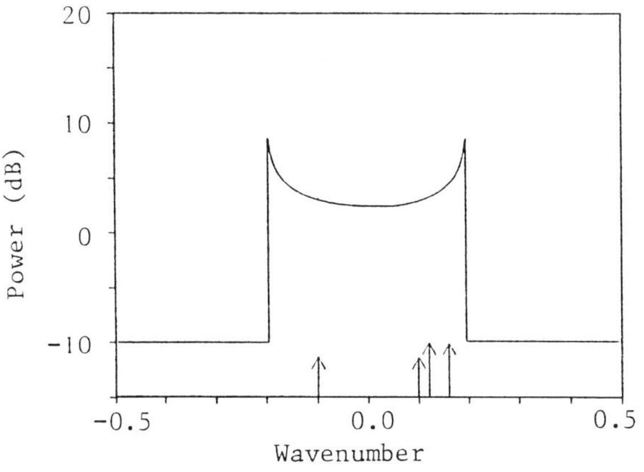

Fig. 1 True Spectrum

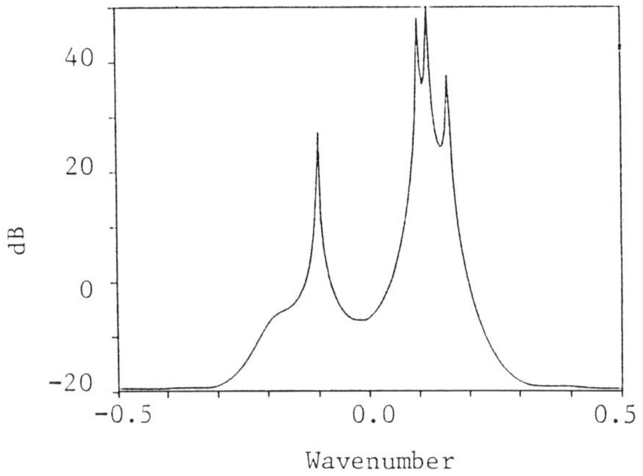

Fig. 2 MUSIC spectrum (exact background)

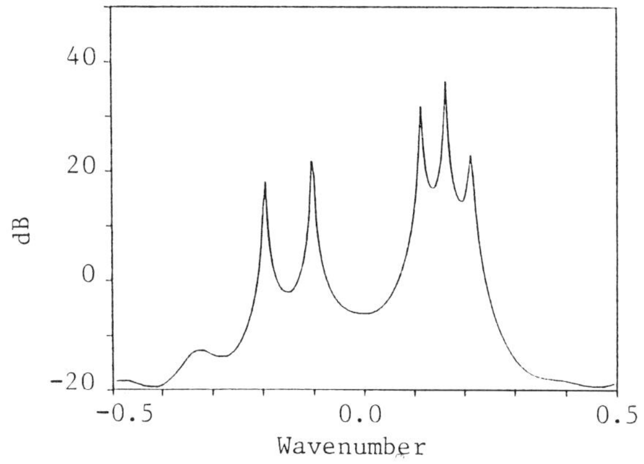

Fig. 3 MUSIC spectrum (False background)

Figure 4 shows the spectrum obtained from the method proposed in this paper. In this case the covariance matrix was sampled and formed from 100 data sets. The four signals have now been resolved and are in approximately the correct directions, even though no assumption other than symmetry has been made about the background noise. The bandlimited maximum entropy technique has also given an accurate representation of the background noise, something which the MUSIC algorithm cannot do.

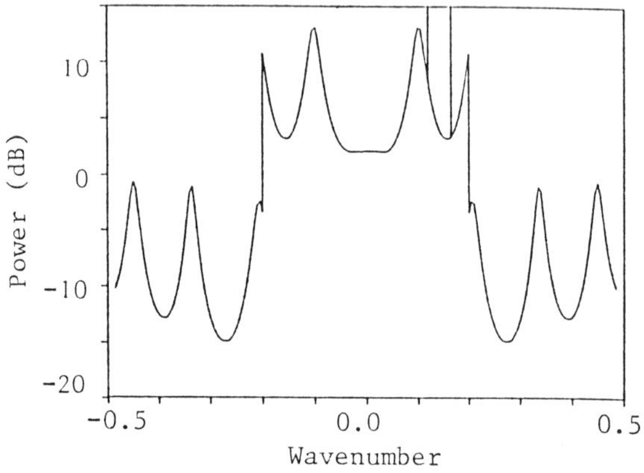

Fig. 4 Discrete and continuous model

A standard MEM solution for the same sampled case is shown in Figure 5, the signals at 0.12 and 0.16 have not been resolved, and the background noise has not been accurately represented.

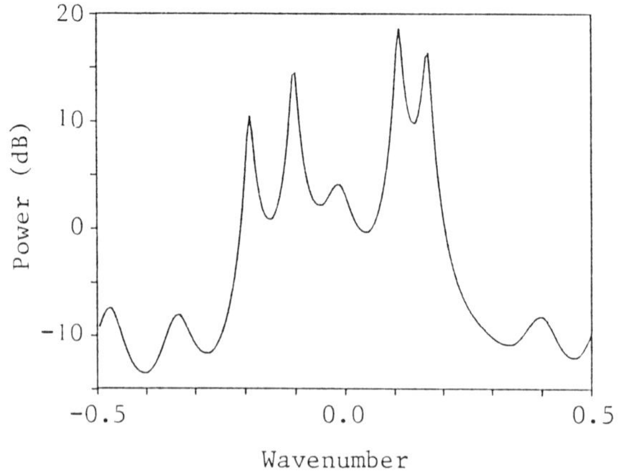

Fig. 5 MEM spectrum

The performance of the Prony method for estimating signal directions and power is shown in Figures 6 and 7. The estimated signal directions (x axis) are plotted against the estimated signal powers (y axis) for 100 sets of sampled data (each obtained from a 100 sample vectors). Figure 6 shows these estimates for data sampled from the covariance matrix considered throughout this section. The estimates for direction are generally good with most points clustered at the correct signal power level of -1dB. Increasing the isotropic noise (3dB) and sensor noise (-7dB) while leaving the signal powers unchanged obviously makes estimation more difficult. This is illustrated by Figure 7, which exhibits inferior performance to those estimates of Figure 6, but two signals are still discernible.

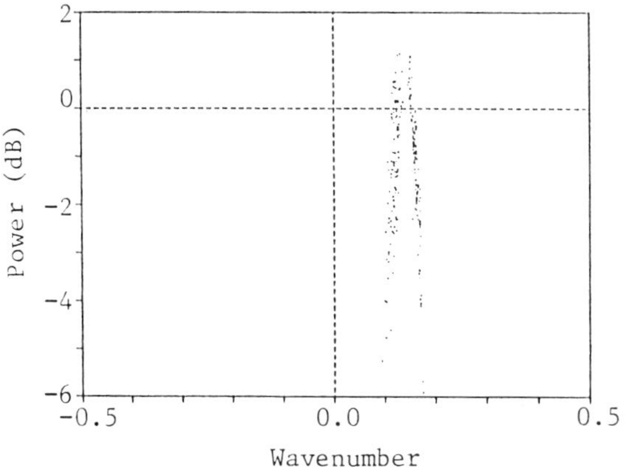

Fig. 6 Discrete Prony estimates

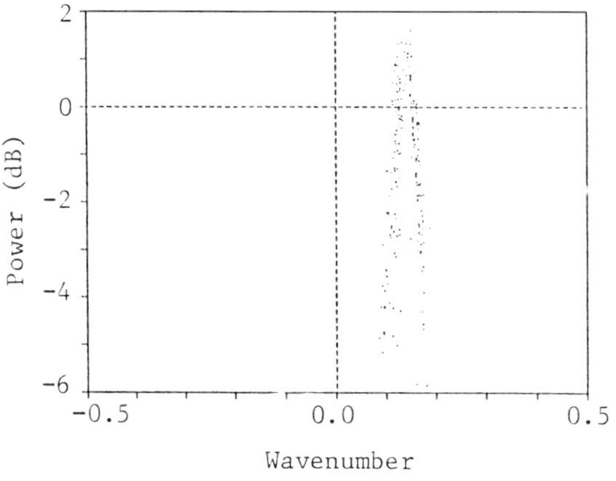

Fig. 7 Prony estimates at lower SNR

By ignoring the real part of the covariance matrix in order to estimate the signal parameters, some prior information has not been included. It is important to quantify how much affect this has. It is not feasible to use the Cramer-Rao lower bound because some signals have not been resolved and the background must be assumed unknown. Instead we compare the solutions to a full maximum likelihood solution [5].

The final estimates for the discrete signal powers and directions as obtained in Figure 7 are used as starting values for a full maximum likelihood solution, with an unknown background noise covariance matrix (constrained to be positive definite).

Eleven observations from Figure 7, which yielded positive definite estimates for R_o are shown in Figure 8. Figure 9 shows the maximum likelihood estimates for signal directions and powers using the estimates of Figure 8 as starting values. No significant improvement is seen in Figure 9, indicating convergence to a local maximum.

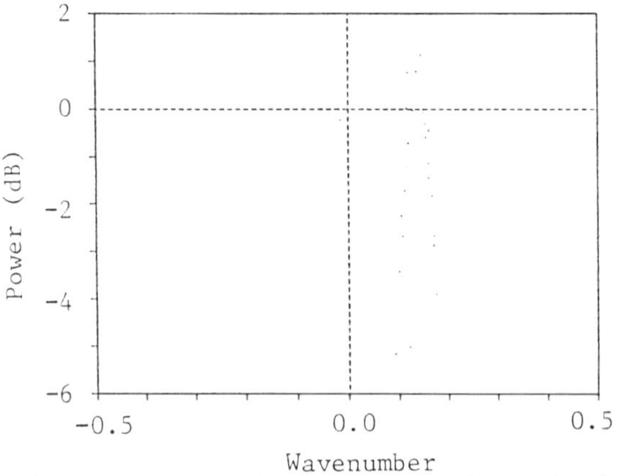

Fig. 8 11 positive definite starting solutions

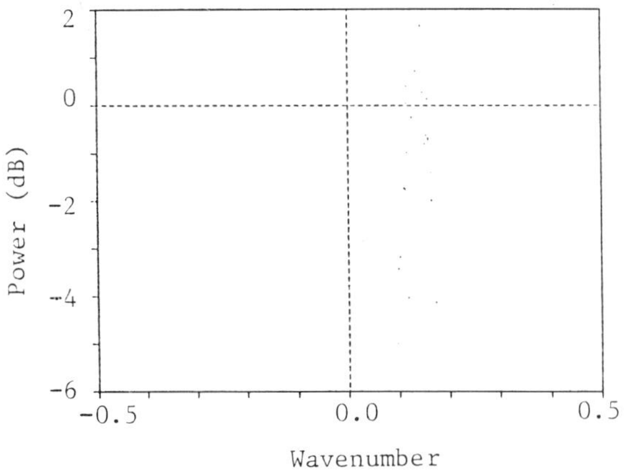

Fig. 9 Maximum Likelihood solutions

6. CONCLUSIONS

A technique has been proposed which is able to accurately reconstruct signal parameters without prior knowledge of the background noise field other than symmetry. It has been compared to eigenvector methods which do not yield correct results with an unknown background even when an exact covariance matrix is available.

The estimates of signal power and direction have been shown to be near full maximum likelihood estimates, which is indicative of a good estimate.

7. REFERENCES

[1] Schmidt, R., (1979) "Multiple source location and signal parameter estimation', *Proc. RADC Spectrum Estimation Workshop,* p. 243.

[2] Hildebrand, F.B., (1956) "Introduction to Numerical Analysis", McGraw-Hill.

[3] Burg, J.P., (1967) "Maximum Entropy Spectral Analysis", Presented at 37th Annual Meeting Soc. Exploration Geophysicists Oklahoma City.

[4] Farrier, D.R., and Jeffries, D.J., (1985) "Maximum Entropy Processing of Band Limited Spectra", *Proc. IEE Part F,* pp. 491-504.

[5] Mardia, K.V., Kent, J.T. and Bibby, J.M., (1979) "Multivariate Analysis" Academic Press.

ADAPTIVE PARALLEL AVERAGING

G.R. Barnes
(University of Louisville, USA)

and

V. Shvartsman
(Electronic Design and Research, Inc)

1. SUMMARY OF RESULTS

This paper discusses qualitative properties of an adaptive parallel processor which has proved useful in the detection of a randomly occurring pulse-like wave. No assumptions are made about the shape of the pulse, or the distribution of noise. Detection and discrimination of the pulse from similar noise induced fluctuations is considered.

2. BACKGROUND

The problem which motivated the development of this system was the noninvasive detection on a beat-by-beat basis of electrical activity of the His-Purkinje system of the heart. This problem involved a very low signal amplitude to noise power ratio (SNR). The noise was environmental background noise (60 Hz), bioelectrical activity, motion artifact from the patient and instrumentation noise generated by the electrode sensors and signal processing system itself. The His bundle signal itself was somewhat unpredictable in shape and location, especially for a population of patients with heart irregularities. The initial effort in this problem focused on the detection of the presence of the signal. Its discrimination from accompanying noise preceding the start of the His Purkinje activity was also considered to be important. This noise made it difficult to determine the time interval between the onset of His Purkinje activity and later events.

The original problem resulted in the formulation of a signal model as a randomly occurring pulse of unknown, and possibly variable, shape or duration. The signal of interest is accompanied by noise with unknown characteristics. The noise power is large relative to signal amplitude so that the randomly occurring noise contaminated signal pulse is indistinguishable from the noise induced fluctuations preceding and following its occurrence. This model makes serial averaging of successive time frames undesirable, and possibly counterproductive as it averages out variations in shape or position which may be of interest.

Signal averaging does have the effect of attenuating noise so the accumulation of information from simultaneous multiple channels was considered. Signal averaging increases signal to noise rms power only by reducing the noise rms power in the denominator at a rate proportional to the square root of the number of channels. The presence of relatively high noise power would require a large number of channels. The fact that averaging acts linearly on the signal-plus-noise combination is useful since it will not distort the system's response to signal as it reduces the effect of noise. However, if the primary concern is the detection or location of a signal hidden in noise and the discrimination of the signal as a pulse of power distinct from the similar fluctuations due to noise when no signal pulse is present, then distortion which emphasizes signal and attenuates noise is acceptable and in some sense desirable.

The average of instantaneous signal uses only the instantaneous amplitudes on the multiple channels to form its output. The signal average ignores other information such as agreement in polarity or the natural generalization of polarity, which is to exceed a threshold value. When no signal is present on multiple, independent channels the periods of coincidence of predominant subsets of channels tend to be short lived, depending on the threshold level, of course. When signal is present it becomes easier for larger subsets of channels to exceed a preset threshold level. These observations were considered in the design of the system described below.

3. SYSTEM DESCRIPTION

A simple block diagram of the system is provided in figure 1.

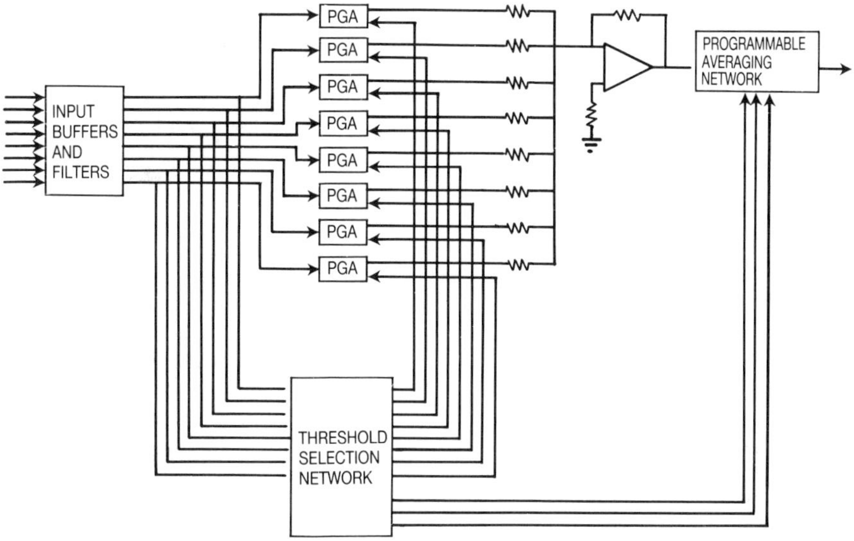

Fig. 1 Block diagram of the Adaptive Parallel Averaging System. The PGA's are programmable gain amplifiers.

The input signal on each channel is subjected to a threshold test. The results of these individual threshold tests are then examined as yes or no responses. If at least k of n channels pass the threshold test then the channels which pass are selected for amplification. Channels which do not pass the threshold test are not amplified. All channels are then used to form a weighted sum, which itself is amplified. If less than k of n channels exceed the threshold then no channel is amplified prior to the formation of a (uniformly weighted) sum which is not amplified. The simultaneous analysis of the signal behaviour on all channels is essentially instantaneous, and is performed continuously. The system continually adapts the assignment of weights to each channel and the decision to amplify the weighted sum output by selecting a "best" subset of channels and analyzing their joint behaviour.

The system response is governed both by the instantaneous signal amplitudes relative to the threshold tests and by the length of the period of coincidence that the select subset continues to satisfy the threshold test. When the input signal channels carry only noise, the periods for which each channel exceeds the threshold (if they do at all) tend not to be "in phase". This results in a few noise generated spikes of relatively short duration which can be filtered out by a low pass filter. This last comment about the pulse phase may become clear if one considers a model for the output of a narrow band pass filter when the input is a sinusoidal wave contaminated by noise. The signal on each channel is modelled by

$$E = R\cos(\omega t + \Theta)$$

where the amplitude envelope R and phase shift Θ are random variables. The quantities R and Θ are independent in the noise only situation. The distribution of these random variables depends on the amplitude A of the sinusoidal wave and the noise rms power p. The joint density of R and Θ and the marginal density of R were studied extensively by Rice [1]. The marginal density of Θ was previously given in [2]. We repeat the formulae for these densities here for completeness. The joint density of R and Θ is given by

$$h(R,\Theta) = \frac{R}{2\pi\sigma^2}\exp -\left(\frac{A^2 + R^2 - 2AR\cos\Theta}{2\sigma^2}\right).$$

The marginal density of the envelope R is given by

$$f(R) = \frac{R^2}{\sigma^2}\exp\left(-\frac{R^2 + A^2}{2\sigma^2}\right) I_0\left(\frac{RA}{\sigma^2}\right)$$

I_0 is the zero-order Bessel function of the first kind with complex argument. The marginal density of the phase shift Θ is given by

$$g(\Theta) = \frac{1}{2\pi}\exp\left\{\frac{-A^2}{2\sigma^2}\right\} \times$$

$$\left[1 + \left(\frac{A\cos\Theta}{\sigma}\right)\exp\left(\frac{A^2\cos^2\Theta}{2\sigma^2}\right)\sqrt{2\pi}\left(1 - \Phi\left(\frac{A\cos\Theta}{\sigma}\right)\right)\right]$$

The graphs of the two marginal densities are shown in Figures 2 and 3. Note that the phase shift changes from a uniformly distributed random variable to one clustered about zero as the signal amplitude to noise rms increases.

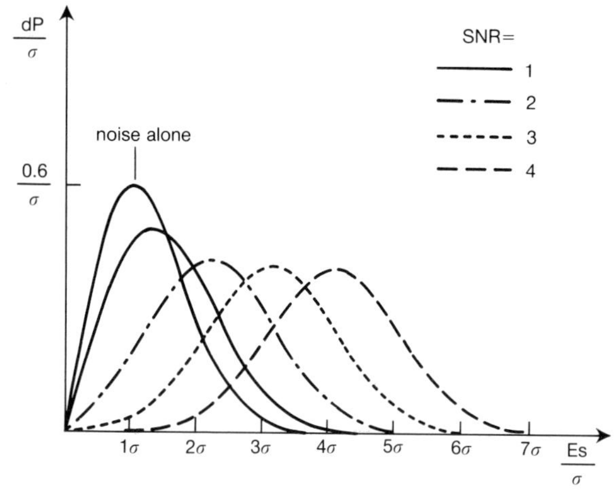

Fig. 2 Density of the envelope of noise alone and a signal plus noise for several signal amplitude-to-noise rms ratios.

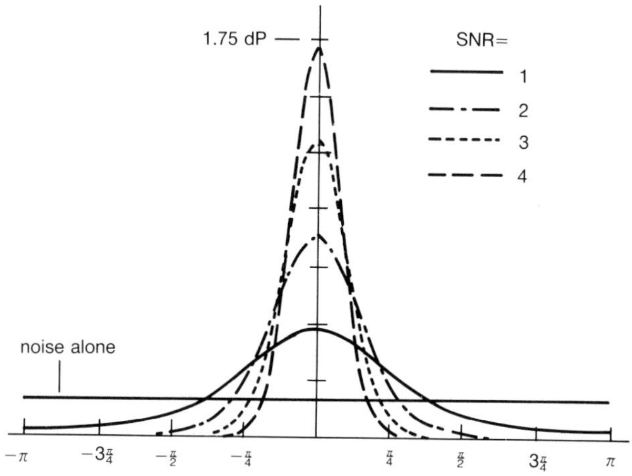

Fig. 3 Density of the phase shift for several signal amplitude-to-noise rms ratios.

We now consider a simplified mathematical model of the system. An assignment of weights to channels satisfying the threshold test and to the weighted sum output will occur if at least k of n channels satisfy the threshold criterion. Let p be the probability that a channel satisfies the threshold test at some fixed instant of time. Under the assumption that the noise on each channel is independent and identically distributed the probability of detection is obtained from the binomial distribution.

The value of p depends upon the threshold value selected, and the presence or absence of signal. The threshold may be selected either by a constant false alarm criterion or by a minimum probability of detection at a given signal-amplitude-to-noise power level.

4. DISCUSSION OF SYSTEM CHARACTERISTICS AND OPERATION

In practice there is a trade-off between signal acquisition and noise reduction. The threshold is adjusted to the level necessary to obtain an acceptable probability of detecting the signal. The acceptability of a given probability of detection depends on the noise clutter on the output channel making it difficult to recognize the randomly occurring signal pulses.

The system behaves more like a detector than a filter, although some filtering of signal from noise is accomplished by the weighted sum. The system was designed to solve a detection and discrimination problem, not to filter noise from the signal pulse itself. It is desirable but not critical that noise accompanying the signal pulse be suppressed. It is more important that noise levels making recognition of the low magnitude signal pulse be reduced, and that noise is the noise between signal pulses. If one knew precisely when the randomly occurring pulse was to occur, then this reduction of noise between pulses could be accomplished, theoretically but not practically due to transient response, by switching the system off between signal pulses. This is almost how the system is behaving. The constant polling of the channels for threshold coincidence and the amplification or attenuation of the averaged channel output based on the result of this poll is an attempt to provide the randomly occurring signal pulse with its own synchronization point.

The system is nonlinear. Clearly, the superposition of two signals, neither of which individually satisfies threshold requirements, may form a sum which does so. The system weighting and amplification would be activated for the sum but not for either component.

By heavily and uniformly weighting the subset of channels exceeding the thresholds, we are, in effect, creating a sum which behaves like the average, conditioned on the terms all being above the threshold value. When some channels do not satisfy the threshold test, the weighted average is similar to a trimmed mean. In this case the weighted average is formed from a sum of the largest (or smallest when the signals agree in negative polarity) order statistics. Since the most negative values from channels failing the threshold test are not included in the sum, a biased estimate of the signal is obtained.

5. EXPERIMENTAL RESULTS

Several working prototypes of this type of system have been constructed and used successfully. Versions based on 4-of-4 channels and at least 4-of-5 channels sampling have been used to detect the electrical activity of the His-Purkinje system of the heart from the body surface on a beat-by-beat basis [2,3]. A phototype having eight channels and using an at-least-6-of-8 algorithm has been developed and is currently under laboratory evaluation.

Figures 4 and 5 show tracings obtained from the eight channel adaptive parallel averaging system when pulses from a generator were mixed with high levels of noise. Lead one is the generator output. Lead 2 is the signal on one channel after the generator output was noise contaminated. Lead 3 is the output of a simple averaging circuit for the eight channels. Lead 4 is the output of the system described in this paper. In both figures 4 and 5 the single channel had a bandpass filter on lead 2. In figure 4 neither the average network on lead 3 nor the system output was subjected to any significant frequency band limitation, i.e., both went through a very wide bandpass filter. In figure 5 both leads 3 and 4 were subjected to moderate bandpass limitation of 20-120 hz. Observe that the eight channel average does improve the signal to noise ratio so that the signal begins to differentiate from noise. However, even with filtering it is difficult to locate the exact start of the signal pulse. The eight channel adaptive parallel averaging system clearly and easily differentiates the start of the signal pulse and the added filtration in figure 5 had little effect. Note that the signal average did a better job of reducing noise during the pulse as evidenced by the consistent pulse amplitudes, but did not do as well as the adaptive parallel averaging system at reducing the activity between generator pulses.

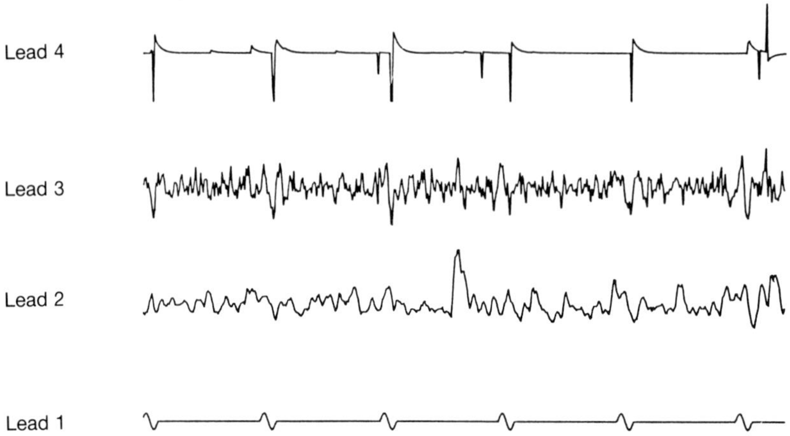

Fig. 4 Lead 1 - Generator output. Lead 2 - Single channel of generator plus noise. Lead 3 - eight channel averaging. Lead 4 - Adaptive parallel averaging utilizing an at-least-six-of-eight algorithm.

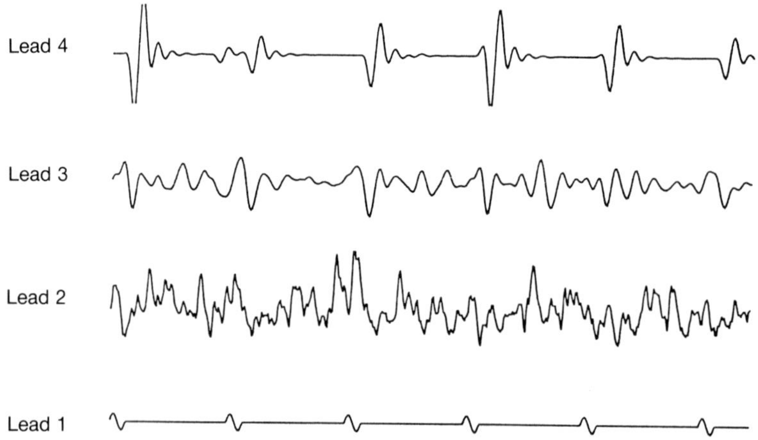

Fig. 5 Lead 1 - Generator output. Lead 2 - Single channel of generator plus noise. Lead 3 - Eight channel averaging. Lead 4 - Adaptive parallel averaging utilizing an at-least-six-of-eight algorithm. Leads 2, 3, and 4 have been 20-120 Hz band-pass filtered.

6. CONCLUSIONS

The system described in this paper was designed to reliably detect randomly occurring pulses which do not have a known shape or duration. It was designed to operate in a heavily noise contaminated environment and specifically in an environment in which the noise and signal have the same frequency characteristics. The discrimination of signal pulse from similar but not concurrent noise pulses, not the separation of the signal from simultaneously occurring noise is what the system does best. The adaptive parallel averaging system has been shown to work well in the environment, and in doing the task for which it was designed. Other methods should be used if more is known about either the shape or location of the signal or the distribution of noise.

7. REFERENCES

[1] Rice, S.O., "Mathematical Analysis of Random Noise", (1944 and 1945), *"Bell Systems Technical Journal"*, Vols 23 and 24, pp. 292-332 and pp. 1-100.

[2] Shvartsman, V., Barnes, G.R. et al, (1982), "Multichannel Signal Processing Bassed on Logic Averaging", *IEEE Trans. Bio. Eng.*, vol. BME-29, pp. 531-536.

[3] Flowers, N.C., Shvartsman, V., et al, (1981), "Surface recording of His-Purkinje activity on an every beat basis without digital averaging", *Circulation*, vol 63, pp. 948-952.